CDX Learning Systems

We support ASE program certification through

Automotive Electricity and Electronics

David M. Jones
Instructor, Ivy Tech
Fort Wayne, Indiana

Kirk VanGelder
ASE Certified Master Automotive Technician & L1 & G1
Technology Educators of Oregon – President
Certified Automotive Service Instructor
Vancouver, Washington

JONES & BARTLETT
LEARNING

World Headquarters
Jones & Bartlett Learning
5 Wall Street
Burlington, MA 01803
978-443-5000
info@jblearning.com
www.jblearning.com

Jones & Bartlett Learning books and products are available through most bookstores and online booksellers. To contact Jones & Bartlett Learning directly, call 800-832-0034, fax 978-443-8000, or visit our website, www.jblearning.com.

Substantial discounts on bulk quantities of Jones & Bartlett Learning publications are available to corporations, professional associations, and other qualified organizations. For details and specific discount information, contact the special sales department at Jones & Bartlett Learning via the above contact information or send an email to specialsales@jblearning.com.

Copyright © 2018 by Jones & Bartlett Learning, LLC, an Ascend Learning Company

All rights reserved. No part of the material protected by this copyright may be reproduced or utilized in any form, electronic or mechanical, including photocopying, recording, or by any information storage and retrieval system, without written permission from the copyright owner.

The content, statements, views, and opinions herein are the sole expression of the respective authors and not that of Jones & Bartlett Learning, LLC. Reference herein to any specific commercial product, process, or service by trade name, trademark, manufacturer, or otherwise does not constitute or imply its endorsement or recommendation by Jones & Bartlett Learning, LLC and such reference shall not be used for advertising or product endorsement purposes. All trademarks displayed are the trademarks of the parties noted herein. *CDX Automotive: Electricity and Electronics* is an independent publication and has not been authorized, sponsored, or otherwise approved by the owners of the trademarks or service marks referenced in this product.

There may be images in this book that feature models; these models do not necessarily endorse, represent, or participate in the activities represented in the images. Any screenshots in this product are for educational and instructive purposes only. Any individuals and scenarios featured in the case studies throughout this product may be real or fictitious, but are used for instructional purposes only.

Production Credits
General Manager: Douglas Kaplan
Executive Publisher: Vernon Anthony
Content Services Manager: Kevin Murphy
Senior Vendor Manager: Sara Kelly
Marketing Manager: Amanda Banner
VP, Manufacturing and Inventory Control: Therese Connell
Composition and Project Management: Integra Software Services Pvt. Ltd.
Cover Design: Scott Moden
Rights & Media Specialist: Robert Boder
Media Development Editor: Shannon Sheehan
Cover Image (Title Page): © sumbul/E+/Getty
Printing and Binding: LSC Communications
Cover Printing: LSC Communications

Library of Congress Cataloging-in-Publication Data unavailable at time of printing.

6048

Printed in the United States of America
20 19 18 17   10 9 8 7 6 5 4 3 2 1

Source: NATEF Program Accreditation Standards, 2013, National Automotive Technicians Education Foundation (NATEF).

BRIEF CONTENTS

CHAPTER 1	Strategy-Based Diagnostics	1
CHAPTER 2	Safety	23
CHAPTER 3	Basic Tools and Precision Measuring	51
CHAPTER 4	Fasteners and Thread Repair	113
CHAPTER 5	Principles of Electrical Systems	137
CHAPTER 6	Sources and Effects of Electricity	155
CHAPTER 7	Ohm's Law and Circuits	167
CHAPTER 8	Electrical Components	181
CHAPTER 9	Electronic Components	203
CHAPTER 10	Digital Multimeter Use and Circuit Testing Procedures	215
CHAPTER 11	Wires and Wiring Harnesses	263
CHAPTER 12	Electrical Testing Procedures	289
CHAPTER 13	Batteries	299
CHAPTER 14	Starting Systems	329
CHAPTER 15	Charging Systems	359
CHAPTER 16	Lighting System Fundamentals	383
Appendix A	2017 NATEF Automobile Accreditation Task List Correlation Guide	405
Glossary		421
Index		429

CONTENTS

CHAPTER 1 Strategy-Based Diagnostics.......1
Introduction 2
Vehicle Service History...................... 2
Strategy-Based Diagnostic Process............. 5
Documenting the Repair..................... 16
Ready for Review 20
Key Terms 21
Review Questions........................... 21
ASE Technician A/Technician B Style Questions 22

CHAPTER 2 Safety........................23
Introduction 24
Personal Safety 24
Shop Safety 31
Hazardous Materials Safety................... 44
First Aid Principles 46
Ready for Review 47
Key Terms 48
Review Questions........................... 49
ASE Technician A/Technician B Style Questions 49

CHAPTER 3 Basic Tools and Precision Measuring..............................51
Introduction 52
General Safety Guidelines..................... 52
Basic Hand Tools 58
Precision Measuring Tools.................... 88
Cleaning Tools and Equipment 105
Ready for Review 108
Key Terms 109
Review Questions.......................... 111
ASE Technician A/Technician B Style Questions 111

CHAPTER 4 Fasteners and Thread Repair..................................113
Introduction 114
Threaded Fasteners and Torque 114
Fastener Standardization.................... 115
Bolts, Studs, and Nuts 116
Threadlocker and Antiseize 123
Screws.................................... 124
Torque-to-Yield and Torque Angle 125

How to Avoid Broken Fasteners............... 128
Thread Repair............................. 128
Ready for Review 134
Key Terms 134
Review Questions.......................... 134
ASE Technician A/Technician B Style Questions 135

CHAPTER 5 Principles of Electrical Systems...............................137
Introduction 138
Electrical Fundamentals..................... 138
Volts, Amps, and Ohms 141
Electrical Circuits 142
Semiconductors 143
Direct Current and Alternating Current 144
Power (Source or Feed) and Ground........... 146
Continuity, Open, Short, and High Resistance (Voltage Drop) 146
Ready for Review 149
Key Terms 151
Review Questions.......................... 151
ASE Technician A/Technician B Style Questions 152

CHAPTER 6 Sources and Effects of Electricity...........................155
Introduction 156
Sources of Electricity 156
Effects of Electricity......................... 160
Ready for Review 162
Key Terms 163
Review Questions.......................... 163
ASE Technician A/Technician B Style Questions 164

CHAPTER 7 Ohm's Law and Circuits........167
Introduction 168
Ohm's Law................................ 168
Circuits 172
Ready for Review 176
Key Terms 177
Review Questions.......................... 178
ASE Technician A/Technician B Style Questions 178

CHAPTER 8 Electrical Components 181
Introduction 182
Switches 182
Fuses, Fusible Links, and Circuit Breakers........ 184
Flash Can/Control.................... 186
Relays and Relay Control Circuits 186
Solenoids.................... 189
Motors 190
Ignition Coils and Transformers 193
Resistors 195
Capacitors.................... 198
Ready for Review 199
Key Terms 201
Review Questions.................... 201
ASE Technician A/Technician B Style Questions 202

CHAPTER 9 Electronic Components........ 203
Introduction 204
Electronic Components 204
Transistors.................... 207
Control Modules.................... 208
Integrated Circuits 209
Microprocessors 209
Microcontrollers.................... 210
Speed Control Circuits.................... 210
Ready for Review 211
Key Terms 212
Review Questions.................... 213
ASE Technician A/Technician B Style Questions 213

CHAPTER 10 Digital Multimeter Use and Circuit Testing Procedures........... 215
Introduction 216
DMM Fundamentals 216
DMM Uses 220
Measuring Volts, Ohms, and Amps.................... 223
DVOM/DMM Testing Procedures 226
Current and Resistance Exercises 232
Series Circuit Exercises.................... 239
Parallel Circuit Exercises.................... 246
Understand Circuit Types 251
Locating Opens, Shorts, Bad Grounds, and High Resistance 256
Ready for Review 259
Key Terms 260
Review Questions.................... 260
ASE Technician A/Technician B Style Questions 261

CHAPTER 11 Wires and Wiring Harnesses ... 263
Introduction 264
Wire Fundamentals.................... 264
Wiring Diagram Fundamentals 270
Wire Maintenance and Repair 275
Ready for Review 285
Key Terms 286
Review Questions.................... 287
ASE Technician A/Technician B Style Questions 287

CHAPTER 12 Electrical Testing Procedures. ..289
Introduction 290
Graphing Multimeters.................... 290
Oscilloscopes 294
Ready for Review 296
Key Terms 297
Review Questions.................... 297
ASE Technician A/Technician B Style Questions 298

CHAPTER 13 Batteries 299
Introduction 300
What Is a Battery?.................... 300
Battery Charging and Discharging Cycles.......... 302
Lead Acid, Gel Cell, and AGM Batteries 306
Battery Testing Procedure.................... 310
Ready for Review 325
Key Terms 327
Review Questions.................... 327
ASE Technician A/Technician B Style Questions 328

CHAPTER 14 Starting Systems 329
Introduction 330
Engine Starting (Cranking) System 330
Starter Motor Construction.................... 333
Starter Motor Operation 337
Starter Drives and the Ring Gear 340
Starting System Procedures 343
Ready for Review 354
Key Terms 355
Review Questions.................... 356
ASE Technician A/Technician B Style Questions 356

CHAPTER 15 Charging Systems............ 359
Introduction 360
Charging System Theory.................... 360
Alternator Principles.................... 360
Hybrid Vehicle Charging Systems 369

Charging System Procedures 370
Ready for Review 379
Key Terms 381
Review Questions......................... 381
ASE Technician A/Technician B Style Questions 381

CHAPTER 16 Lighting System Fundamentals...383
Introduction 384
Types of Lamps 384
Types and Styles of Lighting Systems 388
Lighting Systems Procedures and Peripheral
    Systems 397
Ready for Review 400
Key Terms 402
Review Questions......................... 402
ASE Technician A/Technician B Style
    Questions 403

Appendix A   2017 NATEF Automobile
    Accreditation Task List Correlation Guide...... 405
Glossary 421
Index 429

NOTE TO STUDENTS

This book was created to help you on your path to a career in the transportation industry. Employability basics covered early in the text will help you get and keep a job in the field. Essential technical skills are built in cover to cover and are the core building blocks of an advanced technician's skill set. This book also introduces "strategy-based diagnostics," a method used to solve technical problems correctly on the first attempt. The text covers every task the industry standard recommends for technicians, and will help you on your path to a successful career.

As you navigate this textbook, ask yourself, "What does a technician need to know and be able to do at work?"

This book is set up to answer that question. Each chapter starts by listing the technicians' tasks that are covered within the chapter. These are your objectives. Each chapter ends by reviewing those things a technician needs to know. The content of each chapter is written to explain each objective. As you study, continue to ask yourself that question. Gauge your progress by imagining yourself as the technician. Do you have the knowledge, and can you perform the tasks required at the beginning of each chapter? Combining your knowledge with hands-on experience is essential to becoming a Master Technician.

During your training, remember that the best thing you can do as a technician is learn to learn. This will serve you well because vehicles keep advancing, and good technicians never stop learning.

Stay curious. Ask questions. Practice your skills, and always remember that one of the best resources you have for learning is right there in your classroom… your instructor.

Best wishes and enjoy!
The CDX Automotive Team

ACKNOWLEDGMENTS

▶ Editorial Board

Keith Santini
Addison Trail High School
　Addison, Illinois

Merle Saunders
Nyssa, Oregon

Tim Dunn
Sydney, New South Wales
　Australia

▶ Contributors

Jerry Clemons
Elizabethtown Community and Tech College

Robert Farro
Moraine Valley Community College

Dale Henry
East Mississippi Community College

Paul Kelley
Cypress College

Daniel M. Kolasinski
Milwaukee Area Technical College

Steve Levin
Columbus State Community College

James Martin
Tarrant County College

Jeffrey Rehkopf
Florida State College

Ronald Strzalkowski
Baker College of Flint

Joseph Wagner
Joliet Junior College

Dan Warning
Joliet Junior College

Mike Wichtendahl
Kansas City Kansas Community College

Daniel L. Wooster
Gateway Technical College Horizon Center

CHAPTER 1
Strategy-Based Diagnostics

NATEF Tasks

- **N01001** Review vehicle service history.
- **N01002** Demonstrate use of the three C's (concern, cause, and correction).
- **N01003** Identify information needed and the service requested on a repair order.
- **N01004** Research vehicle service information including fluid type, vehicle service history, service precautions, and technical service bulletins.

Knowledge Objectives

After reading this chapter, you will be able to:

- **K01001** Describe the purpose and use of vehicle service history.
- **K01002** Demonstrate an understanding of the active listening process.
- **K01003** Demonstrate an understanding of the strategy-based diagnosis process.
- **K01004** Describe step one of the strategy-based diagnosis.
- **K01005** Describe step two of the strategy-based diagnosis.
- **K01006** Describe step three of the strategy-based diagnosis.
- **K01007** Describe step four of the strategy-based diagnosis.
- **K01008** Describe step five of the strategy-based diagnosis.
- **K01009** Explain how the three Cs are applied in repairing and servicing vehicles.
- **K01010** Describe the information and its use within a repair order.

Skills Objectives

After reading this chapter, you will be able to:

- **S01001** Use service history in the repair and service of vehicles.
- **S01002** Complete a repair order.

You Are the Automotive Technician

A regular customer brings his 2014 Toyota Sienna into your shop, complaining of a "clicking" noise when he turns the steering wheel. You ask the customer further questions and learn that the clicking happens whenever he turns the wheel, especially when accelerating. He tells you he has just returned from vacation with his family and has probably put 300 miles (482 kms) on the car during their trip.

1. What additional questions should you ask the customer about his concern, the clicking noise he hears when turning?
2. How would you verify this customer's concern?
3. What sources would you use to begin gathering information to address this customer's concern?
4. Based on what you know this far about the customer's concern, what systems might be possibly related to this customer's concern?

▶ Introduction

The overall vehicle service involves three major components. Those pieces are gathering information from the customer, the strategy-based diagnostic process, and documenting the repair. The flow of the overall service can be seen below.

1. Initial information gathering is often completed by a service advisor (consultant) and should contain details about the customer concern and pertinent history.
2. Verifying the customer concern begins the strategy-based diagnostic process. Technicians will complete this step to ensure that a problem exists and that their repair eliminated it.
3. Researching the possible cause will provide a list of possible faults. The technician will expand this list as testing continues.
4. Testing will focus on the list of possibilities. Technicians will start with broad, simple tests that look at an entire system or group of components. Testing will progressively become more narrowly focused as it pinpoints an exact cause.
5. Repairs will be made using suggested tools and recommended procedures. This is done to ensure a reliable repair and that manufacturer requirements are met.
6. Repairs must always be verified. This confirms that the technician has completed the diagnosis accurately and completely.

The repair must be documented. The technician has been doing this all along. When the customer concern is recorded, the tests are recorded, and the final repair procedure recorded, the repair has been documented.

▶ Vehicle Service History

K01001 Describe the purpose and use of vehicle service history.

N01001 Review vehicle service history.

S01001 Use service history in the repair and service of vehicles.

Service history is a complete list of all the servicing and repairs that have been performed on a vehicle (**FIGURE 1-1**). The scheduled service history can be recorded in a service booklet or owner's manual that is kept in the glove compartment. The service history can provide valuable information to technicians conducting repairs. It also can provide potential new owners of used vehicles an indication of how well the vehicle was maintained. A vehicle with a regular service history is a good indication that all of the vehicle's systems have been well maintained, and the vehicle will often be worth more during resale. Most manufacturers store all service history performed in their dealerships (based on the VIN) on a corporate server that is accessible from any of their dealerships. They also use this vehicle service history when it comes to evaluating warranty claims. A vehicle that does not have a complete service history may not be eligible for warranty claims. Independent shops generally keep records of the repairs they perform. However, if a vehicle is repaired at multiple shops, repair history is much more difficult to track and, again, may result in a denial of warranty claims.

Vehicle service history can be very valuable to the technician. This history is typically retrieved from service records kept by the shop, dealer network, **original equipment manufacturer (OEM)**, or **aftermarket** service center. This information often contains a list of services performed on a vehicle and the date and mileage at which they were completed. Not all service history contains the same information. Some histories may only contain repair information, while others include every customer concern and maintenance task performed. This information can be very helpful when diagnosing a concern. Service history may help technicians diagnose a vehicle and can also be used to prevent costly duplicated repairs.

Service history can also be used to guide repairs. Records of vehicle service history may indicate that the customer has recently been in for service and now has returned with a new concern. This all-too-common situation is usually found to be caused by error during the previous service. When working on a vehicle that has returned after a recent repair, the previous technician's work (whoever that may be) should be inspected meticulously.

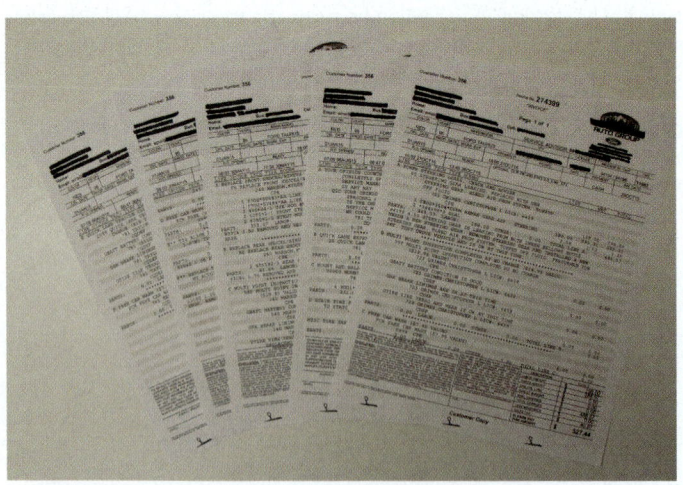

FIGURE 1-1 Print outs of completed repair order as saved in the online repair order system.

The service history may also show that the customer is returning for the same issue due to a component failure. The history might indicate when the component was installed, help the customer get their vehicle repaired, and help the shop to get paid under the component warranty. A vehicle that returns more than once for the same repair could be an indicator that an undiagnosed problem is causing these failures. The service history allows technicians to determine if the vehicle has been well maintained. This can be extremely useful when a technician suspects that lack of maintenance may be the cause of the problem.

The vehicle's service history helps technicians determine what maintenance needs to be performed, and therefore helps customers save money over time by preventing future costly repairs. Routine maintenance is essential on today's modern automobile and prevents premature failures due to contamination and component wear.

Today's vehicles also require regular software updates. There are many advanced computer systems on modern vehicles. From time to time, updates will be available to fix a bug or glitch in the computer programming. These updates are often designed to eliminate a customer concern, improve owner satisfaction, or increase vehicle life. This is very similar to an update for your PC or mobile device. Service history will indicate to the technician that the vehicle may need an update. The technician will inspect the vehicle's computer system and perform any needed updates as necessary.

Service history can also be used to keep customers safe. Occasionally, manufacturers may need to recall a vehicle for service due to a safety concern that has been identified for a vehicle (**FIGURE 1-2**). This means that the manufacturer has found that the potential exists for a dangerous situation to occur, and the vehicle must be serviced to eliminate it. Depending on the nature of the problem, recalls can be mandatory and required by law, or manufacturers may voluntarily choose to conduct a recall to ensure the safe operation of the vehicle or minimize damage to their business or product image. The service history would be used to verify that the vehicle is subject to the recall and has or has not had the recall service completed. The technician would perform the service, update the service history, and return the vehicle to the customer.

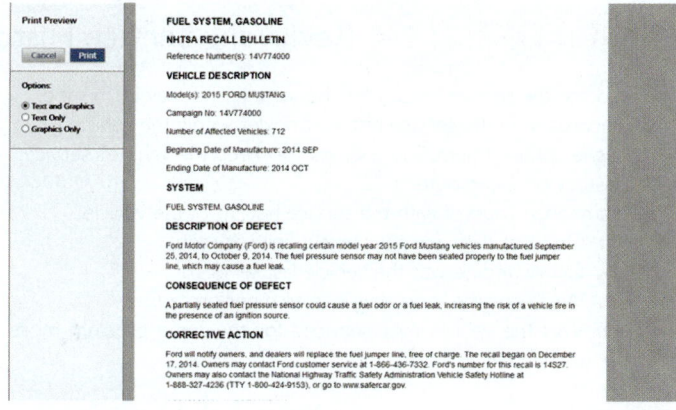

FIGURE 1-2 Recall notice example.

▶ **TECHNICIAN TIP**

Technicians and service advisors should check the vehicle service history against the manufacturer's service maintenance schedule to determine if the vehicle is due for scheduled maintenance. The maintenance schedule is a guide that indicates what service is due when; it can be found in the manufacturer's service information and often in the owner's manual. Keeping the vehicle well maintained can avoid a failure that strands the customer on the roadside.

Applied Science

AS-11 Information Processing: The technician can use computer databases to input and retrieve customer information for billing, warranty work, and other record-keeping purposes.

Dealership service departments have access to databases run by manufacturers for the purposes of accessing warranty information, tracking vehicle servicing and warranty repair history, and logging warranty repair jobs for payment by the manufacturer. When a customer presents their vehicle for a warranty repair, the customer service department staff begin by consulting the database to confirm that the vehicle is within its warranty period and that the warranty has not been invalidated for any reason. Once it is confirmed that the vehicle is still under a valid warranty, the repair order will be passed to the workshop for diagnosis and repair. Any parts required for the warranty repair must be labeled by the technician and stored for possible recall by the manufacturer.

For example, a young man comes in complaining that his vehicle is "running rough." The customer service staff confirms that the vehicle is nine months old and only has 14,500 miles (approx. 23,000 km), so it is within the manufacturer's 3-year/100,000 mile (160,000 km) warranty period. They check the manufacturer's database to confirm that the vehicle's warranty has not been invalidated before handing the repair order onto the workshop. Then a technician diagnoses the fault as a defective ignition coil and fills out a warranty parts form.

Once the repair has been completed and the parts labeled, the warranty parts form and any repair order paperwork is passed back to administrative staff for processing. Processing will include billing the manufacturer for the correct, pre-approved amount of time, logging the repair on the database for payment, and ensuring that all documentation is correct for auditing purposes.

Warranty Parts Form	
Customer concern: Vehicle running rough	Vehicle Information
Cause: #6 ignition coil open circuit on primary winding	VIN: IG112345678910111
Correction: Replaced #6 ignition coil	RO Number: 123456
Parts description: #6 ignition coil	Date of repair: 10/04/2016

SKILL DRILL 1-1 Reviewing Service History

1. Locate the service history for the vehicle. This may be in shop records or in the service history booklet within the vehicle glove compartment. Some shops may keep the vehicle's service history on a computer.
2. Familiarize yourself with the service history of the vehicle.
   a. On what date was the vehicle first serviced?
   b. On what date was the vehicle last serviced?
   c. What was the most major service performed?
   d. Was the vehicle ever serviced for the same problem more than once?
3. Compare the vehicle service history to the manufacturer's scheduled maintenance requirements, and list any discrepancies.
   a. Have all the services been performed?
   b. Have all the items been checked?
   c. Are there any outstanding items?

To review the vehicle service history, follow the steps in **SKILL DRILL 1-1**.

Active Listening Skills

K01002 Demonstrate an understanding of the active listening process.

▶ **TECHNICIAN TIP**

A vehicle's service history is valuable for several reasons:

- It can provide helpful information to the technician when performing repairs.
- It allows potential new owners of the vehicle to know how well the vehicle and its systems were maintained.
- Manufacturers use the history to evaluate warranty claims.

Depending on the size of a shop, the first point of contact for the customer is the **service advisor** or consultant. This person answers the phone, books customer work into the shop, fills out repair orders, prices repairs, invoices, keeps track of work being performed, and builds customer relations with the goal of providing a high level of customer support. The service advisor also serves as a liaison between the customer and the technician who is working on the vehicle. A service advisor or consultant may advance to become a service manager. In smaller shops, a technician may perform these duties.

When the customer brings his or her vehicle in for service, the service advisor or technician should ask for more information than just the customer concern. It is important to let the customer speak while you use active listening skills to gather as many pertinent details as possible. Active listening means paying close attention to not only the customer's words, but also to their tone of voice and body language. Maintain eye contact with the customer throughout your conversation and nod to show you understand and are paying attention. Do not interrupt. Wait for the customer to finish speaking before responding, then ask open-ended questions to verify that you have heard the complaint clearly and understand the problem. An open-ended question is one that cannot be answered with a yes or no, but instead requires the customer to provide you with more information about the problem (**FIGURE 1-3**). If the shop is noisy, try to find a quieter location in which to speak with the customer. Excellent communication helps ensure that all relevant information is collected. It also makes a good first impression with customers; they are likely to feel that they were listened to and cared for.

Politely use open-ended questions to ask about any symptoms the customer may have noticed, such as:

- Under what circumstances does the concern occur or not occur?
- What unusual noises do you hear (e.g., squeaks, rattles, clunks, and other noises)?
- What odd smells or fluid leaks have you noticed?
- What recent work, service, or accessories have been added to the vehicle?
- What other recent changes or experiences have you had with the vehicle?
- What other systems seem to be operating improperly?

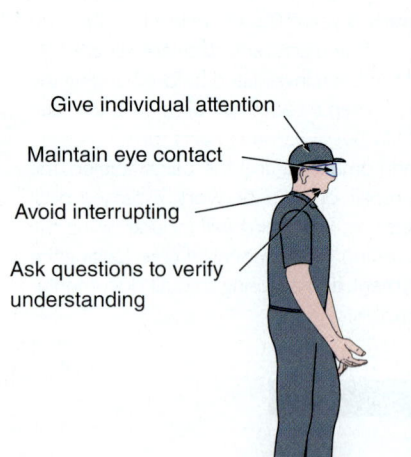

FIGURE 1-3 The active listening process.

Although problems may seem unrelated initially, when multiple systems fail at the same time, the issues are frequently related. Open-ended questions can provide valuable information to the technician who is performing the diagnosis.

▶ Strategy-Based Diagnostic Process

Diagnostic problems can be very challenging to identify and correct in a timely and efficient manner. Technicians will find that having a plan in place ahead of time will vastly simplify the process of logically and systematically (strategically) solving problems. The plan should be simple to remember and consistent in its approach; yet it must work for the entire range of diagnostic problems that technicians will encounter. In this way, technicians will have one single plan to approach any diagnostic situation they may encounter, and will be confident in their ability to resolve it. This problem solving plan is called the **Strategy-Based Diagnostic Process**.

The strategy-based diagnostic process is focused on fixing problems correctly the first time. It is a scientific process of elimination, which is much the same process as a medical doctor uses for their diagnosis. It begins with identifying the customer's concern and ends with confirming that the problem has been resolved. The purpose of the problem-solving process is twofold: to provide a consistent road map for technicians as they address customer concerns that require diagnosis, and to ensure that customer concerns are resolved with certainty.

This process simplifies the problem-solving portion of the repair, making the job easier for the technician; it prevents technicians from having to work on the same job more than once; and it all but eliminates customer comebacks. While repeat customers are good for business, a customer coming back with the same problem is not. The customer is likely to be upset and the technician is likely to be working for free. In order to avoid this scenario, it is imperative to address customer concerns correctly the first time.

Proper diagnosis is important to consumers and to the federal government. Federal and state law protects consumers against the purchase of vehicles with significant persistent defects. Technicians are held to a standard of reasonable repair times and limited visits for the same concern. Although the law varies from state to state, this means technicians must not return a vehicle to a customer without addressing the customer's original concern. Also, technicians cannot make the vehicle unavailable to the customer for a long period while the vehicle is being repaired. The purpose of the state and federal laws is to protect consumers buying new vehicles.

Failure to comply with the state and federal law can be very expensive for the dealership and manufacturer. Although most state laws hold the manufacturer directly responsible, dealerships are also hurt by a loss in sales revenue, a loss in repair revenue, and irreparable damage to their customer and sometimes manufacturer relationships. Many state laws hold the manufacturer responsible for full purchase price, incurred loan fees, installed accessories, and registration and similar government charges. This can be a heavy cost on top of the value of the vehicle itself.

K01003 Describe each step in strategy-based diagnosis.

▶ **TECHNICIAN TIP**

Technicians need to do their best to find the issue and resolve it; otherwise, the vehicle may be required to be bought back from the customer, costing the dealership and manufacturer significant money.

Need for the Strategy-Based Diagnostic Process

Finding the source of every customer concern can prove to be a challenge. Novice technicians frequently struggle with diagnostics situations. Even some veteran technicians have difficulty tackling diagnosis on some new technologies. However, if the strategy for solving a problem is generally the same every time, this greatly simplifies the process. Hopefully, by applying a strategy-based diagnostic process, technicians will resolve challenging customer concerns 100% of the time in an efficient manner.

Customer comebacks occur when the customer picks up the vehicle after service, only to bring it back shortly thereafter with the same concern. This situation is understandably upsetting to the customer. Typically, the end result is wasted labor time and a loss in shop productivity. The customer is left with one of the following impressions:

- The work was not performed;
- The shop is incompetent;
- Or, worse yet, the shop was trying to scam the customer.

▶ **TECHNICIAN TIP**

The diagnostic process makes the technician's job easier by providing a step-by-step strategy to solving the problem. It also answers the question: "Now what do I do?" As even the toughest job becomes easier, technicians will find their rate of diagnostic success increasing.

Customer comebacks are usually caused by one of two avoidable reasons:

1. The customer concern is misinterpreted or misunderstood. This results in the technician "fixing" a problem that does not exist or missing a problem altogether.
2. The technician failed to verify that the original concern was resolved. Technicians are often hurried; some will forget to ensure that the repair they had performed actually fixed the original customer concern.

Use of the strategy-based diagnostic process enables the technician and shop to make more money and satisfy more customers. This is a win-win situation for all involved. Using the strategy-based diagnostic process requires starting at the beginning and following it through to the end every time (**FIGURE 1-4**). This systematic approach will ensure the best results for each diagnostic situation.

Step 1: Verify the Customer's Concern

K01004 Describe step one of the strategy-based diagnosis.

The first step in the diagnostic process is to verify the customer's concern. This step is completed for two main purposes:

- To verify that the vehicle is not operating as designed
- To guarantee that the customer's concern is addressed

Failure to complete this step may result in wasted time, wasted money, and, worst of all, an unhappy customer. The customer is probably not an experienced automotive technician. For this reason, the customer does not always accurately verbalize the problem that may be occurring. Therefore, it is very important that you have a complete understanding of the customer's concern before beginning the diagnosis. This will enable you to know with certainty that you have actually resolved the original concern after repairing the vehicle and before returning it to the customer. During this step, you may perform several of the following tasks, depending on the customer concern.

First, ask the customer to demonstrate the concern, if possible. This may necessitate a test drive (**FIGURE 1-5**). The customer should be encouraged to drive the vehicle while you ride along as a passenger and gather symptoms and details about the concern. Seeing the customer recreate the concern in real time will often provide some much needed context to the problem. Having the customer demonstrate the concern is ideal in most situations, though not always possible. In the event that the customer is not present, you must do your best to recreate the concern on your own based on the information obtained from the customer. With or without the customer present, be sure to document in writing any details about the scenario in which the concern arises.

Next, make sure that the customer concern doesn't fall outside the range of normal operation of the component or system. The manufacturer's service information provides

SAFETY TIP

There should be limits to recreating the customer concern. Technicians need to be careful when riding as a passenger with the customer or driving an unfamiliar customer vehicle. Technicians have died during test drives due to customers' driving or their own driving of unfamiliar vehicles. The purpose of the test drive is to verify the concern or its repair. It is not an opportunity for a thrill ride. Customers and their vehicles should be treated with respect. Additionally, customers should respect the technician. If a customer asks a technician to verify a concern in an unsafe situation, such as a high rate of speed, the technician should decline. This is for both safety and liability reasons.

FIGURE 1-4 The strategy-based diagnostic process.

FIGURE 1-5 Ask the customer to describe the concern.

system descriptions and expected operations; technicians can use these details, provided in the owner's manual or in the vehicle service manual, to become familiar with the system and then explain its operation to the customer. Especially on new cars with many amenities, customers may not be familiar with the controls and subsequent operation. This can cause a customer to bring a vehicle in for service unnecessarily, due to unfamiliarity with the system controls. Many shops use online service (shop) manuals where you can quickly access any information related to the customer's concern (**FIGURE 1-6**). Checking to make sure that the concern is really a fault, and not a normal operation, will avoid unnecessary diagnosis time. This is also an opportunity to provide excellent customer service by demonstrating the features and their controls to the customer.

Conducting a quick visual inspection to look for obvious faults can be very helpful (**FIGURE 1-7**). However, it does not replace the need for testing and is absolutely not intended as a shortcut to the diagnostic process. With that said, the visual inspection can provide valuable information that may speed up the testing processes. The visual inspection provides an opportunity for a quick safety check by the technician and may help to avoid some potentially dangerous situations during service.

While visual inspections can be very valuable, technicians must be careful not to jump to conclusions based on what they see. For example, a customer comes in with an illuminated and flashing overdrive light on the control panel. The technician has seen this problem before and it was caused by a bad solenoid pack in the transmission. If the technician decides that this problem is also caused by a bad solenoid pack, this determination is one that was reached solely on conjecture; no actual test was performed. Although the flashing light might indicate a fault with the solenoid pack, steps in the diagnostic process should never be skipped. This guess can lead to a very costly mistake when it is discovered that the new solenoid pack does not, in fact, fix the problem. In reality, the wiring harness to the transmission is frayed and shorting out. Had the technician performed a test, the cause of the customer concern could have been confirmed or denied before a solenoid pack was put in unnecessarily. While the visual inspection is very valuable, tests must always be performed to compare suspected faults against the expectations and specifications defined in the service information.

When recreating the customer concern, the technician should operate the system in question in all practically available modes. System operation should be checked to see if there are other symptoms that may have gone unnoticed by the customer. These other symptoms can be very valuable when determining which tests to perform; they could save the technician significant time during the diagnosis. When recreating the customer concern, it is important to check the entire system for symptoms and related faults.

Recreating **intermittent faults** can be a challenge. Intermittent symptoms often stem from a component or system that is failing or one where the nature of the fault is not yet clear. In these situations, the aforementioned check of system operation can prove to be highly valuable, as it may uncover previously unnoticed but consistent symptoms. Attempting to repair

FIGURE 1-6 A technician researching service information.

FIGURE 1-7 Performing a visual inspection.

an intermittent fault without consistent symptoms, data, or diagnostic trouble codes (DTCs) is a gamble, because a technician cannot be certain that the actual problem is isolated. This means that there would be no way to confirm with certainty that a repair was effective. The fault could appear again as soon as the vehicle is returned to the customer. To avoid such a situation, look for symptoms, data, or DTCs that are repeatable or consistent. Intermittent diagnosis may require the use of an oscilloscope (a specialized tool for looking at electrical waveforms), or a "wiggle" test (as the name implies, a test instrument is monitored as the electrical or vacuum harness is manipulated by hand). This can verify the customer concern and remove some of the challenge from the diagnosis of an intermittent fault.

Lastly, but notably, save DTCs and freeze frame data. **Freeze frame data** refers to snapshots that are automatically stored in a vehicle's power train control module (PCM) when a fault occurs; this is only available on vehicles model year 1996 and newer. Intermittent faults may be found by reviewing data stored just before, during, and after the fault occurred, similar to an instant replay. When working with computer controlled systems, it is very important to save the recorded data. It may become necessary to erase this information from the computer, though that should generally be avoided. This information is absolutely critical when the technician is trying to answer the questions, "When did this happen?" and "What was going on at the time?"

What will step one look like? When information is gathered and recorded for step one, it should contain the customer concern, any symptoms, and any retrieved DTCs. View the following example from a vehicle that has no reverse. The technician verified the customer concern and recorded:

1. Vehicle will not move when shifted into reverse
2. Vehicle operates normally in all forward gears in OD, D, L2, and L1
3. Current code P0868

Notice that the technician in this example verified and recorded the customer concern. The technician also tried other functions in the system. Specifically, the technician drove the vehicle and tested the other gears in each of the gear ranges and then recorded the results. The DTC data was also retrieved from the control module and recorded. Although it was short and concise, the information will be very useful in the next step.

Step 2: Researching Possible Faults and Gathering Information

The second step in the diagnostic process is to research possible faults that may be related to the customer's concern. The goal of this step is to create a list of possible faults. The list is created based on the information gathered in step one. The list will later be narrowed down by the tests performed in step three until the cause of the concern has been confirmed.

Before testing can begin, a technician must know what possible faults need to be tested. Researching possible faults should begin broadly. Especially when diagnosing electrical and electronic systems, this step should begin at the system level and work down to individual components. For example, if a vehicle engine cranks, but will not start, a technician would list these familiar possible faults: Air, Fuel, Ignition, Compression, and Security. These possible faults are not single components, but rather they are systems. This is where a diagnosis should begin. Starting a diagnosis by listing the dozens of components for each system will make the job unreasonably time intensive. However, once a test determines that there is a fault within a specific system, the list should be expanded to encompass that particular system's subsystems and components. This systematic elimination starts broadly and narrows, allowing technicians to work more efficiently.

In the second step of the diagnostic process, the technician creates a list to help focus their tests. The list may aid in a simple process of elimination by testing one possibility after the next. The list can also start broadly and narrow as testing continues. When starting a list, it may look similar to the following:

1. Air
2. Fuel

▶ **TECHNICIAN TIP**

All too often, the customer does not have symptoms to share and their only concern is that the malfunction indicator is illuminated. In this situation, the data stored in the computer is invaluable. Record it and do not clear it out unless directed to do so in the manufacturer's service procedure. Even then, you should capture the information before clearing the memory.

K01005 Describe step two of the strategy-based diagnosis.

N01004 Research vehicle service information including fluid type, vehicle service history, service precautions, and technical service bulletins.

3. Ignition
4. Compression
5. Security

This list is broad and starts at the system level. As you'll soon see in the next step, the technician would eliminate possible faults with a test that is focused on analyzing the whole system. When a system is located with a fault, in the ignition system for example, the list would become more specific:

1. Spark Plug
2. Coil on Plug
3. CKP
4. CMP
5. Sensor Triggers
6. Harness
7. Control Module

The technician would again focus his or her testing on the list, seeking to eliminate possible faults until one is confirmed, repaired, and verified.

Several great sources of information are available for researching possible faults, although the best source of information is usually the manufacturer's service information system. These systems are typically found online; however, some manufactures still publish paper service manuals. The manufacturer's service information contains definitions for diagnostic trouble codes, system description and operation, electrical wiring diagrams, diagnostic steps, repair procedures, and much more. Fault diagnosis should almost always begin with the factory service information.

Other resources for identifying faults can be used in conjunction with the factory service information. As previously discussed, the vehicle service history can provide valuable insight into the past maintenance or lack thereof. It can also provide information about recent or repeated repairs. **Technical service bulletins (TSBs)** are service notifications and procedures sent out by the manufacturers to dealer groups alerting technicians about common issues with a particular vehicle or group of vehicles (**FIGURE 1-8**). Some aftermarket sources also exist for the pattern failures addressed by TSBs (**FIGURE 1-9**). Additionally, both original equipment manufacturers (OEM) and aftermarket technician support services offer hotlines, or call-in support, that specifically provide technical support to professional technicians. Some of these hotlines offer subscriptions to searchable web-based components. These resources do not guarantee a repair; that is still the responsibility of the technician. However, all of the sources mentioned here can be a huge help as technicians research possible faults.

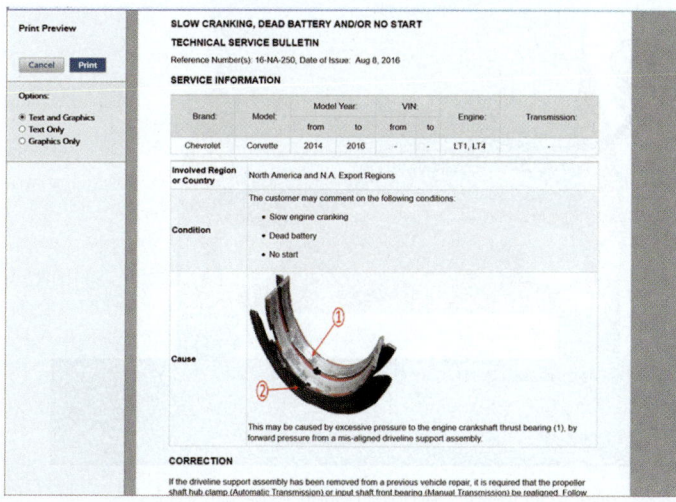

FIGURE 1-8 Technical service bulletin.

FIGURE 1-9 Aftermarket source.

K01006 Describe step three of the strategy-based diagnosis.

▶ **TECHNICIAN TIP**

A repair should never be performed unless the possible fault has been verified through testing. Do not let a possible fault become a possible mistake. In some cases, the list of possible faults can be found in the service information, but many times the technician will need to produce the list based on the concern, the information gathered, and the results of the research.

▶ **TECHNICIAN TIP**

The test description must provide enough information that someone could repeat the test with the same result. This is very important!

While these resources are essential, the list of possible faults is just that: a list of possible faults. A technician must always be aware that steps in the diagnostic process cannot be skipped.

Step 3: Focused Testing

Step three of the diagnostic process involves focused testing. In this step technicians use their testing skills to eliminate possible faults from the list they created in step two. Steps two and three work together; testing will start at a system level and work down to subsystems, then finally to individual components. The idea of focused testing should be to eliminate as many potential faults as possible with each test.

Focused testing is intended to eliminate possible causes with certainty. Each time a test is performed, the following three pieces of information must be recorded:

- a test description
- an expectation
- a result

These can be recorded on the repair order, electronic service record, or on an extra sheet of paper. Test records must be kept handy because they will become part of the documented record for this repair.

The three pieces of test information are recorded carefully for several reasons. Having an expectation before a test is performed makes each test objective and effective. The expectation is what the result is compared against, in order to determine if the vehicle passed or failed.

Many manufacturers, both original equipment and aftermarket, require that documented test results be submitted with each warranty claim. If the technician fails to document his or her work, the manufacturer will not pay the claim. The result is that the shop is out money for the parts and service, and the technician will not be paid for their work. Be sure to document the work properly (**FIGURE 1-10**).

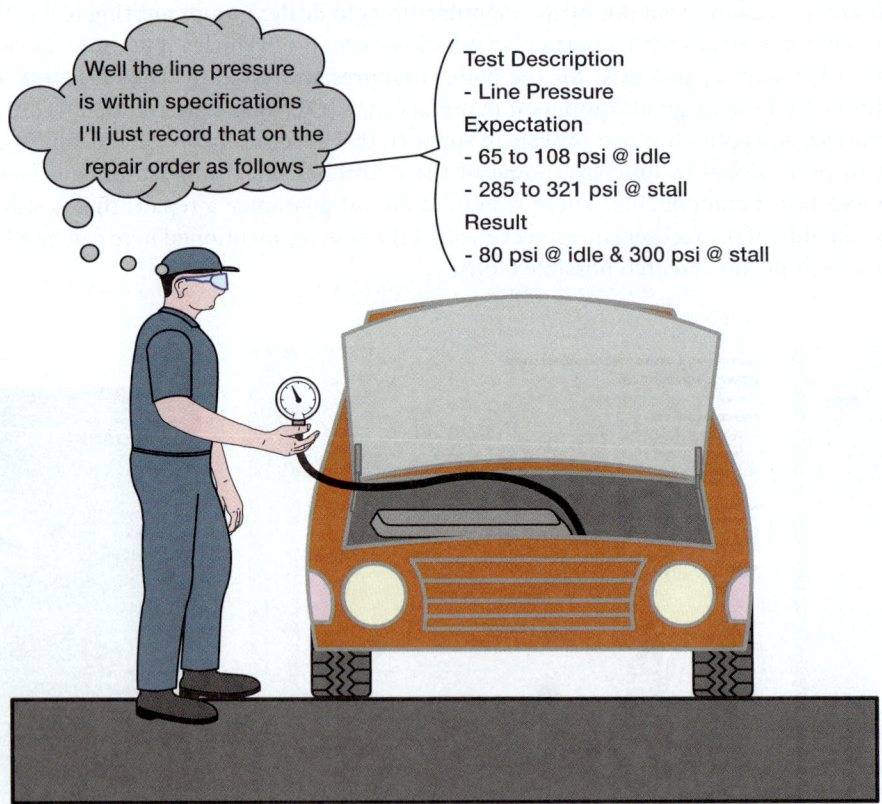

FIGURE 1-10 The test record should include the test description, expectation or specification, and the result of measurement.

1. The test description is not long, or even a complete sentence; it is simply a brief description. It allows the reader to know what test was performed and on what component or system. The test description should be accurate enough that the reader could repeat the test with the same result.
2. The expectation should describe the expected result as if the system is operating normally. The expectation could come directly from the system specifications listed in the manufacturer's service information or from system description and operation.
3. The result is the third part that must be recorded for each test. This information should accurately reflect what happened when the test was performed.

In summary, the testing is focused on isolating a fault or faults from the list of potential faults, and the results are compared to the expectation.

Testing should begin broadly and simply. Consider the following example: A light bulb circuit is suspected of having a fault. If the light bulb is easily accessible, the first test might be to check the voltage drop (i.e., voltage used to push current through the bulb). If the result of the voltage drop measurement is as expected (i.e., within specification), then the problem is in the bulb or socket. In this test, the technician is able to check the integrity of the entire electrical circuit with one test. If the result of the measurement is outside of the expectation (i.e., out of specification), the technician would know that the bulb is not the source of the problem. Further testing would isolate the problem to the ground or power side of the circuit.

The technician in the example performed a simple test with an easy expectation. The test allowed the technician to quickly determine the state of operation for the entire system/circuit and move on. If a fault had been found, then the technician would have isolated the cause of the customer's concern to that particular system/circuit and would need to perform further testing to isolate the cause to a particular component. To do that, the technician would use the service information to determine what components comprise the system and adjust the list from step two to take into account the new information. Then testing would continue.

The next test might measure voltage supply at the bulb (i.e., available voltage). In this way, the technician would be testing the power supply, the conductors, and the switch (assuming a power-side switched circuit). The technician would have an expectation for the circuit voltage and compare his or her result to this expected voltage. As we saw earlier, the technician is testing more than one component with a single test, thereby operating in an efficient manner.

This strategy—starting with broad, simple tests and moving to more complicated, pinpoint tests—makes efficient use of the technician's time while still effectively testing the possible faults.

1. A technician is investigating a customer concern of "no heat from the dash." The technician's investigation might begin with a simple list.
    a. Engine cooling system
    b. HVAC duct and controls
2. The technician would then eliminate one or the other and expand the list. The technician might verify coolant level and temperature at the inlet and outlet of the heater core. The HVAC components controlling and delivering warm air could then be used to expand the list for the next round of testing.
    a. Doors and ducts
    b. Cables
    c. Servos
    d. HVAC control head
    e. Blower motor
    f. Harness
    g. In-cabin filter or debris
3. Notice that the technician has moved from broad system tests to individual components or component groups. The technician's test continues to become more specific as the possibilities are narrowed down.

FIGURE 1-11 Select tests that have simple expectations and are easy to perform.

▶ **TECHNICIAN TIP**

When selecting tests, it is not a bad idea to choose those tests that might look at components of both systems (e.g., voltage drop on a shared electrical ground), but DO NOT attempt to test for both faults at once. While multiple faults within a companion or the same system often turn out to be related, they should be isolated and tested separately.

▶ **TECHNICIAN TIP**

When performing tests for an inspection under warranty, it is absolutely necessary to follow the manufacturers' guidelines.

Technicians commonly encounter vehicles with more than one customer concern. When these concerns both originate from the same or companion systems, technicians are inclined to search for one cause to both problems. Unfortunately, trying to diagnose two faults at once can quickly become problematic and confusing. Instead, select the easier customer concern and follow it through to the end. If both problems were caused by the same faults, then both were fixed. If they were caused by two separate faults, the technician is no worse off for having fixed one concern.

When selecting tests to perform, remember that they should be simple and easy (**FIGURE 1-11**). Except when following service procedures, you should select tests that have simple expectations, are easy to perform, and provide you with the maximum amount of information. This means simple tests that inspect an entire system or circuit are ideal ways to begin testing. Simple tests have expectations and results that are quickly understood and interpreted. They are short and involve basic tools and access to areas that are comfortable to reach.

When selecting tests, prioritize your testing. First choose tests that can be performed quickly and simply, even if they do not test an entire circuit. If a preferred test is in a difficult place to access, move to another test and come back to it, if needed. The answer may be found in the meantime and the time-consuming test can be avoided. Simple and easy tests are ideal, but they must be measurable or objective.

Yes, a visual inspection is a simple and valuable test, but a technician must determine what the issue is in an objective manner, with help from the service information. A guess based only on appearance is insufficient. If the service information says, "cracks in the serpentine belt indicate that it needs replaced," the belt can be visually and objectively (yes or no) tested. The belt will either have the indicated wear or it will not. If the service information states, "Chain deflection cannot exceed 0.75," then the deflection can be measured and compared to the specification. As testing continues, it may become necessary to use advanced tests, sophisticated equipment, additional time, or tests in areas difficult to access. Keeping initial testing simple and easy will produce the quickest, most reliable, and effective results.

When testing, use the recommended procedures and equipment. Manufacturers frequently recommend a particular procedure when testing one of their systems.

Failure to follow the specified service procedure can result in the warranty claim being denied by the manufacturer. In that case, both the shop and the technician lose money. Manufacturers may recommend a certain procedure because of the way their system is designed or monitored. Technicians must also be very careful to perform tests safely (**FIGURE 1-12**).

Beyond the mechanical dangers posed by automobiles, many of today's vehicles have dangerously high fluid pressures and deadly high voltage. It is of the utmost importance for the safety of the technician, and those working in the area, that safety procedures are always followed.

Proper test equipment and procedures are intended to test a particular component or system without causing any damage. Improper equipment or test procedures can create a second fault in the system being tested; making the technician's job even more difficult. For example, front probing an electrical terminal with the lead of a DMM can cause the terminal to spread or deform. This can create an intermittent high resistance or open within the circuit that was not there prior to the technician's test. Using the recommended equipment and procedures will help to ensure warranty claims are approved, people are safe, and testing goes smoothly.

When performing repairs, look beyond the obvious for the root cause. This simple suggestion can avoid customer comebacks. Novice technicians frequently have problems with misdiagnosing fuse-related issues. For example, a technician diagnoses a blown fuse as the cause of the customer concern. While replacing the fuse may have fixed the immediate fault, the technician did not look beyond the obvious. What causes a fuse to blow? Low

FIGURE 1-12 Always perform all tests safely.

resistance and increased amperage cause a fuse to blow. However, the technician did not test for one of these faults and the vehicle is likely to return with the same customer concern and the same blown fuse.

In another example of incomplete reasoning, a technician diagnoses a leaky transmission cooler line. The line is chaffed and leaking. This cooler line runs along the frame rail; the inner and outer tie rods are immediately below. The technician diagnoses the vehicle while it is on a lift and the suspension is unloaded (increasing the distance between the hose and steering linkage). The technician should have looked for the root cause of the chaffing, but instead the vehicle and customer come back some time later for the same concern. The technician notices several broken clips that held the flexible line into place on the frame rail. In both cases, the technician will work for free to repair the same vehicle, because time was not taken to ask the question: "Did something else cause this failure?" Testing must be focused beyond the obvious to identify the root cause of the problem and consequently avoid customer comebacks.

In summary, focused testing has several key elements. It picks up the possible faults identified in step two and begins testing each one broadly, narrowing down to more specific tests. Focused testing requires accurately documenting the tests performed, including a test description, expectation, and result, each and every time a test is performed. It should also be performed in a safe and proper manner, following manufacturers' guidelines and safety protocols. Focused testing is a safe, accurate, and repeatable method for isolating possible faults.

Step 4: Performing the Repair

The fourth step of the diagnostic process is to perform the repair. Although performing the repair is often the most straightforward step in the process, technicians must still avoid making several common mistakes. The following tips will help you to perform an effective and reliable repair.

K01007 Describe step four of the strategy-based diagnosis.

Use Proper Service Procedures

Manufacturers will often indicate what procedures are appropriate for their vehicles and components. Many design features and component materials require certain procedures be used and others avoided. Following the manufacturer's service information can prevent premature failure of the repair (FIGURE 1-13). For example, repair methods that are safe around the home may be unacceptable in the automotive industry. The use of twist-on wire connectors can create an unreliable and potentially dangerous electrical situation when used in a vehicle. Additionally, warranties, both original equipment and aftermarket, rely on the technicians' adherence to the manufacturer's service information. If technicians fail to do so, the warranty claim can go unpaid and the shop will lose money. Therefore, it is important for reliable repairs and warranty reimbursements that technicians follow the service information when performing repairs.

Use the Correct Tool for the Job

Failure to use the correct tool can lead to a customer comeback and injury to the technician. Proper tool selection is essential. If you are ever in doubt, refer to the manufacturer's service information. Improper tool use or selection can damage the component being installed or other components around it. For example, a technician may choose to install a pump busing with a hammer instead of using the recommended press and bushing driver. This incorrect tool selection can easily lead to misalignment, or damage to the bushing, pump, or torque converter. Using the wrong tool (or the right tool in the wrong manner) can also damage the tool and potentially injure those in the area. For example, if a technician is using a hardened chrome socket on an impact wrench, the socket may shatter, sending shrapnel flying. Using the correct tool for the job will produce better work and ensure the safety of the technician.

MAINTENANCE/SPECIFICATIONS

CHANGING YOUR WIPERS

The wiper arms can be manually moved when the ignition is disabled. This allows for ease of blade replacement and cleaning under the blades.

1. Disable the ignition before removing the blade.
2. Pull the arm away from the glass.
3. Left leading edge retaining block to release the blade. Swing the blade, away from with the arm, to remove it.
4. Swing the new blade toward the arm and snap it into place. Replace the retaining block at the leading edge of the wiper arm. Lower the wiper arm back to the windshield. The wiper arms will automatically return to their normal position the next time the ignition is enabled.

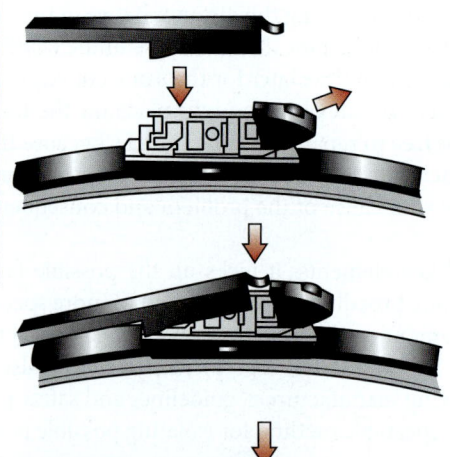

Refresh wiper blades at least twice a year for premium performance.

Poor preforming wipers quality can be improved by cleaning the blades and the windshield. See *Windows and wiper blades* in the *Cleaning* chapter.

To extend the life of wiper blades, scrape off the ice on the windshield BEFORE turning on the wipers. The ice has many sharp edges and will damage and shred the cleaning edge of your wiper blade.

FIGURE 1-13 A typical shop manual page has a task description broken into steps and diagrams or pictures to aid the technician.

Take Time to Perform the Repair Properly

Because technicians are frequently paid by the job, or a flat rate, rather than paid hourly, it is possible for technicians to feel a rush to complete their current job. Rushing increases the likelihood of a mistake. If a mistake occurs, the customer will come back with the vehicle and the technician will work for free to repair the mistake. For example, if a technician replaces a water pump and fills the coolant without bleeding the system, a potentially damaging situation can occur. The trapped gas can affect the flow of coolant and create a hot spot in the cylinder head. This can lead to warning lights, poor performance, and possible engine damage. Take a little extra time to ensure that the work is performed correctly, with the right tools and the proper service procedures. Taking time to perform the repair will ensure fewer "comebacks" and more satisfied customers.

Make Sure the Customer Approves of the Repair

This may seem trivial, but it is extremely important. Most states' laws protect consumers by preventing unauthorized services from being charged or performed. This means that technicians cannot just repair a vehicle and charge the customer for the cost incurred. If the customer is paying, shops must receive a customer's approval prior to performing repairs.

> ▶ **TECHNICIAN TIP**
>
> It is very important to quote accurately and wait for approval before performing repairs on a customer's vehicle.

Check for Updates Prior to the Repair

It is also good practice to check for updated parts and software/firmware before performing a repair. It is possible that manufacturers have become aware of a problem with a particular component or software version and issued a software update or produced an updated component. When performing repairs it is a good idea to check for these sorts of updates (often found in TSBs), because it may prevent a customer comeback. Software updates are often downloaded from the manufacturer's website. For hard parts, the best resource is frequently the respective dealership's parts department.

Technical service bulletins also provide information related to unexpected problems, updated parts, or changes to a repair procedure on a particular vehicle system, part, or component. The typical TSB contains step-by-step procedures and diagrams on how to identify if there is a fault and perform an effective repair. Shops typically keep TSBs in a central location, or you may look them up online. Compare the information contained in the TSB with that of the shop manual. Note the differences and, if necessary, copy the TSB to take with you to perform the repair.

Pay Attention to Details

Performing the repair is straightforward but requires attention to detail. There are several things to keep in mind. Proper service procedures can be located in the manufacturer's service information. The correct tool for the job will lessen injury and ensure reliability. Use the necessary time to make sure that the repair was completed correctly. Document your work. These tips can greatly improve the likelihood of a successful repair, but the process does not stop with the repair.

Step 5: Verify the Repair

The most important step of the strategy-based diagnostic process is verifying the repair. The reason that this is the most important step is straightforward. The vehicle would never have been in the shop if the customer did not have a concern. If the technician fails to address the original concern, the customer may view the trip as ineffective, a waste of their time and money. Even when a valid repair that makes the vehicle safer and more reliable was performed, the customer will still be unsatisfied if his or her original concern was not addressed. For example, a customer brings the vehicle into the shop for a sticky glove box latch. The technician identifies and repairs a dangerous brake line leak, but fails to fix the glove box. Some customers may view this trip to the shop as unsuccessful because it failed to fix their original issue. When verifying the repair, technicians must always double check their work. This is a valuable confirmation that the repair performed did fix the identified problem.

K01008 Describe step five of the strategy-based diagnosis.

There are several ways to verify a repair, but generally, the simplest method is the best method. For example, a customer is concerned that the wipers stop moving when the switch is moved into the high position. In step one, the technician will verify that the customer concern and fault exist by turning the wiper switch to all positions. Then the technician uses the wiring diagram (step 2) to diagnose a fault (step 3) within the wiring harness and repairs it (step 4). The technician could then verify the repair by performing the last diagnostic test (from step 3) again. In most cases, the repair would be confirmed if the results had changed and were within expectation/specification.

But what if there was a second problem affecting the wipers such as worn brushes in the motor, or the wiper linkage fell off of the pivot on one side? The customer would still have issues with the wipers and would likely to be unhappy with the repair. So while performing the last diagnostic test (step 3) is a valid verification method, it is not foolproof. An easier method exists: simply return to the process used in step one to verify the customer concern. If the repair has eliminated the problem, the technician should now be able to turn the wiper switch to all positions (step 1) and confirm normal operation. Be certain to perform the same inspections used to verify the customer concern in step one after the repair is performed. This may include checking the entire system operation, not just a single function. This method of verifying that the customer concern is resolved is usually best in most scenarios because it is simple and it is exactly what the customer will do to check your work.

However, sometimes verifying the repair requires a more complicated means of verification. A common concern that falls into this scenario is as follows: The customer brings their vehicle in with a concern that the MIL (malfunction indicator lamp) or "check engine" light is illuminated. In this scenario, *NEVER* verify the repair by simply checking to see that the light is off. While this is what the customer will do to check your work, the failure of the MIL to light can often be misleading and result in a comeback for the exact same problem. This can occur because the MIL is illuminated when tests run by the computers in the vehicle fail. The computers are constantly running tests, but some tests require very specific conditions before they can be run and, hence, fail. Due to the requirements for the conditions to be right, simply checking to see if the light is illuminated is an inadequate method of verification.

For more complicated computer-controlled systems, the best method of verification is checking the test results stored on the vehicle's computer. This option will require an electronic scan tool that communicates with the vehicle's computer, along with a high level of diagnostic experience and service information to verify that the concern has been fully resolved. If the communication option is not available, the second best method of verification is repeating the last diagnostic test performed (in step 3) and confirming that the result has changed to now match the expectation/specification. Complicated computer-controlled systems require that the technician do more than verify the customer concern is eliminated. The technician will have to repeat a diagnostic test (step 3) or view test results stored on the vehicle's computer (this is the preferred method) in order to verify that the repair was effective.

Step five of the strategy-based diagnostic process is the most important. A vehicle should never be returned to a customer without this step completed.

▶ **TECHNICIAN TIP**

The job is not complete until you have verified that the repair resolved the customer's concern.

▶ Documenting the Repair

The first two components, gathering information from the customer and the strategy-based diagnostic process, have already been described; this section discusses documentation. The repair is documented for several reasons: accurate vehicle history, returns or comebacks, and OEM or aftermarket warranties. Keeping accurate service records will help technicians to know what services and repairs have been performed on a vehicle when it needs any future services. This can be invaluable during the diagnostic process and can also help service advisors and technicians identify what maintenance or recall work still needs to be completed. Documenting the repair also helps technicians in the event that a vehicle returns, now or in the future, with the same customer concern or fault. This can help to identify defective parts or common problems.

Warranty work is another reason that all repairs must be documented. Whether the repair is submitted to an original equipment manufacturer or to an aftermarket warranty company for reimbursement, the repair must be well documented. Warranty clerks will review the repair order to ensure that proper testing and repair procedures have been followed. Technicians must document their work to ensure that the shop, and in turn the technician, get paid for the work performed.

Finally, documenting the work provides the shop with a record that the work was initiated and completed. This is important in case the vehicle is later involved in an accident or other mishap and the shop is involved in a lawsuit. It is important to have the customer sign or initial, depending on shop policy, the repair order, to verify that the customer accepted the repair.

The Three Cs of Documentation

When documenting a repair, technicians need to remember the 3 Cs: concern, cause, and correction (FIGURE 1-14).

K01009 Explain how the three Cs are applied in repairing and servicing vehicles.

N01002 Demonstrate use of the three C's (concern, cause, and correction).

Concern

The main focus of the 3 Cs is the customer *concern*, which is also the focus of step one of the diagnostic process. Often the concern is documented on the repair order prior to the technician receiving the vehicle. If this is the case, the technician who works on the vehicle should take time to fully understand the concern, read the repair order, and possibly talk further with the customer to understand the nature of the problem. Think through the problem and develop a strategy to attack it. Other symptoms and diagnostic troubles codes are some examples of other information that should be included in the "concern."

Cause

The second C in the 3 Cs is *cause,* which details the cause of the customer concern. This correlates to the documentation done in step three of the diagnostic process. The technician should document any tests that they perform with enough detail that they can be repeated, as well as specifications/expectations, and results. This goes for all tests, even the simple ones.

Correction

The technician should then document the last C, the correction. This must include the procedure used as well as a brief description of the correction. This information comes from the fourth step of the diagnostic process. When documenting the repair order, technicians should include the customer concern and symptoms (DTCs are symptoms); brief descriptions of tests; expectations; and results, along with the procedure and repair that were performed. The technician should also include all parts that were replaced as well, and noted if they were new or used, OEM, or aftermarket.

Other Parts of Documentation

Additional service recommendations should also be documented on the repair order. While working on the vehicle, technicians should also be mindful of other work that may need to be performed. Technicians are obligated to make the customer aware of safety concerns that require attention. Customers may be unaware of a potential hazard or lack of maintenance. Bringing this to the attention of the customer right away can help the technician, as well as the customer. For example, if the technician is already working on the vehicle, they would not have to remove the vehicle from the service bay, bring in a new vehicle, and start all over. Repairing multiple issues in one trip to the service bay makes good use of the technician's time. It also improves customer relations by bringing the customer's attention to problems and thereby preventing possible failures.

FIGURE 1-14 The 3Cs of documenting the repair.

CHAPTER 1 Strategy-Based Diagnostics

For example, a technician may be changing the fluid in a transmission and notice that the brake friction pads are extremely low. Bringing this to the attention of the customer can result in additional work for the technician and save the customer from a potentially more costly repair. For this reason, technicians should also note safety issues and maintenance items on the repair order.

Repair Order

K01010 Describe the information and its use within a repair order.

N01003 Identify information needed and the service requested on a repair order.

S01002 Complete a repair order.

A **repair order** is a key document used to communicate with both your customers and coworkers. Thoroughly document the information provided by the customer on the repair order; every bit of it may be helpful during the diagnosis (**FIGURE 1-15**). If you are not typing this information, make sure your handwriting is clear and easy for others to read. Unfortunately, if documentation of a complaint is not done well, the technician could be led on a much longer diagnostic path, wasting everyone's time. It can also be a time-consuming process for the diagnosing technician to make contact with the customer in order to get more information that was missed the first time. From time to time, it may be inevitable that the customer will need to be contacted for further inquiry after the initial visit. However, carefully gathering information from the customer on their initial visit will save time, prevent inconveniencing the customer, and aid in the diagnostic process.

To complete a repair order using the 3 Cs, follow the steps in **SKILL DRILL 1-2**.

▶ TECHNICIAN TIP

A repair order is a legal document that can be used as evidence in the event of a lawsuit. Always make sure the information you enter on a repair order is complete and accurate. The information required on a repair order includes: date; customer's name, address, and phone number; vehicle's year, make, model, color, odometer reading, and VIN; and description of the customer's concern. Store repair orders in a safe place, such as in a fireproof filing cabinet or electronically on a secure computer network. Finally, to prevent future complications, it is a good idea to have the customer sign or initial the repair order, indicating that they understand and agree to the needed repair. Having the customer's signature will help prevent the shop from being held liable in an accident involving the vehicle later.

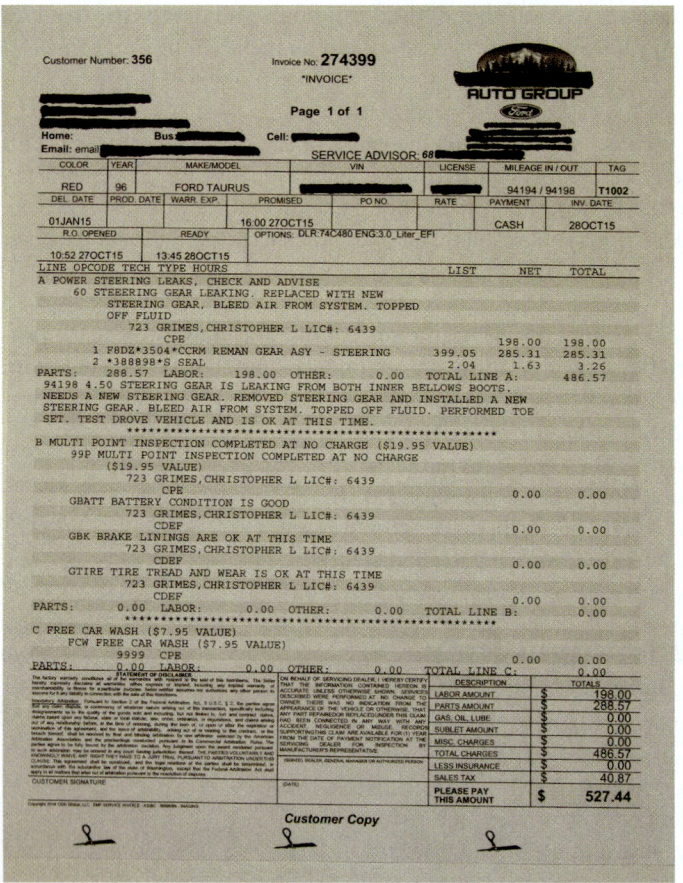

FIGURE 1-15 A repair order.

Applied Communications

AC-23: Repair Orders: The technician writes a repair order containing customer vehicle information, customer complaints, parts and materials used (including prices), services performed, labor hours, and suggested repairs/maintenance.

A repair order is a legal contract between the service provider and the customer. It contains details of the services to be provided by you and the authorization from the customer. To make sure everyone understands clearly what is involved, a repair order should contain information about the following aspects of the repair.

- Your company or service providers: The service provider section contains the company name, address, and contact details; the name of a service advisor who is overseeing the job; and the amount of time the service technician will have to service the vehicle.
- The customer: The customer section contains the customer's name, address, and contact phone numbers.
- The customer's vehicle: The vehicle section includes details about the vehicle to be serviced. Check the vehicle's license plate before starting work. The license plate numbers are usually unique within a country. You should also record information about the vehicle's make, model, and color. This information will make it easier for you to locate the vehicle on the parking lot. You may also need to know the manufacture date of the vehicle to be able to order the right parts. The odometer reading and the date will help keep track of how much distance the vehicle travels and the time period between each visit to the shop. The VIN is designed to be unique worldwide and contains specific information about the vehicle. Many shops do a "walk-around" with the customer to note any previous damage to the vehicle and to look for any obvious faults such as worn tires, rusted-out exhaust pipes, or torn wiper blades.
- The service operations: This section contains the details of the service operations and parts.
- The first part is the service operation details. For example, the vehicle is in for a 150,000 mile (240,000-km) service, which can be done in 3 hours and results in approximately $300 of labor costs. The information about the chargeable labor time to complete a specific task can be found in a labor guide manual. In some workplaces, this information is built into the computer system and will be automatically displayed.
- The second part of this section is the details of parts used in the service, including the descriptions, quantities, codes, and prices. The codes for each service and part are normally abbreviations that are used for easy reference in the shop. Some shops may have their own reference code system.
- As you do the vehicle inspection, you may discover other things that need replaced or repaired. These additional services can be recorded in another section. It is essential that you check with the customer and obtain their approval before carrying out any additional services.
- The parts requirements: This section lists the parts required to perform the repair.

Some repair orders also contain accounting information so they can be used as invoices.

SKILL DRILL 1-2 Completing a Repair Order

1. Greet the customer.
2. Locate a repair order used in your shop and obtain or verify the customer's name, address, and phone number.
3. Obtain details about the vehicle, including the year, make, model, color, odometer reading, and VIN.
4. Ask the customer to tell you more about the *concern* by using open-ended questions, such as "When does problem occur?" "At what speed(s)?" "How do you experience the problem?" "How long has this been occurring?" "How many passengers do you typically carry?" Type or clearly write the customer's responses on the repair order.
5. Ask the customer about other changes with the vehicle, such as recent work, or recent travel. Type or clearly write the customer's responses on the repair order.
6. Remembering the lessons learned regarding the proper diagnostic process, begin to verify the customer's *concern* by first performing a visual inspection.
7. If you see nothing unusual during your visual inspection, continue to verify the customer's concern by conducting a road test of the vehicle. The customer may ride along, if possible, to help identify the issue as it occurs, or you may conduct the test by yourself. Following the test drive, after verifying the customer's concern, record it on the repair order.
8. The second step of the diagnostic process is to research the possible faults, and gather information. Access the vehicle service history to determine if the vehicle has experienced a similar problem in the past, requires a routine service maintenance, or has been serviced recently. Document this information, if applicable, on the repair order.
9. Conduct research by accessing various sources of information related to the vehicle, such as the vehicle service manual or the owner's manual. Check to see if a TSB related to the issue exists. As part of the process, rule out the possibility that the customer's concern is a normal operation of the vehicle.
10. Now that you have your broad list of possible faults related to the concern, begin step three, focused testing. Choose one of the possible broad faults you identified in step two. Now refer to the service manual to locate information that matches the concern. Service manuals usually contain diagnostic charts to aid in the focused testing process.
11. Conduct a test and record its description, your expectation, and the result on the repair order or another piece of paper. Continue to check each possible fault until you identify the *cause* of the concern.
12. Once you have identified the fault, you're ready for step four, performing the repair. You would inform the customer of your finding and obtain his or her approval to make the repair. Pending customer approval, you would then follow proper safety procedures and use the manufacturer's guidelines to correct the problem, being sure to use the correct tools and taking the time to complete the job properly.
13. Once you've made the repair, you are ready for step five, verifying the repair. The simplest way to verify that you have addressed and corrected the customer's concern is to repeat the test drive. Take the vehicle for a test drive and repeat the tests you initially performed. Is the issue gone? If so, you have verified the repair and can return the vehicle to the customer.
14. Document the correction on the repair order. If the issue is not resolved, you must return to your list of possible faults and continue testing after first alerting the customer that additional work and time will be necessary.

Wrap-Up

Ready for Review

- Service history is typically retrieved from service records kept by the shop, dealer network, original equipment manufacturer (OEM), or aftermarket service center and contains a list of services performed on a vehicle and the date and mileage at which they were completed.
- The service history allows technicians to determine if the vehicle has been well maintained. This can be extremely useful when a technician suspects that lack of maintenance may be the cause of the problem.
- Failure to comply with the state and federal law can be very expensive for the dealership and manufacturer.
- Today's vehicles also require regular software updates made available to fix a bug or glitch in the computer programming. These updates are often designed to eliminate a customer concern, improve owner satisfaction, or increase vehicle life.
- The strategy-based diagnostic process is focused on fixing problems correctly the first time. It begins with identifying the customer's concern and ends with confirming that the problem has been resolved.
- The problem-solving process provides a consistent road map for technicians as they address customer concerns that require diagnosis and to make sure that customer concerns are resolved with certainty.
- Strategy-based diagnosis simplifies the problem-solving portion of the repair, making the job easier for the technician; it prevents technicians from having to work on the same job more than once; and it all but eliminates customer comebacks.
- Customer comebacks are usually caused by the customer concern being misinterpreted or misunderstood or failing to verify that the original concern was resolved.
- The strategy-based diagnostic process begins by gathering preliminary information from the customer and by reviewing the vehicle's service history.
- The first step in the diagnostic process is to verify the customer concern. This step is completed for two main purposes: verify that there is an actual problem present, and guarantee that the customer's concern is addressed.
- Visual inspections can be very valuable, but technicians need to be careful not to jump to conclusions.
- DTC's (Diagnostic Trouble Codes) and freeze frame data should always be saved and recorded on the repair order. Freeze-frame data provide a snapshot of the entire engine data when the DTC occurs, which allows for duplication of the condition so that the DTC can be replicated.
- The second step in the diagnostic process is to research possible faults. The goal of this step is to create a list of possible faults. The list will be created based on the information gathered in step 1 and narrowed down by the tests performed in step 3 until the cause of the concern has been confirmed.
- The best source of information is usually the manufacturer's service information system.
- Technical service bulletins (TSBs) are service notifications and procedures sent out by the manufacturers to dealer groups, alerting technicians to common issues with a particular vehicle or group of vehicles. Some aftermarket sources also exist for the pattern failures addressed by TSBs
- A technician must always be aware that steps in the diagnostic process cannot be skipped. A repair should never be performed unless the possible fault has been verified through testing.
- Step 3 of the diagnostic process involves focused testing, where technicians use their testing skills to eliminate possible faults from the list they created in step two. Steps 2 and 3 work together, because testing starts at a system level and works down to subsystems, then finally to individual components.
- When selecting tests prioritize your testing. First choose tests that can be performed quickly and simply, even if they do not test an entire circuit. If a preferred test is in a very difficult place to access, move to another test and come back to it, if needed.
- Following manufacturers' guidelines and safety protocols keeps technicians safe. Focused testing is a safe, accurate, and repeatable method for isolating possible faults. Once the fault has been isolated, it is time to perform the repair.
- The fourth step of the diagnostic process is to perform the repair. Performing the repair is often the most straightforward step in the process.
- Use proper service procedures when performing a repair. Manufacturers often indicate what procedures are appropriate for their vehicles and components.
- Use the correct tool for the job when performing a repair. Failure to use the correct tool for the job can lead to a customer comeback and injury to the technician.
- Take time to perform the repair properly. Technicians are frequently paid by the job, or flat rate, rather than paid hourly, it is possible for technicians to feel a rush to complete their current job. Rushing increases the likelihood of a mistake and the next time you may pay for it.
- The most important step of the strategy-based diagnostic process is verifying the repair. The reason that this is the most important step is straightforward. The vehicle would never have been in the shop if the customer did not have a concern.
- Verifying the original concern is the best method of double-checking your work and meeting your customers' expectations. The job is not complete until you have verified that the repair resolved the customer's concern.

- Documentation is key to effective and efficient repairs. Keeping all the information available to the service advisor, technician, and the customer allows for a more open dialogue which can limit the confusion of the repair process.
- The repair is documented for several reasons: accurate vehicle history, returns or comebacks, and warranties. Keeping accurate service records will help technicians to know what services and repairs have been performed on a vehicle when it needs any future services.
- When documenting a repair, technicians need to remember the three Cs: concern, cause, and correction.
- When documenting the repair order, technicians should include the customer concern and symptoms (Diagnostic Trouble Codes are symptoms) and a brief description of tests, expectations, and results, along with the procedure and repair that were processed.

Key Terms

Strategy-Based Diagnostic Process A systematic process used to diagnose faults in a vehicle.

service advisor The person at a repair facility that is in charge of communicating with the customer.

service history A complete listing of all the servicing and repairs that have been performed on that vehicle.

repair order The document that is given to the repair technician that details the customer concern and any needed information.

freeze frame data Refers to snapshots that are automatically stored in a vehicle's power train control module (PCM) when a fault occurs (only available on model year 1996 and newer).

technical service bulletin (TSB) Service notifications and procedures sent out by the manufacturers to dealer groups alerting technicians about common issues with a particular vehicle or group of vehicles.

original equipment manufacturer (OEM) The company that manufactured the vehicle.

aftermarket A company other than the original manufacturer that produces equipment or provides services.

intermittent faults A fault or customer concern that you can not detect all of the time and only occurs sometimes.

3 Cs A term used to describe the repair documentation process of 1st documenting the customer concern, 2nd documenting the cause of the problem, and 3rd documenting the correction.

concern Part of the 3Cs, documenting the original concern that the customer came into the shop with. This documentation will go on the repair order, invoice, and service history.

cause Part of the 3Cs, documenting the cause of the problem. This documentation will go on the repair order, invoice, and service history.

correction Part of the 3Cs, documenting the repair that solved the vehicle fault. This documentation will go on the repair order, invoice, and service history.

Review Questions

1. When a vehicle comes in for repair, detailed information regarding the vehicle should be recorded in the:
   a. service booklet.
   b. repair order.
   c. vehicle information label.
   d. shop manual.
2. The service history of the vehicle gives information on whether:
   a. the vehicle was serviced for the same problem more than once.
   b. an odometer rollback has occurred.
   c. the vehicle meets federal standards.
   d. the vehicle has Vehicle Safety Certification.
3. Which of the following steps is the last step in a strategy-based diagnostic process?
   a. Verifying the customer's concern
   b. Researching possible faults
   c. Performing the repair
   d. Verifying the repair
4. When possible, which of the following is the best way to understand the customer's concern?
   a. Asking the customer to guess the cause of the problem.
   b. Asking the customer to suggest a solution to the problem.
   c. Encouraging the customer to demonstrate the problem.
   d. Encouraging the customer to help you fix the problem.
5. The best way to address intermittent faults is to:
   a. look for symptoms, data, or DTCs that are repeatable or consistent.
   b. reverse the steps in the diagnostic process.
   c. ask the customer to bring back the vehicle when the fault occurs.
   d. take it up only when it is covered by warranty.
6. When the technician encounters a vehicle with more than one customer concern, and both originate from companion systems, the technician:
   a. should attempt to test for both faults at once.
   b. need not attempt to fix the second fault.
   c. should never choose those tests that might look at components of both systems.
   d. should isolate the faults and test them separately.
7. Choose the correct statement.
   a. When performing tests for an inspection under warranty, follow your intuition rather than the manufacturers' guidelines.
   b. Researching possible faults should begin with a specific cause in mind.
   c. For hard parts, the best resource is frequently the respective dealership's parts department.
   d. DTCs and freeze frame data need not be captured before clearing the memory.
8. All of the following will happen if the technician fails to document test results *except*:
   a. The manufacturer will not pay the claim.
   b. The shop is out money for the parts and service.

c. The technician will be unable to diagnose the fault.
d. The technician will not be paid for his or her work.

9. All of the following statements with respect to the 3 Cs are true *except*:
   a. Customer concern is documented on the repair order prior to the technician receiving the vehicle.
   b. The second C in the 3 Cs refers to the cause of the customer's concern.
   c. Technicians should note safety issues and maintenance items on the repair order.
   d. Additional service recommendations should never be documented on the repair order.

10. Which of the following is not one of the 3 Cs of vehicle repair?
    a. Cause
    b. Cost
    c. Concern
    d. Correction

ASE Technician A/Technician B Style Questions

1. Tech A says that when diagnosing a transmission problem, it is important to first verify the customer concern by taking the vehicle on a road test if possible. Tech B says that you should check for TSBs during the diagnostic process. Who is correct?
   a. Tech A only
   b. Tech B only
   c. Both A and B
   d. Neither A nor B

2. Tech A says that additional service recommendations should be documented on the repair order. Tech B says technicians are obligated to make the customer aware of safety concerns that require attention. Who is correct?
   a. Tech A only
   b. Tech B only
   c. Both A and B
   d. Neither A nor B

3. Tech A says the strategy-based diagnostic process is a scientific process of elimination. Tech B says the strategy-based diagnostic process begins with scanning the vehicle for DTCs. Who is correct?
   a. Tech A only
   b. Tech B only
   c. Both A and B
   d. Neither A nor B

4. Tech A says that manufacturers will often indicate what procedures are appropriate for their vehicles and components. Tech B says that the manufacturer's service information can avoid premature failure of the repair. Who is correct?
   a. Tech A only
   b. Tech B only
   c. Both A and B
   d. Neither A nor B

5. Tech A says that technicians are frequently paid by the job, or flat rate, rather than paid hourly. Tech B says rushing the repair is best for the customer, so they get their vehicle back quickly. Who is correct?
   a. Tech A only
   b. Tech B only
   c. Both A and B
   d. Neither A nor B

6. Tech A says that it is only necessary for dealerships to check for updated parts and software/firmware before performing a repair. Tech B says that it is possible that manufacturers have become aware of a problem with a particular component or software version and have issued a software update. Who is correct?
   a. Tech A only
   b. Tech B only
   c. Both A and B
   d. Neither A nor B

7. Tech A says that the customer concern is the focus of step 1 of the diagnostic process. Often the concern is documented on the repair order prior to the technician receiving the vehicle. Tech B says the technician who works on the vehicle should take time to fully understand the concern, read the repair order, and possibly talk further with the customer to understand the nature of the problem. Who is correct?
   a. Tech A only
   b. Tech B only
   c. Both A and B
   d. Neither A nor B

8. Tech A says that a repair order is only used in the shop and will be discarded when the vehicle is complete. Tech B says that the repair order is a legal document and could be used in a court. Who is correct?
   a. Tech A only
   b. Tech B only
   c. Both A and B
   d. Neither A nor B

9. Two technicians are discussing a transmission problem. Tech A says that it is important to test drive the vehicle because there may actually be no issue with the vehicle and the customer complaint is actually a normal operational characteristic of the transmission. Tech B says you should always check TSBs before performing any service of the transmission because the manufacturer may have updated a component. Who is correct?
   a. Tech A only
   b. Tech B only
   c. Both A and B
   d. Neither A nor B

10. Tech A says that experience will allow you to skip many of the steps of the diagnostic process because you will be familiar with the transmission. Tech B says that skipping steps of the diagnostic process can cause issues to be missed, or misdiagnosis of the problem. Who is correct?
    a. Tech A only
    b. Tech B only
    c. Both A and B
    d. Neither A nor B

CHAPTER 2
Safety

NATEF Tasks

- **N02001** Comply with the required use of safety glasses, ear protection, gloves, and shoes during lab/shop activities.
- **N02002** Identify and wear appropriate clothing for lab/shop activities.
- **N02003** Identify general shop safety rules and procedures.
- **N02004** Utilize proper ventilation procedures for working within the lab/shop area.
- **N02005** Identify the location and the types of fire extinguishers and other fire safety equipment; demonstrate knowledge of the procedures for using fire extinguishers and other fire safety equipment.
- **N02006** Identify the location of the posted evacuation routes.
- **N02007** Locate and demonstrate knowledge of safety data sheets (SDS).

Knowledge Objectives

After reading this chapter, you will be able to:

- **K02001** Describe the personal safety equipment and precautions for the workplace.
- **K02002** Describe the different kinds of hand protection.
- **K02003** Understand why it is important to wear headgear.
- **K02004** Describe the types of ear protection.
- **K02005** Describe the types of breathing devices.
- **K02006** Describe the types of eye protection.
- **K02007** Describe proper lifting techniques.
- **K02008** Comply with safety precautions in the workplace.
- **K02009** Describe how OSHA rules and the EPA impact the automotive workplace.
- **K02010** Explain how shop policies, procedures, and safety inspections make the workplace safer.
- **K02011** Describe the importance of demonstrating a safe attitude in the workplace.
- **K02012** Identify workplace safety signs and their meanings.
- **K02013** Describe the standard safety equipment.
- **K02014** Maintain a safe air quality in the workplace.
- **K02015** Describe appropriate workplace electrical safety practices.
- **K02016** Prevent fires in the workplace.
- **K02017** Identify hazardous environments and the safety precautions that should be applied.
- **K02018** Identify the proper method to clean hazardous dust safely.
- **K02019** Explain the basic first aid procedures when approaching an emergency.

Skills Objectives

After reading this chapter, you will be able to:

- **S02001** Maintain a clean and orderly workplace.
- **S02002** Use information in an SDS.
- **S02003** Properly dispose of used engine oil and other petroleum products.

You Are the Automotive Technician

It's your first day on the job, and you are asked to report to the main office, where your new supervisor gives you your PPE. Before you can begin working on the shop floor, you are given training on the proper use of PPE. Here are some of the questions you must be able to answer.

1. Which type of gloves should be worn when handling solvents and cleaners?
2. Why must safety glasses be worn at all times in the shop?
3. Why should rings, watches, and jewelry never be worn in the shop?
4. When should hearing protection be worn?
5. For what types of tasks should a face shield be worn?
6. Why must hair be tied up or restrained in the shop?
7. Which type of eye protection should be worn when using or assisting a person using an oxyacetylene welder?

K02001 Describe the personal safety equipment and precautions that should be used in the workplace.

N02001 Comply with the required use of safety glasses, ear protection, gloves, and shoes during lab/shop activities.

N02002 Identify and wear appropriate clothing for lab/shop activities.

▶ Introduction

Motor vehicle servicing is one of the most common vocations worldwide. Hundreds of thousands of shops service millions of vehicles every day. That means at any given time, many people are conducting automotive servicing, and there is great potential for things to go wrong. It is up to you and your workplace to make sure all work activities are conducted safely. Accidents are not caused by properly maintained tools; accidents are generally caused by people.

▶ Personal Safety

Personal safety is not something to take lightly. Accidents cause injury and death every day in workplaces across the world (**FIGURE 2-1**). Even if accidents don't result in death, they can be very costly in lost productivity, disability, rehabilitation, and litigation costs. Because workplace safety affects people and society so heavily, government has an interest in minimizing workplace accidents and promoting safe working environments. The primary federal agency for workplace safety is the Occupational Safety and Health Administration (OSHA). States have their own agencies that administer the federal guidelines as well as create additional regulations that apply to their state.

Personal protective equipment (PPE) is equipment used to block the entry of hazardous materials into the body or to protect the body from injury. PPE includes clothing, shoes, eye protection, face protection, head protection, hearing protection, gloves, masks, and respirators (**FIGURE 2-2**). Before you undertake any activity, consider all potential hazards and select the correct PPE based on the risk associated with the activity. For example, if you are going to change hydraulic brake fluid, put on some impervious gloves to protect your skin from chemicals.

Protective Clothing

Protective clothing includes items like shirts, vests, pants, shoes, and gloves. These items are your first line of defense against injuries and accidents, and clothing appropriate for the task must be worn when performing any work. Always make sure protective clothing is kept clean and in good condition. You should replace any clothing that is not in good condition, as it is no longer able to fully protect you. Types of protective clothing materials and their uses are as follows:

- Paper-like fiber: Disposable suits made of this material provide protection against dust and splashes.
- Treated wool and cotton: Adapts well to changing workplace temperatures. Comfortable and fire resistant. Protects against dust, abrasion, and rough and irritating surfaces.

FIGURE 2-1 Accidents are costly.

- Duck: Protects employees against cuts and bruises while they handle heavy, sharp, or rough materials.
- Leather: Often used against dry heat and flame.
- Rubber, rubberized fabrics, neoprene, and plastics: Provides protection against certain acids and other chemicals.

Source: PPE Assessment, Occupational Safety & Health Administration, U.S. Department of Labor.

Always wear appropriate work clothing. Whether this is a one-piece coverall/overall or a separate shirt and pants, the clothes you work in should be comfortable enough to allow you to move, without being loose enough to catch on machinery (**FIGURE 2-3**). The material must be flame retardant and strong enough that it cannot be easily torn. A flap must cover buttons or snaps. If you wear a long-sleeve shirt, the cuffs must be close fitting, without being tight. Pants should not have cuffs so that hot debris cannot become trapped in the fabric.

Always wash your work clothes separately from your other clothes to prevent contaminating your regular clothes. Start a new working day with clean work clothes, and change out of contaminated clothing as soon as possible. It is a good idea to keep a spare set of work clothes in the shop in case the ones you are wearing become overly dirty or a toxic or corrosive fluid is spilled on them.

The proper footwear provides protection against items falling on your feet, chemicals, cuts, abrasions, punctures, and slips. They also provide good support for your feet, especially when working on hard surfaces like concrete. The soles of your shoes must be acid and slip resistant, and the uppers must be made from a puncture-proof material such as leather.

Some shops and technicians prefer safety shoes with a steel toe cap to protect the toes. Always wear shoes that comply with your local shop standards.

FIGURE 2-2 Personal protective equipment (PPE) includes clothing, shoes, safety glasses, hearing protection, masks, and respirators.

▶ **TECHNICIAN TIP**

Each shop activity requires specific clothing, depending on its nature. Research and identify what specific type of clothing is required for every activity you undertake. Wear appropriate clothing for the activity you will be involved in, according to the shop's policies and procedures.

FIGURE 2-3 A. One-piece coverall. **B.** Shirt and pants.

K02002 Describe the different kinds of hand protection.

Hand Protection

Hands are a very complex and sensitive part of the body, with many nerves, tendons, and blood vessels. They are susceptible to injury and damage. Nearly every activity performed on vehicles requires the use of your hands, which provides many opportunities for injury. Whenever required, wear gloves to protect your hands. There are many types of gloves available, and their applications vary greatly as you will see below. It is important to wear the correct type of glove for the various activities you perform. In fact, when working on rotating equipment, it may be necessary for you to remove your gloves so that they don't get caught in the machinery and pull you into it.

Heavy-duty and impenetrable chemical gloves should always be worn when using solvents and cleaners. They should also be worn when working on batteries. Chemical gloves should extend to the middle of your forearm to reduce the risk of chemicals splashing onto your skin (**FIGURE 2-4**). Always inspect chemical gloves for holes or cracks before using them, and replace them when they become worn. Some chemical gloves are also slightly heat resistant. This type of chemical glove is suitable for use when removing radiator caps and mixing coolant.

Leather gloves protect your hands from burns when welding or handling hot components (**FIGURE 2-5**). You should also use them when removing steel from a storage rack and when handling sharp objects. When using leather gloves for handling hot components, be aware of the potential for **heat buildup**. Heat buildup occurs when the leather glove can no longer absorb or reflect heat, and heat is transferred to the inside of the leather glove. At this point, the leather gloves' ability to protect you from the heat is reduced, and you need to stop work, remove the leather gloves, and allow them to cool down before continuing to work. Also avoid picking up very hot metal with leather gloves, because it causes the leather to harden, making it less flexible during use. If very hot metal must be moved, it would be better to use an appropriate pair of pliers.

Light-duty gloves should be used to protect your hands from exposure to greases and oils (**FIGURE 2-6**). Light-duty gloves are typically disposable and can be made from a few different materials, such as nitrile, latex, and even plastic. Some people have allergies to these materials. If you have an allergic reaction when wearing these gloves, try using a glove made from a different material.

Cloth gloves are designed to be worn in cold temperatures, particularly during winter, so that cold tools do not stick to your

FIGURE 2-4 Chemical gloves should extend to the middle of your forearm to reduce the risk of chemical burns.

FIGURE 2-5 Leather gloves protect your hands from burns when welding or handling hot components.

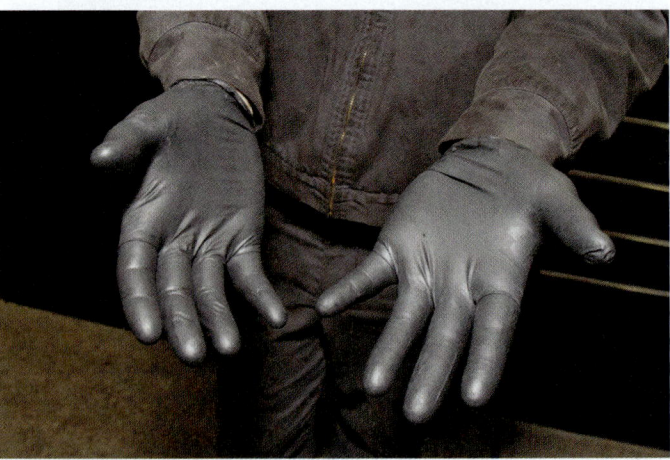

FIGURE 2-6 Light-duty gloves should be used to protect your hands from exposure to greases and oils.

FIGURE 2-7 Cloth gloves work well in cold temperatures, particularly during winter, so that cold tools do not stick to your skin.

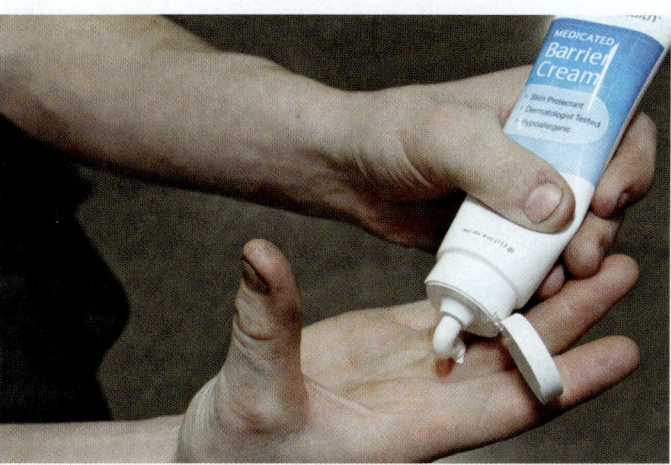

FIGURE 2-8 Barrier cream helps prevent chemicals from being absorbed into your skin and should be applied to your hands before you begin work.

skin (**FIGURE 2-7**). Over time, cloth gloves accumulate dirt and grime, so you need to wash them regularly. Regularly inspect cloth gloves for damage and wear, and replace them when required. Cloth gloves are not an effective barrier against chemicals or oils, so never use them for that purpose.

Barrier cream looks and feels like a moisturizing cream, but it has a specific formula to provide extra protection from chemicals and oils. Barrier cream prevents chemicals from being absorbed into your skin and should be applied to your hands before you begin work (**FIGURE 2-8**). Even the slightest exposure to certain chemicals can lead to dermatitis, a painful skin irritation. Never use a standard moisturizer as a replacement for proper barrier cream. Barrier cream also makes it easier to clean your hands because it can prevent fine particles from adhering to your skin.

When cleaning your hands, use only specialized hand cleaners, which protect your skin. Your hands are porous and easily absorb liquids on contact. Never use solvents such as gasoline or kerosene to clean your hands, because they can be absorbed into the bloodstream and remove the skin's natural protective oils.

Headgear

Headgear includes items like hairnets, caps, and hard hats. These help protect you from getting your hair caught in rotating machinery and protect your head from knocks or bumps. For example, a hard hat can protect you from bumping your head on vehicle parts when working under a vehicle that is raised on a hoist. Head wounds tend to bleed a lot, so hard hats can prevent the need for visiting an emergency room for stitches.

Some technicians wear a cap to keep their hair clean when working under vehicles, or to contain hair that reaches a shirt collar. Some caps are designed specifically with additional padding on the top to provide extra protection against bumps. If hair is longer than can be contained in a cap, then technicians can either use a ponytail holder or hairnet (**FIGURE 2-9**).

When in a workshop environment, watches, rings, necklaces, and dangling earrings and other jewelry present a number of hazards. They can get caught in rotating machinery, and because they are mainly constructed from metal, they can conduct electricity. Imagine leaning over a running engine with a dangling necklace; it could get caught in the fan belt and pull you into the rotating parts if it doesn't break; not only will it get destroyed but it could seriously injure you. A ring or watch could inadvertently short out an electrical circuit, heat up quickly and severely burn you, or cause a spark that might make the battery explode. A ring can also get caught on moving parts, breaking the finger bone or even ripping the finger out of the hand (**FIGURE 2-10**). To be safe, always remove watches, rings, and jewelry before starting work. Not only is it safer to remove these items but your valuables will not get damaged or lost.

K02003 Understand why it is important to wear headgear.

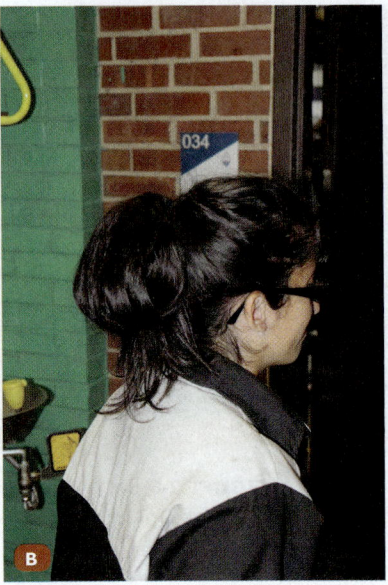

FIGURE 2-9 Containing hair. **A.** Ball cap. **B.** Pony tail.

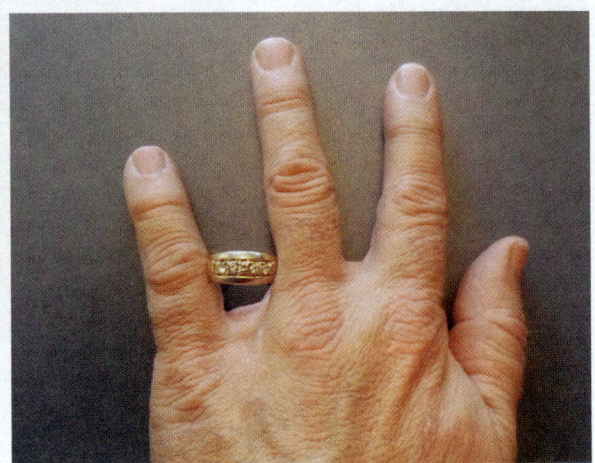

FIGURE 2-10 Finger missing because the wedding ring caught on rotating machinery.

Ear Protection

K02004 Describe the types of ear protection.

Ear protection should be worn when sound levels exceed 85 decibels, when you are working around operating machinery for any period of time, or when the equipment you are using produces loud noise. If you have to raise your voice to be heard by a person who is 2' (50 mm) away from you, then the sound level is about 85 decibels or more. Ear protection comes in two forms: One type covers the entire outer ear, and the other is fitted into the ear canal (**FIGURE 2-11**). Generally speaking, the in-the-ear style has higher noise-reduction ratings. If the noise is not excessively loud, either type of protection will work. If you are in an extremely loud environment, you will want to verify that the option you choose is rated high enough.

Breathing Devices

K02005 Describe the types of breathing devices.

Dust and chemicals from your workspace can be absorbed into the body when you breathe. When working in an environment where dust is present or where the task you are performing will produce dust, you should always wear an appropriate form of breathing device. When working in an environment where chemical vapors are present, you should always wear the proper respirator. There are two types of breathing devices: disposable dust masks and respirators.

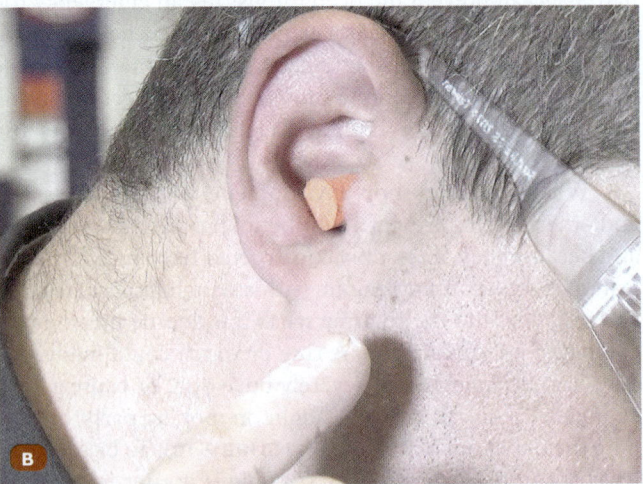

FIGURE 2-11 Ear protection comes in two forms: **A.** One type covers the entire outer ear. **B.** The other type is fitted into the ear canal.

A disposable dust mask is made from paper with a wire-reinforced edge that is held to your face with an elastic strip. It covers your mouth and nose and is disposed of at the completion of the task. This type of mask should only be used as a dust mask and should not be used if chemicals, such as paint solvents, are present in the atmosphere. It should also not be used when working around asbestos dust as the asbestos particles are too small for the filter to remove them, allowing then to be inhaled deeply into the lungs where their sharp tips pierce the lung's lining and become trapped. Over time, these create scar tissue in the lungs and can potentially cause cancer or other life-threatening diseases. Dust masks and respirators should fit securely on your face to minimize leaks around the edges. This can be especially difficult to prevent if you have a beard.

The **respirator** has removable cartridges that can be changed according to the type of contaminant being filtered. Always make sure the cartridge is the correct type for the contaminant in the atmosphere. For example, when chemicals are present, use the appropriate chemical filter in your respirator. The cartridges should be replaced according to the manufacturer's recommended replacement schedule to ensure their effectiveness. To be completely effective, the respirator mask must make a good seal onto your face (**FIGURE 2-12**).

FIGURE 2-12 To be completely effective, the respirator mask must make a good seal onto your face.

In some situations where the environment either contains too high a concentration of hazardous chemicals or a lack of oxygen, a fresh air respirator must be used. This device pumps a supply of fresh air to the mask from an outside location (**FIGURE 2-13**). Being aware of the environment you are working in allows you to determine the proper respirator or fresh air supply system.

Eye Protection

Eyes are very sensitive organs, and they need to be protected against damage and injury. There are many things in the workshop environment that can damage or injure eyes, such as high-velocity particles coming from a grinder or high-intensity light coming from a welder. In fact, the American National Standards Institute (ANSI) reports that 2000 workers per day suffer on-the-job eye injuries. Always select the appropriate eye protection for the work you are undertaking. Sometimes this may mean that more than one type of protection is required. For example, when grinding, you should wear a pair of safety glasses underneath your face shield for added protection.

K02006 Describe the types of eye protection.

The most common type of eye protection is a pair of **safety glasses**, which must be clearly marked with "Z87.1." Safety glasses have built-in side shields to help protect your eyes from the side. Approved safety glasses should be worn whenever you are in a workshop. They are designed to help protect your eyes from direct impact or debris damage (**FIGURE 2-14**).

FIGURE 2-13 Fresh air respirator.

FIGURE 2-14 Safety glasses are designed to protect your eyes from direct impact or debris damage.

SAFETY TIP

You might be tempted to take your safety glasses off while you are doing a nonhazardous task in the shop like a former student of mine. He was doing paperwork in his stall while his best friend was driving pins out of the tracks of a bulldozer. Unfortunately, the head of the punch his friend was using was mushroomed. And on one hit, a fragment broke off, flew across the stall, and hit the student in the eye, blinding him permanently. So always wear safety glasses while in a work area, even if you aren't working.

▶ TECHNICIAN TIP

Each lab/shop activity requires at least the safe use of safety glasses, clothing, and shoes, depending on its nature. Research and identify whether any additional safety devices are required for every activity you undertake.

SAFETY TIP

Be aware that the ultraviolet radiation can burn your skin like a sunburn, so wear the appropriate welding apparel to protect yourself from this hazard.

The only time they should be removed is when you are using other eye protection equipment. Prescription and tinted safety glasses are also available. Tinted safety glasses are designed to be worn outside in bright sunlight conditions. Never wear them indoors or in low-light conditions because they reduce your ability to see clearly. For people who wear prescription glasses, there are three acceptable options that OSHA makes available:

- Prescription spectacles, with side shields and protective lenses meeting requirements of ANSI Z87.1
- Goggles that can fit comfortably over corrective eyeglasses without disturbing their alignment
- Goggles that incorporate corrective lenses mounted behind protective lenses

Safety goggles don't provide as much eye protection as safety glasses, but they do have added protection against harmful chemicals that may splash up behind the lenses of safety glasses (**FIGURE 2-15**). Goggles also provide additional protection from foreign particles. Safety goggles must be worn when servicing air-conditioning systems or any other system that contains pressurized gas. Goggles can sometimes fog up when in use; if this occurs, use one of the special antifog cleaning fluids or cloths to clean them.

A full face shield gives you added protection from sparks or chemicals over safety glasses alone (**FIGURE 2-16**). The clear mask of the face shield allows you to see all that you are doing and helps protect your eyes and face from chemical burns should there be any splashes or battery explosions. It is also recommended that you use a full face shield combined with safety goggles when using a bench or angle grinder.

The light from a welding arc is very bright and contains high levels of ultraviolet radiation. So wear a **welding helmet** when using, or assisting a person using, an electric welder. The lens on a welding helmet has heavily shaded glass to reduce the intensity of the light from the welding arc, allowing you to see the task you are performing more clearly (**FIGURE 2-17**).

Lenses come in a variety of ratings depending on the type of welding you are doing; always make sure you are using a properly rated lens for the welder you are using. The remainder of the helmet is made from a durable material that blocks any other light, which can burn your skin similar to a sunburn, from reaching your face. It also protects you from welding sparks. Photosensitive welding helmets that darken automatically when an arc is struck are also available. Their big advantage is that you do not have to lift and lower the helmet by hand while welding.

Gas welding goggles can be worn instead of a welding mask when using or assisting a person using an oxyacetylene welder (**FIGURE 2-18**). The eyepieces are available in heavily shaded versions, but not as shaded as those used in an electric welding helmet. There is much less ultraviolet radiation from an oxyacetylene flame, so a welding helmet is not required. However, the flame is bright enough to damage your eyes, so always use goggles of the correct shade rating.

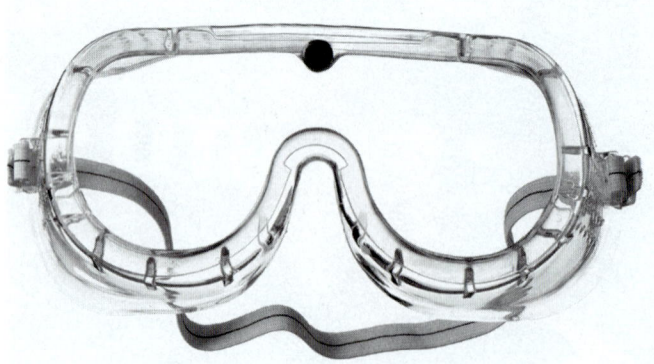

FIGURE 2-15 Safety goggles provide much the same eye protection as safety glasses, but with added protection against any harmful fluid that may find its way behind the lenses.

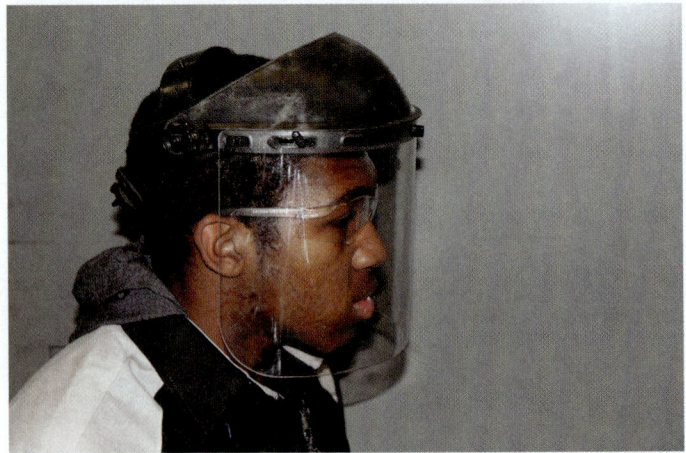

FIGURE 2-16 Full face shield.

FIGURE 2-17 The lens on a welding helmet has heavily tinted glass to reduce the intensity of the light from the welding tip, allowing you to see what you are doing.

FIGURE 2-18 Gas welding goggles can be worn instead of a welding helmet when using or assisting a person using an oxyacetylene welder.

Lifting

Whenever you lift something, there is always the possibility of injury; however, by lifting correctly, you reduce the chance of injuring yourself or others. Before lifting anything, you can reduce the risk of injury by breaking down the load into smaller quantities, asking for assistance if required, or possibly using a mechanical device to assist the lift. If you have to bend down to lift something, you should bend your knees to lower your body; do not bend over with straight legs because this can damage your back (**FIGURE 2-19**). Place your feet about shoulder width apart, and lift the item by straightening your legs while keeping your back as straight as possible.

▶ Shop Safety

The work environment can be described as anywhere you work. The condition of the work environment plays an important role in making the workplace safer. A safe work environment goes a long way toward preventing accidents, injuries, and illnesses. There are many ways to describe a safe work environment, but generally it would contain a well-organized shop layout, use of shop policies and procedures, safe equipment, safety equipment, safety training, employees who work safely, good supervision, and a workplace culture that supports safe work practices. Conversely, a shop that is cluttered with junk, poorly lit, and full of safety hazards is unsafe (**FIGURE 2-20**).

K02007 Describe proper lifting techniques.

SAFETY TIP

Never lift anything that is too heavy for you to comfortably lift, and always seek assistance if lifting the object could injure you. Always err on the side of caution.

K02008 Comply with safety precautions in the workplace.

N02003 Identify general shop safety rules and procedures.

FIGURE 2-19 Prevent back injuries when lifting heavy objects by crouching with your legs slightly apart, standing close to the object, positioning yourself so that the center of gravity is between your feet.

FIGURE 2-20 **A.** Relatively safe shop. **B.** Relatively unsafe shop.

Applied Science

AS-1: Safety: The technician follows all safety regulations and applicable procedures while performing the task.

Using a bench grinder to grind down a steel component is a simple everyday activity in shops. In terms of its potential safety implications, it carries significant risk of injury. Before beginning the task, you must ensure that the machinery is safe and ready to use. Inspect and/or adjust the guards/shields, grinding wheels, electrical cord, etc. From a personal perspective, you must make sure your clothing is suitable and safe for the task. Clothing cannot be loose, as it may get caught in the grinder; it must be made of flame-retardant material because of the risk of ignition from sparks. Long hair must be tied back or contained within a hairnet or cap because of the risk of it getting caught in rotating machinery. Various items of PPE are required to safely carry out this task: safety goggles to protect against foreign objects entering the eyes, ear protection to guard against hearing damage due to excessive noise, steel-capped boots to prevent injury from falling heavy objects, and heavy-duty gloves to protect against skin contact with sharp or hot components.

K02009 Describe how OSHA rules and the EPA impact the automotive workplace.

OSHA and EPA

OSHA stands for the **Occupational Safety and Health Administration (OSHA)**. It is a U.S. government agency that was created to provide national leadership in occupational safety and health, and it works toward finding the most effective ways to help prevent worker fatalities and workplace injuries and illnesses. OSHA has the authority to conduct workplace inspections and, if required, fine employers and workplaces if they violate its regulations and procedures. For example, a fine may be imposed on the employer or workplace if a worker is electrocuted by a piece of faulty machinery that has not been regularly tested and maintained.

OSHA standard 29 CFR 1910.132 requires employers to assess the workplace to determine if hazards are present, or are likely to be present, which necessitate the use of PPE. In addition, OSHA regulations require employers to protect their employees from workplace hazards such as machines, work procedures, and hazardous substances that can cause injury. To do that, employers must institute all feasible **engineering and work practice controls** to eliminate and reduce hazards before using PPE to protect against hazards. This means that the employers have a responsibility not only to assess safety issues but to eliminate or reduce those that can be mitigated, and then provide PPE and training for those hazards that cannot be mitigated. At the same time, if you are injured, you are the one suffering the consequences, such as being out of work for a period of time, permanent disability, or death. So it is in your best interest to take responsibility for your own safety while on the job. Engineering controls means the employer has physically changed the machine or work environment to prevent employee exposure to the potential hazard. An example might be adding a ventilation system to an area where solvent tanks are used. Work practice controls means the employer changes the way employees do their jobs in order to reduce exposure to the hazard. An example of this might be requiring employees to use a brake wash station when working on brake systems.

EPA stands for the **Environmental Protection Agency**. This federal government agency deals with issues related to environmental safety. The EPA conducts research and monitoring, sets standards, conducts workplace inspections, and holds employees and companies legally accountable in order to keep the environment protected. Shop activities need to comply with EPA laws and regulations by ensuring that waste products are disposed of in an environmentally responsible way, chemicals and fluids are correctly stored, and work practices do not contribute to damaging the environment.

Applied Science

AS-2: Environmental Issues: *The technician develops and maintains an understanding of all federal, state, and local rules and regulations regarding environmental issues related to the work of the automobile technician.*

You need to keep up to date with local laws and regulations regarding environmental issues. There are large fines associated with disregarded environmental regulations. To remain informed, you can log on to local state websites to check the latest laws and regulations. The U.S. Environmental Protection Agency (EPA) website also has up-to-date information on environmental regulations (www.epa.gov/lawsregs/). More information can be found on the specific regulations for the automotive industry at www.epa.gov/lawsregs/sectors/automotive.html. This site has a full listing of laws and regulations, compliance measures, and enforcement tactics. For additional laws that are enforced by your local state environmental agencies, go to www.epa.gov/epahome/state.htm.

AS-3: Environmental Issues: *The technician uses such things as government impact statements, media information, and general knowledge of pollution and waste management to correctly use and dispose of products that result from the performance of a repair task.*

When you complete a job, you must be able to identify what to do with any waste products created from the repair task. This could be as simple as knowing where and how to recycle cardboard boxes or as complex as knowing what to do with brake components that may contain asbestos. Normally, there is a table posted in the garage that details the correct measures for disposing of and storing waste material. A sample table follows:

Component	Material/Parts	Removal and Safety Information	Recommended Storage
Air-conditioning gases (refrigerant)	R12, R134a	Use approved evacuation and collection equipment required. *Note:* Requires AC license.	Use approved storage containers—reused or recycled.
Batteries	Plastic, rubber, lead, sulfuric acid	Avoid contact between sulfuric acid and your skin, clothing, or eyes.	Store off the ground in a covered area for collection by recycler.
Brake fluid	Diethylene and polyethylene glycol-monoalkyl ethers	These are corrosive and highly toxic to the environment. Drain into pan or tray.	Store in a drum in a covered area for collection by a licensed operator.
Brake shoes and pads pre-2004	Asbestos	Fibers are dangerous if inhaled.	Put in a plastic bag in a sealed container for collection by contractor.
Coolant	Phosphoric acid, hydrazine ethylene glycol, alcohols	Radiator coolant can be toxic to the environment.	Store in sealed drums in a covered area for recycling or collection by a licensed operator.
Coverings for plastic parts, and plastic bags and containers for parts shipping	Plastic-made components	Plastic components that can be recycled bear a recycling symbol: this symbol has a number inside telling the recycling company what the product is made of.	If the plastic container has a recycle code, then put it in the recycle bin. If it does not, then put in general waste. Plastic oil containers cannot be recycled.
Fuel	Unleaded, diesel	Avoid fumes. Fire hazard; keep well away from ignition sources. Siphon from tank to avoid spillage.	Store in a drum in a covered area.
Metal	Brake discs, housings made of metal (gearbox/engine case and components), metal cuttings	Some metal products can be heavy, so lift with care.	All metal components can be recycled; keep waste metal in a separate recycle bin for sale or disposal.
Oil	Engine, transmission, and differential oils	Fire hazard; keep well away from ignition sources.	Store in a drum/container in a covered area for collection by a licensed operator.

(Continues)

Applied Science (Continued)

Component	Material/Parts	Removal and Safety Information	Recommended Storage
Oil filters	Steel paper fiber	—	Drain filter, then crush and store in a leakproof drum for collection.
Parts boxes and paperwork	Paper, cardboard	—	Store in a recycling bin to be taken away for recycling.
Tires	Rubber, steel, fabric	Keep away from ignition sources.	Store in a fenced area for collection by recycler.
Trim, plastic fittings, and seats	Plastic, metal, cloth	—	Store racked or binned for reuse, sale, or recycling.
Tubes and rubber components	Rubber hoses, mounts, etc.	Keep away from ignition sources.	Store in a collection bin, to be collected and recycled.
Undeployed airbags	Plastics, metals, igniters, explosives	Recommended specific training on airbags before attempting removal. Handle with care; accidental deployment can cause serious harm. If unit is to be scrapped, ensure that it is safely deployed first.	Store face up in a secure area.

K02010 Explain how shop policies, procedures, and safety inspections make the workplace safer.

Shop Policies, Procedures, and Safety Inspections

Shop policies and procedures are a set of documents that outline how tasks and activities in the shop are to be conducted and managed. They also ensure that the shop operates according to OSHA and EPA laws and regulations. A **policy** is a guiding principle that sets the shop direction, and a **procedure** is a list of the steps required to get the same result each time a task or activity is performed (**FIGURE 2-21**). An example of a policy is an OSHA document for the shop that describes how the shop complies with legislation, such as a sign simply saying, "Safety glasses must be worn at all times in the shop." An example of a procedure is a document that describes the steps required to safely use the vehicle hoist.

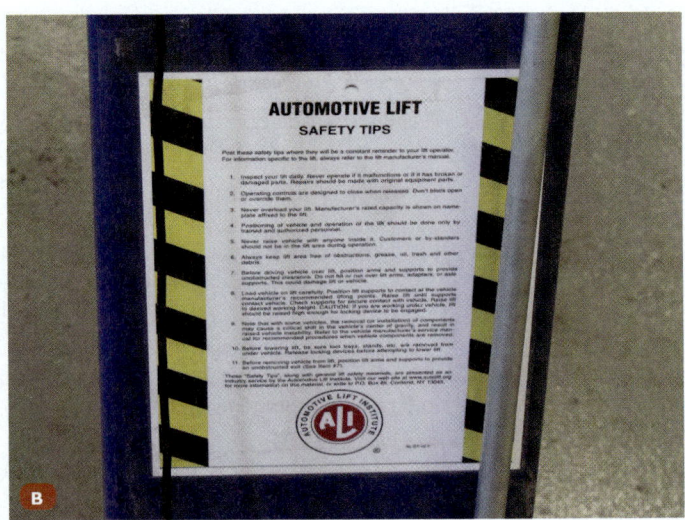

FIGURE 2-21 **A.** Policy. **B.** Procedure.

Each shop has its own set of policies and procedures and a system in place to make sure the policies and procedures are regularly reviewed and updated. Regular reviews ensure that new policies and procedures are developed and old ones are modified in case something has changed. For example, if the shop moves to a new building, then a review of policies and procedures will ensure that they relate to the new shop, its layout, and equipment. In general, the policies and procedures are written to guide shop practice; help ensure compliance with laws, statutes, and regulations; and reduce the risk of injury. Always follow your shop policies and procedures to reduce the risk of injury to your coworkers and yourself and to prevent damage to property.

FIGURE 2-22 Shop safety inspection using a checklist.

It is everyone's responsibility to know and follow the rules. Locate the general shop rules and procedures for your workplace. Look through the contents or index pages to familiarize yourself with the contents. Discuss the policy and the shop rules and procedures with your supervisor. Ask questions to ensure that you understand how the rules and procedures should be applied and your role in making sure they are followed.

Shop safety inspections are valuable ways of identifying unsafe equipment, materials, or activities so they can be corrected to reduce the risk of accidents or injuries. The inspection can be formalized by using inspection sheets to check specific items, or they can be general walk-arounds where you consciously look for and document problems that can be corrected (**FIGURE 2-22**).

Here are some of the common things to look for:
Items blocking emergency exits or walkways

- Poor safety signage
- Unsafe storage of flammable goods
- Tripping or slipping hazards
- Faulty or unsafe equipment or tools
- Missing equipment guards
- Misadjusted bench grinder tool rests
- Missing or expired fire extinguishers
- Clutter, spills, unsafe shop practices
- People not wearing the correct PPE

Formal and informal safety inspections should be held regularly. For example, an inspection sheet might be used weekly or monthly to formally evaluate the shop, and informal inspections might be held daily to catch issues that are of a more immediate nature. Never ignore or put off a safety issue.

Housekeeping and Orderliness

Good housekeeping is about always making sure the shop and your work surroundings are neat and kept in good order. This habit pays off in a couple of ways. It makes the shop a safer place to work by not allowing clutter or spills to accumulate. It also makes the shop more efficient if everything is kept neat and orderly, by making it easier to find needed tools and equipment. Trash and liquid spills should be quickly cleaned up; tools need to be cleaned and put away after use; spare or removed parts need to be stored or disposed of correctly; and generally everything needs to have a safe place to be kept.

S02001 Maintain a clean and orderly workplace.

You should carry out good housekeeping practices while working, not just after a job is completed. For example, throw trash in the garbage can as it accumulates, clean up spills when they happen, and put tools away when you are finished working with them. It is also good practice to periodically perform a deep clean of the shop so that any neglected areas are taken care of.

K02011 Describe the importance of demonstrating a safe attitude in the workplace.

Safe Attitude

Develop a safe attitude toward your work. You should always think "safety first" and then act safely (**FIGURE 2-23**). Think ahead about what you are doing, and put in place specific measures to protect yourself and those around you. For example, you could ask yourself the following questions:

- What could go wrong?
- What measures can I take to ensure that nothing goes wrong?
- What PPE should I use?
- Have I been trained to use this piece of equipment?
- Is the equipment I'm using safe?

Answering these questions and taking appropriate action before you begin will help you work safely. If you don't know the answers to those questions, don't work! Stop and ask someone with more experience than yourself, or your supervisor. Also, don't count on others for your safety. You alone are primarily responsible for your own safety.

Also remember that the time to make a good decision is before an accident happens. Once the accident occurs, you will likely have very little to no control over the outcome. In fact, there are very few true accidents that happen all by themselves. Accidents are almost always caused by poor decisions or poor actions. Almost all accidents can be prevented with a little more thought, training, and practice.

Don't Underestimate the Dangers

Because vehicle servicing and repair are so commonplace, it is easy to overlook the many potential risks related to this field. Think carefully about what you are doing and how you are doing it. Think through the steps, trying to anticipate things that may go wrong and

FIGURE 2-23 Always think "safety first."

taking steps to prevent them. Also be wary of taking shortcuts. In most cases, the time saved by taking a shortcut is nothing compared to the time spent recovering from an accident.

Accidents and Injuries Can Happen at Any Time

There is the possibility of an accident occurring whenever work is undertaken. For example, fires and explosions are a constant hazard wherever there are flammable fuels. Electricity can kill quickly as well as cause painful shocks and burns. Heavy equipment and machinery can easily cause broken bones or crush fingers and toes. Hazardous solvents and other chemicals can burn or blind as well as contribute to many kinds of illness. Oil spills and tools left lying around can cause slips, trips, and falls. Poor lifting and handling techniques can cause chronic strain injuries, particularly to your back.

FIGURE 2-24 Clean up spills immediately.

Slip, Trip, and Fall Hazards

Slip, trip, and fall hazards are ever present in the shop, and they can be caused by trash, tools and equipment, or liquid spills being left lying around. Always be on the lookout for hazards that can cause slips, trips, or falls. Floors and steps can become slippery, so they should be kept clean and have antislip coatings applied to them. High-visibility strips with antislip coatings can be applied to the edge of step treads to reduce the hazard. Clean up liquid spills immediately, and mark the area with wet floor signs until the floor is dry (**FIGURE 2-24**). Make sure the workshop has good lighting, so hazards are easy to spot, and keep walkways clear from obstruction. Think about what you are doing and make sure the work area is free of slip, trip, and fall hazards as you work.

Accidents and Injuries Are Avoidable

Almost all accidents are avoidable or preventable by taking a few precautions. Think of nearly every accident you have witnessed or heard about. In most cases someone made a mistake. Whether caused by horseplay, neglecting maintenance on tools or equipment, or using tools improperly, these instances lead to injury. Most of these accidents can be prevented if people follow policies and develop a "safety first" attitude. By following regulations and safety procedures, you can make your workplace much safer. Learn and follow all of the correct safety procedures for your workplace. Always wear the right PPE, and stay alert and aware of what is happening around you.

Think about what you are doing, how you are doing it, and its effect on others. You also need to know what to do in case of an emergency. Document and report all accidents and injuries whenever they happen, and take the proper steps to make sure they never happen again.

Signs

Always remember that a shop is a hazardous environment. To make people more aware of specific shop hazards, legislative bodies have developed a series of safety signs. These signs are designed to give adequate warning of an unsafe situation. Each sign has four components:

K02012 Identify workplace safety signs and their meanings.

- Signal word: There are three signal words—danger, warning, and caution. Danger indicates an immediately hazardous situation, which, if not avoided, will result in death or serious injury. Danger is usually indicated by white text with a red background. Warning indicates a potentially hazardous situation, which, if not avoided, could result in death or serious injury. The sign is usually in black text with an orange or yellow background. Caution indicates a potentially hazardous situation, which, if not avoided, may result in minor or moderate injury. It may also be used to alert against unsafe practices. This sign is usually in black text with a yellow background (**FIGURE 2-25**).

FIGURE 2-25 Signs. **A.** Danger is usually indicated by white text on a red background. **B.** Warning is usually in black text with an orange background. **C.** Caution is usually in black text with a yellow background.

- Background color: The choice of background color also draws attention to potential hazards and is used to provide contrast so the letters or images stand out. For example, a red background is used to identify a definite hazard; yellow indicates caution for a potential hazard. A green background is used for emergency-type signs, such as for first aid, fire protection, and emergency equipment. A blue background is used for general information signs.
- Text: The sign will sometimes include explanatory text intended to provide additional safety information. Some signs are designed to convey a personal safety message.
- Pictorial message: In symbol signs, a pictorial message appears alone or is combined with explanatory text. This type of sign allows the safety message to be conveyed to people who are illiterate or who do not speak the local language.

Safety Equipment

K02013 Describe the standard safety equipment.

Shop safety equipment includes items such as:

- Handrails: Handrails are used to separate walkways and pedestrian traffic from work areas. They provide a physical barrier that directs pedestrian traffic and also offer protection from vehicle movements.
- Machinery guards: Machinery guards and yellow lines prevent people from accidentally walking into the operating equipment or indicate that a safe distance should be kept from the equipment.

- Painted lines: Large, fixed machinery such as lathes and milling machines present a hazard to the operator and others working in the area. To prevent accidents, a machinery guard or a painted yellow line on the floor usually borders this equipment.
- Sound-insulated rooms: Sound-insulated rooms are usually used when operating equipment makes a lot of noise, for example a chassis dynamometer. A vehicle operating on a dynamometer produces a lot of noise from its tires, exhaust, and engine. To protect other shop users from the noise, the dynamometer is usually placed in a sound-insulated room, keeping shop noise to a minimum.
- Adequate ventilation: Exhaust gases and chemical vapors are serious health hazards in the shop. Whenever a vehicle's engine is running, toxic gases are emitted from its exhaust. To prevent an excess of toxic gas buildup, a well-ventilated work area is needed, as well as a method of directly venting the vehicle's exhaust to the outside. It may only take a minute or two for a poorly running vehicle to fill the shop with enough carbon monoxide to affect people's health. Chemical vapors are also a hazard and need to be vented outside.
- Doors and gates: Doors and gates are used for the same reason as machinery guards and painted lines. A doorway is a physical barrier that can be locked and sealed to separate a hazardous environment from the rest of the shop or a general work area from an office or specialist work area.
- Temporary barriers: In the day-to-day operation of a shop, there is often a reason to temporarily separate one work bay from others. If a welding machine or an oxyacetylene cutting torch is in use, it may be necessary to place a temporary screen or barrier around the work area to protect other shop users from welding flash or injury.

Air Quality

Managing air quality in shops helps protect you from potential harm and also protects the environment. There are many shop activities, stored liquids, and other hazards that can reduce the quality of air in shops. Some of these are listed here:

K02014 Maintain a safe air quality in the workplace.

- Dangerous fumes from running engines
- Welding (gas and electric)
- Painting
- Liquid storage areas
- Air conditioning servicing
- Dust particles from brake servicing

Running engines produce dangerous exhaust gases including carbon monoxide (CO) and carbon dioxide (CO_2). Carbon monoxide in small concentrations can kill or cause serious injuries. Carbon dioxide is a greenhouse gas, and vehicles are a major source of carbon dioxide in the atmosphere. Exhaust gases also contain hydrocarbons (HC) and oxides of nitrogen (NOx). These gases can form smog and also cause breathing problems for some people.

Carbon monoxide in particular is extremely dangerous, as it is odorless and colorless and can build up to toxic levels very quickly in confined spaces. In fact, it doesn't take very much carbon monoxide to pose a danger. The maximum exposure limit is regulated by the following agencies:

- OSHA permissible exposure limit (PEL) is 50 parts per million (ppm) of air for an eight-hour period.
- National Institute for Occupational Safety and Health has established a recommended exposure limit of 35 ppm for an eight-hour period.

The reason the PEL is so low is because carbon monoxide attaches itself to red blood cells much more easily than oxygen does, and it never leaves the blood cell. This prevents the blood cells from carrying as much oxygen, and if enough carbon monoxide has been inhaled, it effectively asphyxiates the person. Always follow the correct safety precautions when running engines indoors or in a confined space, including over service pits,

FIGURE 2-26 Exhaust hoses should be vented to where the fumes will not be drawn back indoors.

as gases can accumulate there. The best solution when running engines in an enclosed space (shop) is to directly connect the vehicle's exhaust pipe to an exhaust extraction system hose that ventilates the fumes to the outside air. The extraction system should be vented to where the fumes will not be drawn back indoors, to a place well away from other people and other premises (**FIGURE 2-26**).

Do not assume that an engine fitted with a catalytic converter can be run safely indoors; it cannot. Catalytic converters are fitted into the exhaust system to help control exhaust emissions through chemical reaction. They require high temperatures to operate efficiently and are less effective when the exhaust gases are relatively cool, such as when the engine is only idling or being run intermittently. A catalytic converter can never substitute for adequate ventilation or exhaust extraction equipment. In fact, even if the catalytic converter were working at 100% efficiency, the exhaust would contain large amounts of carbon dioxide and very low amounts of oxygen, neither of which conditions can sustain life.

Proper Ventilation

N02004 Utilize proper ventilation procedures for working within the lab/shop area.

Proper ventilation is required for working in the shop area. The key to proper ventilation is to ensure that any task or procedure that may produce dangerous or toxic fumes is recognized, so measures can be put in place to provide adequate ventilation. Ventilation can be provided by natural means, such as by opening doors and windows to provide air flow for low-exposure situations. However, in high-exposure situations, such as vehicles running in the shop, a mechanical means of ventilation is required; an example is an exhaust extraction system. Parts cleaning areas or areas where solvents and chemicals are used should also have good general ventilation, and if required, additional exhaust hoods or fans should be installed to remove dangerous fumes. In some cases, such as when spraying paint, it may be necessary to use a personal respirator in addition to proper ventilation.

> ▶ **TECHNICIAN TIP**
>
> Before starting a vehicle in the shop, make sure the correct ventilation equipment is connected properly and turned on. Also, if the system uses rubber hoses, make sure they aren't kinked.

Electrical Safety

K02015 Describe appropriate workplace electrical safety practices.

Many people are injured by electricity in shops. Poor electrical safety practices can cause shocks and burns as well as fires and explosions. Make sure you know where the electrical shutoffs, or panels for your shop, are located. All circuit breakers and fuses should be clearly labeled so that you know which circuits and functions they control (**FIGURE 2-27**).

In the case of an emergency, you may need to know how to shut off the electricity supply to a work area or to your entire shop. Keep the circuit breaker and/or electrical panel covers closed to keep them in good condition, prevent unauthorized access, and prevent accidental contact with the electricity supply. It is important that you do not block or obstruct access to this electrical panel; keep equipment and tools well away so emergency access is not hindered. In some localities, 3' (0.91 m) of unobstructed space must be maintained around the panel at all times.

There should be a sufficient number of electrical receptacles in your work area for all your needs. Do not connect multiple appliances to a single receptacle with a simple double adapter. If necessary, use a multi-outlet safety strip that has a built-in overload cutout feature. Electric receptacles should be at least 3' (0.91 m) above floor level to reduce the risk of igniting spilled fuel vapors or other flammable liquids.

If you need to use an extension cord, make sure it is made of flexible wiring—not the stiffer type of house wiring—and that it is fitted with a ground wire. The cord should be neoprene-covered, as this material resists oil damage.

Always check it for cuts, abrasions, or other damage. Be careful how you place the extension cord, so it does not cause a tripping hazard. Also avoid rolling equipment or vehicles over it, as doing so can damage the cord. Never use an extension cord in wet conditions or around flammable liquids. Portable electric tools that operate at 240 volts are often sources of serious shock and burn accidents. Be particularly careful when using these items. Always inspect the cord for damage and check the security of the attached plug before connecting the item to the power supply. Use 110-volt or lower voltage tools if they are available. All electric tools must be equipped with a ground prong or be **double insulated**.

If electrical cords don't have the ground prong or aren't double insulated, do not use them. Never use any high-voltage tool in a wet environment. In contrast, air-operated tools cannot give you an electric shock, because they operate on air pressure instead of electricity, and therefore, they are safer to use in a wet environment.

Portable shop lights/droplights have been the cause of many accidents over the years. Incandescent bulbs get extremely hot and can cause burns. They are also prone to shatter, which can cut skin and damage eyes or cause fires and electric shock. In fact, in some places incandescent portable shop lights cannot be used in automotive shops and must be replaced with less hazardous lights.

One such light is the fluorescent droplight. Although this type of light stays much cooler than incandescent lights, it still can shatter, causing cuts or damage to eyes. Because of this, droplights should be fully enclosed in a clear, insulating case. They may also contain mercury, which becomes dispersed when the bulb is shattered, creating a hazardous-materials situation.

The safest portable shop light to come on the market is the LED (light-emitting diode) shop light. It uses much lower voltage, and the LED is much less prone to shattering, so it is much safer to use. It also uses a small amount of electricity to produce a large amount of light, so many of them are cordless. They may also include a magnetic base, so the light can be attached to any steel surface and then adjusted to shine where needed.

FIGURE 2-27 All electrical switches and fuses should be clearly labeled so that you know which circuits and functions they control.

Preventing Fires

The danger of a gasoline fire is always present in an automotive shop. Most automobiles carry a fuel tank, often with large quantities of fuel on board, which is more than sufficient to cause a large, very destructive, and potentially explosive fire.

In fact, 1 gallon of gasoline has the same amount of energy as 20 sticks of dynamite. So take precautions to make sure you have the correct type and size of extinguishers on hand for a potential fuel fire. Make sure you clean up spills immediately and avoid ignition sources, like sparks, near flammable liquids or gases. Being aware of the following topics will help you know how to minimize the risk of fires in the shop.

Extinguishing Fires

Three elements must be present at the same time for a fire to occur: fuel, oxygen, and heat (**FIGURE 2-28**). The secret of firefighting involves the removal of at least one of these elements. If a fire occurs in the shop, it is usually the oxygen or the heat that is removed to extinguish the fire. For example, a fire blanket when applied correctly removes the oxygen, and a water extinguisher removes heat from the fire. Fire extinguishers are used to extinguish the majority of small fires in a shop. Never hesitate to call the fire department if you cannot extinguish a fire quickly and safely.

In the United States, there are five classes of fire:

- Class A fires involve ordinary combustibles such as wood, paper, or cloth.
- Class B fires involve flammable liquids or gaseous fuels.

K02016 Prevent fires in the workplace.

▶ **TECHNICIAN TIP**

Try out the new LED flashlights or cordless LED shop lights. Good ones provide a lot of light, and the batteries last a long time. It is not unusual to hear about an LED flashlight that got dropped behind a toolbox while it was on, and was still illuminated days later.

N02005 Identify the location and the types of fire extinguishers and other fire safety equipment; demonstrate knowledge of the procedures for using fire extinguishers and other fire safety equipment.

FIGURE 2-28 Fire triangle.

FIGURE 2-29 Traditional labels on fire extinguishers often incorporate a shape as well as a letter.

▶ **TECHNICIAN TIP**

It's not a matter of "if" but "when" a shop is going to have a fire. When shops do catch on fire, the fire can grow quickly out of control. So preventing a fire is the best action you can take. If a fire starts, react appropriately. In some cases, you just need to get out safely. In other cases, it may be safe to try to fight the fire quickly before it spreads too far. Always be aware of the potential for a fire, and plan ahead by thinking through the task you are about to undertake. Know where firefighting equipment is kept and how it works.

N02006 Identify the location of the posted evacuation routes.

- Class C fires involve electrical equipment.
- Class D fires involve combustible metals such as sodium, titanium, and magnesium.
- Class K fires involve cooking oil or fat.

Fire extinguishers are marked with pictograms depicting the types of fires that the extinguisher is approved to fight (**FIGURE 2-29**):

- Class A: Green triangle
- Class B: Red square
- Class C: Blue circle
- Class D: Yellow pentagram
- Class K: Black hexagon

Unless the fire is very small, always sound the alarm before attempting to fight a fire. If you cannot fight the fire safely, leave the area while you wait for backup. You need to size up the fire before you make the decision to fight it with a fire extinguisher, by identifying what sort of material is burning, the extent of the fire, and the likelihood of it spreading. Also, if the fire is in electrical wires or equipment, make sure you won't be electrocuted while trying to extinguish it. To operate a fire extinguisher, follow the acronym for fire extinguisher use: PASS (pull, aim, squeeze, sweep).

- Pull out the pin that locks the handle at the top of the fire extinguisher to prevent accidental use. Carry the fire extinguisher in one hand, and use your other hand to:
- Aim the nozzle at the base of the fire. Stand about 8–12 feet (2.4–3.7 m) away from the fire, and
- Squeeze the handle to discharge the fire extinguisher. Remember that if you release the handle on the fire extinguisher, it will stop discharging.
- Sweep the nozzle from side to side at the base of the fire.

Continue to watch the fire. Although it may appear to be extinguished, it may suddenly reignite. Portable fire extinguishers only operate for about 10–25 seconds before they are empty. So use them effectively (**FIGURE 2-30**).

If the fire is indoors, you should be standing between the fire and the nearest safe exit. If the fire is outside, you should stand facing the fire, with the wind on your back, so that the smoke and heat are being blown away from you. If possible, get an assistant to guide you and inform you of the fire's progress. Again, make sure you have a means of escape, should the fire get out of control. When you are certain that the fire is out, report it to your supervisor. Also report what actions you took to put out the fire. Once the circumstances of the fire have been investigated and your supervisor or the fire department has given you the all clear, clean up the debris and submit the used fire extinguisher for inspection and service.

Fire blankets are designed to smother a small fire. They are ideal for use in situations where a fire extinguisher could cause damage. For example, if there is a small fire under the hood of a vehicle, a fire blanket might be able to smother the fire without running the risk of getting powder from the fire extinguisher down the intake system. They are also very useful in putting out a fire on a person. The fire blanket can be thrown around the person, smothering any fire.

Obtain a fire blanket and study the use instructions on the packaging. If instructions are not provided, research how to use a fire blanket, or ask your supervisor. You may require instruction from an authorized person in using the fire blanket. If you do use a fire blanket, make sure you return the blanket for use or, if necessary, replace it with a new one.

Evacuation Routes

Evacuation routes are a safe way of escaping danger and gathering in a prearranged safe place where everyone can be accounted for in the event of an emergency. It is important to have more than one evacuation route in case any single route is blocked during the

FIGURE 2-30 To operate a fire extinguisher, follow PASS. **A.** Pull. **B.** Aim. **C.** Squeeze. **D.** Sweep.

emergency. Your shop may have an evacuation procedure that clearly identifies the evacuation routes (**FIGURE 2-31**).

Often the evacuation routes are marked with colored lines painted or taped on the floors. Exits should be highlighted with signs that may be illuminated, and should never be chained closed or obstructed (**FIGURE 2-32**).

Always make sure you are familiar with the evacuation routes for the shop. Before conducting any task, identify which route you will take if an emergency occurs.

> ▶ **TECHNICIAN TIP**
>
> Never place anything in the way of evacuation routes, including equipment, tools, parts, cleaning supplies, or vehicles.

FIGURE 2-31 Your shop may have an evacuation procedure that clearly identifies the evacuation routes.

FIGURE 2-32 Exits should be marked, clear of obstructions, and not chained closed.

N02007 Locate and demonstrate knowledge of safety data sheets (SDS).

S02002 Use information in an SDS.

▶ Hazardous Materials Safety

A **hazardous material** is any material that poses an unreasonable risk of damage or injury to persons, property, or the environment if it is not properly controlled during handling, storage, manufacture, processing, packaging, use and disposal, or transportation. These materials can be solids, liquids, or gases. Most technicians use hazardous materials daily, such as cleaning solvents, gasket cement, brake fluid, and coolant. In fact, there are likely to be hundreds of hazardous materials in a typical shop. Hazardous materials must be properly handled, labeled, and stored in the shop.

Safety Data Sheets

Hazardous materials are used daily and may make you very sick if they are not used properly. **Safety data sheets (SDS)** contain detailed information about hazardous materials to help you understand how they should be safely used, any health effects relating to them, how to treat a person who has been exposed to them, and how to deal with them in a fire situation. SDS can be obtained from the manufacturer of the material. The shop should have an SDS for each hazardous substance or dangerous product. In the United States, it is required that workplaces have an SDS for every chemical that is on site.

Safety data sheets (SDS) were formerly called material safety data sheets (MSDS). In 2012, OSHA changed the requirements for the hazard communication system (HCS) to conform to the United Nations Globally Harmonized System of Classification and Labeling of Chemicals (GHS). The MSDS needed to change its name and its format to fit the new standards. Whereas the original MSDS had 8 sections, safety data sheets are required to have 16 sections that provide additional details and make it easier to find specific data when needed. In addition, GHS requires all employers to train their employees in the new chemical labeling requirements and the new format for the safety data sheets.

Whenever you deal with a potentially hazardous product, you should consult the SDS to learn how to use that product safely. If you are using more than one product, make sure you consult all the SDS for those products. Be aware that certain combinations of products can be more dangerous than any of them separately.

SDS are usually kept in a clearly marked binder and should be regularly updated as chemicals come into the workplace. As of June 1, 2015, the HCS requires new SDS to be in a uniform format, and include the section numbers, the headings, and associated information under the headings below:

Section 1, Identification, includes product identifier; manufacturer or distributor name, address, and phone number; emergency phone number; recommended use; restrictions on use.

Section 2, Hazard(s) identification, includes all hazards regarding the chemical; required label elements.

Section 3, Composition/information on ingredients, includes information on chemical ingredients; trade secret claims.

Section 4, First-aid measures, includes important symptoms/effects, acute, delayed; required treatment.

Section 5, Firefighting measures, lists suitable extinguishing techniques, equipment; chemical hazards from fire.

Section 6, Accidental release measures, lists emergency procedures; protective equipment; proper methods of containment and cleanup.

Section 7, Handling and storage, lists precautions for safe handling and storage, including incompatibilities.

Section 8, Exposure controls/personal protection, lists OSHA's permissible exposure limits (PELs); threshold limit values (TLVs); appropriate engineering controls; personal protective equipment (PPE).

Section 9, Physical and chemical properties, lists the chemical's characteristics.

Section 10, Stability and reactivity, lists chemical stability and possibility of hazardous reactions.

Section 11, Toxicological information, includes routes of exposure; related symptoms, acute and chronic effects; numerical measures of toxicity.

Section 12, Ecological information*

Section 13, Disposal considerations*

Section 14, Transport information*

Section 15, Regulatory information*

Section 16, Other information, includes the date of preparation or last revision.

*Note: Since other Agencies regulate this information, OSHA will not be enforcing Sections 12 through 15 (29 CFR 1910.1200(g)(2)).

Source: https://www.osha.gov/Publications/HazComm_QuickCard_SafetyData.html.

To identify information found on an SDS, follow the steps in **SKILL DRILL 2-1**.

Cleaning Hazardous Dust Safely

Toxic dust is any dust that may contain fine particles that could be harmful to humans or the environment. If you are unsure as to the toxicity of any particular dust, then you should always treat it as toxic and take the precautions identified in the SDS or shop procedures. Brake and clutch dust are potential toxic dusts that automotive shops must manage. The dust is made up of very fine particles that can easily spread and contaminate an area. One of the more common sources of toxic dust is inside drum brakes and manual transmission bell housings. It is a good idea to avoid all dust if possible, whether it is classified as toxic or not. If you do have to work with dust, never use compressed air to blow it from components or parts, and always use PPE such as face masks, eye protection, and gloves.

After completing a service or repair task on a vehicle, there is often dirt and dust left behind. The chemicals present in this dirt usually contain toxic chemicals that can build up and cause health problems. To keep the levels of dirt and dust to a minimum, clean it up immediately after the task is complete. The vigorous action of sweeping causes the dirt and dust to rise; therefore, when sweeping the floor, use a soft broom that pushes, rather than flicks, the dirt and dust forward. Create smaller piles and dispose of them frequently. Another successful way of cleaning shop dirt and dust is to use a water hose. The

K02018 Identify the proper method to clean hazardous dust safely.

SKILL DRILL 2-1 Identifying Information in an SDS

1. Once you have studied the information on the container label, find the SDS for that particular material. Always check the revision date to ensure that you are reading the most recent update.
2. Note the chemical and trade names for the material, its manufacturer, and the emergency telephone number to call.
3. Find out why this material is potentially hazardous. It may be flammable, it may explode, or it may be poisonous if inhaled or touched with your bare skin. Check the **threshold limit values (TLVs)**. The concentration of this material in the air you breathe in your shop must not exceed these figures. There could be physical symptoms associated with breathing harmful chemicals. Find out what will happen to you if you suffer overexposure to the material, either through breathing it or by coming into physical contact with it. This helps you to take safety precautions, such as wearing eye, face, or skin protection, or a mask or respirator, while using the material, or like washing your skin afterward.
4. Note the flash point for this material so that you know at what temperature it may catch fire. Also note what kind of fire extinguisher you would use to fight a fire involving this material. The wrong fire extinguisher could make the emergency even worse.
5. Study the reactivity for this material to identify the physical conditions or other materials that you should avoid when using this material. It could be heat, moisture, or some other chemical.
6. Find out what special precautions you should take when working with this material. These include personal protection for your skin, eyes, and lungs, and proper storage and use of the material.
7. Be sure to refresh your knowledge of your SDS from time to time. Be confident that you know how to handle and use the material and what action to take in an emergency, should one occur.

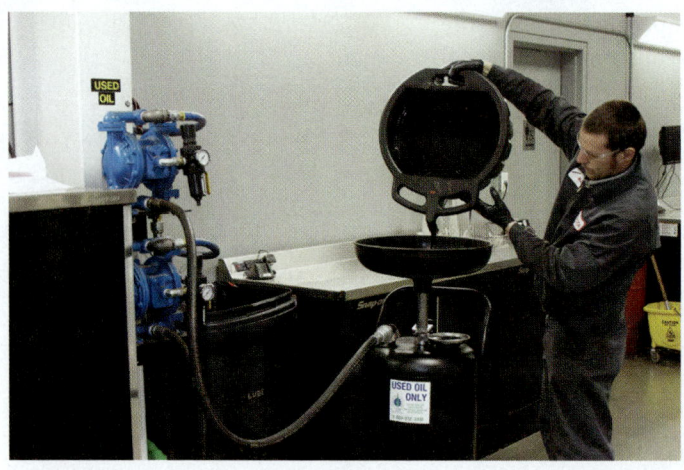

FIGURE 2-33 Used oil and fluids often contain dangerous chemicals and need to be safely recycled or disposed of in an environmentally friendly way.

wastewater must be caught in a settling pit and not run into a storm water drain. Many shops also have floor scrubbers that use a water/soap solution to clean the floor. These shops usually vacuum up the dirty water and store it in a tank until it can be disposed of properly.

Various tools have been developed to clean toxic dust from vehicle components. The most common one is the brake wash station. It uses an aqueous solution to wet down and wash the dust into a collection basin. The basin needs periodic maintenance to properly dispose of the accumulated sludge. This tool is probably the simplest way to effectively deal with hazardous dust because it is easy to set up, use, and store. One system that used to be approved for cleanup of hazardous dust is no longer approved for that purpose. It is called a high-efficiency particulate air (HEPA) dust collection system and used a HEPA filter to trap very small particles. But these types of hazardous dust collection systems are no longer approved.

Used Engine Oil and Fluids

S02003 Properly dispose of used engine oil and other petroleum products.

Used engine oil and fluids are liquids that have been drained from the vehicle, usually during servicing operations. Used oil and fluids often contain dangerous chemicals and impurities such as heavy metals that need to be safely recycled or disposed of in an environmentally friendly way (**FIGURE 2-33**).

There are laws and regulations that control the way in which they are to be handled and disposed. The shop will have policies and procedures that describe how you should handle and dispose of used engine oil and fluids. Be careful not to mix incompatible fluids such as used engine oil and used coolant. Doing so makes the hazardous materials very much more expensive to dispose of.

Generally speaking, petroleum products can be mixed together. Follow your local, state, and federal regulations when disposing of waste fluids. Used engine oil is a hazardous material containing many impurities that can damage your skin. Coming into frequent or prolonged contact with used engine oil can cause dermatitis and other skin disorders, including some forms of cancer. Avoid direct contact as much as possible by always using impervious gloves and other protective clothing, which should be cleaned or replaced regularly. Using a barrier-type hand lotion also helps protect your hands and makes cleaning them much easier. Always follow safe work practices, which minimize the possibility of accidental spills. Keeping a high standard of personal hygiene and cleanliness is important so that you get into the habit of washing off harmful materials as soon as possible after contact. If you have been in periodic contact with used engine oil, you should regularly inspect your skin for signs of damage or deterioration. If you have any concerns, consult your doctor.

▶ First Aid Principles

K02019 Explain the basic first aid procedures when approaching an emergency.

The following information is designed to provide you with an awareness of basic first aid principles and the importance of first aid training courses. You will find general information about how to take care of someone who is injured. However, this information is only a guide. It is not a substitute for training or professional medical assistance. Always seek professional advice when tending to an injured person. One of the best things you can do is take a first aid class and periodic refresher courses. These courses help you stay current on first aid practices as they change. They also help remind you of the importance of preventing accidents so that first aid is less likely to be needed.

First aid is the immediate care given to an injured or suddenly ill person. Learning first aid skills is valuable in the workplace in case an accident or medical emergency arises. First aid courses are available through many organizations, such as the Emergency Care and Safety Institute (ECSI). It is strongly advised that you seek out a certified first aid course and become certified in first aid. The following information highlights some of the principles of first aid.

In the event of an accident, the possibility of injury to the rescuer or further injury to the victim must be evaluated. This means the first step in first aid is to survey the scene. While doing this, try to determine what happened, what dangers may still be present, and the best actions to take. Only remove the injured person from a dangerous area if it is safe for you to do so. When dealing with electrocution or electrical burns, make sure the electrical supply is switched off before attempting any assistance. Always perform first aid techniques as quickly as is safely possible after an injury. When breathing or the heart has stopped, brain damage can occur within four to six minutes. The degree of brain damage increases with each passing minute, so make sure you know what to do, and do it as quickly as is safe to do so.

Prompt care and treatment before the arrival of emergency medical assistance can sometimes mean the difference between life and death. The goals of first aid are to make the immediate environment as safe as possible, preserve the life of the patient, prevent the injury from worsening, prevent additional injuries from occurring, protect the unconscious, promote recovery, comfort the injured, prevent unnecessary delays in treatment, and provide the best possible immediate care for the injured person. When attending to an injured victim, always send for assistance. Make sure the person who stays with the injured victim is more experienced in first aid than the messenger. If you are the only person available, request medical assistance as soon as reasonably possible.

When you approach the scene of an accident or emergency, do the following:

1. **Danger:** Survey the scene to make sure there are no other dangers, and assist only if it is safe to do so.
2. **Response:** Check to see if the victim is responsive and breathing. If responsive, ask the victim if he or she needs help. If the victim does not respond, he or she is unresponsive.
3. **Send for help:** Have a bystander call 9-1-1. Always have the person with the most first aid experience stay with the victim. If alone, call 9-1-1 yourself.
4. **Airway:** Open the airway by tilting the victim's head back and lifting the chin.
5. **Breathing:** Check for normal breathing by placing one of your hands on the victim's chest, and lean down so your ear is near the person's mouth. Listen and watch for breathing. If the victim is breathing, monitor the victim for further issues. If the victim is not breathing normally, start CPR.
6. **Circulation/CPR (cardiopulmonary resuscitation):** If no pulse is present, start chest compressions at the rate of approximately 100 per minute. You can either give sequences of 30 compressions to two breaths, or just compressions if you are reluctant to give mouth to mouth resuscitation. Repeat the compression and breath cycles, or compression only cycles, until an automated external defibrillator (AED) is available or emergency medical system (EMS) personnel arrive.
7. **Defibrillation:** If a pulse is still not present once an AED arrives, expose the victim's chest and turn on the AED. Attach the AED pads. Ensure that no one touches the victim. Follow the audio and visual prompts from the AED. If no shock is advised, resume CPR immediately (five sets of 30 compressions and two breaths, or compressions by themselves). If a shock is advised, do not touch the victim and give one shock. Or shock as advised by AED. Follow the directions given by the AED.

▶ **TECHNICIAN TIP**

- Some vehicle components, including brake and clutch linings, contain asbestos, which, despite having very good heat properties, is toxic. Asbestos dust causes lung cancer. Complications from breathing the dust may not show until decades after exposure.
- Airborne dust in the shop can also cause breathing problems such as asthma and throat infections.
- Never cause dust from vehicle components to be blown into the air. It can stay floating for many hours, meaning that other people will breathe the dust unknowingly.
- Wear protective gloves whenever using solvents.
- If you are unfamiliar with a solvent or a cleaner, refer to the SDS for information about its correct use and applicable hazards.
- Always wash your hands thoroughly with soap and water after performing repair tasks on brake and clutch components.
- Always wash work clothes separately from other clothes so that toxic dust does not transfer from one garment to another.
- Always wear protective clothing and the appropriate safety equipment.

▶ **TECHNICIAN TIP**

Three important rules of first aid:

1. Know what you must not do.
2. Know what you must do.
3. If you are not sure what procedures to follow, send for trained medical assistance.

▶ Wrap-Up

Ready for Review

▶ Your employer is responsible for maintaining a safe work environment; you are responsible for working safely.
▶ Always wear the correct personal protective equipment, such as gloves or hearing protection.
▶ Personal protective equipment (PPE) protects the body from injury but must fit correctly and be task appropriate.
▶ Work clothing should be clean, loose enough for movement, but not baggy, and should also be flame-retardant.
▶ Footwear should be acid- and slip-resistant and made of puncture-proof material.
▶ Headgear can protect your head from bumps and should hold long hair in place.

- Hand protection includes chemical gloves, leather gloves, light-duty gloves, general-purpose cloth gloves, and barrier cream.
- Hazardous chemicals and oils can be absorbed into your skin.
- Wear ear protection if the sound level is 85 decibels or above.
- Breathing devices include disposable dust masks and respirators.
- Forms of eye protection are safety glasses, welding helmet, gas welding goggles, full face shield, and safety goggles.
- You may need two types of eye protection for some tasks.
- Before starting work, remove all jewelry and watches, and make sure your hair is contained.
- Lifting correctly or seeking assistance helps prevent back injuries.
- Thinking "safety first" will lead to acting safely.
- Accidents and injuries can be avoided by safe work practices.
- OSHA is a federal agency that oversees safe workplace environments and practices.
- The EPA monitors and enforces issues related to environmental safety.
- Shop policies and procedures are designed to ensure compliance with laws and regulations, create a safe working environment, and guide shop practice.
- Shop safety inspections ensure that safety policies and procedures are being followed.
- Safety includes keeping a clean shop with everything put where it belongs and all spills cleaned up.
- Safety signs include a signal word, background color, text, and a pictorial message.
- Shop safety equipment includes handrails, machinery guards, painted lines, soundproof rooms, adequate ventilation, gas extraction hoses, doors and gates, and temporary barriers.
- Air quality is an important safety concern.
- All shops require proper ventilation.
- Carbon monoxide and carbon dioxide from running engines can create a hazardous work environment.
- Electrical safety in a shop is important to prevent shocks, burns, fires, and explosions.
- Portable electrical equipment should be the proper voltage and should always be inspected for damage.
- Use caution when plugging in or using a portable shop light.
- Fuels and fuel vapors are potential fire hazards.
- Use fuel retrievers when draining fuel, and have a spill response kit nearby.
- Fuel, oxygen, and heat must all be present for fire to occur.
- Types of fires are classified as A, B, C, D, or K, and fire extinguishers match them accordingly.
- Do not fight a fire unless you can do so safely.
- Operating a fire extinguisher involves the PASS method: pull, aim, squeeze, and sweep.
- Every shop should mark evacuation routes; always know the evacuation route for your shop.
- Identify hazards and hazardous materials in your work environment.
- Safety data sheets contain important information on each hazardous material in the shop.
- Using a soap and water solution is the safest method of cleaning dust or dirt that may be toxic.
- Used engine oil and fluids must be handled and disposed of properly.
- First aid involves providing immediate care to an ill or injured person.
- Do not perform first aid if it is unsafe to do so.

Key Terms

barrier cream A cream that looks and feels like a moisturizing cream but has a specific formula to provide extra protection from chemicals and oils.

double insulated Tools or appliances that are designed in such a way that no single failure can result in a dangerous voltage coming into contact with the outer casing of the device.

ear protection Protective gear worn when the sound levels exceed 85 decibels, when working around operating machinery for any period of time, or when the equipment you are using produces loud noise.

engineering and work practice controls Systems and procedures required by OSHA and put in place by employers to protect their employees from hazards.

Environmental Protection Agency (EPA) Federal government agency that deals with issues related to environmental safety.

first aid The immediate care given to an injured or suddenly ill person.

gas welding goggles Protective gear designed for gas welding; they provide protection against foreign particles entering the eye and are tinted to reduce the glare of the welding flame.

hazardous material Any material that poses an unreasonable risk of damage or injury to persons, property, or the environment if it is not properly controlled during handling, storage, manufacture, processing, packaging, use and disposal, or transportation.

headgear Protective gear that includes items like hairnets, caps, or hard hats.

heat buildup A dangerous condition that occurs when the glove can no longer absorb or reflect heat, and heat is transferred to the inside of the glove.

Occupational Safety and Health Administration (OSHA) Government agency created to provide national leadership in occupational safety and health.

personal protective equipment (PPE) Safety equipment designed to protect the technician, such as safety boots, gloves, clothing, protective eyewear, and hearing protection.

policy A guiding principle that sets the shop direction.

procedure A list of the steps required to get the same result each time a task or activity is performed.

respirator Protective gear used to protect the wearer from inhaling harmful dusts or gases. Respirators range from single-use disposable masks to types that have replaceable cartridges. The correct types of cartridge must be used for the type of contaminant encountered.

safety data sheet (SDS) A sheet that provides information about handling, use, and storage of a material that may be hazardous.

safety glasses Safety glasses are protective eye glasses with built-in side shields to help protect your eyes from the front and side. Approved safety glasses should be worn whenever you are in a workshop. They are designed to help protect your eyes from direct impact or debris damage.

threshold limit value (TLV) The maximum allowable concentration of a given material in the surrounding air.

toxic dust Any dust that may contain fine particles that could be harmful to humans or the environment.

welding helmet Protective gear designed for arc welding; it provides protection against foreign particles entering the eye, and the lens is tinted to reduce the glare of the welding arc.

Review Questions

1. Neoprene gloves are ideal for protection from:
   a. dust.
   b. heavy materials.
   c. changing workplace temperatures.
   d. certain acids and other chemicals.
2. You can use a disposable dust mask to protect yourself against:
   a. asbestos dust.
   b. brake dust.
   c. brake fluid vapors.
   d. paint solvents.
3. When bending down to lift something, you should:
   a. bend over with straight legs.
   b. put your feet together.
   c. bend your knees.
   d. bend your back.
4. Which of the following works toward finding the most effective ways to help prevent worker fatalities and workplace injuries and illnesses?
   a. EPA
   b. FDA
   c. OSHA
   d. ASA
5. The list of steps required to get the same result each time a task or activity is performed is known as the:
   a. shop policy.
   b. shop regulations.
   c. shop procedure.
   d. shop system.
6. The signal word that indicates a potentially hazardous situation, which, if not avoided, could result in death or serious injury is:
   a. danger.
   b. caution.
   c. warning.
   d. hazardous.
7. Which of these should not be used in a shop?
   a. Incandescent bulbs
   b. Extension cords
   c. Low-voltage tools
   d. Air-operated tools
8. To operate a fire extinguisher, follow the acronym for fire extinguisher use: PASS, which stands for:
   a. Press, aim, squeeze, sweep.
   b. Pull, aim, shake, sweep.
   c. Press, aim, shake, sweep.
   d. Pull, aim, squeeze, sweep.
9. Which of these should never be used to remove dust?
   a. Floor scrubbers
   b. Water hoses
   c. Compressed air
   d. Brake wash stations
10. When disposing of used liquids:
    a. federal regulations need not be followed.
    b. incompatible liquids can be mixed together.
    c. impervious gloves are not needed.
    d. petroleum products can be mixed together.

ASE Technician A/Technician B Style Questions

1. Tech A says that personal protective equipment (PPE) does not include clothing. Tech B says that the PPE used should be based on the task you are performing. Who is correct?
   a. Tech A
   b. Tech B
   c. Both A and B
   d. Neither A nor B
2. Tech A says that rings, watches, and other jewelry should be removed prior to working. Tech B says that you should always wear cuffed pants when working in a shop. Who is correct?
   a. Tech A
   b. Tech B
   c. Both A and B
   d. Neither A nor B
3. Tech A says that one use of barrier creams is to make cleaning your hands easier. Tech B says that hearing protection only needs to be worn by people operating loud equipment. Who is correct?
   a. Tech A
   b. Tech B
   c. Both A and B
   d. Neither A nor B

4. Tech A says that tinted safety glasses can be worn when working outside. Tech B says that welding can cause a sunburn. Who is correct?
   a. Tech A
   b. Tech B
   c. Both A and B
   d. Neither A nor B
5. Tech A says that exposure to solvents may have long-term effects. Tech B says that accidents are almost always avoidable. Who is correct?
   a. Tech A
   b. Tech B
   c. Both A and B
   d. Neither A nor B
6. Tech A says that after an accident you should take measures to avoid it in the future. Tech B says that it is okay to block an exit for a shop if you are actively working. Who is correct?
   a. Tech A
   b. Tech B
   c. Both A and B
   d. Neither A nor B
7. Tech A says that both OSHA and the EPA can inspect facilities for violations. Tech B says that a shop safety rule does not have to be reviewed once put in place. Who is correct?
   a. Tech A
   b. Tech B
   c. Both A and B
   d. Neither A nor B
8. Tech A says that a safety data sheet (SDS) contains information on procedures to repair a vehicle. Tech B says that you only need an SDS after an accident occurs. Who is correct?
   a. Tech A
   b. Tech B
   c. Both A and B
   d. Neither A nor B
9. Tech A says that one approved way to clean dust off brakes is with compressed air. Tech B says that some auto parts may contain asbestos. Who is correct?
   a. Tech A
   b. Tech B
   c. Both A and B
   d. Neither A nor B
10. Tech A says that if you are unsure of what personal protective equipment (PPE) to use to perform a job, you should just use what is nearby. Tech B says that air tools are less likely to shock you than electrically powered tools. Who is correct?
    a. Tech A
    b. Tech B
    c. Both A and B
    d. Neither A nor B

CHAPTER 3

Basic Tools and Precision Measuring

NATEF Tasks

- **N03001** Demonstrate safe handling and use of appropriate tools.
- **N03002** Utilize safe procedures for handling of tools and equipment.
- **N03003** Identify standard and metric designation.
- **N03004** Demonstrate proper use of precision measuring tools (i.e., micrometer, dial-indicator, dial-caliper).
- **N03005** Demonstrate proper cleaning, storage, and maintenance of tools and equipment.

Knowledge Objectives

After reading this chapter, you will be able to:

- **K03001** Describe the safety procedures to take when handling and using tools.
- **K03002** Describe how to properly lockout and tag-out faulty equipment and tools.
- **K03003** Describe typical tool storage methods.
- **K03004** Identify tools and their usage in automotive applications.
- **K03005** Describe the type and use of wrenches.
- **K03006** Describe the type and use of sockets.
- **K03007** Describe the type and use of torque wrenches.
- **K03008** Describe the type and use of pliers.
- **K03009** Describe the type and use of cutting tools.
- **K03010** Describe the type and use of Allen wrenches.
- **K03011** Describe the type and use of screwdrivers.
- **K03012** Describe type and use of magnetic pickup tools and mechanical fingers.
- **K03013** Describe the type and use of hammers.
- **K03014** Describe the type and use of chisels.
- **K03015** Describe the type and use of punches.
- **K03016** Describe the type and use of pry bars.
- **K03017** Describe the type and use of gasket scrapers.
- **K03018** Describe the type and use of files.
- **K03019** Describe the type and use of clamps.
- **K03020** Describe the type and use of taps and dies.
- **K03021** Describe the type and use of screw extractors.
- **K03022** Describe the type and use of pullers.
- **K03023** Describe the type and use of flaring tools.
- **K03024** Describe the type and use of riveting tools.
- **K03025** Describe the type and use of measuring tapes.
- **K03026** Describe the type and use of steel rulers.
- **K03027** Describe the type and use of outside, inside, and depth micrometers.
- **K03028** Describe the type and use of telescoping gauges.
- **K03029** Describe the type and use of split ball gauges.
- **K03030** Describe the type and use of dial bore gauges.
- **K03031** Describe the type and use of vernier calipers.
- **K03032** Describe the type and use of dial indicators.
- **K03033** Describe the type and use of straight edges.
- **K03034** Describe the type and use of feeler gauges.
- **K03035** Describe the importance of proper cleaning and storage of tools.

Skills Objectives

There are no Skills Objectives for this chapter.

You Are the Automotive Technician

After finishing work on the last vehicle of the day, you are required to return your workstation back to order. You clean, inspect, and return tools and equipment to their designated place. You wipe up any spills, according to the shop procedure, and clear the floor of any debris, to avoid slips and falls. During your workspace inspection, you determine that the insulation on the droplight cord is frayed, and there are some tools that need to be cleaned and put away.

1. What needs to happen with the droplight?
2. What should you do to with a micrometer before storing it?
3. How do you check a micrometer for accuracy?
4. Describe a double flare and how it is different from a single flare.

Introduction

In this chapter, we explore a variety of tool and basic shop equipment topics that are fundamental to your success as an automotive technician. They provide the means for work to be undertaken on vehicles, from lifting to diagnosing, removing, installing, cleaning, and inspecting. Nearly all shop tasks involve the use of some sort of tool or piece of equipment (**FIGURE 3-1**). This makes their purchase, use, and maintenance very important to the overall performance of the shop. In fact, most tool purchases are considered an investment because they generate income when they are used. For example, if you can buy a tool for $100 that saves you two hours of working time every time you use it, you only need to use it a couple of times in order to pay for it. Then, every time after that, it pays you a bonus. This means that you need to treat tools like your own personal moneymakers. One way to do that is to always use tools and equipment in the way they are designed to be used. Don't abuse them. Think about the task at hand, identify the most effective tools to do the task, inspect the tool before using it, use it correctly, clean and inspect it after you use it, and store it in the correct location. Doing all of these things ensures that your tools will be available the next time you use them and that they will last a long time.

General Safety Guidelines

Although it is important to be trained on the safe use of tools and equipment, it is even more critical to have a safe attitude. A safe attitude will help you avoid being involved in an accident. Students who think they will never be involved in an accident will not be as aware of unsafe situations as they should be. And that can lead to accidents. So while we are covering the various tools and equipment you will encounter in the shop, pay close attention to the safety and operation procedures. Tools are a technician's best friend, but if used improperly, they can injure or kill (**FIGURE 3-2**).

Work Safe and Stay Safe

Whenever using tools, always think safety first. There is nothing more important than your personal safety. If tools (both hand and power) are used incorrectly, you can potentially injure yourself and others. Always follow equipment and shop instructions, including the use of recommended personal protective equipment (PPE). Accidents only take a second or two to happen but can take a lifetime to recover from. You are ultimately responsible for your safety, so remember to work safe and stay safe.

Safe Handling and Use of Tools

Tools must be safely handled and used to prevent injury and damage. Always inspect tools prior to use, and never use damaged tools. Check the manufacturer and the shop procedures, or ask your supervisor if you are uncertain about how to use tools. Inspect and clean tools when you are finished using them. Always return tools to their correct storage location.

To safely handle and use appropriate tools, follow **SKILL DRILL 3-1**.

Safe Procedures for Handling Tools and Equipment

Some tools are heavy or awkward to use, so seek assistance if required, and use correct manual handling techniques when using tools. To utilize safe procedures for handling tools and equipment, follow the steps in **SKILL DRILL 3-2**.

Tool Usage

Tools extend our abilities to perform many tasks; for example, jacks, stands, and hoists extend our ability to lift and hold heavy objects. Hand tools extend our ability to perform

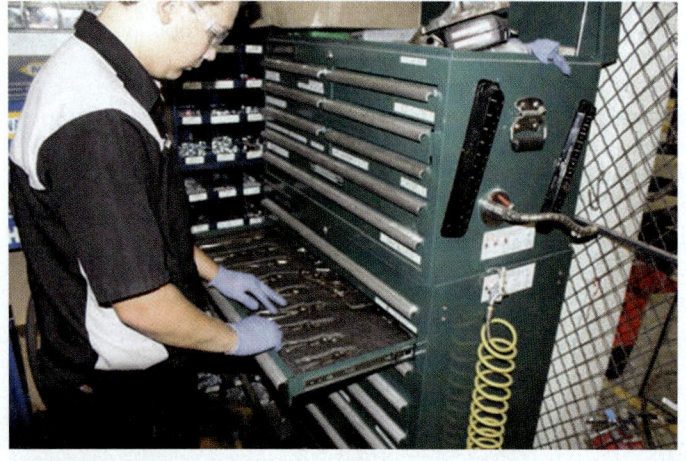

FIGURE 3-1 Technicians rely on tools to perform work.

▶ **TECHNICIAN TIP**

A sticker on the dash of a work truck I once used read: "Treat me well—your paycheck depends on it. If I can't work, neither can you!" That goes for all tools and equipment. Your ability to do work as a technician will either be enhanced or reduced by the condition of the tools and equipment you use.

N03001 Demonstrate safe handling and use of appropriate tools.

N03002 Utilize safe procedures for handling of tools and equipment.

K03001 Describe the safety procedures to take when handling and using tools.

▶ **TECHNICIAN TIP**

Using tools can make you much more efficient and effective in performing your job. Without tools, it would be very difficult to carry out vehicle repairs and servicing. It is also the reason that many technicians invest well over $20,000 in their personal tools. If purchased wisely, tools can help you perform more work in a shorter amount of time, thereby making you more money. So think of your tools as an investment that pays for themselves over time.

General Safety Guidelines 53

FIGURE 3-2 Tools used improperly can injure or kill.

SKILL DRILL 3-1 Safe Handling and Use of Tools

1. Select the correct tool(s) to undertake tasks.

2. Inspect tools prior to use to ensure they are in good working order. If tools are faulty, remove them from service according to shop procedures.

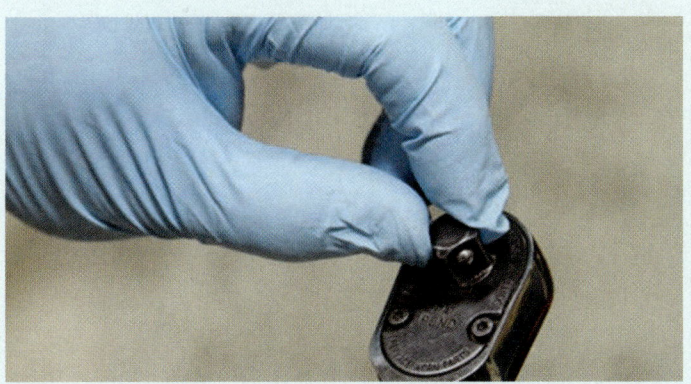

3. Clean tools prior to use if necessary.

4. Use tools to complete the task while ensuring manufacturer and shop procedures are followed. Always use tools safely to prevent injury and damage.

CHAPTER 3 Basic Tools and Precision Measuring

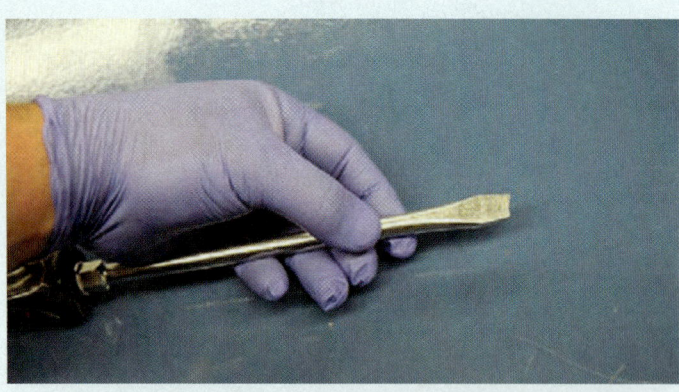

5. Ensure tools are clean and in good working order after use. Report and tag damaged tools, and remove them from service, following shop procedures.

6. Return tools to correct storage locations.

SKILL DRILL 3-2 Safe Procedures for Handling Tools and Equipment

1. Seek assistance if tools and equipment are too heavy or too awkward to be managed by a single person.

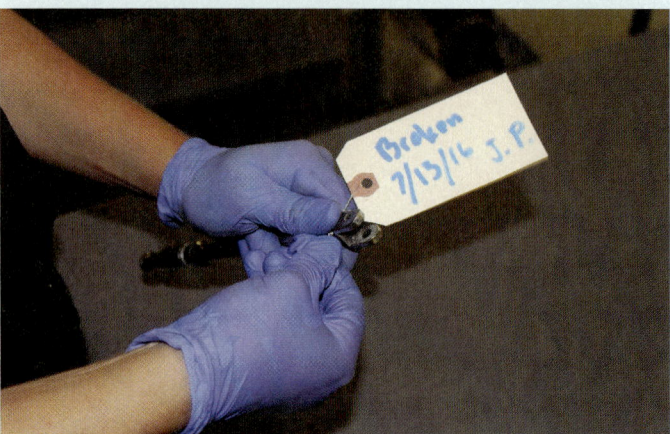

2. Inspect tools and equipment for possible defects before starting work. Report and/or tag faulty tools and equipment according to shop procedures.

Continued

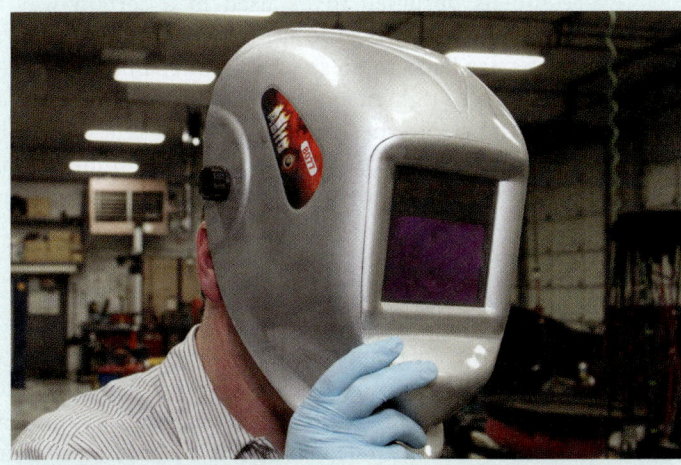

3. Select and wear appropriate PPE for the tools and equipment being used.

4. Use tools and equipment safely.

5. Check tools for faults after using them and report and/or tag faulty tools and equipment according to shop procedures.

6. Clean and return tools and equipment to correct storage locations when tasks are completed.

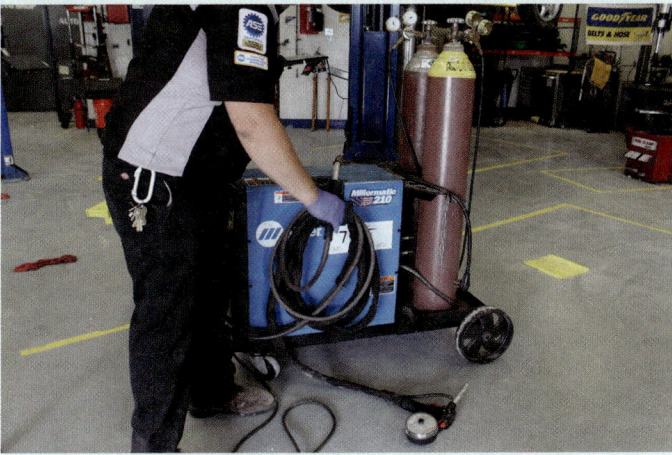

fundamental tasks like gripping, turning, tightening, measuring, and cutting (**FIGURE 3-3**). Electrical meters enable us to measure things we cannot see, feel, or hear; and power and air tools multiply our strength by performing tasks quickly and efficiently. As you are working, always think about what tool can make the job easier, safer, or more efficient. As you become familiar with more tools, your productivity, quality of work, and effectiveness will improve.

FIGURE 3-3 Tools extend our abilities.

SAFETY TIP

Standardized lockout/tag-out procedures are a mandatory part of workplace safety regulations in most countries. Familiarize yourself with your local legislation and with the specific lockout/tag-out practices that apply in your workplace.

K03003 Describe typical tool storage methods.

Every tool is designed to be used in a certain way to do the job safely. It is critical to use a tool in the way it was designed to be used and to do so safely. For example, a screwdriver is designed to tighten and loosen screws, not to be used as a chisel. Ratchets are designed to turn sockets, not to be used as a hammer. Think about the task you are undertaking, select the correct tools for the task, and use each tool as it was designed.

Lockout/Tag-Out

Lockout/tag-out is an umbrella term that describes a set of safety practices and procedures that are intended to reduce the risk of technicians inadvertently using tools, equipment, or materials that have been determined to be unsafe or potentially unsafe, or that are in the process of being serviced. An example of lockout would be physically securing a broken, unsafe, or out-of-service tool so that it cannot be used by a technician. In many cases, the item is also tagged out so it is not inadvertently placed back into service or operated. An example of tag-out would be affixing a clear and obvious label to a piece of equipment that describes the fault found, the name of the person who found the fault, and the date that the fault was found, and that warns not to use the equipment (**FIGURE 3-4**).

Tool Storage

Typically, technicians have a selection of their own tools, which include various hand tools, air tools, measuring tools, and electrical meters (**FIGURE 3-5**). Often they add to their toolbox over time. These tools are kept in a toolbox that can be rolled around as needed but is typically kept in the technician's work stall. The shop usually has a selection of specialty tools and equipment that is available for technicians to use on a shared basis. These tools are located in specific areas around the shop, typically in a centralized location so they are relatively easy to access (**FIGURE 3-6**). They include specialized manufacturer tools, such as pullers or installation tools; high-cost tools such as factory scan tools; and tools that are not portable, such as hoists and tire machines. But because they are shared by everyone in the shop, it is critical that they are inspected for damage and put in their proper storage space after each use so they will be available to the other technicians.

FIGURE 3-4 A. An example of lockout would be physically locking out a tool or piece of equipment so that it cannot be accessed and used by someone who may be unaware of the potential danger of doing so. **B.** An example of tag-out would be affixing a clear and obvious label to a piece of equipment that describes the fault found and that warns not to use the equipment.

FIGURE 3-5 Typical technician's toolbox.

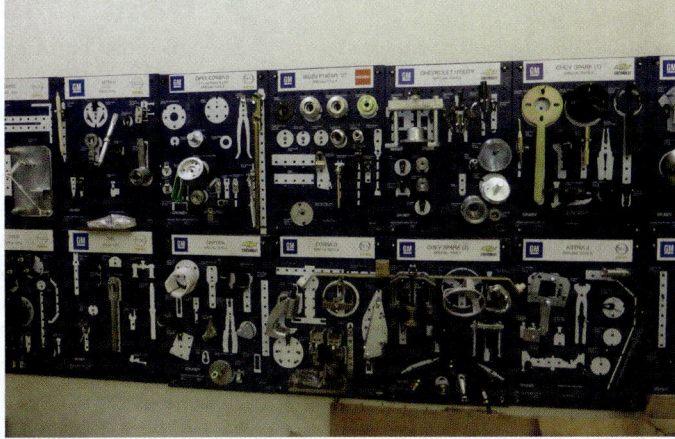
FIGURE 3-6 Manufacturer's special tools.

To identify tools and their usage in automotive applications, follow these steps:

1. Create a list of tools in your toolbox and identify their application for automotive repair and service.
2. Look through the shop's tool storage areas, and create a list of the tools found in each storage area; identify their application for automotive repair and service.

Standard and Metric Designations

Many tools, measuring instruments, and fasteners come in United States customary system (USCS) sizes, more commonly referred to as "standard," or in metric sizes. Tools and measuring instruments can be identified as standard or metric by markings identifying their size on tools or the increments on a measuring tool. **Fasteners** bought new have their designation identified on the packaging. Other fasteners may have to be measured by a ruler or **vernier caliper** to identify their designation. Manufacturer's charts showing thread and fastener sizing assist in identifying standard or metric sizing.

N03003 Identify standard and metric designation.

To identify standard and metric designation, follow these steps:

1. Examine the component, tool, or fastener to see if any marking identifies it as standard or metric. Manufacturer specifications and shop manuals can be referred to and may identify components as standard or metric (**FIGURE 3-7**).
2. If no markings are available, use measuring devices to gauge the size of the item and compare thread and fastener charts to identify the sizing. Inch-to-metric conversion charts assist in identifying component designation (**FIGURE 3-8**).

FIGURE 3-7 Manufacturer specifications and shop manuals can be referred to and may identify components as standard or metric.

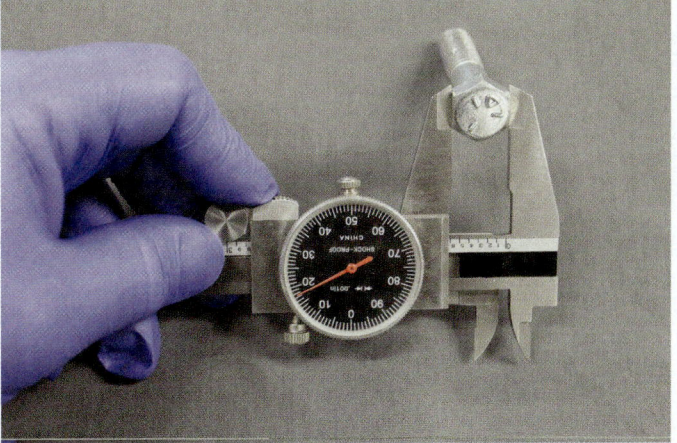
FIGURE 3-8 If no markings are available, use measuring devices to gauge the size of the item and compare thread and fastener charts to identify the sizing.

K03004 Identify tools and their usage in automotive applications.

▶ **TECHNICIAN TIP**

Invest in quality tools. Because tools extend your abilities, poor-quality tools affect the quality and quantity of your work. Price is not always the best indicator of quality, but it plays a role. As you learn the purpose and function of the tools in this chapter, you should be able to identify high-quality tools versus poor-quality tools by looking at them, handling them, and putting them to work.

K03005 Describe the type and use of wrenches.

▶ Basic Hand Tools

Like all tools, hand tools extend our ability to do work. Hand tools come in a variety of shapes, sizes, and functions (**FIGURE 3-9**). A large percentage of your personal tools will be hand tools. Over the years, manufacturers have introduced new fasteners, wire harness terminals, quick-connect fittings for fuel and other lines, and additional technologies that require their own different types of hand tools. This means that technicians need to continually add updated tools to their toolbox.

Speaking of toolboxes, technicians need to invest in a quality toolbox to keep their tools secure, so toolboxes should have enough capacity to hold all current and future tools. They should be built to last because a toolbox is in continuous use throughout the workday. A toolbox should have drawers that can handle the weight of the tools in it. Plus it should open and close easily; drawer slides that use bearings make them much easier to open and close (**FIGURE 3-10**). The toolbox should also be equipped with an easy-to-use locking system to secure the tools when they aren't being used.

Wrenches

Wrenches are used to tighten and loosen nuts and **bolts**, which are two types of fasteners. There are three commonly used wrenches: the **box-end wrench**, the **open-end wrench**, and the **combination wrench** (**FIGURE 3-11**). The box-end wrench fits fully around the head of the bolt or nut and grips each of the six points at the corners, just like a socket. This is exactly the sort of grip needed if a nut or bolt is very tight, and makes the box-end wrench less likely to round off the points on the head of the bolt than the open-end wrench. The ends of box-end wrenches are bent or offset, so they are easier to grip and have different-sized heads at each end (**FIGURE 3-12**). One disadvantage of the box-end wrench is that it can be awkward to use once the nut or bolt has been loosened a bit, because you have to lift it off the head of the fastener and move it to each new position.

The open-end wrench is open on the end, and the two parallel flats only grip two points of the fastener. Open-end wrenches usually either have different-sized heads on each end of the wrench, or they have the same size, but with different angles (**FIGURE 3-13**). The head is at an angle to the handle and is not bent or offset, so it can be flipped over and used on both sides. This is a good wrench to use in very tight spaces as you can flip it over and get a new angle so the head can catch new points on the fastener. Although an open-end wrench often gives the best access to a fastener, if the fastener is extremely

FIGURE 3-9 Hand tools.

FIGURE 3-10 Toolbox drawers need to open and close easily.

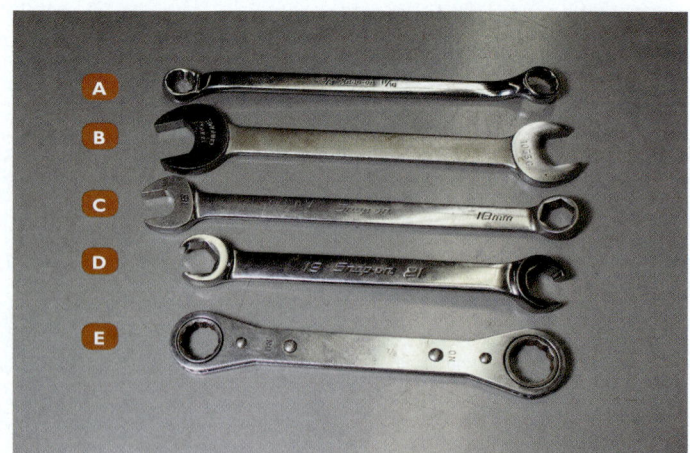

FIGURE 3-11 A. Box-end wrench. **B.** Open-end wrench. **C.** Combination wrench. **D.** Flare nut wrench. **E.** Ratcheting box-end wrench.

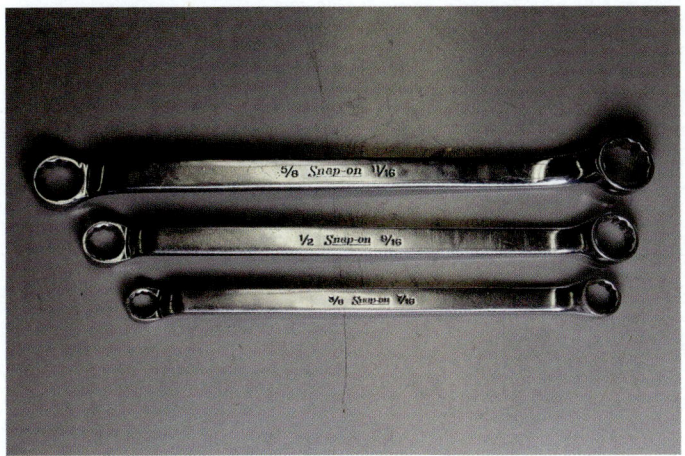

FIGURE 3-12 Box-end wrenches.

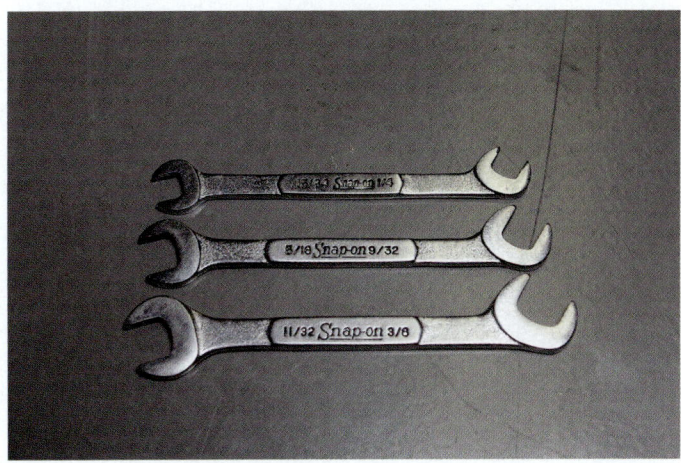

FIGURE 3-13 Types of open-end wrenches.

tight, the open-end wrench should not be used, as this type of wrench only grips two points. If the jaws flex slightly, the wrench can suddenly slip when force is applied. This slippage can round off the points of the fastener. The best way to approach a tight fastener is to use a box-end wrench to break the bolt or nut free, then use the open-end wrench to finish the job. The open-end wrench should only be used on fasteners that are no more than firmly tightened.

The combination wrench has an open-end head on one end and a box-end head on the other end (**FIGURE 3-14**). Both ends are usually of the same size. That way the box-end wrench can be used to break the bolt loose, and the open end can be used for turning the bolt. Because of its versatility, this is probably the most popular wrench for technicians.

A variation on the open-end wrench is the **flare nut wrench**, also called a flare tubing wrench, or line wrench. It gives a better grip than the open-end wrench because it grabs all six points of the fastener, not two (**FIGURE 3-15**). However, because it is open on the end, it is not as strong as a box-end wrench. The partially open sixth side lets the wrench be placed over the tubing or pipe so the wrench can be used to turn the tube fittings. Do not use the flare nut wrench on extremely tight fasteners as the jaws may spread, damaging the nut.

One other open-end wrench is the open-end adjustable wrench or crescent wrench. This wrench has a movable jaw that can be adjusted by turning an adjusting screw to fit any fastener within its range (**FIGURE 3-16**). It should only be used if other wrenches are not available because it is not as strong as a fixed wrench, so it can slip off and damage the head of tight bolts or nuts. Still, it is a handy tool to have because it can be adjusted to fit most any fastener size.

A **ratcheting box-end wrench** is a useful tool in some applications because it does not require removal of the tool to reposition it. It has an inner piece that fits over and grabs the points on the fastener and is able to rotate within the outer housing. A ratcheting mechanism lets it rotate in one direction and lock in the other direction. In some cases, the wrench just needs to be flipped over to be used in the opposite direction. In other cases, it has a lever that changes the direction from clockwise to counterclockwise (**FIGURE 3-17**). Just be careful to not overstress this tool by using it to tighten or loosen very

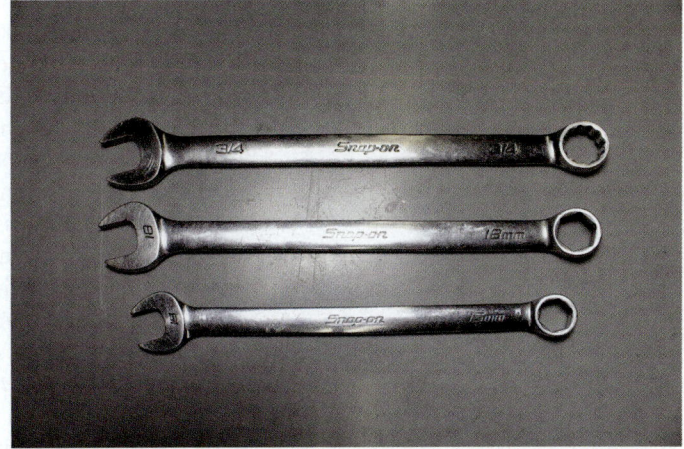

FIGURE 3-14 Combination wrenches.

FIGURE 3-15 Flare nut wrenches.

FIGURE 3-16 Open-end adjustable wrench.

FIGURE 3-17 Ratcheting box-end wrench.

> ▶ **TECHNICIAN TIP**
>
> Wrenches (which are also known as spanners in some countries) will only do a job properly if they are the right size for the given nut or bolt head. The size used to describe a wrench is the distance across the flats of the nut or bolt. There are two systems in common use—standard (in inches) and metric (in millimeters). Each system provides a range of sizes, which are identified by either a fraction, which indicates fractions of an inch for the standard system, or a number, which indicates millimeters for the metric system. In most cases, standard and metric tools cannot be used interchangeably on a particular fastener. Even though a metric tool fits a standard sized fastener, it may not grip as tight, which could cause it to round the head of the fastener off.

tight fasteners, as the outer housing is not very strong. There is also a ratcheting open-end wrench, but it uses no moving parts. One of the sides is partially removed so that only the bottom one-third remains to catch a point on the bolt (**FIGURE 3-18**). When it is used, the normal side works just like a standard open-end wrench. The shorter side of the open-end wrench catches the point on the fastener so it can be turned. When moving the wrench to get a new bite, the wrench is pulled slightly outward, disengaging the short side while leaving the long side to slide along the faces of the bolt. The wrench is then rotated to the new position and pushed back in so the short side engages the next point. This wrench, like other open-end wrenches, is not designed to tighten or loosen tight fasteners, but it does work well in blind places where a socket or ratcheting box-end wrench cannot be used.

Specialized wrenches such as the **pipe wrench** grip pipes and can exert a lot of force to turn them (**FIGURE 3-19**). Because the handle pivots slightly, the more pressure put on the handle to turn the wrench, the more the grip tightens. The jaws are hardened and serrated, and increasing the pressure also increases the risk of marking or even gouging metal from the pipe. The jaw is adjustable, so it can be threaded in or out to fit different pipe sizes. Also, they come in different lengths, allowing you to increase the leverage applied to the pipe.

A specialized wrench called an **oil filter wrench** grabs the filter and gives you extra leverage to remove an oil filter when it is tight. These wrenches are available in various designs and sizes (**FIGURE 3-20**), and some are adjustable to fit many filter sizes. Also note that an oil filter wrench should be used *only* to remove an oil filter, never to install it. Almost all oil filters should be installed and tightened by hand.

FIGURE 3-18 Ratcheting open-end wrench.

FIGURE 3-19 Pipe wrench.

FIGURE 3-20 Oil filter wrenches.

Using Wrenches Correctly

Choosing the correct wrench for a job usually depends on two things: how tight the fastener is and how much room there is to get the wrench onto the fastener, and then to turn it. When being used, it is always possible that a wrench could slip. Before putting a lot of tension on the wrench, try to anticipate what will happen if it does slip. If possible, it is usually better to pull a wrench toward you than to push it away (**FIGURE 3-21**). If you have to push, use an open palm to push, so your knuckles won't get crushed if the wrench slips. If pulling toward yourself, make sure your face is not close to your hand or in line with your pulling motion. Many technicians have punched themselves in the face when the wrench slipped.

Sockets

Sockets are very popular because of their adaptability and ease of use (**FIGURE 3-22**). Sockets are a good choice where the top of the fastener is reasonably accessible. The **socket** fits onto the fastener snugly and grips it on all six corners, providing the type of grip needed on any nut or bolt that is extremely tight. Sockets come in a variety of configurations, and technicians usually have a lot of sockets so they can get in a multitude of tight places. Individual sockets fit a particular size nut or bolt, so they are usually purchased in sets.

K03006 Describe the type and use of sockets.

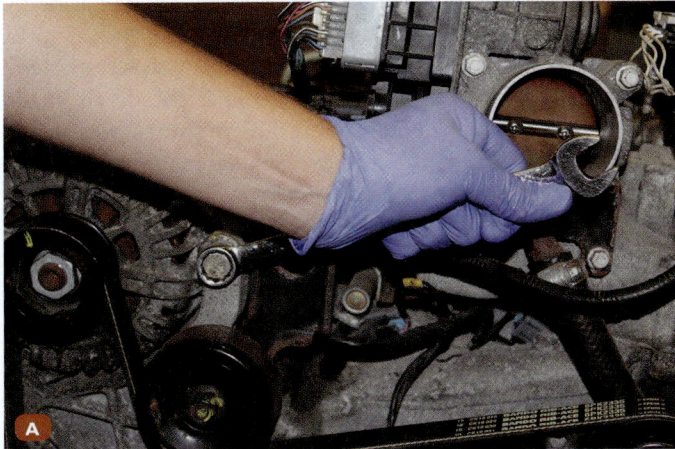

FIGURE 3-21 A. Pulling on a wrench is generally better than pushing on a wrench. **B.** Pushing regularly results in bruised or broken knuckles.

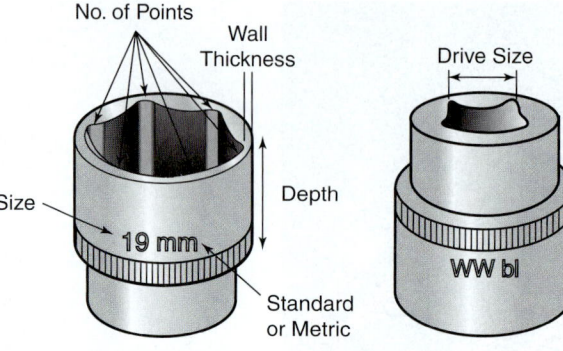

FIGURE 3-22 The anatomy of a socket.

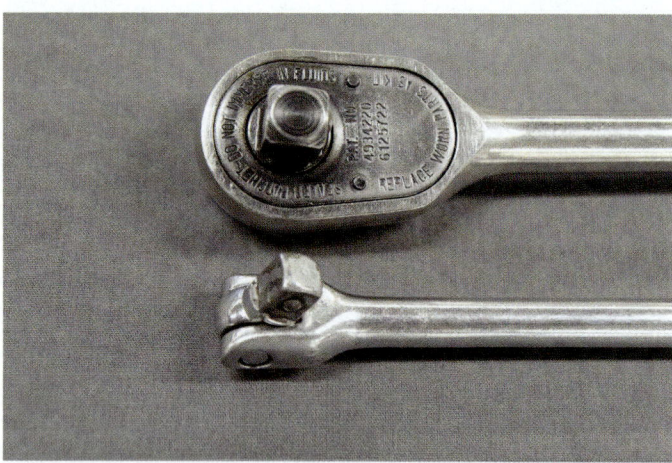

FIGURE 3-23 Sockets are designed to fit a matching drive on a ratchet.

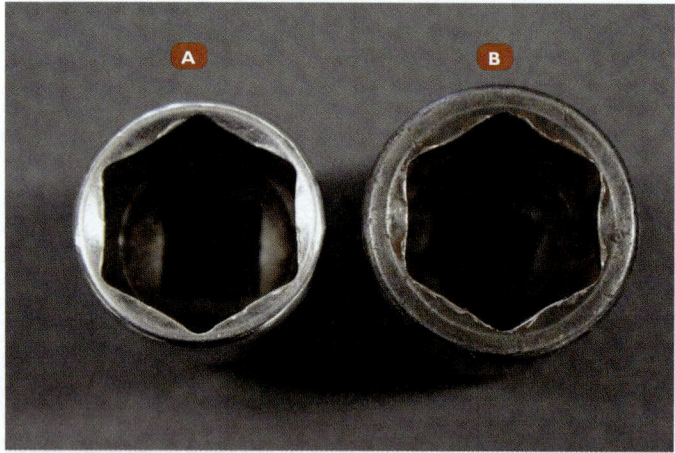

FIGURE 3-24 **A.** Standard wall socket. **B.** Impact socket.

▶ **TECHNICIAN TIP**

Because sockets are usually purchased in sets, with each set providing a slightly different capability, you can see why technicians could easily have several hundred sockets in their toolbox. Having a variety of sockets allows the technician to do jobs easier and quicker, which makes them an investment. Turning a socket requires a handle (**FIGURE 3-27**).

Sockets are classified by the following characteristics:

- Standard or metric
- Size of drive used to turn them: 1/2" (12.7 mm), 3/8" (9.525 mm), and 1/4" (6.35 mm) are most common; 1" (25.4 mm) and 3/4" (19.05 mm) are less common.
- Number of points: 6 and 12 are most common; 4 and 8 are less common.
- Depth of socket: Standard and deep are most common; shallow is less common.
- Thickness of wall: Standard and impact are most common; thin wall is less common.

Sockets are built with a recessed square drive that fits over the square drive of the ratchet or other driver (**FIGURE 3-23**). The size of the drive determines how much twisting force can be applied to the socket. The larger the drive, the larger the twisting force. Small fasteners usually only need a small torque, so having too large of a drive may make it so the socket cannot gain access to the bolt. For fasteners that are really tight, an impact wrench exerts a lot more torque on a socket than turning it by hand does. Impact sockets are usually thicker walled than standard wall sockets and have six points so they can withstand the forces generated by the impact wrench as well as grip the fastener securely (**FIGURE 3-24**).

Six- and 12-point sockets fit the heads of hexagonal-shaped fasteners. Four- and 8-point sockets fit the heads of square-shaped fasteners (**FIGURE 3-25**). Because 6-point and 4-point sockets fit the exact shape of the fastener, they have the strongest grip on the fastener, but they only can fit on the fastener in half as many positions as a 12- or 8-point socket, making them harder to fit onto the fastener in places where the ratchet handle is restricted.

Another factor in accessing a fastener is the depth of the socket. If a nut is threaded quite a way down a stud or bolt, then a standard length socket will not fit far enough over the stud to reach the nut (**FIGURE 3-26**). In this case, a deep socket will usually reach the nut.

The most common socket handle, the **ratchet**, makes easy work of tightening or loosening a nut where not a lot of pressure is involved. It can be set to turn in either direction and does not need much room to swing. It is built to be convenient, not super strong, so too much pressure could strip the ratchet mechanism. For heavier tightening or loosening, a breaker bar gives the most leverage. When that is not available, a **sliding T-handle** may be more useful. With this tool, both hands can be used, and the position of the T-piece is adjustable to clear any obstructions when turning it. The connection between the socket and the accessory is made by a square drive. The larger the drive, the heavier and bulkier the socket will be. The 1/4" (6.35 mm) drive is for small work in

FIGURE 3-25 **A.** Six- and 12-point sockets. **B.** Four- and 8-point sockets.

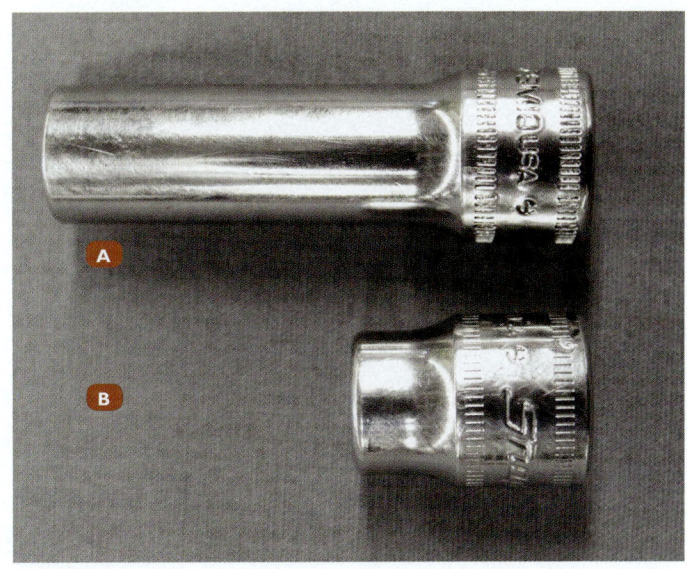

FIGURE 3-26 **A.** Deep socket. **B.** Standard length socket.

difficult areas. The 3/8" (9.525 mm) drive accessories handle a lot of general work where torque requirements are not too high. The 1/2" (12.7 mm) drive is required for all-around service. The 3/4" (19.05 mm) and 1" (25.4 mm) drives are required for large work with high torque settings.

Many fasteners are located in positions where access can be difficult. There are many different lengths of extensions available to allow the socket to be on the fastener while extending the drive point out to where a handle can be attached. Because extensions come in various lengths, they can be connected together to get just the right length needed for a particular situation.

If an object is in the way of getting a socket on a fastener, a flexible joint is used to apply the turning force to the socket through an angle. This can allow you to still turn the socket even though you are no longer directly in line with the fastener. There are four common types of flexible joints: U-joint style, wobble extension, cable extension, and flex socket (**FIGURE 3-28**). The flex

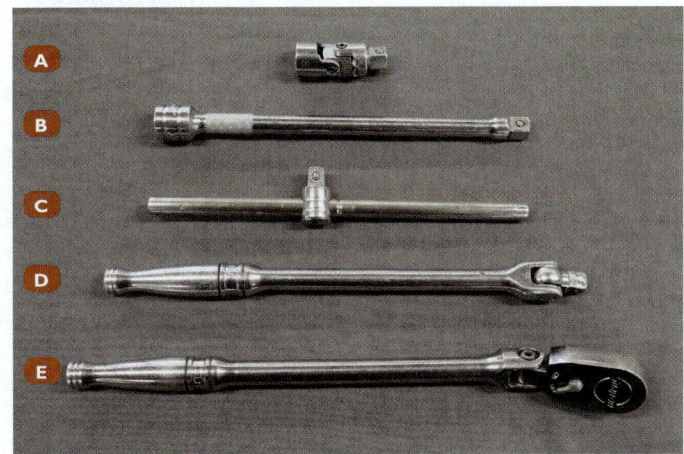

FIGURE 3-27 Tools to turn sockets. **A.** Universal joint. **B.** Extension. **C.** T-handle. **D.** Breaker bar. **E.** Ratchet.

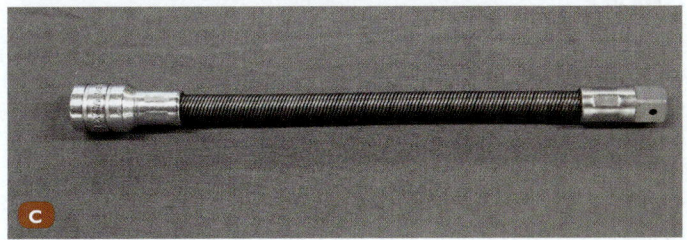

FIGURE 3-28 Flexible extensions. **A.** U-joint style. **B.** Wobble extension style. **C.** Cable extension style. **D.** Flex socket style.

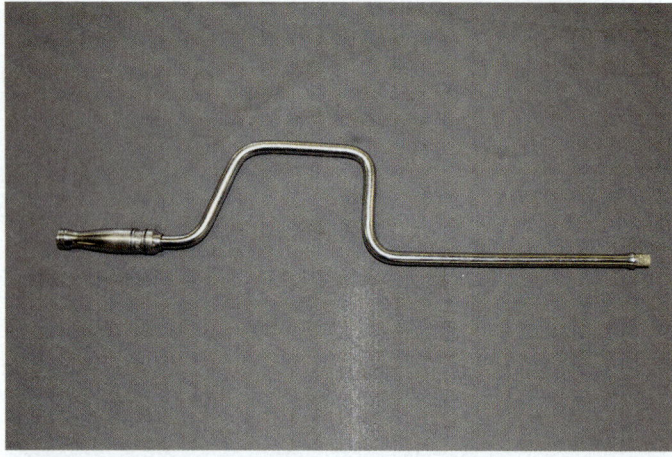

FIGURE 3-29 Speed brace.

FIGURE 3-30 Lug wrench.

socket has the universal built into it, so it's overall length is shorter, making it able to get into tighter spaces. In some situations, you may need to use more than one flexible joint to get around objects that are in the way. This is especially true when removing some bell housing bolts on some transmissions.

A **speed brace** or speeder handle is the fastest way to spin a fastener on or off a thread by hand, but it cannot apply much torque to the fastener; therefore, it is mainly used to remove a fastener that has already been loosened or to run the fastener onto the thread until it begins to tighten (**FIGURE 3-29**).

A **lug wrench** has special-sized lug nut sockets permanently attached to it. One common model has four different-sized sockets, one on each arm (**FIGURE 3-30**). Never hit or jump on a lug wrench when loosening lug nuts. If the lug wrench will not remove them, you should use an impact wrench. The impact wrench provides a hammering effect in conjunction with rotation to help loosen tight fasteners. *Never* use an impact wrench to tighten lug fasteners. Torque all lug fasteners to the proper torque with a properly calibrated torque wrench. And do it in the proper sequence.

Torque and Torque Wrenches

K03007 Describe the type and use of torque wrenches.

Bolt tension is what keeps a bolt from loosening and what causes it to hold parts together with the proper clamping force. **Torque** is the twisting force used to create bolt tension so that surfaces are clamped together with the proper force. The torque value is the amount of twisting force applied to a fastener by the torque wrench. A foot-pound is described as the amount of twisting force applied to a shaft by a perpendicular lever 1 ft (0.30m) long with a weight of 1 lb (0.45 kg) placed on the outer end (**FIGURE 3-31**). A torque value of 100 ft-lb (13.83 m-kg) is the same as a 100 lb (45.36 kg) weight placed at the end of a 1' (0.30m) long lever. A torque value of an inch-pound is 1 lb (0.45 kg) placed at the end of a 1" (25.40 mm) long lever. This means that 12 in-lb equals 1 ft-lb (0.14 m-kg) and vice versa. A newton meter (Nm) is described as the amount of twisting force applied to a shaft by a perpendicular lever 1 meter long with a force of 1 newton applied to the outer end. A torque value of 100 Nm is the same as applying a 100-newton force to the end of a 1 m long lever. One ft-lb is equal to 1.35 Nm. So torque is the measurement of twisting force.

FIGURE 3-31 Torque is the measurement of twisting force. 1 ft-lb of torque.

Torque Charts

Torque specifications for bolts and nuts in vehicles are usually contained within service information. Bolt, nut, and stud manufacturers also produce torque charts, which contain the information you need to determine the maximum torque of bolts or nuts. For example, most charts include the bolt diameter, threads per inch, grade, and maximum torque setting for both dry and lubricated bolts and nuts (**FIGURE 3-32**).

A lubricated bolt and nut reach their maximum clamping force at a lower torque setting than if they are dry. In practice, most torque specifications call for the nuts and bolts to have dry threads prior to tightening. There are some exceptions, so close examination of the torque specification is critical. Also remember that the bolt manufacturer's torque chart is a maximum recommended torque, not necessarily the torque required by the vehicle manufacturer for the specific application that the bolt is used for.

Torque Wrenches

A **torque wrench** is also known as a tension wrench (**FIGURE 3-33**) and is used to tighten fasteners to a predetermined torque. The drive on the end fits any socket and accessory of the same drive size found in ordinary socket sets. Although manufacturers do not specify torque settings for every nut and bolt, when they do, it is important to follow the specifications. For example, manufacturers specify a torque for head bolts.

The torque specified ensures that the bolt provides the proper clamping pressure and will not come loose, but will not be so tight as to risk breaking the bolt or stripping the threads. The torque value is specified in foot-pounds (ft-lb), inch-pounds (in-lb), or newton meters (N m).

Torque wrenches come in various types: beam style, clicker, dial, and electronic (**FIGURE 3-34**). The simplest and least expensive is the beam-style torque wrench, which uses a spring steel beam that flexes under tension. A smaller fixed rod then indicates the amount of torque on a scale mounted to the bar. The amount of deflection of the bar coincides with the amount of torque on the scale. One drawback of this design is that you have to be positioned directly above the scale so you can read it accurately. That can be a problem when working under the hood of a vehicle.

The clicker-style torque wrench uses an adjustable clutch inside that slips (clicks) when the preset torque is reached. You can set it for a particular torque on the handle

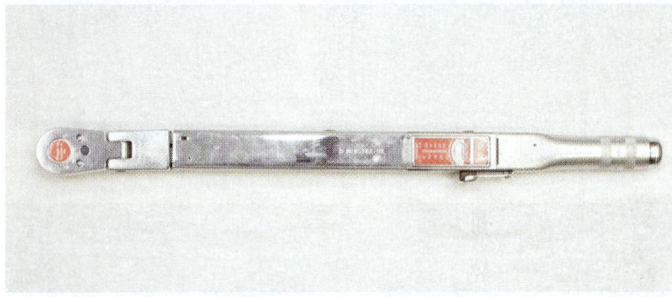

FIGURE 3-32 Bolt torque chart.

FIGURE 3-33 A torque wrench.

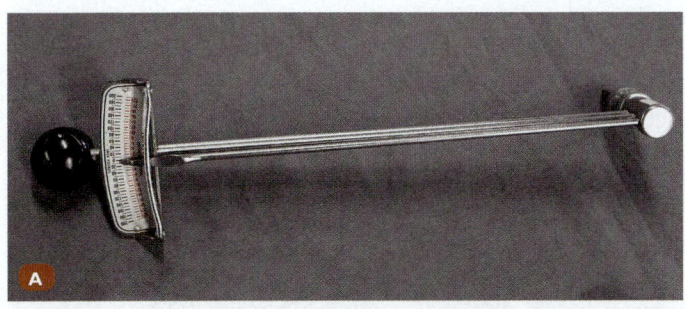

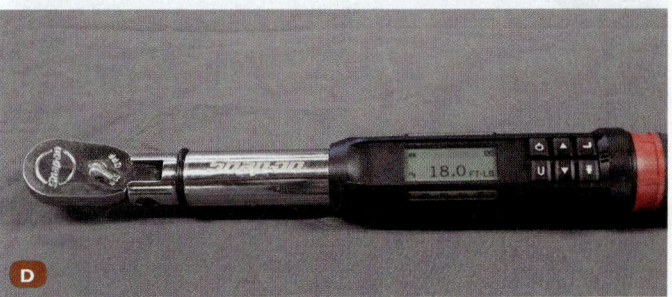

FIGURE 3-34 Torque wrenches. **A.** Beam style. **B.** Clicker style. **C.** Dial. **D.** Electronic.

(**FIGURE 3-35**). As the bolt is tightened, once the preset torque is reached, the torque wrench clicks. The higher the torque, the louder the click; the lower the torque, the quieter the click. Be careful when using this style of torque wrench, especially at lower torque settings. It is easy to miss the click and over-tighten, break, or strip the bolt. Once the torque wrench clicks, stop turning it, as it will continue to tighten the fastener if you turn it past the click point.

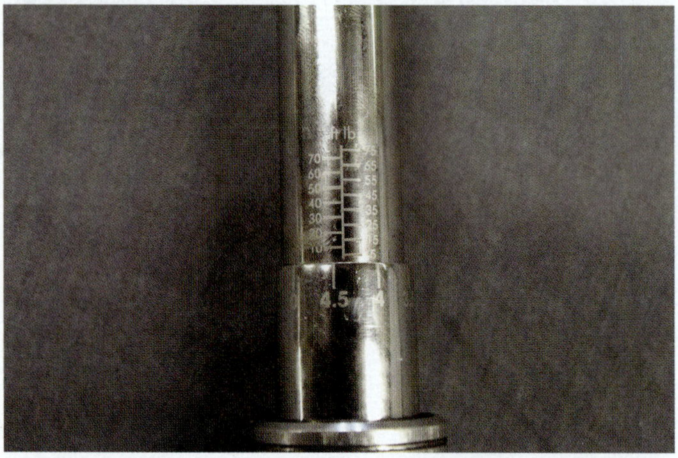

The dial torque wrench turns a dial that indicates the torque based on the torque being applied. Like the beam-style torque wrench, you have to be able to see the dial to know how much torque is being applied (**FIGURE 3-36**). Many dial torque wrenches have a movable indicator that is moved by the dial and stays at the highest reading. That way you can double-check the torque achieved once the torque wrench is released. Once the proper torque is reached, the indicator can be moved back to zero for the next fastener being torqued.

The digital torque wrench usually uses a spring steel bar with an electronic strain gauge to measure the amount of torque being applied. The torque wrench can be preset to the desired torque. It will then display the torque as the fastener is being tightened (**FIGURE 3-37**). When it reaches the preset torque, it usually gives an audible signal, such as a beep. This makes it useful in situations where a scale or dial cannot be read.

Torque wrenches fall out of calibration over time or if they are not used properly, so they should be checked and calibrated on a periodic basis (**FIGURE 3-38**). This can be performed in the shop if the proper calibration equipment is available, or the torque wrench can be sent to a qualified service center. Most quality torque wrench manufacturers provide a recalibration service for their customers.

Using Torque Wrenches

The torque wrench is used to apply a specified amount of torque to a fastener. There are various methods used by torque wrenches to indicate that the correct torque has been reached. Some give an audible signal, such as a click or a beep, whereas others give a visual signal such as a light or a pin moving or clicking out. Some provide a scale and needle that must be observed while you are torquing the fastener. To help ensure that the proper amount of torque gets from the torque wrench to the bolt, support the head

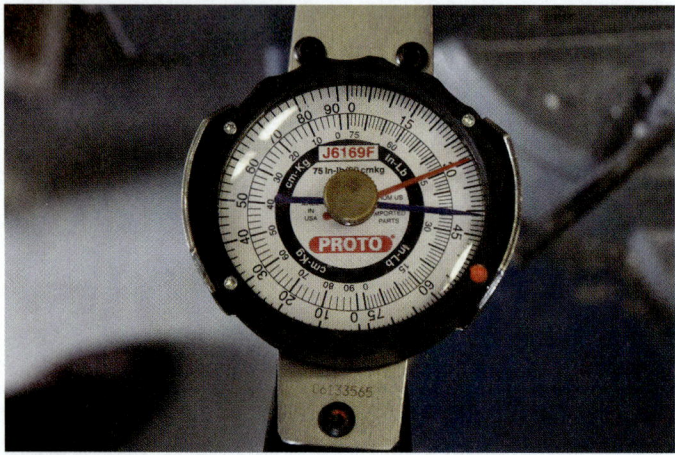

FIGURE 3-35 Torque setting scale on the handle.

FIGURE 3-36 Dial torque wrench reading torque.

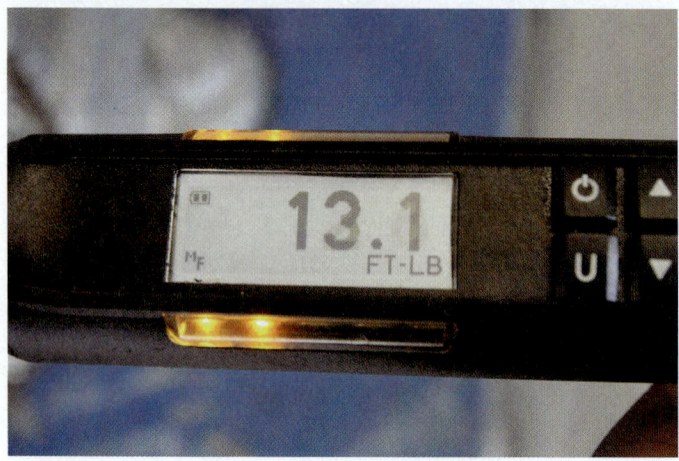

FIGURE 3-37 Digital torque wrench displaying torque.

FIGURE 3-38 Checking torque wrench calibration.

of the torque wrench with one hand (**FIGURE 3-39**). When using a torque wrench, it is best not to use extensions. Extensions make it harder to support the head, which can end up absorbing some of the torque. If possible, use a deep socket instead.

Pliers

Pliers are a hand tool designed to hold, cut, or compress materials (**FIGURE 3-40**). They are usually made out of two pieces of strong steel joined at a fulcrum point, with jaws and cutting surfaces at one end and handles designed to provide leverage at the other. There are many types of pliers, including slip-joint, combination, arc joint, needle-nose, and flat-nose (**FIGURE 3-41**).

Quality **combination pliers** are one of the most commonly used pliers in a shop. They pivot together so that any force applied to the handles is multiplied in the strong jaws. Most combination pliers are designed so they have surfaces to both grip and cut. Combination pliers offer two gripping surfaces, one for gripping flat objects and one for gripping rounded objects, and one or two pairs of cutters. The cutters in the jaws should be used for softer materials that will not damage the blades. The cutters next to the pivot can shear through hard, thin materials, like steel wire or pins.

Most pliers are limited by their size in what they can grip. Beyond a certain point, the handles are spread too wide, or the jaws cannot open wide enough, but **arc joint pliers** overcome that limitation with a moveable pivot. Often, these are called Channel-locks™ after the company that first made them. These pliers have parallel jaws that allow you to increase or decrease the size of the jaws by selecting a different set of channels (**FIGURE 3-42**). They are useful for a wider grip and a tighter squeeze on parts too big for conventional pliers.

Another type of pliers is **needle-nose pliers**, which have long, pointed jaws and can reach into tight spots or hold small items that other pliers cannot (**FIGURE 3-43**). For example, they can pick up a small bolt that has fallen into a tight spot. **Flat-nose pliers** have an end or nose that is flat and square; in contrast, combination pliers have a rounded end. A flat nose makes it possible to bend wire or even a thin piece of sheet steel accurately along a straight edge. **Diagonal cutting pliers** are used for cutting wire or cotter pins (**FIGURE 3-44**). Diagonal cutters are the most common cutters in the toolbox, but they should not be used on hard or heavy-gauge materials because the cutting surfaces will be damaged. End cutting pliers, also called **nippers**, have a cutting edge at right angles to their length (**FIGURE 3-45**). They are designed to cut through soft metal objects sticking out from a surface.

Snap ring pliers have metal pins that fit in the holes of a snap ring (**FIGURE 3-46**). Snap rings can be of the internal or external type. If internal, then internal snap ring pliers compress the snap ring so it can be removed from and installed in its internal groove. If external, then external snap ring pliers are used to expand the snap ring so it can

K03008 Describe the type and use of pliers.

SAFETY TIP

When applying pressure to pliers, make sure your hands are not greasy, or they might slip. Select the right type and size of pliers for the job. As with most tools, if you have to exert almost all your strength to get something done, then you are using either the wrong tool or the wrong technique. If the pliers slip, you will get hurt. At the very least, you will damage the tool and what you are working on. Pliers get a lot of hard use in the shop, so they do get worn and damaged. If they are worn or damaged, they will be inefficient and can be dangerous. Always check the condition of all shop tools on a regular basis, and replace any that are worn or damaged.

FIGURE 3-39 It may be necessary to support the torque wrench when using extensions.

FIGURE 3-40 Pliers are used for grasping and cutting. These are slip-joint pliers.

68 CHAPTER 3 Basic Tools and Precision Measuring

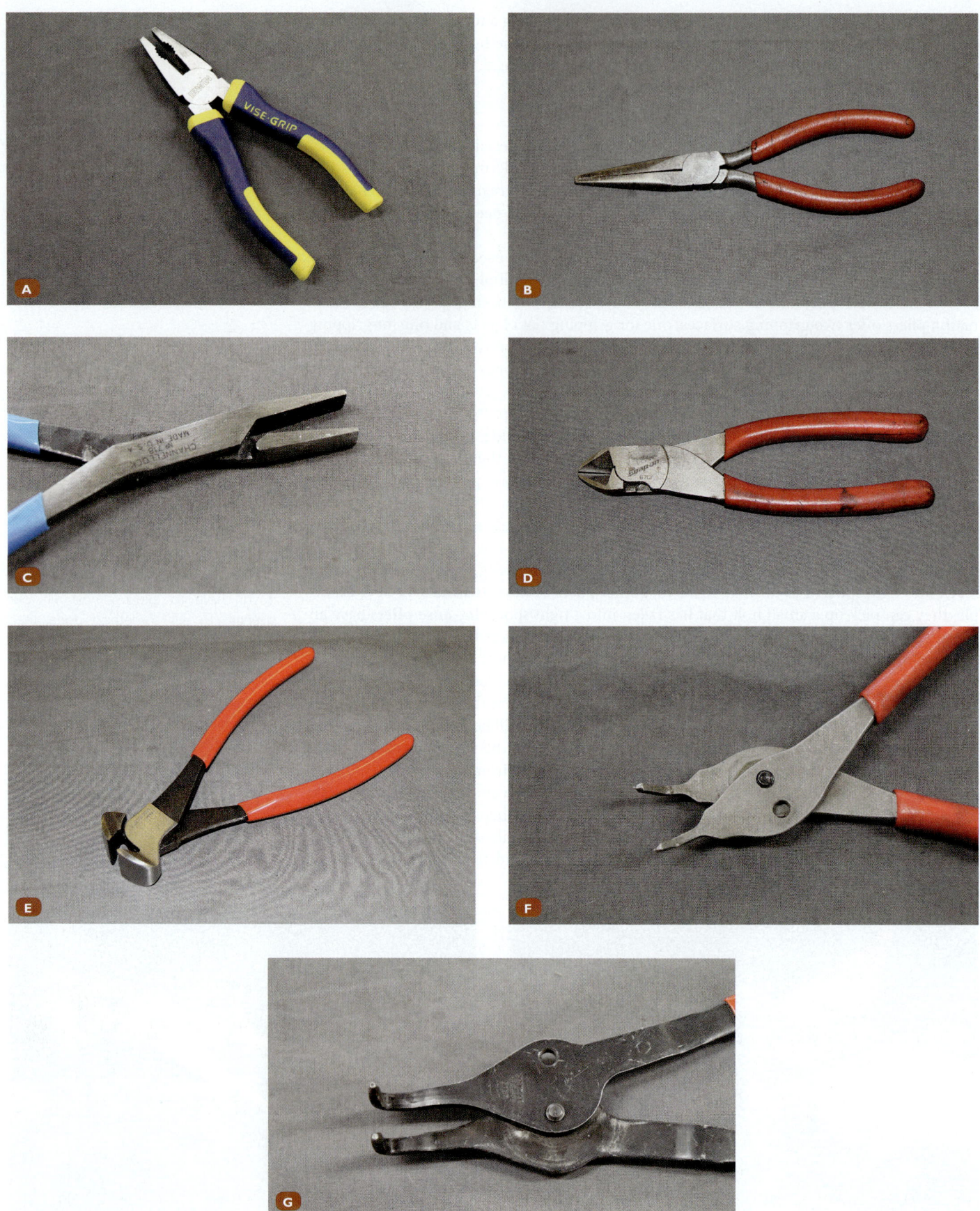

FIGURE 3-41 **A.** Combination pliers. **B.** Needle-nose pliers. **C.** Flat-nose pliers. **D.** Diagonal cutting pliers. **E.** Nippers. **F.** Internal snap ring pliers. **G.** External snap ring pliers.

Basic Hand Tools 69

FIGURE 3-42 Arc joint pliers.

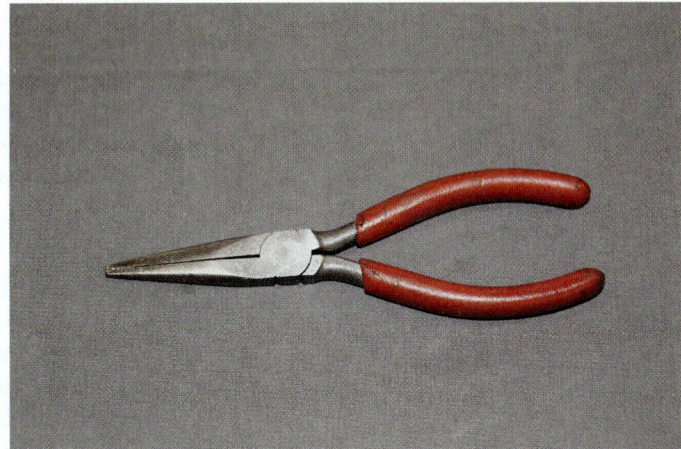
FIGURE 3-43 Needle-nose pliers.

FIGURE 3-44 Diagonal side cutters.

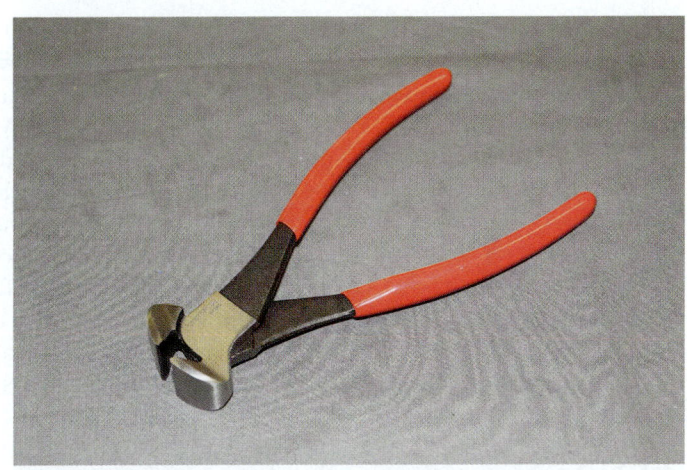
FIGURE 3-45 Nippers, or end cutting, pliers.

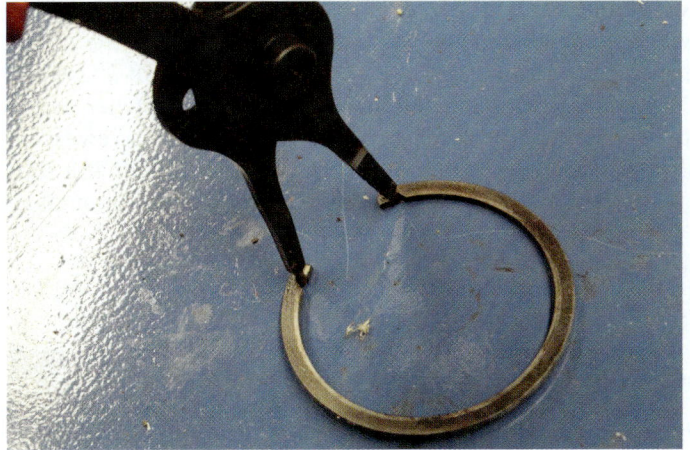
FIGURE 3-46 Snap ring pliers and snap ring.

FIGURE 3-47 Locking pliers.

remove and install the snap ring in its external groove. Always wear safety glasses when working with snap rings, as the snap rings can easily slip off the snap ring pliers and fly off at tremendous speeds, possibly causing severe eye injuries.

Locking pliers, also called vice grips, are general-purpose pliers used to clamp and hold one or more objects (**FIGURE 3-47**). Locking pliers are helpful by freeing up one or more of your hands when you are working, because they can clamp something and lock themselves in place to hold it. They are also adjustable, so they can be used for a variety of tasks. To clamp an object with locking pliers, put the object between the jaws, turn the screw until the handles are almost closed, then squeeze them together to lock them shut. You can increase or decrease the gripping force with the adjustment screw. To release them, squeeze the release lever, and they should open right up.

Cutting Tools

K03009 Describe the type and use of cutting tools.

Bolt cutters cut heavy wire, non-hardened rods, and bolts (**FIGURE 3-48**). Their compound joints and long handles give the leverage and cutting pressure that is needed for heavy gauge materials. **Tin snips** are the nearest thing in the toolbox to a pair of scissors. They can cut thin sheet metal, and lighter versions make it easy to follow the outline of gaskets. Most snips come with straight blades, but if there is an unusual shape to cut, there is a pair with left- or right-handed curved blades. **Aviation snips** are designed to cut soft metals. They are easy to use because the handles are spring-loaded in the open position and double pivoted for extra leverage.

Allen Wrenches

K03010 Describe the type and use of Allen wrenches.

Allen wrenches, sometimes called Allen or hex keys, are tools designed to tighten and loosen fasteners with Allen heads (**FIGURE 3-49**). The Allen head fastener has an internal hexagonal recess that the Allen wrench fits in snugly. Allen wrenches come in sets, and there is a correct wrench size for every Allen head. They give the best grip on a screw or bolt of all the drivers, and their shape makes them good at getting into tight spots. Care must be taken to make sure the correct size of Allen wrench is used, or else the wrench and/or socket head will be rounded off. The traditional Allen wrench is a hexagonal bar with a right-angle bend at one end. They are made in various metric and standard sizes. As their popularity has increased, so too has the number of tool variations. Now Allen sockets are available, as are T-handle Allen keys (**FIGURE 3-50**).

K03011 Describe the type and use of screwdrivers.

Screwdrivers

The correct screwdriver to use depends on the type of slot or recess in the head of the screw or bolt, and how accessible it is (**FIGURE 3-51**). Most screwdrivers cannot grip as securely as wrenches, so it is very important to match the tip of the screwdriver exactly with the slot or recess in the head of a fastener. Otherwise, the tool might slip, damaging the fastener or the tool and possibly injuring you. When using a screwdriver, always check where the screwdriver blade can end up if it slips off the head of the screw. Many technicians who have not taken

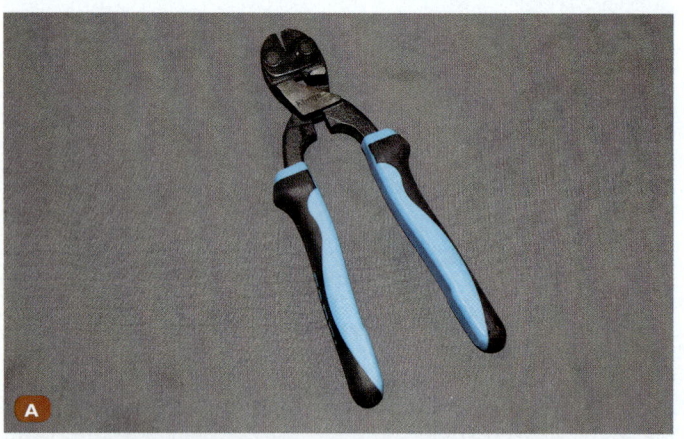

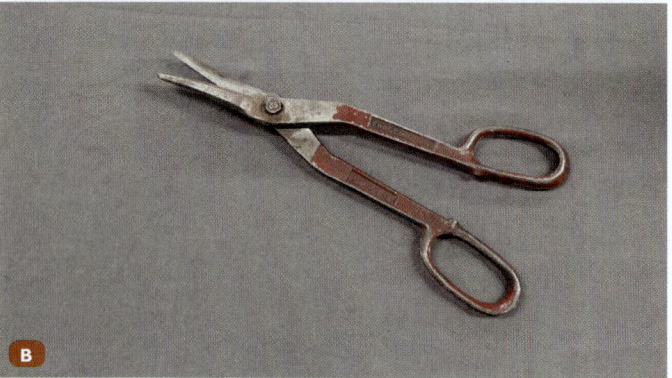

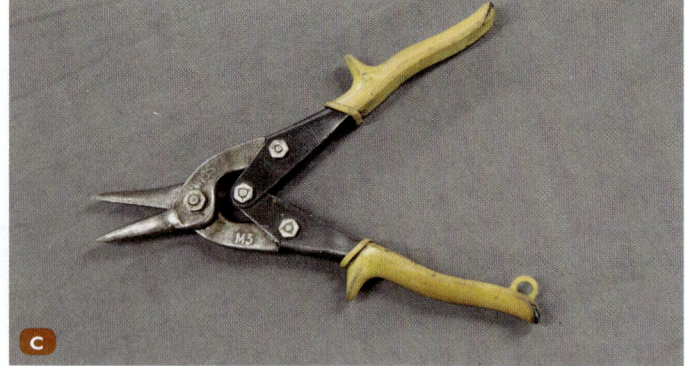

FIGURE 3-48 A. Bolt cutters. **B.** Tin snips. **C.** Aviation snips.

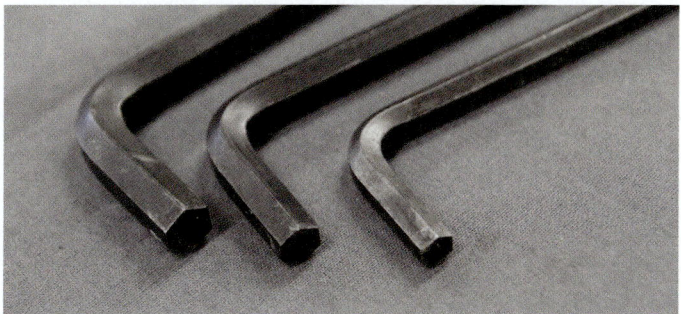

FIGURE 3-49 Typical Allen wrench head.

FIGURE 3-50 **A.** Allen socket. **B.** T-handle Allen wrench.

this precaution have stabbed a screwdriver into or through their hand, which can be painful as well as become infected or damage nerves.

The most common screwdriver has a flat tip, or blade, which gives it the name **flat blade screwdriver**. The blade should be almost as wide and thick as the slot in the fastener so that twisting force applied to the screwdriver is transferred right out to the edges of the head, where it has most effect. The blade should be a snug fit in the slot of the screw head. Then the twisting force is applied evenly along the sides of the slot. This guards against the screwdriver suddenly chewing a piece out of the slot and slipping just when the most force is being exerted. Flat blade screwdrivers come in a variety of sizes and lengths, so find the right one for the job. If viewed from the side, the blade should taper slightly to the very end where the flat tip fits into the slot. If the tip of the blade is not clean and square, it should be reshaped or replaced.

When you use a flat blade screwdriver, support the shaft with your free hand as you turn it (but keep it behind the tip). This helps keep the blade square in the slot and centered. Screwdrivers that slip are a common source of damage and injury in shops. A screw or bolt with a cross-shaped recess requires a **Phillips head screwdriver** or a Pozidriv screwdriver (**FIGURE 3-52**). The cross-shaped slot holds the tip of the screwdriver securely on the head. The Phillips tip fits a tapered recess, whereas the Pozidriv fits into slots with parallel sides in the head of the screw. Both a Phillips and a Pozidriv screwdriver are less likely to slip sideways, because the point is centered in the screw, but again the screwdriver must be the

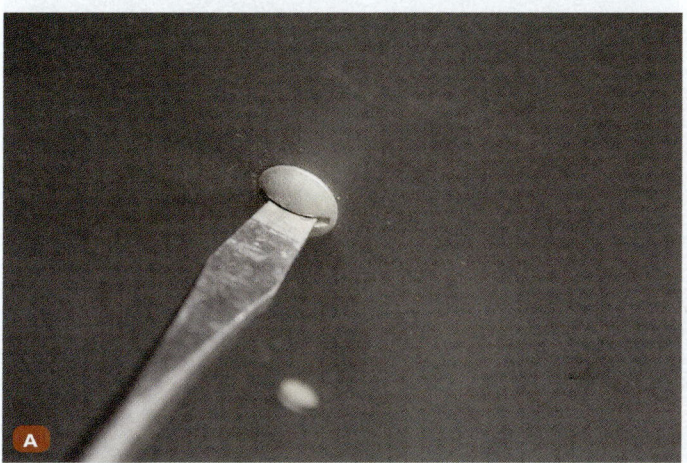

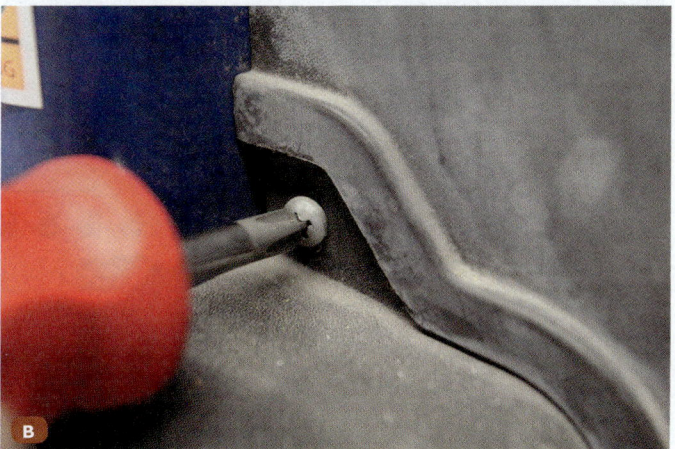

FIGURE 3-51 **A.** Slotted screw and screw driver. **B.** Phillips screw and screw driver.

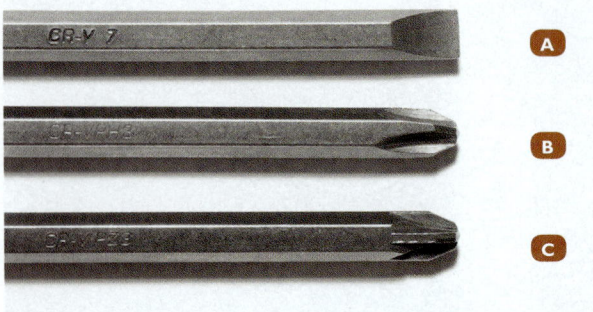
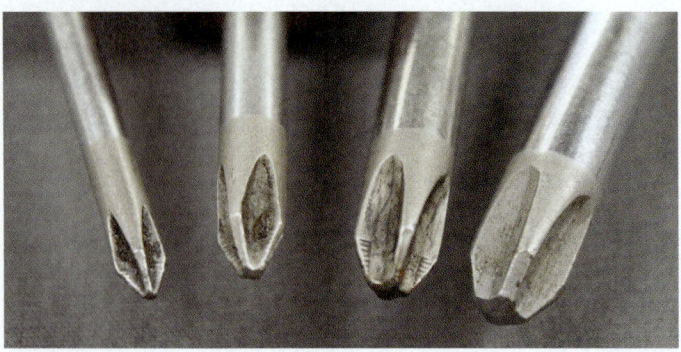

FIGURE 3-52 **A.** Flat blade screwdrivers. **B.** Phillips screwdriver. **C.** Pozidriv screwdriver.

FIGURE 3-53 Four sizes of Phillips screwdrivers.

right size. The fitting process is simplified with these two types of screwdrivers because four sizes are enough to fit almost all fasteners with this sort of screw head (**FIGURE 3-53**).

The **offset screwdriver** fits into spaces where a straight screwdriver cannot and is useful where there is not much room to turn it (**FIGURE 3-54**). The two tips look identical, but one is set at 90 degrees to the other. This is because sometimes there is only room to make a quarter turn of the driver. Thus the driver has two blades on opposite ends so that offset ends of the screwdriver can be used alternately.

The **ratcheting screwdriver** is a popular screwdriver handle that usually comes with a selection of removable flat and Phillips tips. It has a ratchet inside the handle that turns the blade in only one direction, depending on how the slider is set. When set for loosening, a screw can be undone without removing the tip of the blade from the head of the screw. When set for tightening, a screw can be inserted just as easily.

An **impact driver** is used when a screw or a bolt is rusted/corroded in place or over-tightened, and needs a tool that can apply more force than the other members of this family. Screw slots can easily be stripped with the use of a standard screwdriver. The force of the hammer forcing the bit into the screw while turning it makes it more likely the screw

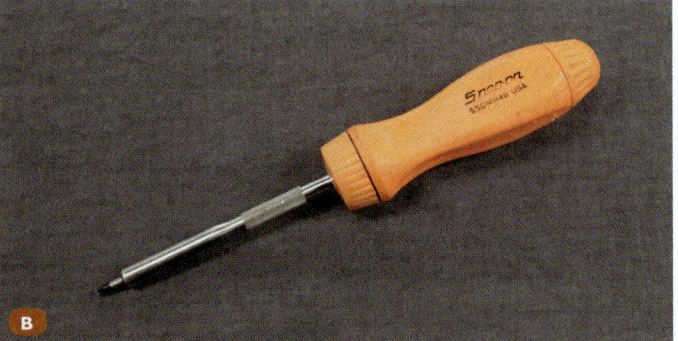

FIGURE 3-54 **A.** Offset screwdriver. **B.** Ratcheting screwdriver. **C.** Impact driver.

will break loose. The impact driver accepts a variety of special impact tips. Choose the right one for the screw head, fit the tip in place, and then tension it in the direction it has to turn. A sharp blow with the hammer breaks the screw free, and then it can usually be unscrewed normally.

Magnetic Pickup Tools and Mechanical Fingers

Magnetic pickup tools and **mechanical fingers** are very useful for grabbing items in tight spaces (**FIGURE 3-55**). A magnetic pickup tool typically is a telescoping stick that has a magnet attached to the end on a swivel joint. The magnet is strong enough to pick up screws, bolts, sockets, and other ferrous (containing iron, making it magnetic) metals. For example, if a screw is dropped into a tight crevice where your fingers cannot reach, a magnetic pickup tool can be used to extract it.

Mechanical fingers are also designed to extract or insert objects in tight spaces. Because they actually grab the object, they can pick up nonmagnetic objects, which makes them handy for picking up rubber or plastic parts. They use a flexible body and come in different lengths, but typically are about 12–18" (305–457 mm) long. They have expanding grappling fingers on one end to grab items, and the other end has a push mechanism to expand the fingers and a retracting spring to contract the fingers.

Hammers

Hammers are a vital part of the shop tool collection, and a variety are commonly used (**FIGURE 3-56**). The most common hammer in an automotive shop is the **ball-peen (engineer's) hammer**. Like most hammers, its head is hardened steel. A punch or a chisel can be driven with the flat face. Its name comes from the ball peen or rounded face. This end is usually used for flattening or **peening** a rivet. The hammer should always match the size of the job, and it is usually better to use one that is too big than too small.

Hitting chisels with a **steel hammer** is fine, but sometimes you only need to tap a component to position it. A steel hammer might mark or damage the part, especially if it is made of a softer metal, such as aluminum. In such cases, a soft-faced hammer should normally be used for the job. Soft-faced hammers range from very soft with rubber or plastic heads to slightly harder with brass or copper.

When a large chisel needs a really strong blow, it is time to use a **sledgehammer**. The sledgehammer is like a small mallet, with two square faces made of high carbon steel. It is the heaviest type of hammer that can be used one-handed. The sledgehammer is used in conjunction with a chisel to cut off a bolt where corrosion has made it impossible to remove the nut.

A **dead blow hammer** is designed not to bounce back when it hits something. A rebounding hammer can be dangerous or destructive. A dead blow hammer can be made

K03012 Describe type and use of magnetic pickup tools and mechanical fingers.

SAFETY TIP

The hammer you use depends on the part you are striking. Hammers with a metal face should almost always be harder than the part you are hammering. Never strike two hardened tools together, as this can cause the hardened parts to shatter.

K03013 Describe the type and use of hammers.

▶ TECHNICIAN TIP

It may be challenging getting the magnet down inside some areas because the magnet wants to keep sticking to other objects. One trick in this situation is to roll up a piece of paper so that a tube is created. Stick that down into the area of the dropped part, then slide the magnet down the tube, which helps it get past magnetic objects. Once the magnet is down, you may want to remove the roll of paper. Just remember two things: First, patience is important when using this tool, and second, don't drop anything in the first place!

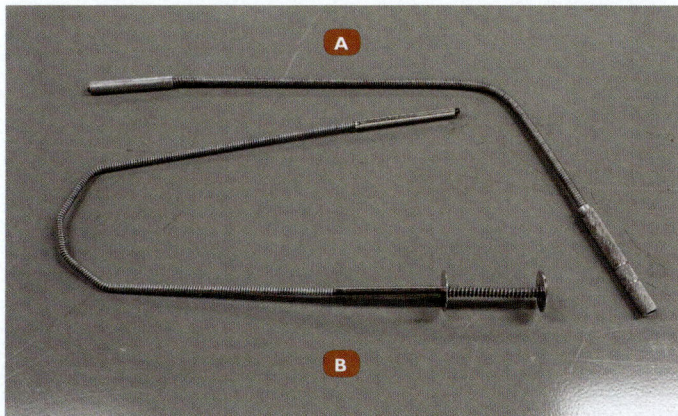

FIGURE 3-55 A. Magnetic pickup tools. **B.** Mechanical fingers.

FIGURE 3-56 A. Ball peen hammer. **B.** Sledge hammer. **C.** Soft-faced hammer. **D.** Dead blow hammer.

> **SAFETY TIP**
>
> When using hammers and chisels, always wear safety glasses. Also be aware that safety glasses by themselves need to be form fitting and must not be allowed to slide down your nose, for them to be effective. It may be prudent to wear either a face shield or safety goggles for extra protection.

K03014 Describe the type and use of chisels.

> **SAFETY TIP**
>
> Chisels and punches are designed with a softer striking end than hammers. Over time, this softer metal "mushrooms," and small fragments are prone to breaking off when hammered. These fragments can cause eye or other penetrative injuries to people in the area. Always inspect chisels and punches for mushrooming, and dress them on a grinder when necessary (**FIGURE 3-59**).

K03015 Describe the type and use of punches.

with a lead head or, more commonly, a hollow polyurethane head filled with lead shot or sand. The head absorbs the blow when the hammer makes contact, reducing any bounce-back or rebounding. This hammer can be used when working on the vehicle chassis or when dislodging stuck parts.

A **hard rubber mallet** is a special-purpose tool and has a head made of hard rubber. It is often used for moving things into place where it is important not to damage the item being moved. For example, it can be used to install a hubcap or to break a gasket seal on an aluminum housing.

Chisels

The most common kind of chisel is a **cold chisel** (**FIGURE 3-57**). It gets its name from the fact it is used to cut cold metals rather than heated metals. It has a flat blade made of high-quality steel and a cutting angle of approximately 70 degrees. The cutting end is tempered and hardened because it has to be harder than the metals it is cutting. The head of the chisel needs to be softer so it will not chip when it is hit with a hammer. Technicians sometimes use a cold chisel to remove bolts whose heads have rounded off.

A variation of the cold chisel is a spring-loaded cold chisel (**FIGURE 3-58**). This chisel works really well in tight spaces where a hammer can't be swung. The chisel is made up of three parts: the chisel, the weighted hammerhead, and a spring in tension, holding the other two components together. It is operated by holding the chisel end against the part you are working on, pulling back on the hammerhead, and allowing the spring to rapidly slam the hammerhead into the end of the chisel. The force of the hammerhead hitting the end of the chisel transfers a lot of energy to the chisel. This type of arrangement is also used on some spring-loaded center punches.

A **cross-cut chisel** is so named because the sharpened edge is across the blade width. This chisel narrows down along the stock, so it is good for getting in grooves. It is used for cleaning out or even making key ways. The flying chips of metal should always be directed away from the user.

Punches

Punches are used when the head of the hammer is too large to strike the object being hit without causing damage to adjacent parts. A punch transmits the hammer's striking power from the soft upper end down to the tip that is made of hardened high-carbon steel. A punch transmits an accurate blow from the hammer at exactly one point, something that cannot be guaranteed using a hammer on its own.

Four of the most common punches are the prick punch, center punch, drift punch, and pin punch (**FIGURE 3-60**). When marks need to be drawn on an object like a steel plate, to help locate a hole to be drilled, a **prick punch** can be used to mark the points so they will not rub off. They can also be used to scribe intersecting lines between given points. The

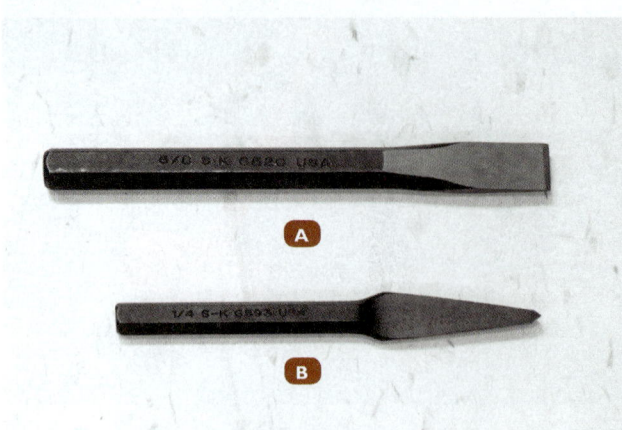

FIGURE 3-57 **A.** Cold chisel. **B.** Cross-cut chisel.

FIGURE 3-58 Spring-loaded chisels.

FIGURE 3-59 Dressing a chisel.

prick punch's point is very sharp, so a gentle tap leaves a clear indentation. The **center punch** is not as sharp as a prick punch and is usually bigger. It makes a bigger indentation that centers a drill bit at the point where a hole is required to be drilled.

Although most center punches are used with a hammer, some center punches operate automatically when the punch is pressed tightly up against the part you are punching. This type of punch has a spring and weighted hammer inside of the back end of the center punch. It is machined in such a way that pushing the center punch against a surface compresses a spring behind a movable weight (**FIGURE 3-61**). When pushed far enough, the weight is released, and the spring forces it against the center rod in the punch, causing it to indent the work surface.

A **drift punch** is also named a starter punch because you should always use it first to get a pin moving. It has a tapered shank, and the tip is slightly hollow so it does not spread the end of a pin and make it an even tighter fit. Once the starter drift has gotten the pin moving, a suitable pin punch will drive the pin out or in. A drift punch also works well for aligning holes on two mating objects, such as a valve cover and cylinder head. Forcing the drift punch in the hole aligns both components for easier installation of the remaining bolts. **Pin punches** are available in various diameters. A pin punch has a long slender shaft that has straight sides. It is used to drive out pins or rivets (**FIGURE 3-62**). A lot of components are either held together or accurately located by pins. Pins can be pretty tight, and a group of pin punches is specially designed to deal with them.

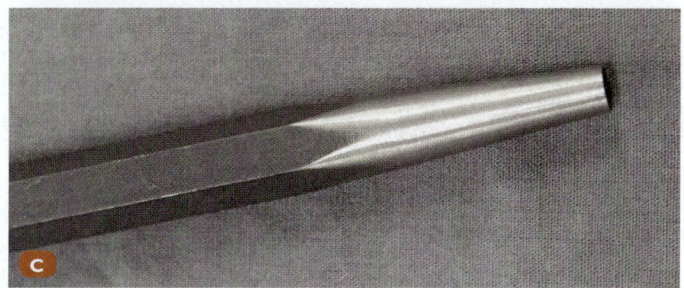
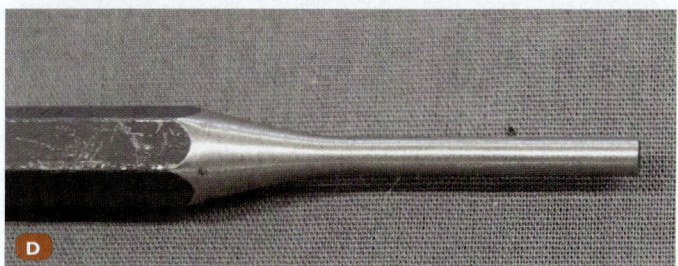

FIGURE 3-60 A. Prick punch. **B.** Center punch. **C.** Drift punch. **D.** Pin punch.

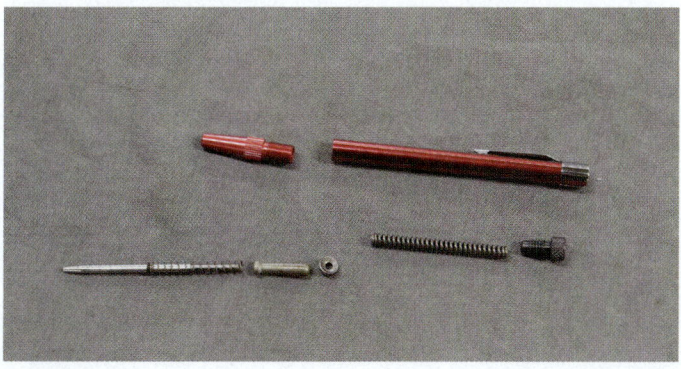

FIGURE 3-61 Internal workings of an automatic center punch.

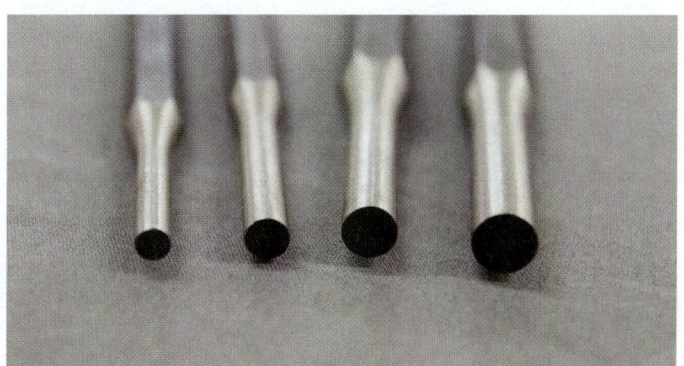

FIGURE 3-62 Various pin punches.

FIGURE 3-63 Wad punch.

FIGURE 3-64 Number and letter punches.

> ▶ **TECHNICIAN TIP**
>
> Many engine components are made of aluminum. Because aluminum is quite soft, it is critical that you use the gasket scraper very carefully so as not to damage the surface. This can be accomplished by keeping the gasket scraper at a fairly flat angle to the surface. Also, the gasket scraper should only be used by hand, not with a hammer. Some manufacturers specify using plastic gasket scrapers only on certain aluminum components such as cylinder heads and blocks.

K03016 Describe the type and use of pry bars.

K03017 Describe the type and use of gasket scrapers.

Special punches with hollow ends are called **wad punches** or **hollow punches** (**FIGURE 3-63**). They are the most efficient tool to make a hole in soft sheet material like shim steel, plastic, and leather, or, most commonly, in a gasket. When being used, there should always be a soft surface under the work, ideally the end grain of a wooden block. If a hollow punch loses its sharpness or has nicks around its edge, it will make a mess instead of a hole.

Numbers and letters, like the engine numbers on some cylinder blocks, are usually made with number and letter punches that come in boxed sets (**FIGURE 3-64**). The rules for using a number or letter punch set are the same as for all punches. The punch must be square with the surface being worked on, not on an angle, and the hammer must hit the top squarely.

Pry Bars

Pry bars are tools constructed of strong metal that are used as a lever to move, adjust, or pry (**FIGURE 3-65**). Pry bars come in a variety of shapes and sizes. Many have a tapered end that is slightly bent, with a plastic handle on the other end. This design works well for applying force to tension belts or for moving parts into alignment. Another type of pry bar is the **roll bar**. One end is sharply curved and tapered, which is used for prying. The other end is tapered to a dull point and is used to align larger holes such as transmission bell housings or engine motor mounts. Because pry bars are made of hardened steel, care should be taken when using them on softer materials, to avoid any damage.

Gasket Scrapers

A **gasket scraper** has a hardened, sharpened blade. It is designed to remove a gasket without damaging the sealing face of the component when used properly (**FIGURE 3-66**). On

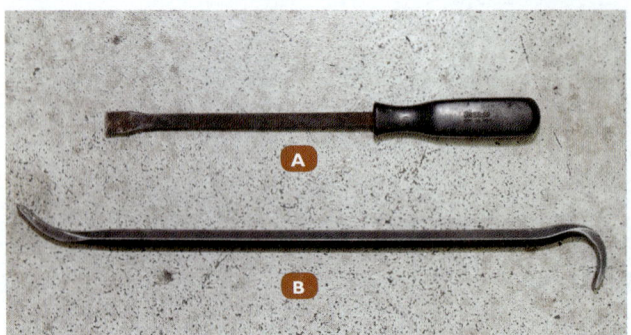

FIGURE 3-65 A. Pry bar. **B.** Roll bar.

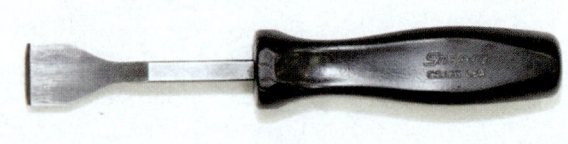

FIGURE 3-66 A gasket scraper.

one end, it has a comfortable handle like a screwdriver handle; on the other end, a blade is fitted with a sharp edge to assist in the removal of gaskets. The gasket scraper should be kept sharp and straight to make it easy to remove all traces of the old gasket and sealing compound. The blades come in different sizes, with a typical size being 1" (25 mm) wide. Whenever you use a gasket scraper, be very careful not to nick or damage the surface being cleaned.

Files

Files are hand tools designed to remove small amounts of material from the surface of a workpiece. Files come in a variety of shapes, sizes, and coarseness depending on the material being worked and the size of the job. Files have a pointed tang on one end that is fitted to a handle. Files are often sold without handles, but they should not be used until a handle of the right size has been fitted. A correctly sized handle fits snugly without working loose when the file is being used. Always check the handle before using the file. If the handle is loose, give it a sharp rap to tighten it up, or if it is the threaded type, screw it on tighter. If it fails to fit snugly, you must use a different-size handle.

What makes one file different from another is not just the shape but how much material it is designed to remove with each stroke. The teeth on the file determine how much material will be removed (**FIGURE 3-67**). Because the teeth face one direction only, the file cuts in one direction only. Dragging the file backward over the surface of the metal only dulls the teeth and wears them out quickly. Teeth on a coarse-grade file are larger, with a greater space between them. A coarse-grade file working on a piece of mild steel removes a lot of material with each stroke, but it leaves a rough finish. A smooth-grade file has smaller teeth cut more closely together. It removes much less material on each stroke, but the finish is much smoother. On many jobs, the coarse file is used first to remove material quickly, and then a smoother file gently removes the remaining material and leaves a smoother finish to the work. The full list of grades in flat files, from rough to smooth, follows:

- **Rough files** have the coarsest teeth, with approximately 20 teeth per inch (25 mm). They are used when a lot of material must be removed quickly. They leave a very rough finish and have to be followed by the use of finer files to produce a smooth final finish.
- **Coarse bastard files** are still a coarse file, with approximately 30 teeth per inch (25 mm), but they are not as course as the rough file. They are also used to rough out or remove material quickly from the job.
- **Second-cut files** have approximately 40 teeth per inch (25 mm) and provide a smoother finish than the rough or coarse bastard file. They are good all-round intermediary files and leave a reasonably smooth finish.
- **Smooth files** have approximately 60 teeth per inch (25 mm) and are a finishing file used to provide a smooth final finish.
- **Dead smooth files** have 100 teeth per inch (25mm) or more and are used where a very fine finish is required.

Some flat files are available with one smooth edge (no teeth), called safe edge files. They allow filing up to an edge without damaging it. Flat files work well on straightforward jobs, but some jobs require special files. A **warding file** is thinner than other files and comes to a point; it is used for working in narrow slots (**FIGURE 3-68**). A **square file** has teeth on all four sides, so you can use it in a square or rectangular hole. A square file can make the right shape for a squared metal key to fit in a slot.

A **triangular file** has three sides. It is triangular, so it can get into internal corners easily. It is able to cut right into a corner without removing material from the sides. **Curved files** are typically either half-round or round. A half-round file has a shallow convex surface that can file in a concave hollow or in

K03018 Describe the type and use of files.

SAFETY TIP

Hands should always be kept away from the surface of the file and the metal that is being worked on. It is easy for skin to get caught between the edge of the file and the piece you are working. This creates a scissor-like condition that can result in severe cuts. Also, filing can produce small slivers of metal that penetrate your skin and can be difficult to remove from a finger or hand.

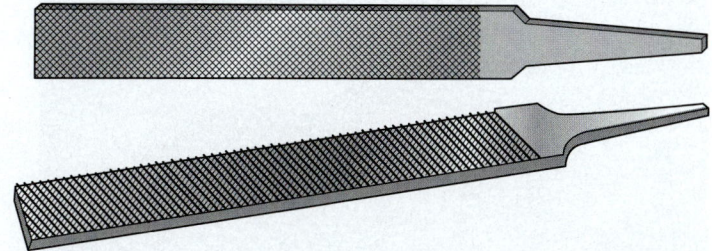

FIGURE 3-67 The teeth on a file determine how much material will be removed from the object being filed.

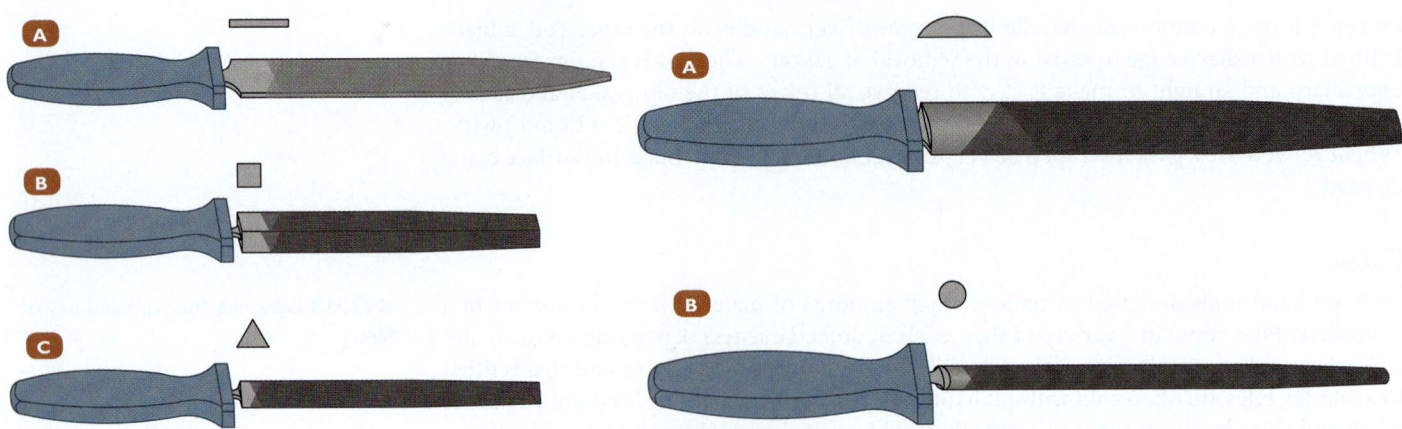

FIGURE 3-68 A. Warding file. B. Square file. C. Triangular file.   FIGURE 3-69 A. Half-round file. B. Round, or rat-tail, file.

an acute internal corner (**FIGURE 3-69**). The fully round file, sometimes called a rat-tail file, can make holes bigger. It can also file inside a concave surface with a tight radius.

The **thread file** cleans clogged or distorted threads on bolts and studs. Thread files come in either standard or metric configurations, so make sure you use the correct file. Each file has eight different surfaces that match different thread dimensions, so the correct face must be used (**FIGURE 3-70**).

Files should be cleaned after each use. If they are clogged, they can be cleaned by using a file card, or file brush (**FIGURE 3-71**). This tool has short steel bristles that clean out the small particles that clog the teeth of the file. Rubbing a piece of chalk over the surface of the file prior to filing makes it easier to clean.

Clamps

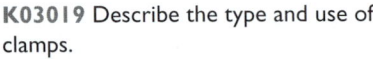

K03019 Describe the type and use of clamps.

There are many types of vices or clamps available (**FIGURE 3-72**). The **bench vice** is a useful tool for holding anything that can fit into its jaws. Some common uses include sawing, filing, or chiseling. The jaws are serrated to give extra grip. They are also very hard, which means that when the vice is tightened, the jaws can mar whatever they are gripping. To prevent this, a pair of soft jaws can be fitted whenever the danger of damage arises. They are usually made of aluminum or some other soft metal or can have a rubber-type surface applied to them.

When materials are too awkward to grip vertically in a plain vice, it may be easier to use an **offset vice**. The offset vice has its jaws set to one side to allow long components to be held vertically (**FIGURE 3-73**). For example, a long threaded bar can be held vertically in an offset vice to cut a thread with a die.

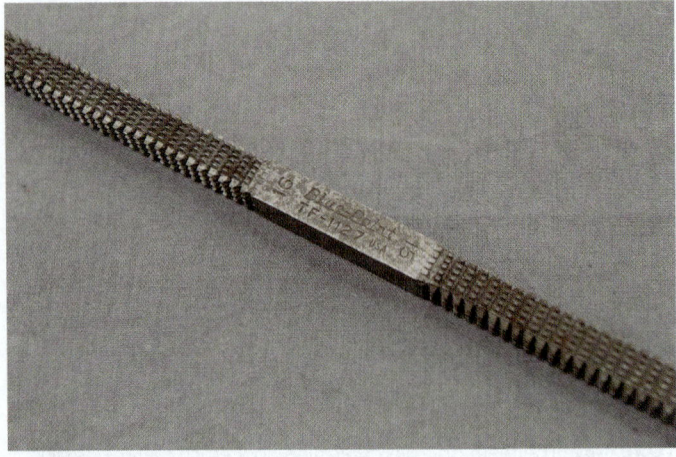

FIGURE 3-70 Thread file.

FIGURE 3-71 File card.

Basic Hand Tools 79

FIGURE 3-72 Bench vice.

FIGURE 3-73 Offset vice being used to hold a pipe.

A **drill vice** is designed to hold material on a drill worktable (**FIGURE 3-74**). The drill worktable has slots cut into it to allow the vice to be bolted down on the table, to hold material securely. To hold something firmly and drill it accurately, the object must be secured in the jaws of the vice. The vice can be moved on the bed until the precise drilling point is located and then tightened down by bolts to hold the drill vice in place during drilling.

The name for the **C-clamp** comes from its shape (**FIGURE 3-75**). It can hold parts together while they are being assembled, drilled, or welded. It can reach around awkwardly shaped pieces that do not fit in a vice. It is also commonly used to retract disc brake caliper pistons. This clamp is portable, so it can be taken to the work.

Taps and Dies

Taps and dies are used to form threads in metal so that they can be fastened together (**FIGURE 3-76**). The tap cuts female threads in a component so a fastener can be screwed into it. The die is used to cut male threads on a bolt so that it can be screwed into the female threads created by the tap. The tap and die are companion tools that create matching threads so that they can both be used to fasten things together.

K03020 Describe the type and use of taps and dies.

Taps

Various types of taps are designed to be used based on what you want to do and the material you are working with. The most common taps are the taper tap, intermediate tap (also called a plug tap), bottoming tap (also called a flat-bottomed tap), and the thread chaser (**FIGURE 3-77**). These are explored in more depth below.

FIGURE 3-74 Drill vice on a drill worktable.

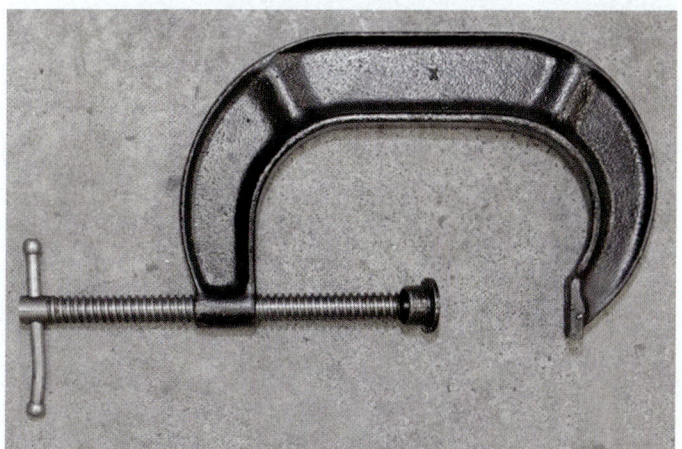

FIGURE 3-75 C-Clamp.

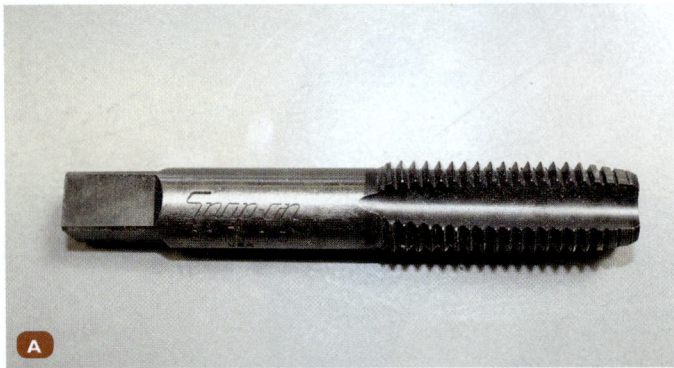

FIGURE 3-76 **A.** Tap. **B.** Die.

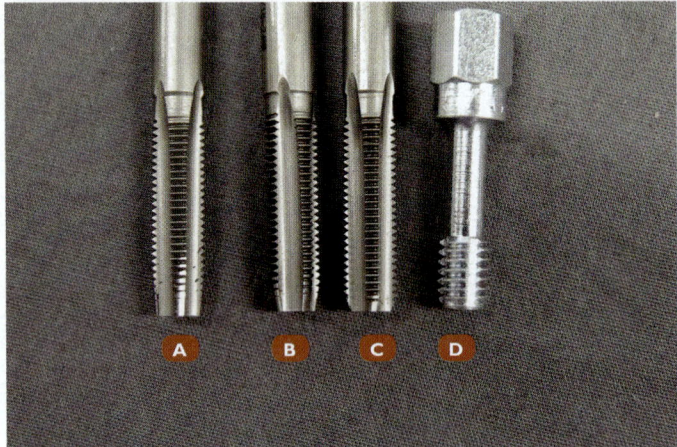

FIGURE 3-77 **A.** Taper tap. **B.** Intermediate tap. **C.** Bottoming tap. **D.** Thread chaser.

Taper Tap

A **taper tap** narrows at the tip, which makes it easier to start straight when cutting threads in a new hole. It also makes it less likely to break the tip off the tap because it removes metal in a less aggressive manner. Taps are very hard, which gives them good wear resistance but also makes them very brittle; they can break off easily. If a tap breaks in the hole, it can be very hard to remove. This makes taper taps good taps to use when starting a hole. Plus, they are easier to get started straight in the hole.

Intermediate Tap

The second type of tap is an **intermediate tap**, also known as a plug tap. It is more aggressive than a taper tap, but not as aggressive as a bottoming tap. This is the most common tap used by technicians. Although it is a bit more aggressive than the taper tap, it can be used as a starter tap for a new hole.

Bottoming Tap

The **bottoming tap** is used when you need to cut threads to the very bottom of a blind hole and in holes that already have threads started that just need to be extended slightly. It has a flat bottom, and the threads are the same all the way to the end. This type of tap is virtually impossible to use in a new, unthreaded hole.

Thread Chaser

Thread chasers are used to clean up the threads of a hole to make sure they are free from debris and dirt. They do not cut threads, but just clean up the existing threads so that the bolt doesn't encounter any excessive resistance.

Dies

A **die** is used to cut external threads on a metal shank or bolt. The threads in a die create the male companion to the female threads that have been cut into the material by a tap. Usually the dies come with a setscrew that allows the user to slightly adjust the size of the die so the threads can be cut to the right fit that matches the threaded hole. Loose-fitting threads strip more easily than they should and are not as secure. Tight-fitting threads increase the torque required to turn the bolt, thereby reducing the clamping force relative to bolt torque. So adjusting thread fit is typically accomplished by adjusting the die. Dies are hardened, which makes them very wear resistant, but very brittle. Die nuts do not have the split like the threading die does because they are used to clean up threads like the thread chasers (**FIGURE 3-78**).

FIGURE 3-78 **A.** Die nut. **B.** Split die.

Proper Use of Taps and Dies

The proper use of taps and dies is a major issue often overlooked in repairing a thread on a bolt or threads in a hole. Improper use of these two items causes the bolt not to thread into the hole properly, or it could even break the tool, which is something to avoid. Both of these tools are made of hardened steel, which makes them very wear resistant but also makes them very brittle. If they are used improperly they have a tendency to facture and break.

Because taps and dies both need to be rotated, special tools are used to turn them. A **tap handle** is used to turn taps. It has a right-angled jaw that matches the squared end of all the taps (**FIGURE 3-79**). The jaws are designed to hold the tap securely, and the handles provide the leverage for the operator to comfortably rotate the tap to cut the thread. To cut a thread in an awkward space, a T-shaped tap handle is very convenient (**FIGURE 3-80**). Its handle is not as long, so it fits into tighter spaces; however, it is harder to turn and to guide accurately.

To cut a brand new thread on a blank rod or shaft, a die held in a **die stock** is used (**FIGURE 3-81**). The die fits into the octagonal recess in the die stock and is usually held in place by a thumb screw. The die may be split so that it can be adjusted more tightly onto the work with each pass of the die as the thread is cut deeper and deeper, until the external thread fits the threaded hole properly.

When tapping a hole, the diameter of the hole is determined by a tap drill chart, which can be obtained from engineering suppliers. This chart shows what hole size has to be drilled and what tap size is needed to cut the right thread for any given bolt size (**FIGURE 3-82**).

Just remember that if you are drilling a 1/4" (6 mm) or larger hole, use a smaller pilot drill first. Once the properly sized hole has been drilled, the taper tap or intermediate tap can be started in the hole. Make sure to use the proper lubricant for the metal you are tapping. Also be sure to start the tap straight. The best way to do that is to start the tap about one turn, stop, and then use a square to check the position of the tap in two places 90 degrees apart. If it isn't perfectly straight, you can usually straighten it while turning it another half turn. Again stop and verify that it is straight in two positions, 90 degrees apart. If the tap is straight, turn the tap about one full turn, and then back it off about a quarter turn. Continue cutting and backing off the tap until either the tap turns easily or you are at the bottom of the hole. Remove the tap. If you are cutting threads in a blind hole, you need to use a bottoming tap to finish the threads off. If not, clean the threaded hole and test-fit a bolt in the hole to check the threads. The bolt should turn smoothly by hand.

Screw Extractors

Screw extractors are devices designed to remove screws, studs, or bolts that have broken off in threaded holes. A common type of extractor uses a coarse left-hand tapered thread formed on its hardened body (**FIGURE 3-83**). Normally, a hole is drilled in the center of the broken screw, and then the extractor is screwed into the hole. The left-hand thread grips the broken part of the bolt and unscrews it. The extractor is marked with two sizes: one showing the size range of screws it is designed to remove, and the other, the size of the hole that needs to be drilled. It is important to carefully drill the hole in the center of the bolt or

FIGURE 3-79 Tap handle.

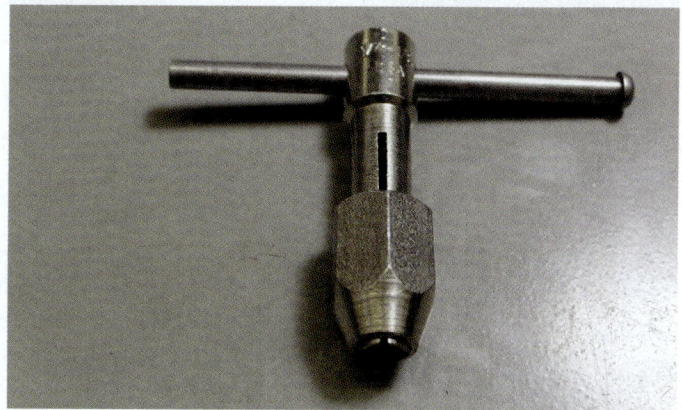

FIGURE 3-80 T-shaped tap handle.

FIGURE 3-81 Die stock.

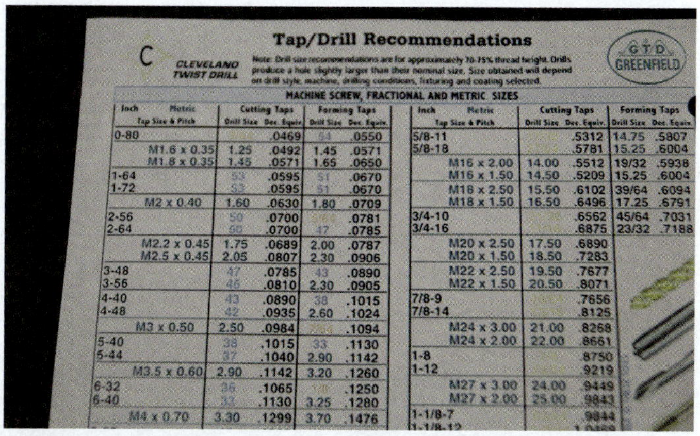

FIGURE 3-82 Tap drill chart.

K03021 Describe the type and use of screw extractors.

FIGURE 3-83 Screw extractors.

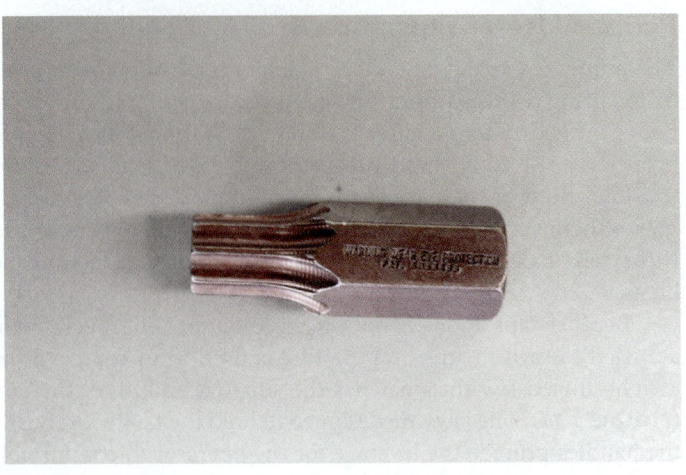

FIGURE 3-84 Straight-sided screw extractor.

stud in case you end up having to drill the bolt out. If you drill the hole off center, you will not be able to drill it out all the way to the inside diameter of the threads, as the hole is offset from center. This makes removal of the bolt much harder.

Some screw extractors use a hardened, tapered square shank that is hammered into a hole drilled into the center of the broken off bolt. The square edges of the screw extractor cut into the bolt and are used to grip it so it can be removed. Another type of screw extractor is the straight-sided, vertically splined, round shaft. The sides of the shaft have straight splines running the length of the extractor (**FIGURE 3-84**). It doesn't taper, so it is strong up and down its length. A straight hole of the correct diameter is drilled into the broken-off bolt. Then the extractor is driven into the newly drilled hole. The vertical splines, being larger than the hole, grab onto the bolt, allowing it to be unthreaded.

Pullers

K03022 Describe the type and use of pullers.

Pullers are a very common universal tool that can be used for removing bearings, bushings, pulleys, and gears (**FIGURE 3-85**). Specialized pullers are also available for specific tasks where a standard puller is not as effective. The most common pullers have two or three legs that grip the part to be removed. A center bolt, called a forcing screw or jacking bolt, is then screwed in, producing a jacking or pulling action, which extracts the part. **Gear pullers** come in a range of sizes and shapes, all designed for particular applications. They consist of three main parts: jaws, a cross-arm, and a forcing screw. There are normally two or three jaws on a puller. They are designed to connect to the component either externally or internally.

FIGURE 3-85 **A.** Puller. **B.** Gear puller.

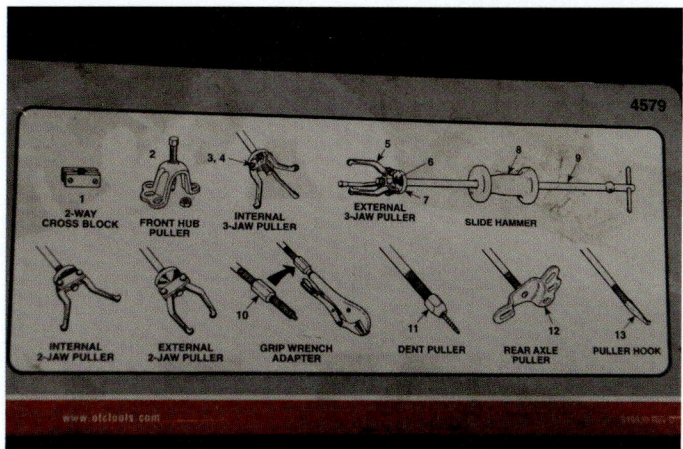

FIGURE 3-86 Interchangeable feet for the gear puller.

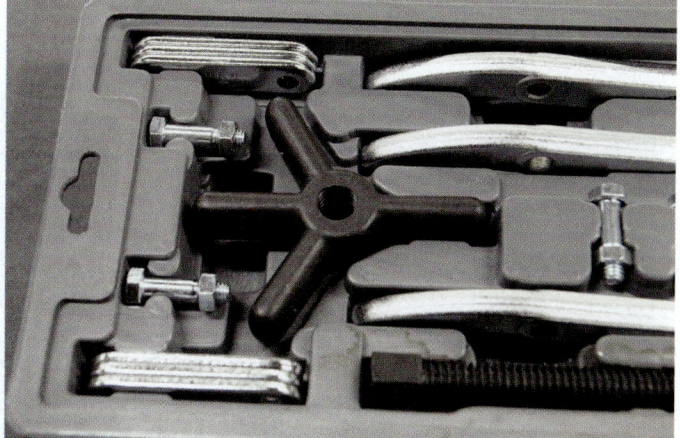

FIGURE 3-87 Four-arm puller.

The **forcing screw** is a long, fine-threaded bolt that is applied to the center of the cross-arm. When the forcing screw is turned, it applies a very large force to the component you are removing. The forcing screw typically has interchangeable feet (**FIGURE 3-86**). A tapered cone style of foot does a good job of centering the puller, but it also creates a very large wedging effect, which can distort the end of the shaft. A flat-style foot is very good for pushing against the end of a shaft, but doesn't center itself. In either situation, the wrong foot size can push on the internal threads of the shaft, damaging them. The cross-arm attaches the jaws to the forcing screw. If the **cross-arm** has four arms, three of the arms are spaced 120 degrees apart. The fourth arm is positioned 180 degrees apart from one arm (**FIGURE 3-87**). This allows the cross-arm to be used as either a two- or a three-arm puller.

Using Gear Pullers

Gear and bearing pullers are designed for hundreds of applications. Their main purpose is to remove a component such as a gear, pulley, or bearing from a shaft; or to remove a shaft from inside a hole. Normally these components have been pressed onto that shaft or into the hole, so removing them requires considerable force. To select, install, and use a gear puller to remove a pulley, follow the steps in **SKILL DRILL 3-3**.

Flaring Tools

A **tube flaring tool** is used to flare the end of a tube so it can be connected to another tube or component. One example of this is where the brake line screws into a wheel cylinder. The flared end is compressed between two threaded parts so that it seals the joint and withstands high pressures. The three most common shapes of flares are the **single flare**, for tubing carrying low pressures like a fuel line; the **double flare**, for higher pressures such as in a brake system; and the ISO flare (sometimes called a bubble flare), which is the metric version used in brake systems (**FIGURE 3-88**).

Flaring tools have two main parts: a set of bars with holes that match the diameter of the tube end that is being shaped, and a yoke that drives a cone into the mouth of the tube (**FIGURE 3-89**). To make a single flare, the end of the tube is placed level with the surface of the top of the flaring bars. With the clamp screw firmly tightened, the feed screw flares the end of the tube. Making a double flare is similar, but an extra step is added before flaring the tube, and more of the tube is exposed to allow for folding the flare over into a double flare. A double flaring button is placed into the end of the tube, and when it is removed after tightening, the pipe looks like a bubble. Placing the cone and yoke over the bubble allows you to force the bubble to fold in on itself, forming the double flare.

SAFETY TIP

- Always wear eye protection when using a gear puller.
- Make sure the puller is located correctly on the workpiece. If the jaws cannot be fitted correctly on the part, then select a more appropriate puller. Do not use a puller that does not fit the job.

K03023 Describe the type and use of flaring tools.

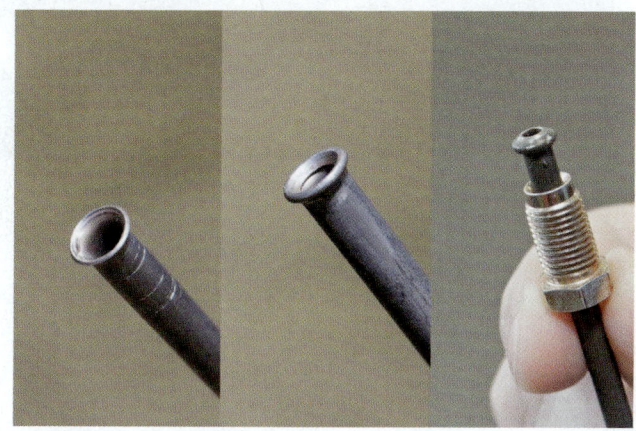

FIGURE 3-88 Single flare, double flare, and ISO flare.

SKILL DRILL 3-3 Using Gear Pullers

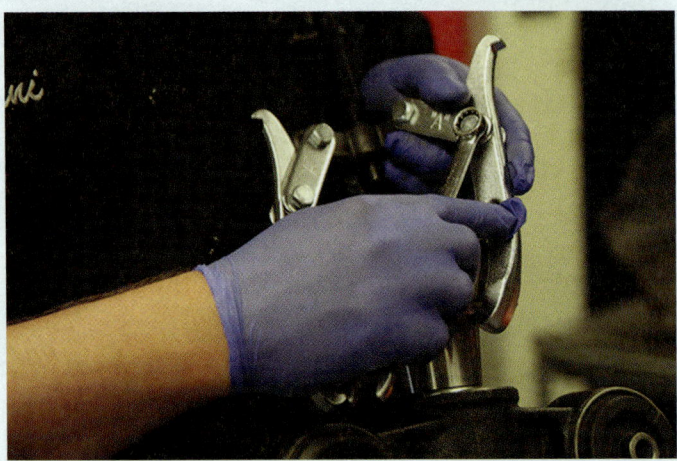

1. Examine the gear puller you have selected for the job. Identify the jaws; there may be two or three of them, and they must fit the part you want to remove. The cross-arm enables you to adjust the diameter of the jaws. The forcing screw should fit snugly onto the part you are removing. Finally, select the right wrench or socket size to fit the nut on the end of the forcing screw.

2. Adjust and fit the puller. Adjust the jaws and cross-arms of the puller so that it fits tightly around the part to be removed. The arms of the jaws should be pulling against the component at close to right angles.

3. Position the forcing screw. Use the appropriate wrench to run the forcing screw down to touch the shaft. Check that the point of the forcing screw is centered on the shaft. If not, adjust the jaws and cross-arms until the point is in the center of the shaft. Also be careful to use the correct foot on the end of the puller.

4. Tighten the forcing screw slowly and carefully onto the shaft. Check that the puller is not going to slip off center or off the pulley. Readjust the puller if necessary.

Continued

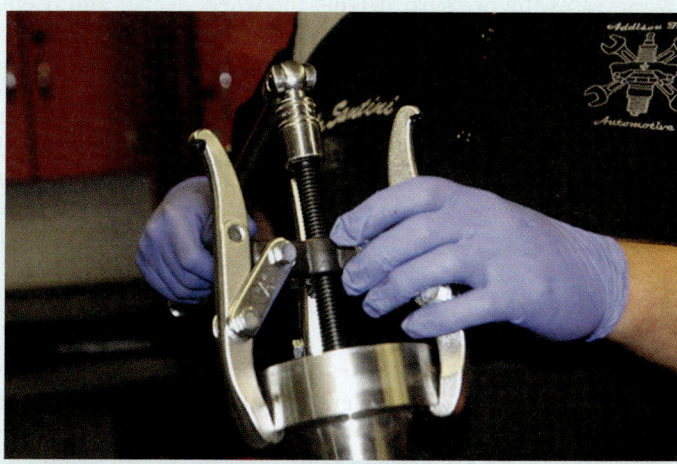

5. If the forcing screw and puller jaws remain in the correct position, tighten the forcing screw, and pull the part off the shaft.

6. You may sometimes have to use a hammer to hit directly on the end of the forcing screw to help break the part loose.

An ISO flare uses a flaring tool made specifically for that type of flare. The process is similar to that of the double-flare but differs in the use of the button, because an ISO flare does not get doubled back on itself. It should resemble a bubble shape when you are finished.

A **tubing cutter** is more convenient and neater than a saw when cutting pipes and metal tubing (**FIGURE 3-90**). The sharpened wheel does the cutting. As the tool turns

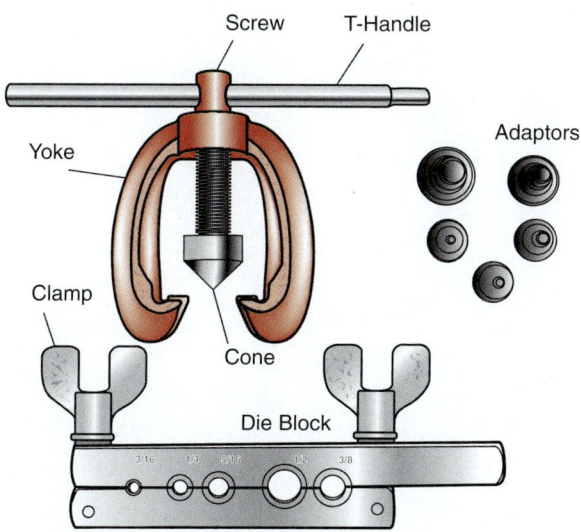

FIGURE 3-89 Components of a flare tool.

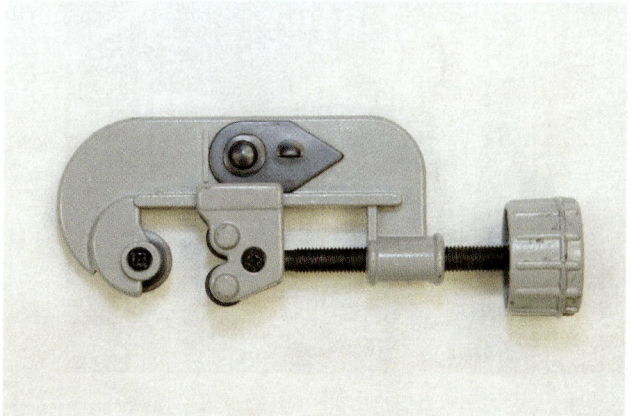

FIGURE 3-90 Tubing cutter.

SAFETY TIP

A flaring tool is used to produce a pressure seal for sealing brake lines and fuel system tubing. Make sure you test the flared joint for leaks before completing the repair; otherwise, the brakes could fail or leaking fuel could catch on fire.

SAFETY TIP

Always wear eye protection when using a flaring tool. Make sure the tool is clean, in good condition, and suitable for the type of material you are going to flare.

K03024 Describe the type and use of riveting tools.

around the pipe, the screw increases the pressure, driving the wheel deeper and deeper through the pipe until it finally cuts through. There is a larger version that is used for cutting exhaust pipes.

Using Flaring Tools

To make a successful flare, it is important to have the correct amount of tube protruding through the tool before clamping. Otherwise, too much of the end will fold over and leave too small of a hole for fluid to pass through. Too little and there won't be enough tube to fold over properly, and the joint won't have full surface contact.

If you are making a double flare or ISO flare, make sure you use the correctly sized button for the tubing size. The button is also used to measure the amount of tube required to protrude from the tool prior to forming it. To prevent the tool from slipping on the tube and ruining the flare, make sure the tool is sufficiently tight around the tube before starting to create the flare.

To use a flaring tool to make a flare in a piece of tubing, follow the steps in **SKILL DRILL 3-4**.

Riveting Tools

There are many applications for blind rivets, and various rivet types and tools may be used to do the riveting. Rivets are used in many places in automotive applications where there is a need for a fastener that doesn't have to be easily removed. Some older window regulators,

SKILL DRILL 3-4 Using Flaring Tools

1. Choose the tube you will use to make the flare, and put the flare nut on the tube before creating the flare.

2. Match the size of the tube to the correct hole in the tubing clamp.

Continued

Basic Hand Tools 87

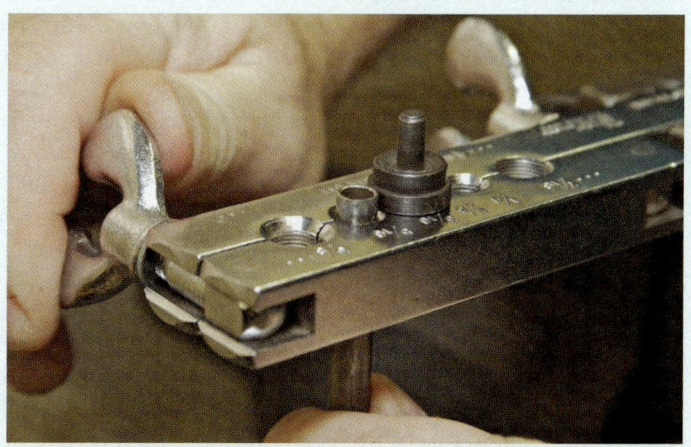

3. Holding the flaring tool, put the tube into the clamp. Position the tube so the correct amount is showing through the tool. If you are conducting a double flare, use the correctly sized button to ensure the proper amount of the tube is sticking up above the top of the clamp. Tighten the two halves of the clamp together using the wing nuts. Make sure the tool is tight enough to clamp the tube so it will not slip.

4. Put the cone and forming tool over the clamp, and turn the handle to make the flare. If you are doing a double flare or ISO flare, place the button in the end of the tube, install the cone and forming tool, and turn the handle to make the bubble. Remove the button from the tube.

5. If this is an ISO flare, inspect it to see if it is properly formed. If it is a double flare, put the cone and forming tool back on the clamp, and tighten the forming tool handle to create the double flare.

6. Remove the forming tool.

7. Remove the tube from the clamp, and check the flare to ensure it is free of burrs and is correctly formed.

CHAPTER 3 Basic Tools and Precision Measuring

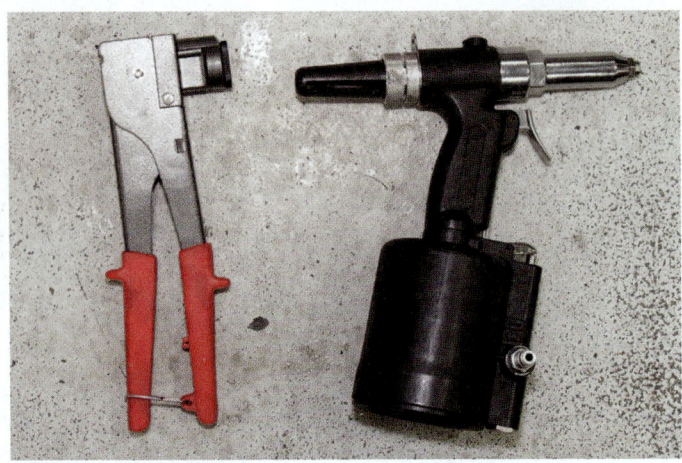

FIGURE 3-91 Pop rivet guns.

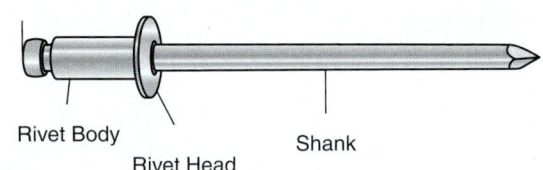

FIGURE 3-92 Anatomy of a rivet.

▶ TECHNICIAN TIP

A rivet is a single-use fastener. Unlike a nut and bolt, which can normally be disassembled and reused, a rivet cannot. The metal shell that makes up a pop rivet is crushed into place so that it holds the parts firmly together. If it ever needs to be removed, it must be drilled out or cut off.

▶ SAFETY TIP

When compressing the rivet handles, be careful not to place your fingers between the handles as they could end up pinching your fingers when the mandrel on the rivet breaks.

N03004 Demonstrate proper use of precision measuring tools (i.e., micrometer, dial-indicator, dial-caliper).

body panels, trim pieces, and even some suspension members use rivets to keep them attached without vibrating loose. **Pop rivet guns** are convenient for occasional riveting of light materials (**FIGURE 3-91**).

A typical pop, or **blind rivet**, has a body that forms the **finished rivet** and a mandrel, which is discarded when the riveting is completed (**FIGURE 3-92**). It is called a blind rivet because there is no need to see or reach the other side of the hole in which the rivet goes to do the work. In some types, the rivet is plugged shut so that it is waterproof or pressure proof.

The rivet is inserted into the riveting tool, which, when squeezed, pulls the end of the **mandrel** back through the body of the rivet. Because the **mandrel head** is bigger than the hole through the body, the body swells tightly against the hole. Finally, the mandrel head will snap off under the pressure and fall out, leaving the rivet body gripping the two pieces of material together.

Using Riveting Tools

Rivet tools are used to join two pieces of metal or other material together—for example, sheet metal that needs to be attached to a stiffening frame. To perform a riveting operation, you need a rivet gun, rivets, a drill, a properly sized drill bit, and the materials to be riveted.

Rivets come in various diameters and lengths for different sizes of jobs and are made of various types of metals to suit the job at hand. When selecting rivets to suit the job, consider the diameter, length, and rivet material. Larger diameter rivets should be used for jobs that require more strength. The rivet length should be sufficient to protrude past the thickness of the materials being riveted by about 1.6 times the diameter of the rivet stem. Typically, you should select rivets that are made from the same material as that being riveted. For example, stainless steel rivets should be used for riveting stainless steel, and aluminum rivets should be used to rivet aluminum.

Pilot holes must be drilled through the metal to be riveted. Ensure that the hole is just large enough for the rivet to comfortably pass through it, but do not make it too large. If the hole is too large, the rivet will be loose and will not hold the materials securely together. When drilling holes for rivets, stay back from the material's edge to ensure that the rivets do not break through the edge of the materials being riveted. A good rule of thumb is to allow at least twice the diameter of the rivet stem as clearance from any edge.

Most rivet tools are capable of riveting various sizes of rivets and have a number of nosepiece sizes to work with different sizes of rivets. Make sure you select the properly sized nosepiece for the rivet you are using.

To use a riveting tool to rivet two pieces of material together, follow the steps in **SKILL DRILL 3-5**.

▶ Precision Measuring Tools

Technicians are required to perform a variety of measurements while carrying out their job. This requires knowledge of what tools are available and how to use them. Measuring tools can generally be classified according to what type of measurements they can make. A measuring tape is useful for measuring longer distances and is accurate to a millimeter or fraction of an inch (**FIGURE 3-93**). A steel rule is capable of accurate measurements on shorter lengths, down to a millimeter or a fraction of an inch. Precision measuring tools are accurate to much smaller dimensions, such as a micrometer, which in some cases can accurately measure down to 1/10,000 of an inch (0.0001") or 1/1000 of a millimeter (0.001 mm).

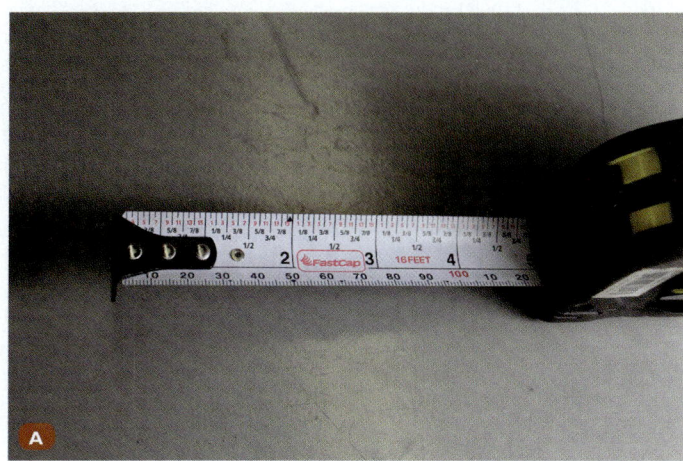 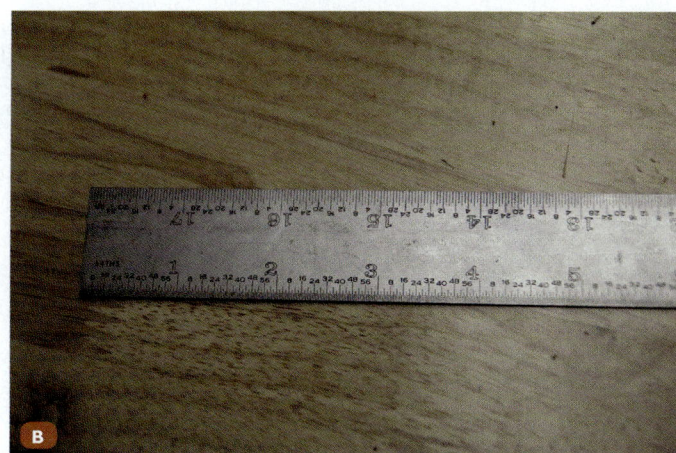

FIGURE 3-93 A. Measuring tape. **B.** Steel rule.

Measuring Tapes

Measuring tapes are a flexible type of ruler and are a common measuring tool. The most common type found in shops is a thin metal strip about 0.5" to 1" (13 to 25 mm) wide that is rolled up inside a housing with a spring return mechanism. Measuring tapes can be of various lengths, with 16' or 25' (5 or 8 m) being very common. The measuring tape

K03025 Describe the type and use of measuring tapes.

SKILL DRILL 3-5 Using Riveting Tools

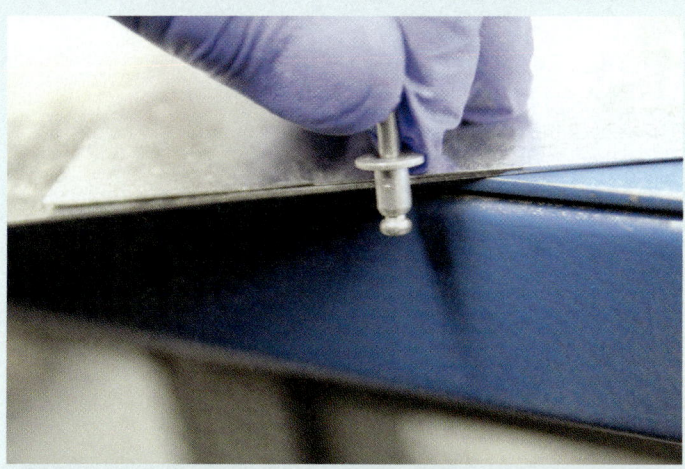

1. Select the correct rivet for the material you are riveting. Make sure the rivet is the correct length.

2. Drill pilot holes in the material to be riveted. Remove any burrs. Ensure that the pilot hole is the correct size—not too large or too small.

Continued

3. Make sure the correctly sized nosepiece for the rivet size is fitted to the rivet tool.

4. Insert the rivet into the gun, and push the rivet through the materials to be riveted. Hold firm pressure while pushing the rivet into the work.

5. Squeeze and release the rivet tool handle to compress the rivet. Continue this process until the rivet stem or shank breaks away from the rivet head.

6. Check the rivet joint to ensure the pieces are firmly held together.

is pulled from the housing to measure items, and a spring return winds it back into the housing. The housing usually has a built-in locking mechanism to hold the extended measuring tape against the spring return mechanism. The hooked end can be placed over the edge of the object you are measuring and pulled against spring tension to take the measurement (**FIGURE 3-94**).

Steel Rulers

K03026 Describe the type and use of steel rulers.

As the name suggests, a **steel rule** is a ruler that is made from steel. Steel rules commonly come in 12" (0.030 m), 24" (0.60 m), and 36" (0.91 m) lengths. They are used like any ruler to measure and mark out items. A steel rule is a very strong ruler, has precise markings, and resists damage. When using a steel rule, you can rest it on its edge so the markings are closer to the material being measured, which helps to mark the work more precisely

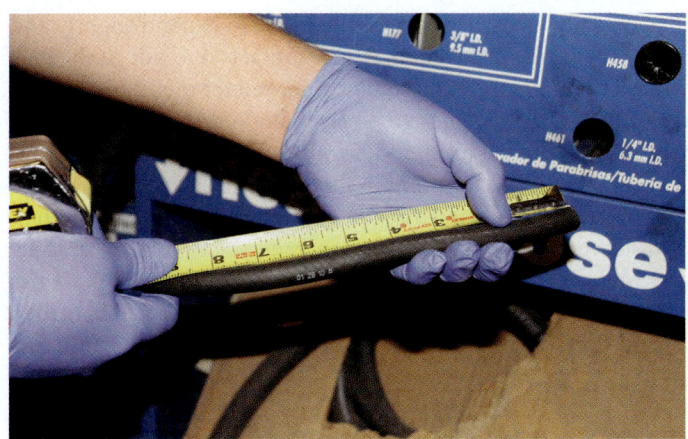
FIGURE 3-94 The hook makes it easy to take a measurement with one hand.

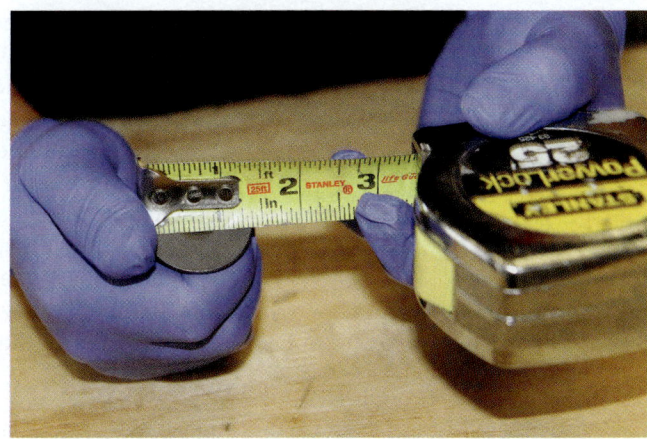
FIGURE 3-95 Tipping a steel rule on its side to get a more accurate reading.

(**FIGURE 3-95**). Always protect the steel rule from damage by storing it carefully; a damaged ruler will not give an accurate measurement. Never take measurements from the very end of a damaged steel rule, as damaged ends may affect the accuracy of your measurements.

Outside, Inside, and Depth Micrometers

Micrometers are precise measuring tools designed to measure small distances and are available in both inch and millimeter (mm) calibrations. Typically they can measure down to a resolution of 1/1000 of an inch (0.001") for a standard micrometer or 1/100 of a millimeter (0.01 mm) for a metric micrometer. Vernier micrometers equipped with the addition of a vernier scale can measure down to 1/10,000 of an inch (0.0001") or 1/1000 of a millimeter (0.001 mm).

K03027 Describe the type and use of outside, inside, and depth micrometers.

The most common types of micrometers are the outside, inside, and depth micrometers (**FIGURE 3-96**). As the name suggests, an **outside micrometer** measures the outside dimensions of an item. For example, it could measure the diameter of a valve stem. The **inside micrometer** measures inside dimensions. For example, the inside micrometer could measure an engine cylinder bore. **Depth micrometers** measure the depth of an item such as how much clearance a piston has below the surface of the block.

The most common micrometer is an outside micrometer and is made up of several parts (**FIGURE 3-97**). The horseshoe-shaped part is the frame. It is built to make sure the micrometer holds its shape. Some frames have plastic finger pads so that body heat is not

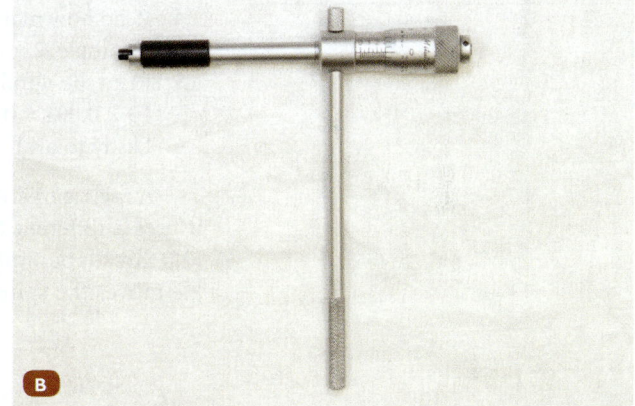

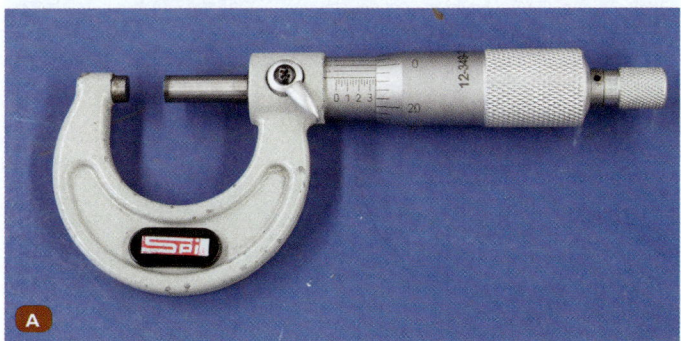
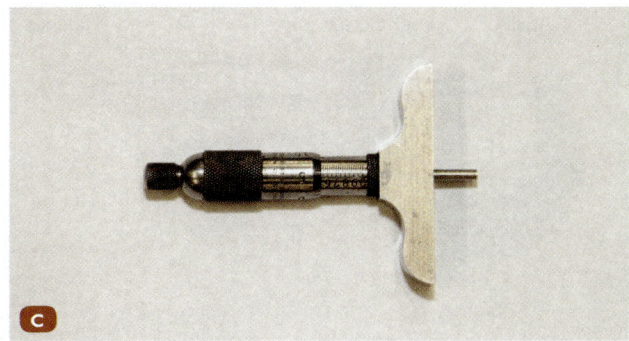
FIGURE 3-96 **A.** Outside micrometer. **B.** Inside micrometer. **C.** Depth micrometer.

TECHNICIAN TIP

The United States customary system (USCS), also called the standard system, and the metric system are two sets of standards for quantifying weights and measurements. Each system has defined units. For example, the standard system uses inches, feet, and yards, whereas the metric system uses millimeters, centimeters, and meters. Conversions can be undertaken from one system to the other. For example, 1 inch is equal to 25.4 millimeters, and 1 foot is equal to 304.8 millimeters. Tools that make use of a measuring system, such as wrenches, sockets, drill bits, micrometers, rulers, and many others, come in both standard and metric measurements. To work on modern vehicles, an understanding of both systems and their conversion is required. Conversion tables can be used to convert from one system to the other. The more you work with both systems, the easier it will be to understand how they relate to each other.

TECHNICIAN TIP

If the end of a rule is damaged, you may be able to measure from the 1" mark and subtract an inch from the measurement.

transferred to the metal frame as easily, as heat can cause the metal to expand slightly and affect the reading. On one end of the frame is the anvil, which contacts one side of the part being measured. The other contact point is the spindle. The micrometer measures the distance between the anvil and spindle, so that is where the part being measured fits.

The measurement is read on the sleeve/barrel and thimble. The sleeve/barrel is stationary and has the linear markings on it. The thimble fits over the sleeve and has the graduated marking on it. The thimble is connected directly to the spindle, and both turn as a unit. Because the spindle and sleeve/barrel have matching threads, the thimble rotates the spindle inside of the sleeve/barrel, and the thread moves the spindle inward and outward. The thimble usually incorporates either a ratchet or a clutch mechanism, which is turned lightly by finger, thus preventing over-tightening of the micrometer thimble when taking a reading.

A lock nut, lock ring, or lock screw is used on most micrometers to lock the thimble in place while you read the micrometer. Standard micrometers use a specific thread of 40 TPI (threads per inch) on the spindle and sleeve. This means that the thimble rotates exactly 40 turns in 1' of travel. Every complete rotation moves the spindle one 40th of an inch, or 0.025" (1 ÷ 40 = 0.025). In four rotations, the spindle moves 0.100" (0.025 × 4 = 0.100). The linear markings on the sleeve show each of the 0.100" marks between 0 and 1 inch as well as each of the 0.025" marks (**FIGURE 3-98**). Because the thimble has graduated marks from 0 to 24 (each mark representing 0.001"), every complete turn of the thimble uncovers another one of the 0.025" marks on the sleeve. If the thimble stops short of any complete turn, it will indicate the number of 0.001" marks past the zero line on the sleeve (**FIGURE 3-99**). So reading a micrometer is as simple as adding up the numbers as shown below.

To read a standard micrometer, perform the following steps (**FIGURE 3-100**):

1. Verify that the micrometer is properly calibrated.
2. Verify what size of micrometer you are using. If it is a 0–1" micrometer, start with 0.000. If it is a 1–2" micrometer, start with 1.000". A 2–3" micrometer would start with 2.000", etc. (To give an example, let's say it is 2.000".)
3. Read how many 0.100" marks the thimble has uncovered (example: 0.300").
4. Read how many 0.025" marks the thimble has uncovered past the 0.100" mark in step 3 (example: 2 × 0.025 = 0.050").
5. Read the number on the thimble that lines up with the zero line on the sleeve (example: 13 × 0.001 = 0.013").
6. Lastly, total all of the individual readings (example: 2.000 + 0.300 + 0.050 + 0.013 = 2.363").

A metric micrometer uses the same components as the standard micrometer. However, it uses a different **thread pitch** on the spindle and sleeve. It uses a 0.5 mm thread pitch (2.0 threads per millimeter) and opens up approximately 25 mm. Each rotation of the thimble moves the spindle 0.5 mm, and it therefore takes 50 rotations of the thimble to move

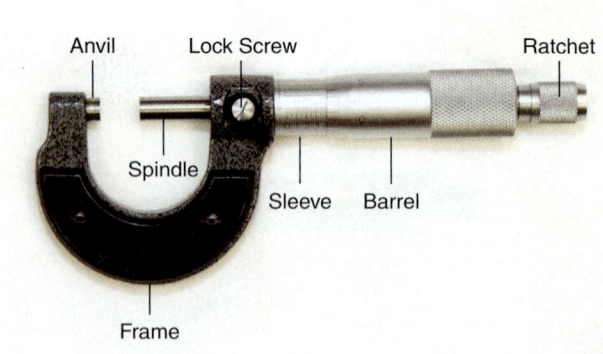

FIGURE 3-97 Parts of an outside micrometer.

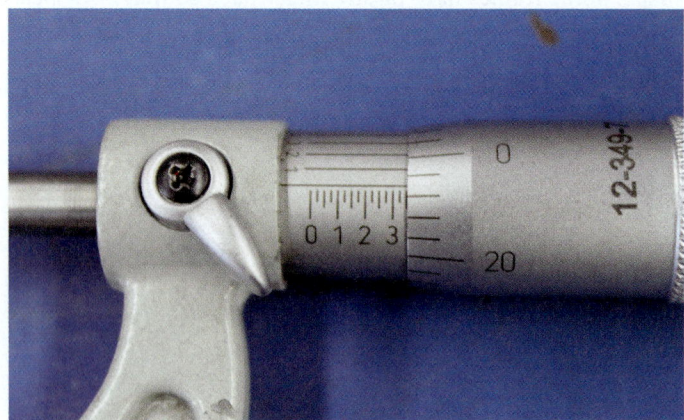

FIGURE 3-98 0.100" and 0.025" markings on the sleeve.

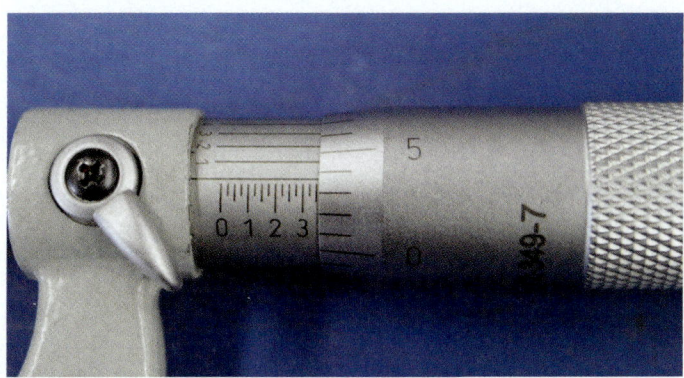

FIGURE 3-99 Markings on the thimble.

> ▶ TECHNICIAN TIP
>
> Micrometers are precision measuring instruments and must be handled and stored with care. They should always be stored with a gap between the spindle and anvil so metal contraction and expansion do not interfere with their calibration.

> ▶ TECHNICIAN TIP
>
> All micrometers need to be checked for calibration (also called "zeroing") before each use. A 0–1" or 0–25 mm outside micrometer can be lightly closed all of the way. If the anvil and spindle are clean, the micrometer should read 0.000, indicating the micrometer is calibrated correctly. If the micrometer is bigger than 1", or 25 mm, then a "standard" is used to verify the calibration. A standard is a hardened, machined rod of a precise length, such as 2", or 50 mm. When inserted in the same-sized micrometer, the reading should be exactly the same as listed on the standard. If a micrometer is not properly calibrated, it should not be used until it is recalibrated. See the tool's instruction manual for the calibration procedure.

the full 25 mm distance. The sleeve/barrel is labeled with individual millimeter marks and half-millimeter marks from the starting millimeter to the ending millimeter, 25 mm away (**FIGURE 3-101**). The thimble has graduated marks from 0 to 49 (**FIGURE 3-102**).

Reading a metric micrometer involves the following steps:

1. Read the number of full millimeters the thimble has passed (To give an example, let's say it is 23.00 mm).
2. Check to see if it passed the 0.5 mm mark (example: 0.50 mm).
3. Check to see which mark on the thimble lines up with or is just passed (example: 37 × 0.01 mm = 0.37 mm).
4. Total all of the numbers (example: 23.00 mm + 0.50 mm + 0.37 mm = 23.87 mm).

If the micrometer is equipped with a vernier gauge, meaning it can read down to 1/10,000 of an inch (0.0001") or 1/1000 of a millimeter (0.001 mm), you need to complete one more step. Identify which of the vernier lines is closest to one of the lines on the thimble (**FIGURE 3-103**). Sometimes it is hard to determine which is the closest, so decide which three are the closest, and then use the center line. At the frame side of the sleeve will be a number that corresponds to the vernier line, numbered 1–0. Add the vernier to the end of your reading. For example: 2.363 + 0.0007 = 2.3637", and 23.77 + 0.007 = 23.777 mm.

For inside measurements, the inside micrometer works on the same principles as the outside micrometer and so does the depth micrometer. The only difference is that the scale on the sleeve of the depth micrometer is backward, so be careful when reading it.

Using Micrometers

To maintain accuracy of measurements, it is important that both the micrometer and the items to be measured are clean and free of any dirt or debris. Also make sure the micrometer is zeroed before taking any measurements. Never over-tighten a micrometer or store it with its measuring surfaces touching, as this may damage the tool and affect its accuracy. When measuring, make sure the item can pass through the micrometer surfaces snugly

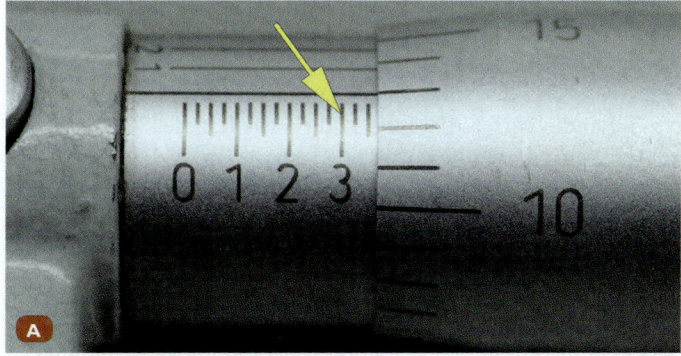
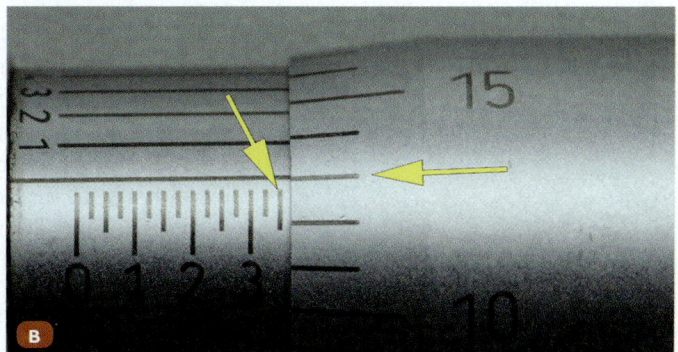

FIGURE 3-100 **A.** Read how many 0.100" marks the thimble has uncovered. **B.** Read the number on the thimble that lines up with the zero line on the sleeve.

FIGURE 3-101 Metric markings on the sleeve.

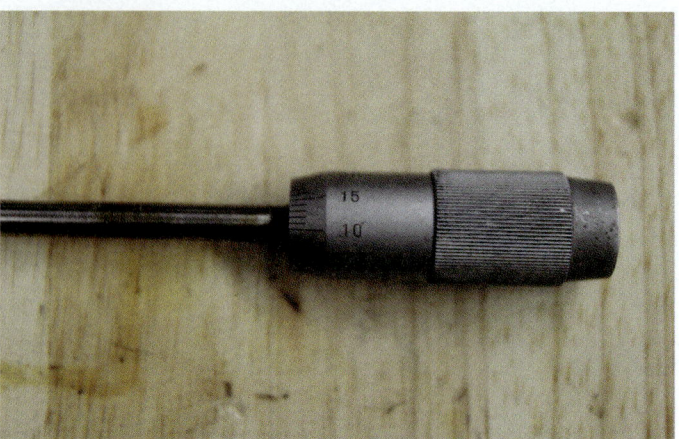

FIGURE 3-102 Markings on the thimble.

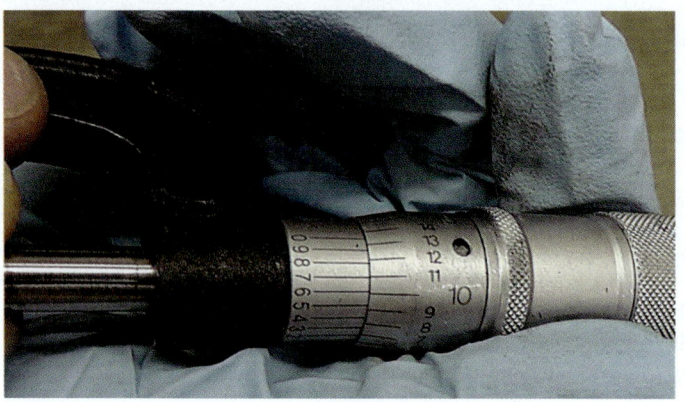

FIGURE 3-103 Vernier scale on a micrometer showing 7 on the sleeve lined up the best with a line on the thimble.

K03028 Describe the type and use of telescoping gauges.

K03029 Describe the type and use of split ball gauges.

K03030 Describe the type and use of dial bore gauges.

and squarely. This is best accomplished by using the ratchet to tighten the micrometer. Always take the measurement a number of times and compare results to ensure you have measured accurately.

To correctly measure using an outside micrometer, follow the steps in **SKILL DRILL 3-6**.

Telescoping Gauges

For measuring distances in awkward spots like the bottom of a deep cylinder, the **telescoping gauge** has spring-loaded plungers that can be unlocked with a screw on the handle so they slide out and touch the walls of the cylinder (**FIGURE 3-104**). The screw then locks them in that position, the gauge can be withdrawn, and the distance across the plungers can be measured with an outside micrometer or calipers to convey the diameter of the cylinder at that point. Telescoping gauges come in a variety of sizes to fit various sizes of holes and bores.

Split Ball Gauges

A **split ball gauge** or small hole gauge is good for measuring small holes where telescoping gauges cannot fit (**FIGURE 3-105**). They use a similar principle to the telescoping gauge, but the measuring head uses a split ball mechanism that allows it to fit into very small holes. Split ball gauges are ideal for measuring valve guides on a cylinder head for wear. A split ball gauge can be fitted in the bore and expanded until there is a slight drag. Then it can be retracted and measured with an outside micrometer.

Dial Bore Gauges

A **dial bore gauge** is used to measure the inside diameter of bores with a high degree of accuracy and speed (**FIGURE 3-106**). The dial bore gauge can measure a bore directly by using telescoping pistons on a T-handle with a dial mounted on the handle. The dial bore gauge combines a telescoping gauge and dial indicator in one instrument. A dial bore gauge determines if the diameter is worn, tapered, or out-of-round according to the manufacturer's specifications. The resolution of a dial bore gauge is typically accurate to 5/10,000 of an inch (0.0005″) or 1/100 of a millimeter (0.01 mm).

Using Dial Bore Gauges

To use a dial bore gauge, select an appropriate-sized adapter to fit the internal diameter of the bore, and install it to the measuring head. Many dial bore gauges also have a fixture to calibrate the tool to the size you desire (**FIGURE 3-107**). The fixture is set to the size desired,

SKILL DRILL 3-6 Using Micrometers

1. Select the correct size of micrometer. Verify that the anvil and spindle are clean and that it is calibrated properly. Clean the surface of the part you are measuring.

2. In your right hand, hold the frame of the micrometer between your pinky, ring finger, and the palm of your hand, with the thimble between your thumb and forefinger.

3. With your left hand, hold the part you are measuring, and place the micrometer over it.

4. Using your thumb and forefinger, lightly tighten the ratchet. It is important that the correct amount of force is applied to the spindle when taking a measurement. The spindle and anvil should just touch the component with a slight amount of drag when the micrometer is removed from the measured piece. Be careful that the part is square in the micrometer so the reading is correct. Try rocking the micrometer in all directions to make sure it is square.

Continued

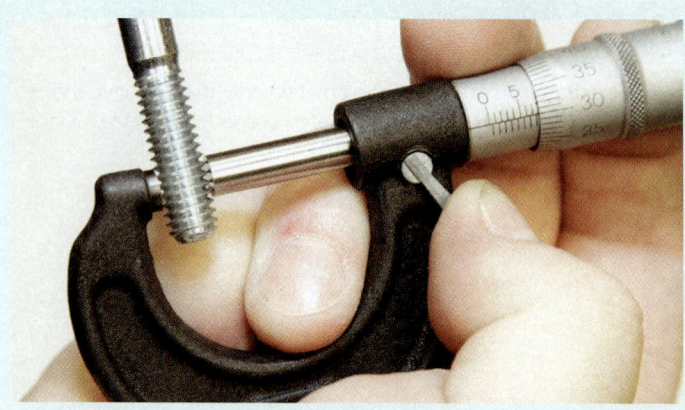

5. Once the micrometer is properly snug, tighten the lock mechanism so the spindle will not turn. Read the micrometer and record your reading.

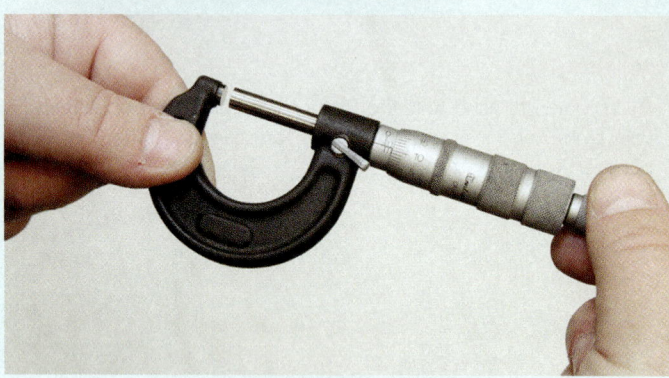

6. When all readings are finished, clean the micrometer, position the spindle so it is backed off from the anvil, and return it to its protective case.

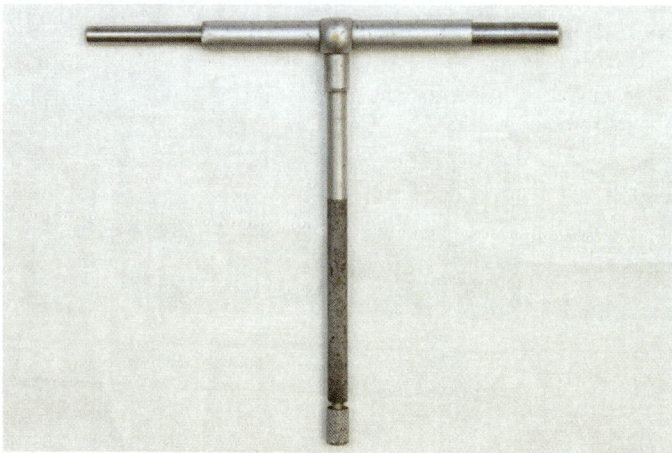

FIGURE 3-104 Telescoping gauge.

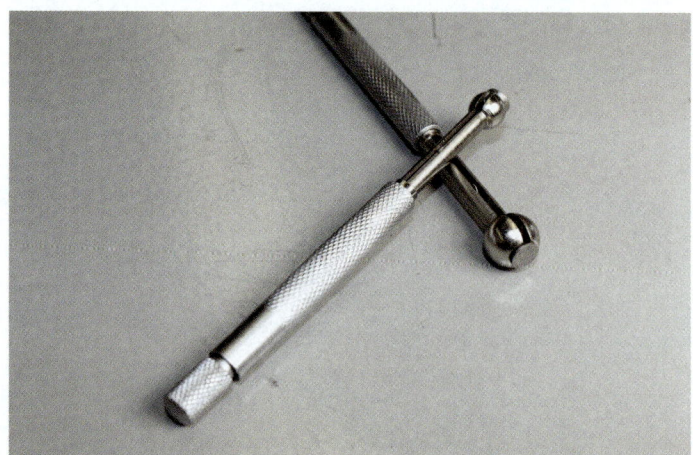

FIGURE 3-105 Split ball gauges.

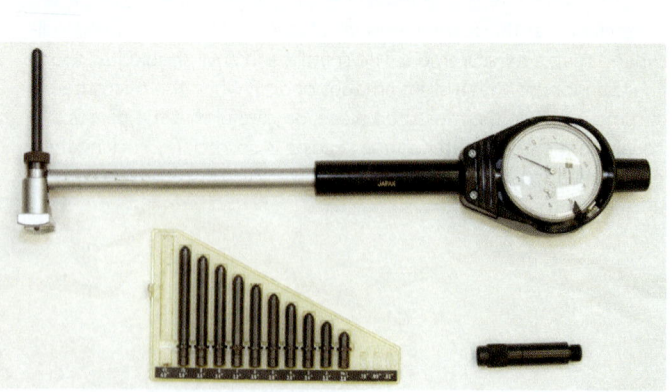

FIGURE 3-106 A dial bore gauge set.

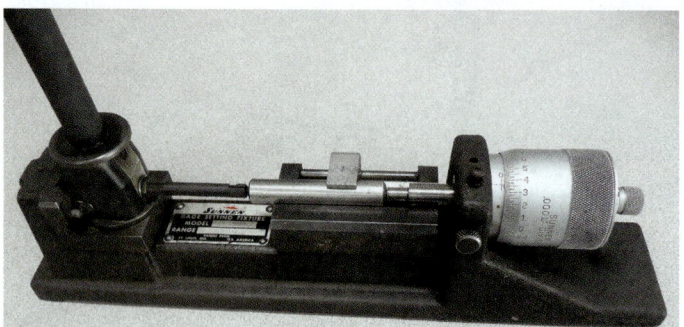

FIGURE 3-107 Dial bore gauge being calibrated to a predetermined size.

and the dial bore gauge is placed in it. The dial bore gauge is then adjusted to the proper reading. Once it is calibrated, the dial bore gauge can be inserted inside the bore to be measured. Hold the gauge in line with the bore, and slightly rock it to ensure it is centered. Read the dial when it is fully centered and square to the bore, to determine the correct measurement. It takes a bit of practice to get accurate readings.

Store a bore gauge carefully in its storage box and ensure the locking mechanism is released while in storage. Bore gauges are available in different ranges of size. It is important to select a gauge with the correct range for the bore you are measuring. When measuring, make sure the gauge is at a 90-degree angle to the bore and read the dial. Always take the measurement a number of times and compare results to ensure you have measured accurately.

To correctly measure using a dial bore gauge, follow the steps in **SKILL DRILL 3-7**.

Vernier Calipers

Vernier calipers are a precision instrument used for measuring outside dimensions, inside dimensions, and depth measurements, all in one tool (**FIGURE 3-108**). They have a graduated bar with markings like a ruler. On the bar, a sliding sleeve with jaws is mounted for taking inside or outside measurements. Measurements on older versions of vernier calipers are taken by reading the graduated bar scales, and fractional measurements are read by comparing the scales between the sliding sleeve and the graduated bar. Technicians often use vernier calipers to measure length and diameters of bolts and pins or the depth of blind holes in housings.

Newer versions of vernier calipers have dial and digital scales (**FIGURE 3-109**). The dial caliper has the main scale on the graduated bar, and fractional measurements are taken from a dial with a rotating needle. These tend to be easier to read than the older versions.

K03031 Describe the type and use of vernier calipers.

SKILL DRILL 3-7 Using Dial Bore Gauges

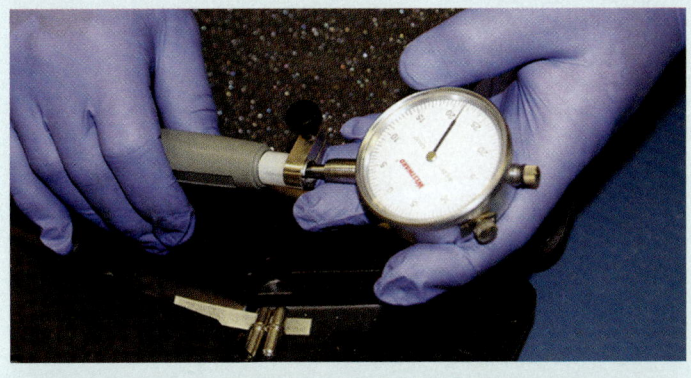

1. Select the correct size of the dial bore gauge you will use, and fit any adapters to it. Check the calibration and adjust it as necessary. Insert the dial bore gauge into the bore. The accurate measurement will be at exactly 90 degrees to the bore. To find the accurate measurement, rock the dial bore gauge handle slightly back and forth until you find the centered position.

2. Check the calibration and adjust it as necessary.

Continued

3. Insert the dial bore gauge into the bore. The accurate measurement will be at exactly 90 degrees to the bore. To find the accurate measurement, rock the dial bore gauge handle slightly back and forth until you find the centered position.

4. Read the dial to determine the bore measurement.

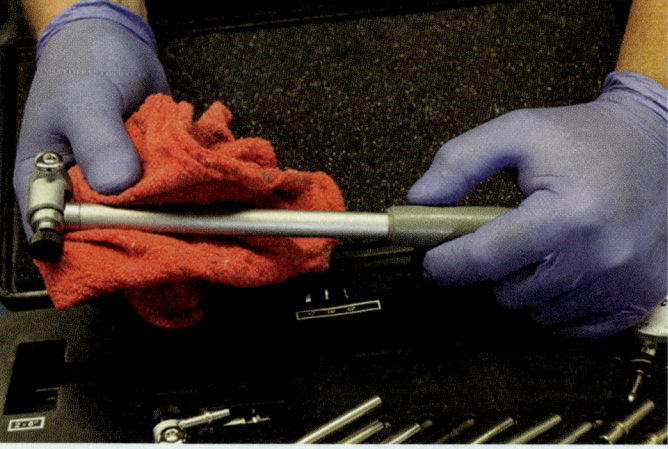

5. Always clean the dial bore gauge, and return it to its protective case when you have finished using it.

More recently, digital scales on vernier calipers have become commonplace (**FIGURE 3-110**). The principle of their use is the same as any vernier caliper; however, they have a digital scale that reads the measurement directly.

Using Vernier Calipers

Always store vernier calipers in a storage box to protect them and ensure the measuring surfaces are kept clean for accurate measurement. If making an internal or external measurement, make sure the caliper is at right angles to the surfaces to be measured. You should always repeat the measurement a number of times and compare results to ensure you have

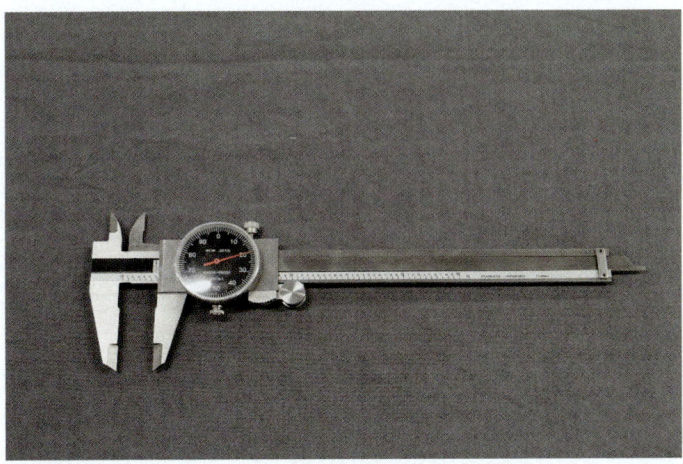

FIGURE 3-108 Vernier calipers can take three types of readings.

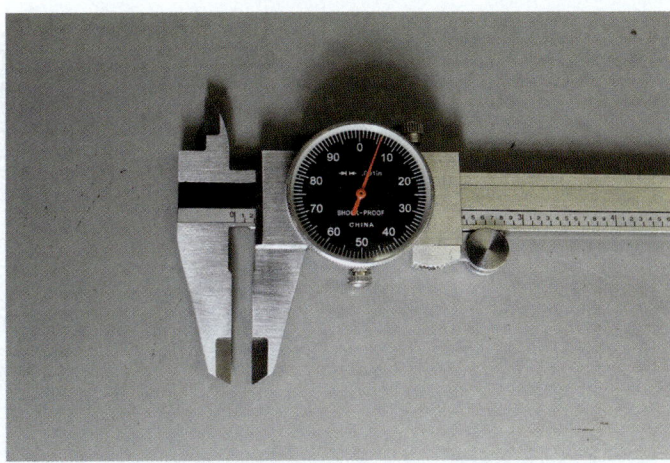

FIGURE 3-109 Dial vernier caliper.

measured accurately. To correctly measure using digital vernier calipers, follow the steps in **SKILL DRILL 3-8**.

Dial Indicators

Dial indicators can also be known as dial gauges, and as the name suggests, they have a dial and needle where measurements are read. They have a measuring plunger with a pointed or rounded contact end that is spring loaded and connected via the housing to the dial needle (**FIGURE 3-111**). The dial accurately displays movement of the plunger in and out as it rests against an object. For example, they can be used to measure the trueness of a rotating disc brake rotor. A dial indicator can also measure how round something is, such as a **crankshaft**, which can be rotated in a set of **V blocks** (**FIGURE 3-112**). If the crankshaft is bent, it will show as movement on the dial indicator as the crankshaft is rotated. The dial indicator senses slight movement at its tip and magnifies it into a measurable swing on the dial. Dial indicators normally have either one or two indicator needles. The large needle indicates the fine reading of thousandths of an inch. If it has a second needle, it will be smaller and indicates the coarse

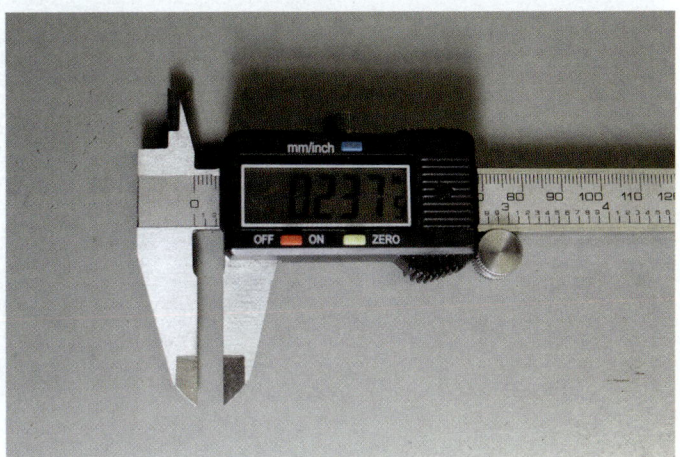

FIGURE 3-110 Digital caliper.

SKILL DRILL 3-8 Using Vernier Calipers

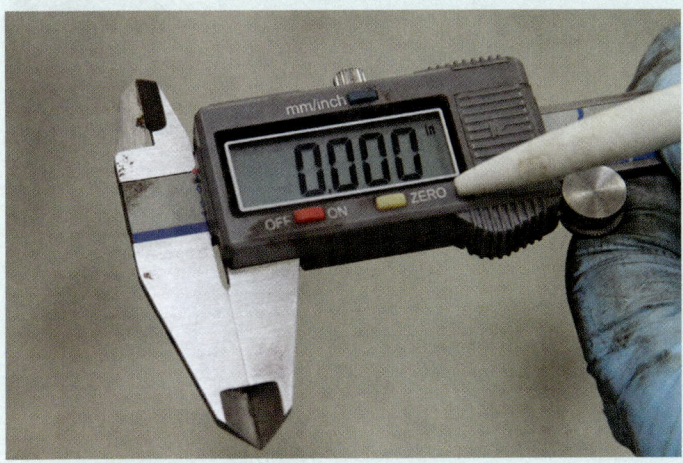

1. Verify that the vernier caliper is calibrated (zeroed) before using it.

Continued

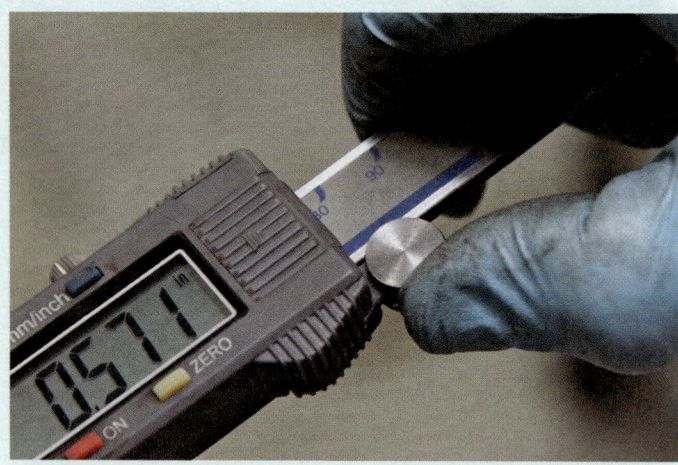

2. Position the caliper correctly for the measurement you are making. Internal and external readings are normally made with the vernier caliper positioned at 90 degrees to the face of the component to be measured. Length and depth measurements are usually made parallel to or in line with the object being measured. Use your thumb to press or withdraw the sliding jaw to measure the outside or inside of the part.

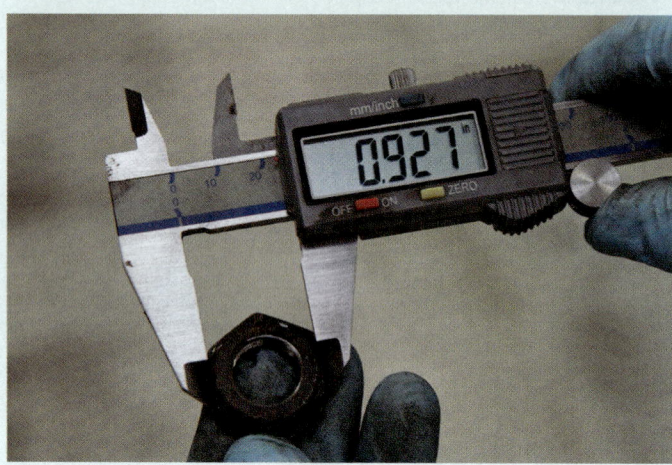

3. Read the scale of the vernier caliper, being careful not to change the position of the moveable jaw. If using a non-digital caliper, always read the dial or face straight on. A view from the side can give a considerable **parallax error**. Parallax error is a visual error caused by viewing measurement markers at an incorrect angle.

reading of tenths of an inch. The large needle is able to move numerous times around the outer scale. One full turn may represent 0.100" or 1 mm. The small inner scale indicates how many times the outer needle has moved around its scale. In this way, the dial indicator is able to read movement of up to 1" or 2 cm. Dial indicators can typically measure with an accuracy of 0.001" or 0.01 mm.

The type of dial indicator you use is determined by the amount of movement you expect from the component you are measuring. The indicator must be set up so that there

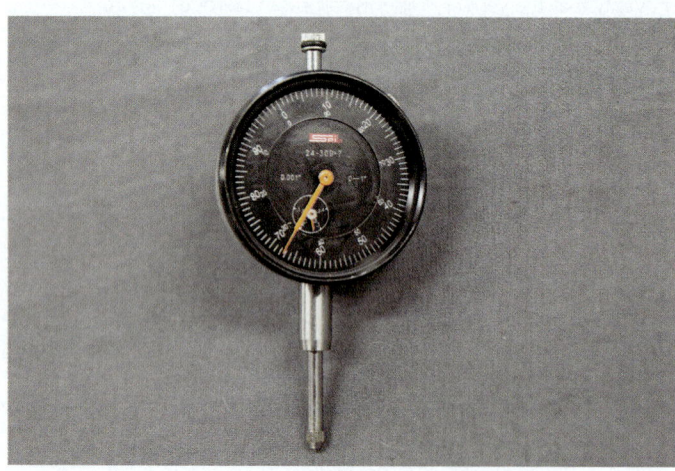

FIGURE 3-111 Dial indicator.

FIGURE 3-112 A dial indicator being used to measure crankshaft runout.

is no gap between the dial indicator and the component to be measured. It also must be set perpendicular and centered to the part being measured. Most dial indicator sets contain various attachments and support arms, so they can be configured specifically for the measuring task.

Using Dial Indicators

Dial indicators are used in many types of service jobs. They are particularly useful in determining runout on rotating shafts and surfaces. Runout is the side-to-side variation of movement when a component is turned. When attaching a dial indicator:

- Keep support arms as short as possible.
- Make sure all attachments are tightened to prevent unnecessary movement between the indicator and the component.
- Make sure the dial indicator plunger is positioned at 90 degrees to the face of the component to be measured.
- Always read the dial face straight on, as a view from the side can give a considerable parallax error.

The outer face of the dial indicator is designed so it can be rotated so that the zero mark can be positioned directly under the pointer. This is how a dial indicator is zeroed. To correctly measure using a dial indicator, follow the steps in **SKILL DRILL 3-9**.

SKILL DRILL 3-9 Using Dial Indicators

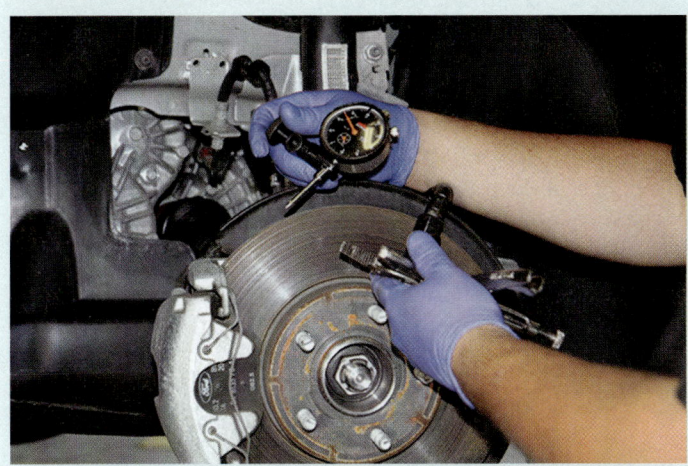

1. Select the gauge type, size, attachment, and bracket that fit the part you are measuring. Mount the dial indicator firmly to keep it stationary.

2. Adjust the indicator so that the plunger is at 90 degrees to the part you are measuring, and lock it in place.

Continued

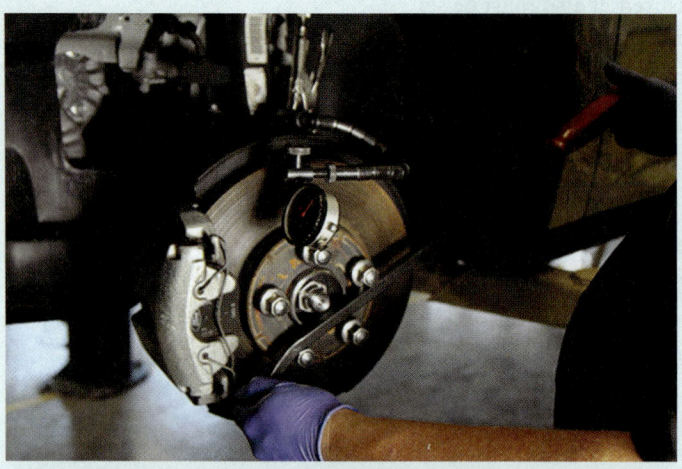

3. Rotate the part one complete turn, and locate the low spot. Zero the indicator.

4. Find the point of maximum height and note the reading. This indicates the runout value.

5. Continue the rotation, making sure the needle does not go below zero. If it does, re-zero the indicator and remeasure the point of maximum variation.

6. Check your readings against the manufacturer's specifications. If the deviation is greater than the specifications allow, consult your supervisor.

CALIPER ILLUS.	Maximum Parallel Variation	Runout Limit	Nominal Thickness	Minimum Machining	Discard Or Und
	NS	.0029	NS	.937	.905
	NS	.0019	NS	.386	.354
	NS			935	90
			NS	1.015	98
			NS	.421	3

Straight Edges

Straight edges are usually made from hardened steel and are machined so that the edge is perfectly straight. A straight edge is used to check the flatness of a surface. It is placed on its edge against the surface to be checked (**FIGURE 3-113**). The gap between the straight edge and the surface can be measured by using feeler gauges. Sometimes the gap can be seen easily if light is shone from behind the surface being checked. Straight edges are often used to measure the amount of warpage the surface of a cylinder head has.

K03033 Describe the type and use of straight edges.

Feeler Gauges

Feeler gauges (also called feeler blades) are used to measure the width of gaps, such as the clearance between valves and rocker arms. Feeler gauges are flat metal strips of varying thicknesses (**FIGURE 3-114**). The thickness of each feeler gauge is clearly marked on each one. They are sized from fractions of an inch or fractions of a millimeter. They usually come in sets with different sizes and are available in standard and metric measurements. Some sets contain feeler gauges made of brass. These are used to take measurements between components that are magnetic. If steel gauges are used, the drag caused by the magnetism would mimic the drag of a proper clearance. Brass gauges are not subject to magnetism, so they work well in that situation. Some feeler gauges come in a bent arrangement to be more easily inserted in cramped spaces. Others come in a stepped version. For example, the end of the gauge might be 0.010" thick, while the rest of the gauge is 0.012" thick. This works well for adjusting valve clearance. If the specification is 0.010", then the 0.010 section can be placed in the gap. If the 0.012 section slides into the gap, then the valve needs to be readjusted. If it stops at the lip of the 0.012" section, then the gap is correct (**FIGURE 3-115**).

K03034 Describe the type and use of feeler gauges.

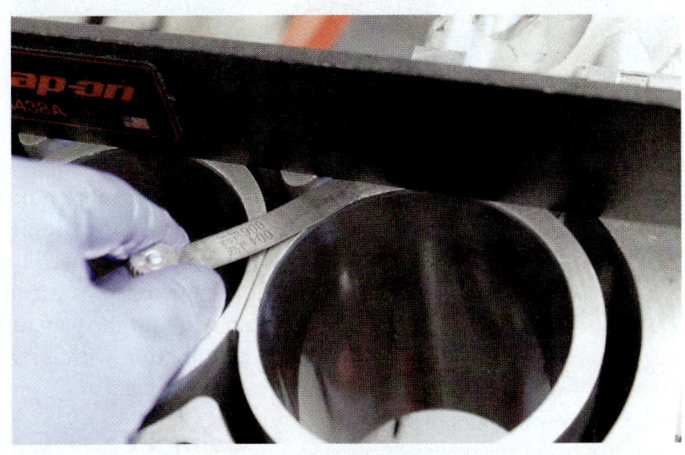

FIGURE 3-113 Straight edge being used to measure the flatness of a surface.

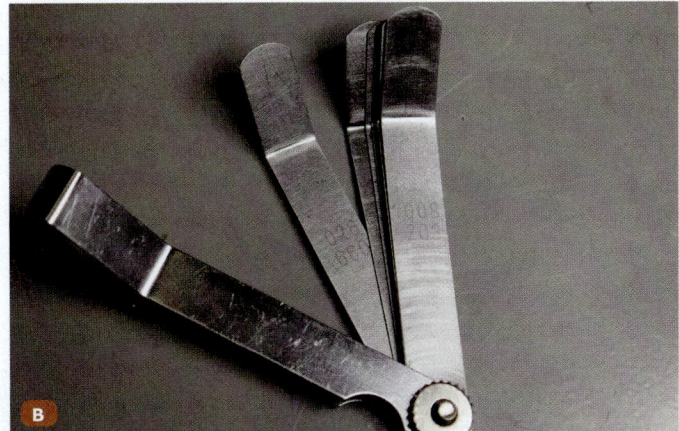

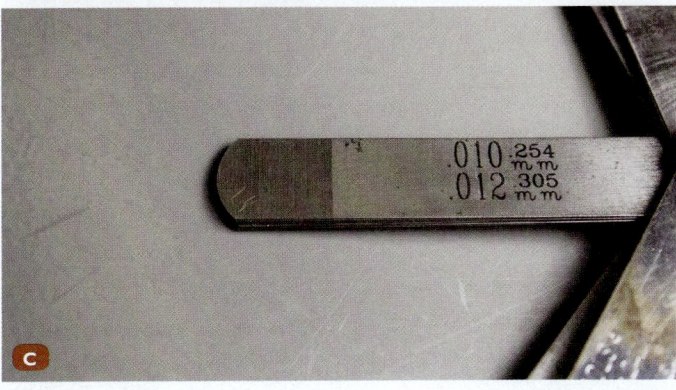

FIGURE 3-114 Feeler blade set. **A.** Straight. **B.** Bent. **C.** Stepped. **D.** Wire gauge.

FIGURE 3-115 Stepped feeler blade being used during a valve adjustment.

FIGURE 3-116 Wire feeler gauge used to check a spark plug gap.

SAFETY TIP
Never use feeler gauges on operating machinery.

SAFETY TIP
Feeler gauges are strips of hardened metal that have been ground or rolled to a precise thickness. They can be very thin and will cut through skin if not handled correctly.

Two or more non-stepped feeler gauges can be stacked together to make up a desired thickness. For example, to measure a thickness of 0.029 of an inch, a 0.017 and a 0.012 feeler gauge could be used together to make up the size. Alternatively, if you want to measure an unknown gap, you can interchange feeler gauges until you find the one or more that fits snugly into the gap and total their thickness to determine the measurement of the gap. In conjunction with a straight edge, they can be used to measure surface irregularities on a cylinder head.

Using Feeler Gauges

If the feeler gauge feels too loose when measuring a gap, select the next size larger, and measure the gap again. Repeat this procedure until the feeler gauge has a slight drag between both parts. If the feeler gauge is too tight, select a smaller size until the feeler gauge fits properly. When measuring a spark plug gap, feeler gauges should not be used because the surfaces of the spark plug electrodes are not perfectly parallel, so it is preferable to use wire feeler gauges (**FIGURE 3-116**).

Wire feeler gauges use accurately machined pieces of wire instead of flat metal strips. To select and use feeler gauge sets, follow the steps in **SKILL DRILL 3-10**.

SKILL DRILL 3-10 Using Feeler Gauges

1. Select the appropriate type and size feeler gauge set for the job you are working on.

2. Inspect the feeler gauges to make sure they are clean, rust-free, and undamaged, but slightly oiled for ease of movement.

Continued

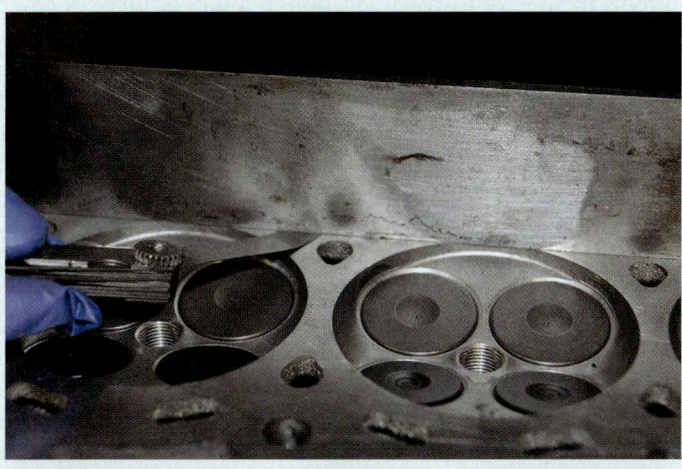

3. Choose one of the smaller wires or blades, and try to insert it in the gap on the part. If it slips in and out easily, choose the next size up. When you find one that touches both sides of the gap and slides with only gentle pressure, then you have found the exact width of that gap.

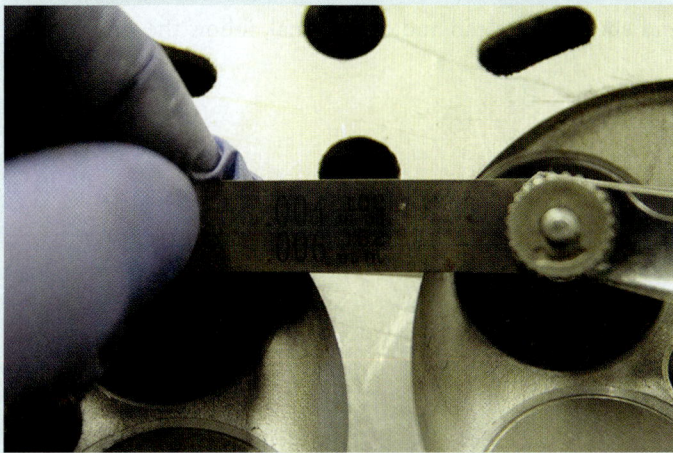

4. Read the markings on the wire or blade, and check these against the manufacturer's specifications for this component. If gap width is outside the tolerances specified, inform your supervisor.

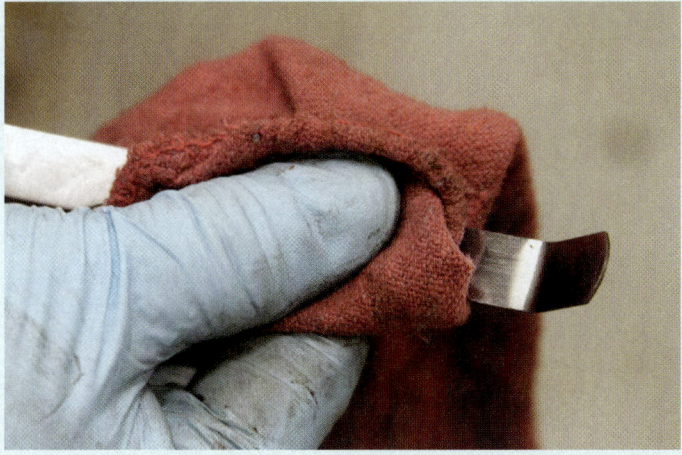

5. Clean the feeler gauge set with an oily cloth before storage to prevent rust.

▶ Cleaning Tools and Equipment

Clean tools and equipment work more safely and efficiently. At the end of each working day, clean the tools and equipment you used and check them for any damage. If you note any damage, tag the tool as faulty and organize a repair or replacement. Electrical current can travel over oily or greasy surfaces. Be sure to keep electrical power tools clean. All shop equipment should have a maintenance schedule. Always complete the tasks described on the schedule at the required time. This helps to keep the equipment in safe working order.

N03005 Demonstrate proper cleaning, storage, and maintenance of tools and equipment.

K03035 Describe the importance of proper cleaning and storage of tools.

Store commonly used tools in an easy-to-reach location. If a tool or piece of equipment is too difficult to return, then it will likely be left on a workbench or on the floor, where it will become a safety hazard. Keep your work area tidy. This will help you work more efficiently and safely. Keep a trash can close to your work area, and place any waste in it as soon as possible. Dispose of liquid and solid waste, such as oils, coolant, and worn components, in the correct manner. Local authorities provide guidelines for waste disposal with fines for noncompliance. When cleaning products lose their effectiveness, they need to be replaced. Refer to the supplier's recommendations for collection or disposal. Do not pour solvents or other chemicals into the sewage system. This is both environmentally damaging and illegal.

Always use chemical gloves when using any cleaning material, because excessive exposure to cleaning materials can damage skin. Also, absorbing some chemicals through the skin over time can cause permanent harm to your body. Some solvents are flammable; never use cleaning materials near an open flame or cigarette. The fumes from cleaning chemicals can be toxic, so wear appropriate respirator and eye protection wherever you are using these products.

To keep work areas and equipment clean and operational, follow the steps in **SKILL DRILL 3-11**.

SAFETY TIP

Do not use flammable cleaners or water on electrical equipment.

SKILL DRILL 3-11 Tool and Equipment Cleaning

1. Clean hand tools. Keep your hand tools in good, clean condition with two types of rags. One rag should be lint-free to clean or handle precision instruments or components. The other should be oily to prevent rust and corrosion.

2. Clean floor jacks. Wipe off any oil or grease on the floor jack, and check for fluid leaks. If you find any, remove the jack from use, and have it repaired or replaced. Occasionally, apply a few drops of lubricating oil to the wheels and a few drops to the posts of threaded jack stands.

Continued

Cleaning Tools and Equipment 107

3. Clean electrical power tools. Keep power tools clean by brushing off any dust and wiping off excess oil or grease with a clean rag. Inspect any electrical cables for dirt, oil, or grease, and for any chafing or exposed wires. With drills, inspect the chuck and lubricate it occasionally with machine oil.

4. Clean air-powered tools. Apply a few drops of oil into the inlet of your air tools every day. Although these tools have no electrical motor, they do need regular lubrication of the internal parts to prevent wear.

5. Clean hoists and heavy machinery. Locate the checklist or maintenance record for each hoist or other major piece of equipment before carrying out cleaning activities.

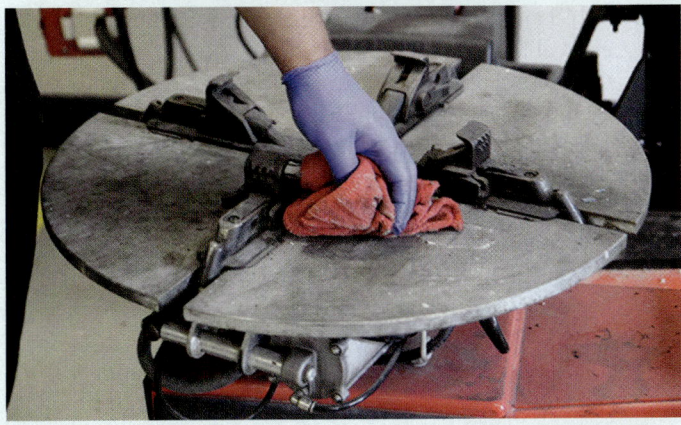

6. You should clean equipment operating mechanisms and attachments of excess oil or grease.

Wrap-Up

Ready for Review

- Tools and equipment should be used only for the task they were designed to do.
- Always have a safe attitude when using tools and equipment.
- Do not use damaged tools; inspect before using, then clean and inspect again before putting them away.
- Lockouts and tag-outs are meant to prevent technicians from using tools and equipment that are potentially unsafe.
- Many tools and measuring instruments have USCS or metric system markings to identify their size.
- Torque defines how much a fastener should be tightened.
- Torque specification indicates the level of tightness each bolt or nut should be tightened to; torque charts list torque specifications for nuts and bolts.
- Torque (or tension) wrenches tighten fasteners to the correct torque specification.
- Torque value—the amount of twisting force applied to a fastener by the torque wrench—is specified in foot-pounds (ft-lb), inch-pounds (in-lb), or newton meters (Nm).
- Torque wrench styles are beam (simplest and least expensive), clicker, dial, and electronic. Each gives an indication of when proper torque is achieved.
- Bolts that are tightened beyond their yield point do not return to their original length when loosened.
- Common wrenches include box end, open end, combination (most popular), flare nut (or flare tubing), open-end adjustable, and ratcheting box end.
- Box-end wrenches can loosen very tight fasteners, but open-end wrenches usually work better once the fastener has been broken loose.
- Use the correct wrench for the situation, so as not to damage the bolt or nut.
- Sockets grip fasteners tightly on all six corners and are purchased in sets.
- Sockets are classified as follows: standard or metric, size of drive used to turn them, number of points, depth of socket, and thickness of wall.
- The most common socket handle is a ratchet; a breaker bar gives more leverage, or a sliding T-handle may be used.
- Fasteners can be spun off or on (but not tightened) by a speed brace or speeder handle.
- Pliers hold, cut, or compress materials; types include slip-joint, combination, arc joint, needle-nose, flat, diagonal cutting, snap ring, and locking.
- Always use the correct type of pliers for the job.
- Cutting tools include bolt cutters, tin snips, and aviation snips.
- Allen wrenches are designed to fit into fasteners with recessed hexagonal heads.
- Screwdriver types include flat blade (most common), Phillips, Pozidriv, offset, ratcheting, and impact.
- The tip of the screwdriver must be matched exactly to the slot or recess on the head of a fastener.
- Magnetic pickup tools and mechanical fingers allow for the extraction and insertion of objects in tight places.
- Types of hammers include ball peen (most common), sledge, mallet, and dead blow.
- Chisels are used to cut metals when hit with a hammer.
- Punches are used to mark metals when hit with a hammer and come in different diameters and different points for different tasks; types of punches include prick, center, drift, pin, ward, and hollow.
- Pry bars can be used to move, adjust, or pry parts.
- Gasket scrapers are designed to remove gaskets without damaging surrounding materials.
- Files are used to remove material from the surface of an automotive part.
- Flat files come in different grades to indicate how rough they are; grades are rough, coarse bastard, second cut, smooth, and dead smooth.
- Types of files include flat, warding, square, triangular, curved, and thread.
- Bench vices, offset vices, drill vices, and C-clamps all hold materials in place while they are worked on.
- Taps are designed to cut threads in holes or nuts; types include taper, intermediate, and bottoming.
- A die is used to cut a new thread on a blank rod or shaft.
- Gear and bearing pullers are designed to remove components from a shaft when considerable force is needed.
- Flaring tools create flares at the end of tubes to connect them to other components; types include single, double, and ISO.
- Rivet tools join together two pieces of metal; each rivet can be used only once.
- Measuring tapes and steel rules are commonly used measuring tools; more precise measuring tools include micrometers, gauges, calipers, dial indicators, and straight edges.
- Micrometers can be outside, inside, or depth.
- Learn to read micrometer measurements on the sleeve/barrel and thimble; always verify the micrometer is properly calibrated before use.
- Gauges are used to measure distances and diameters; types include telescoping, split ball, and dial bore.
- Vernier calipers measure outside, inside, and depth dimensions; newer versions have dial and digital scales.
- Dial indicators are used to measure movement.
- A straight edge is designed to assess the flatness of a surface.
- Feeler blades are flat metal strips that are used to measure the width of gaps.
- Keep work area, tools, and equipment clean and organized.

Key Terms

Allen wrench A type of hexagonal drive mechanism for fasteners.

arc joint pliers Pliers with parallel slip jaws that can increase in size. Also called Channellocks.

aviation snips A scissor-like tool for cutting sheet metal.

ball-peen (engineer's) hammer A hammer that has a head that is rounded on one end and flat on the other; designed to work with metal items.

bench vice A device that securely holds material in jaws while it is being worked on.

blind rivet A rivet that can be installed from its insertion side.

bolt A type of threaded fastener with a thread on one end and a hexagonal head on the other.

bolt cutters Strong cutters available in different sizes, designed to cut through non-hardened bolts and other small-stock material.

bottoming tap A thread-cutting tap designed to cut threads to the bottom of a blind hole.

box-end wrench A wrench or spanner with a closed or ring end to grip bolts and nuts.

C-clamp A clamp shaped like the letter C; it comes in various sizes and can clamp various items.

center punch Less sharp than a prick punch, the center punch makes a bigger indentation that centers a drill bit at the point where a hole is required to be drilled.

cold chisel The most common type of chisel, used to cut cold metals. The cutting end is tempered and hardened so that it is harder than the metals that need to be cut.

combination pliers A type of pliers for cutting, gripping, and bending.

combination wrench A type of wrench that has a box-end wrench on one side and an open end on the other.

crankshaft A vehicle engine component that transfers the reciprocating movement of pistons into rotary motion.

cross-arm A description for an arm that is set at right angles or 90 degrees to another component.

cross-cut chisel A type of chisel for metal work that cleans out or cuts key ways.

curved file A type of file that has a curved surface for filing holes.

dead blow hammer A type of hammer that has a cushioned head to reduce the amount of head bounce.

depth micrometer A measuring device that accurately measures the depth of a hole.

diagonal cutting pliers Cutting pliers for small wire or cable.

dial bore gauge An accurate measuring device for inside bores, usually made with a dial indicator attached to it.

dial indicator An accurate measuring device where measurements are read from a dial and needle.

die Used to cut external threads on a metal shank or bolt.

die stock A handle for securely holding dies to cut threads.

double flare A seal that is made at the end of metal tubing or pipe.

drift punch A type of punch used to start pushing roll pins to prevent them from spreading.

drill vice A tool with jaws that can be attached to a drill press table for holding material that is to be drilled.

fasteners Devices that securely hold items together, such as screws, cotter pins, rivets, and bolts.

feeler gauge A thin blade device for measuring space between two objects.

finished rivet A rivet after the completion of the riveting process.

flare nut wrench A type of box-end wrench that has a slot in the box section to allow the wrench to slip through a tube or pipe. Also called a flare tubing wrench.

flat blade screwdriver A type of screwdriver that fits a straight slot in screws.

flat-nose pliers Pliers that are flat and square at the end of the nose.

forcing screw The center screw on a gear, bearing, or pulley puller. Also called a jacking screw.

gasket scraper A broad sharp flat blade to assist in removing gaskets and glue.

gear pullers A tool with two or more legs and a cross bar with a center forcing screw to remove gears.

hard rubber mallet A special-purpose tool with a head made of hard rubber; often used for moving things into place where it is important not to damage the item being moved.

hollow punch A punch with a center hollow for cutting circles in thin materials such as gaskets.

impact driver A tool that is struck with a hammer to provide an impact turning force to remove tight fasteners.

inside micrometer A micrometer designed to measure internal diameters.

intermediate tap One of a series of taps designed to cut an internal thread. Also called a plug tap.

locking pliers A type of pliers where the jaws can be set and locked into position.

lockout/tag-out A safety tag system to ensure that faulty equipment or equipment in the middle of repair is not used.

lug wrench A tool designed to remove wheel lugs nuts and commonly shaped like a cross.

magnetic pickup tools An extending shaft, often flexible, with a magnet fitted to the end for picking up metal objects.

magnetic pickup tools Useful for grabbing items in tight spaces, it typically is a telescoping stick that has a magnet attached to the end on a swivel joint.

mandrel The shaft of a pop rivet.

mandrel head The head of the pop rivet that connects to the shaft and causes the rivet body to flare.

measuring tape A thin measuring blade that rolls up and is contained in a spring-loaded dispenser.

mechanical fingers Spring-loaded fingers at the end of a flexible shaft that pick up items in tight spaces.

micrometer An accurate measuring device for internal and external dimensions. Commonly abbreviated as "mic."

needle-nose pliers Pliers with long tapered jaws for gripping small items and getting into tight spaces.

nippers (pincer pliers) Pliers designed to cut protruding items level with the surface.

nut A fastener with a hexagonal head and internal threads for screwing on bolts.

offset screwdriver A screwdriver with a 90-degree bend in the shaft for working in tight spaces.

offset vice A vice that allows long objects to be gripped vertically.

oil filter wrench A specialized wrench that allows extra leverage to remove an oil filter when it is tight.

open-end wrench A wrench with open jaws to allow side entry to a nut or bolt.

outside micrometer A micrometer designed to measure the external dimensions of items.

parallax error A visual error caused by viewing measurement markers at an incorrect angle.

peening A term used to describe the action of flattening a rivet through a hammering action.

phillips head screwdriver A type of screwdriver that fits a head shaped like a cross in screws.

pin punch A type of punch in various sizes with a straight or parallel shaft.

pipe wrench A wrench that grips pipes and can exert a lot of force to turn them. Because the handle pivots slightly, the more pressure put on the handle to turn the wrench, the more the grip tightens.

pliers A hand tool with gripping jaws.

pop rivet gun A hand tool for installing pop rivets.

prick punch A pinch with a sharp point for accurately marking a point on metal.

pry bar A high-strength carbon steel rod with offsets for levering and prying.

pullers A generic term to describe hand tools that mechanically assist the removal of bearings, gears, pulleys, and other parts.

punches A generic term to describe a high-strength carbon steel shaft with a blunt point for driving. Center and prick punches are exceptions and have a sharp point for marking or making an indentation.

ratchet A generic term to describe a handle for sockets that allows the user to select direction of rotation. It can turn sockets in restricted areas without the user having to remove the socket from the fastener.

ratcheting box-end wrench A wrench with an inner piece that is able to rotate within the outer housing, allowing it to be repositioned without being removed.

ratcheting screwdriver A screwdriver with a selectable ratchet mechanism built into the handle that allows the screwdriver tip to ratchet as it is being used.

roll bar Another type of pry bar, with one end used for prying and the other end for aligning larger holes, such as engine motor mounts.

screw extractor A tool for removing broken screws or bolts.

single flare A sealing system made on the end of metal tubing.

sledgehammer A heavy hammer, usually with two flat faces, that provides a strong blow.

sliding T-handle A handle fitted at 90 degrees to the main body that can be slid from side to side.

snap ring pliers A pair of pliers for installing and removing snap rings or circlips.

socket An enclosed metal tube commonly with 6 or 12 points to remove and install bolts and nuts.

speed brace A U-shaped socket wrench that allows high-speed operation. Also called a speeder handle.

split ball gauge A measuring device used to accurately measure small holes.

square file A type of file with a square cross section.

steel hammer A hammer with a head made of hardened steel.

steel rule An accurate measuring ruler made of steel.

straight edge A measuring device generally made of steel to check how flat a surface is.

tap A term used to generically describe an internal thread-cutting tool.

taper tap A tap with a tapper; it is usually the first of three taps used when cutting internal threads.

tap handle A tool designed to securely hold taps for cutting internal threads.

telescoping gauge A gauge that expands and locks to the internal diameter of bores; a caliper or outside micrometer is used to measure its size.

thread file A type of file that cleans clogged or distorted threads on bolts and studs.

thread pitch The coarseness or fineness of a thread as measured by either the threads per inch or the distance from the peak of one thread to the next. Metric fasteners are measured in millimeters.

tin snips Cutting device for sheet metal, works in a similar fashion to scissors.

torque Twisting force applied to a shaft that may or may not result in motion.

torque specifications Supplied by manufacturers and describes the amount of twisting force allowable for a fastener or a specification showing the twisting force from an engine crankshaft.

torque wrench A tool used to measure the rotational or twisting force applied to fasteners.

triangular file A type of file with three sides so it can get into internal corners.

tube flaring tool A tool that makes a sealing flare on the end of metal tubing.

tubing cutter A hand tool for cutting pipe or tubing squarely.

V blocks Metal blocks with a V-shaped cutout for holding shafts while working on them. Also referred to as vee blocks.

vernier calipers An accurate measuring device for internal, external, and depth measurements that incorporates fixed and adjustable jaws.

wad punch A type of punch that is hollow for cutting circular shapes in soft materials such as gaskets.

warding file A type of thin, flat file with a tapered end.

wrenches A generic term to describe tools that tighten and loosen fasteners with hexagonal heads.

Review Questions

1. Specialty tools:
   a. should not be shared among technicians.
   b. should have tag-out practices for regular storage.
   c. should be put in the proper storage space after each use.
   d. can be used for purposes they were not designed for.
2. A set of safety practices and procedures that are intended to reduce the risk of technicians inadvertently using tools that have been determined to be unsafe is known as:
   a. lockout.
   b. shop policy.
   c. equipment storage procedure.
   d. PPE maintenance.
3. Which of these wrenches can be awkward to use once the nut or bolt has been loosened a bit?
   a. Open-end
   b. Flare nut
   c. Combination
   d. Box-end
4. Oil filters should be installed using:
   a. an oil filter wrench.
   b. a flare nut wrench.
   c. your hand.
   d. arc joint pliers.
5. When tightening lug fasteners, an impact wrench should:
   a. always be used.
   b. never be used.
   c. be used in difficult-to-access positions.
   d. never be used in high-torque settings.
6. Which of these would you use for a wider grip and a tighter squeeze on parts too big for conventional pliers?
   a. Diagonal cutting pliers
   b. External snap ring pliers
   c. Arc joint pliers
   d. Combination pliers
7. When a large chisel needs a really strong blow, use a:
   a. sledgehammer.
   b. hard rubber mallet.
   c. dead blow hammer.
   d. ball-peen hammer.
8. Which style of torque wrench is the simplest and least expensive?
   a. Clicker
   b. Dial
   c. Beam
   d. Electronic
9. The depth of blind holes in housings can best be measured using a:
   a. measuring tape.
   b. steel rule.
   c. dial bore gauge.
   d. vernier caliper.
10. All of the following are good practices *except*:
    a. replacing cleaning products when they lose their effectiveness.
    b. discarding solvents into the sewage system immediately after use.
    c. using chemical gloves when using any cleaning material.
    d. wearing a respirator when using toxic cleaning chemicals.

ASE Technician A/Technician B Style Questions

1. Tech A says that knowing how to use tools correctly creates a safe working environment. Tech B says that a flare nut wrench is used to loosen very tight bolts and nuts. Who is correct?
   a. Tech A
   b. Tech B
   c. Both A and B
   d. Neither A nor B
2. Tech A says that lockout is designed to secure a vehicle after all work on it has been completed. Tech B says that tag-out is used when a tool is no longer fit for use and identifies what is wrong with it. Who is correct?
   a. Tech A
   b. Tech B
   c. Both A and B
   d. Neither A nor B
3. Tech A says that torque wrenches need to be calibrated periodically to ensure proper torque values. Tech B says that when using a clicker-style torque wrench, keep turning the torque wrench ⅛–¼ turn to make sure the bolt is properly tightened. Who is correct?
   a. Tech A
   b. Tech B
   c. Both A and B
   d. Neither A nor B
4. Tech A says that when using a micrometer, a "standard" is used to hold the part you are measuring. Tech B says that the micrometer spindle should be firmly closed against the anvil prior to storage. Who is correct?
   a. Tech A
   b. Tech B
   c. Both A and B
   d. Neither A nor B

5. Tech A says that a box-end wrench is more likely to round the head of a bolt than an open-end wrench. Tech B says that 6-point sockets and wrenches will hold more firmly when removing and tightening bolts. Who is correct?
   a. Tech A
   b. Tech B
   c. Both A and B
   d. Neither A nor B
6. Tech A says that it is usually better to pull a wrench to tighten or loosen a bolt. Tech B says that pushing a wrench will protect your knuckles if the wrench slips. Who is correct?
   a. Tech A
   b. Tech B
   c. Both A and B
   d. Neither A nor B
7. Tech A says that a feeler gauge is used to measure the diameter of small holes. Tech B says that a feeler gauge and straight edge are used to check surfaces for warpage. Who is correct?
   a. Tech A
   b. Tech B
   c. Both A and B
   d. Neither A nor B
8. Tech A says that a dead blow hammer reduces rebound of the hammer. Tech B says that a dead blow hammer should be used with a chisel to cut the head of a bolt off. Who is correct?
   a. Tech A
   b. Tech B
   c. Both A and B
   d. Neither A nor B
9. Tech A says that gaskets can be removed quickly and safely with a hammer and sharp chisel. Tech B says that extreme care must be used when removing a gasket on an aluminum surface. Who is correct?
   a. Tech A
   b. Tech B
   c. Both A and B
   d. Neither A nor B
10. Tech A says that when using a file, apply pressure to file in the direction of the cut and no pressure when pulling the file back. Tech B says that file cards are used to file uneven surfaces. Who is correct?
    a. Tech A
    b. Tech B
    c. Both A and B
    d. Neither A nor B

CHAPTER 4

Fasteners and Thread Repair

NATEF Tasks

- **N04001** Perform common fastener and thread repair, to include: remove broken bolt, restore internal and external threads, and repair internal threads with thread insert.

Knowledge Objectives

After reading this chapter, you will be able to:
- **K04001** Identify threaded fasteners and describe their use.
- **K04002** Identify standard and metric fasteners.
- **K04003** Describe how bolts are sized.
- **K04004** Describe thread pitch and how it is measured.
- **K04005** Describe bolt grade.
- **K04006** Describe the types of bolt strength required.
- **K04007** Describe nuts and their application.
- **K04008** Describe washers and their applications.
- **K04009** Describe the purpose and application of thread-locking compounds.
- **K04010** Describe screws and their applications.
- **K04011** Describe the torque-to-yield and torque angle.
- **K04012** Describe how to avoid broken fasteners.
- **K04013** Identify situations where thread repair is necessary.

Skills Objectives

- **S04001** Perform common fastener and thread repair.

You Are the Automotive Technician

The new intern was helping you disassemble a leaking water pump. All was going well until one of the bolts broke off while removing it. It is evident that the shank of the bolt had been narrowed considerably due to rust and corrosion over time. You know that removing this bolt could turn the job into a real nightmare if not performed properly and carefully. It is also an opportunity to teach the intern how to perform this important task. As you are thinking about this task, several questions come to mind.

1. If part of the bolt extends past the surface it is threaded into, what can be done to try to remove it without using a bolt extractor?
2. What processes can be tried on bolts that are stuck, before they are broken off?
3. Is it okay to use a replacement bolt of a different grade from the original bolt? Why or why not?
4. What is the purpose of a thread-locking compound? Give an example of when should it be used?
5. What is the purpose of antiseize? Give examples of when it should and shouldn't be used?

Introduction

Fasteners come in two common types: threaded fasteners and non-threaded fasteners (**FIGURE 4-1**). Both types of **fasteners** are designed to secure parts. **Threaded fasteners** are primarily designed to clamp objects together. Non-threaded fasteners are designed to hold parts together, such as preventing a component from falling off a shaft by using C-clips, cotter pins, roll pins, or other retainers. They can also be used to clamp components together, for example, with rivets. Each type of fastener has a different purpose or application, so getting to know the various kinds will help you know how to disassemble and reassemble component assemblies. We cover the major types of threaded and non-threaded fasteners below.

Threaded Fasteners and Torque

K04001 Identify threaded fasteners and describe their use.

Threaded fasteners are designed to secure parts that are under various tension and sheer stresses. The nature of the stresses placed on parts and threaded fasteners depends on their use and location. For example, head bolts withstand tension stresses by clamping the head gasket between the cylinder head and block so that combustion pressures can be contained in the cylinder (**FIGURE 4-2**). The bolts must withstand the very high combustion pressures trying to push the head off the top of the engine block in order to prevent leaks past the head gasket. In this situation, the pressure is trying to stretch the head bolts, which means they are under tension.

Some fasteners must withstand sheer stresses, which are sideways forces trying to shear the bolt in two (**FIGURE 4-3**). An example of fasteners withstanding sheer stresses is wheel lug studs and lug nuts. They clamp the wheel assembly to the suspension system, and the weight of the vehicle tries to sheer the lug studs. If this were to happen, the wheel would fall off the vehicle, likely leading to an accident.

To accomplish their job, fasteners come in a variety of diameters for different sized loads, and hardnesses, which are defined in grades. Threaded fasteners are designed to be tightened to a specified amount depending on several factors such as the job at hand, the grade or hardness of the material they are made from, their size, and the thread type. If a fastener is over-tightened, it could become damaged or could break. If it is under-tightened, it could work loose over time. Torque is a measure of twisting force around an axis, so it is a way of defining how much a fastener should be, or is, tightened to. Manufacturers print **torque specifications** in torque charts for virtually every fastener on a vehicle.

The idea of threaded fasteners has been around since approximately 240 B.C., when Archimedes invented the screw conveyor. The screw design was adapted to a straight pin to attach different materials together. As time progressed, the materials that the screw was made from became stronger, and the sizing became more precise. Whitworth created a thread profile that had a rounded crest and rounded root in England in 1841 (**FIGURE 4-4**). Then in 1864, in the United States, Sellers created a thread profile that had a flat crest and flat root (**FIGURE 4-5**).

In order to create the threaded fasteners we use today, the flat crest and rounded root is used along with a 60-degree pitch angle (**FIGURE 4-6**). Metric and standard size bolts use the same shape of threads, excluding specialty-use bolts, which are made for a particular purpose, such as the lead screw for a vice or gear puller. The square edges of this specialty thread (**square thread**) tend to exert force more in line with the screw, where a V-thread tends to push at a 90-degree angle to the thread, not the screw (**FIGURE 4-7**).

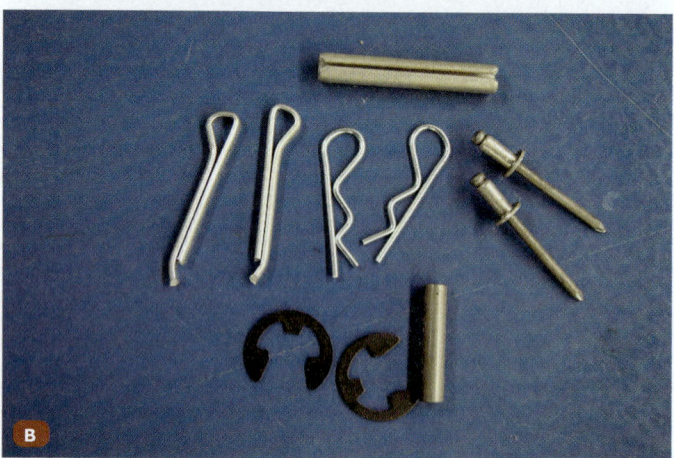

FIGURE 4-1 A. Threaded fasteners. **B.** Non-threaded fasteners.

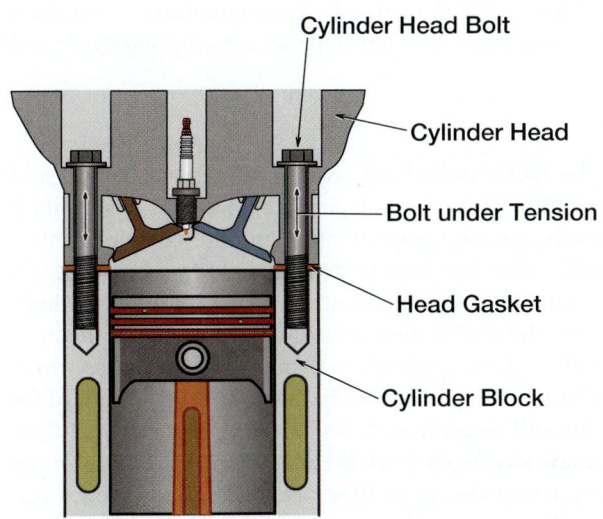

FIGURE 4-2 Bolts clamp parts together.

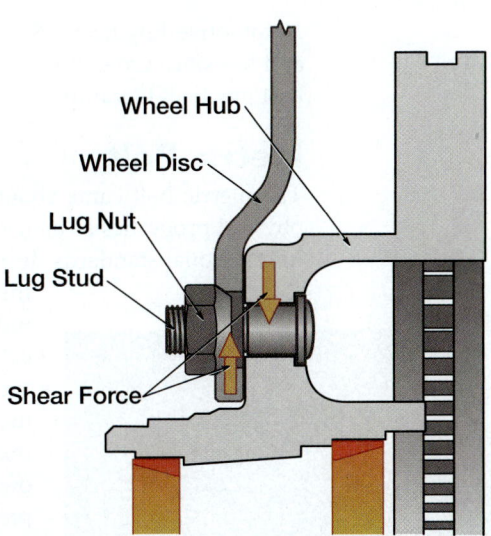

FIGURE 4-3 Bolt under sheer stresses.

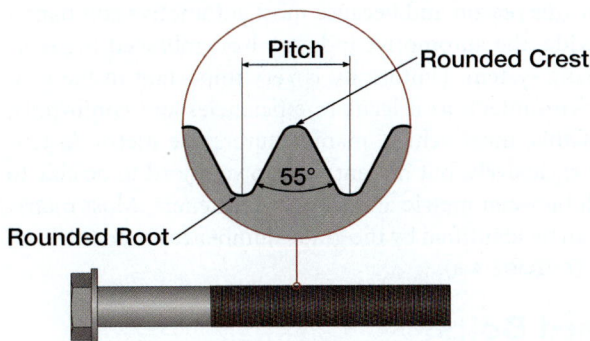

FIGURE 4-4 Whitworth thread.

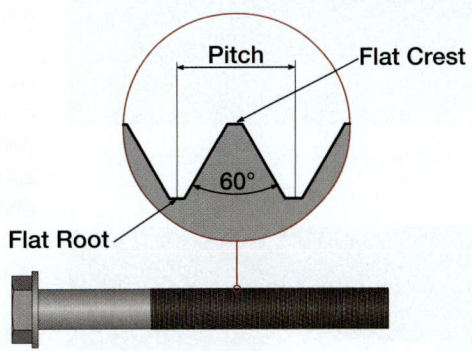

FIGURE 4-5 Sellers thread.

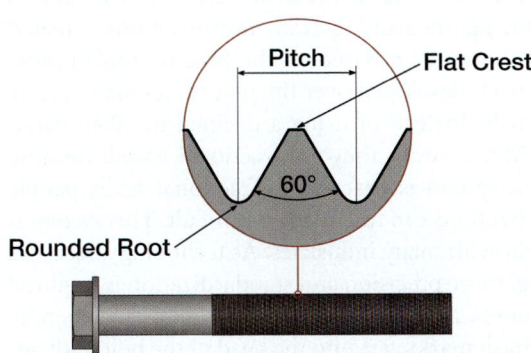

FIGURE 4-6 Current thread.

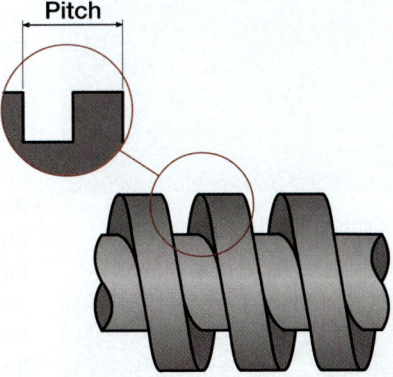

FIGURE 4-7 Square thread as used in a vice.

▶ Fastener Standardization

There are many different groups that monitor and set the standards that make the automotive industry conform to one universally recognized specification set. There are three main groups to know about. The American Society for Testing and Materials (ASTM) is a nongovernmental controlled group that tests and sets standards in all areas of industry. The International Organization for Standardization (ISO) is an independent developer of standards that are created to ensure a reliable, safe, and quality product. The Society of

K04002 Identify standard and metric fasteners.

Automotive Engineers (SAE) was initially started to standardize automotive production and has since grown to encompass a lot of engineering disciplines creating standards and best practices literature.

Metric Bolts

The metric bolt came about from the ISO standard 898, which defines mechanical and physical properties of metric fasteners. The ISO is an independent developer of voluntary international standards. In other words, any one government or group does not control them. The ISO develops standards that apply to many different industries in many different countries, using an unbiased logical approach. The metric decimal system has been around since the late 1600s and is sequential, so people can quickly see which number is larger. Many countries have adopted this standard for their weights and measure systems, but have also retained their own customary standards. Certain industries have required more precision in measurement, so they have adopted the metric system exclusively.

The automotive industry is one that deals in precise measurement and exact specifications in every aspect of automobile design. For this reason and because most automotive companies are worldwide, the automotive industry has embraced the metric measuring system. Uniformity is very important in the production environment, as it leads to efficiencies and conformity. Because of this, most vehicle manufacturers use metric fasteners almost exclusively, but not entirely. So you need to be able to distinguish between metric and standard fasteners. Most metric fasteners can be identified by the grade number cast into the head of the bolt (**FIGURE 4-8**).

FIGURE 4-8 Metric bolts can be identified by the grade number on the top of the bolt head.

FIGURE 4-9 Standard bolts can be identified by the hash marks on top of the bolt head.

FIGURE 4-10 Bolts, studs, and nuts.

Standard Bolts

The standard measurement of bolts is a combination of the Imperial and U.S. customary measurement systems. SAE standard J429 covers the Mechanical and Material Requirements that govern this standard for externally threaded fasteners. It combines the British measuring units and the U.S. measuring units created after the Revolutionary War. Based on the British units of measure, units have been developed over the past millennia to arrive at today's standards. Instead of using a decimal number–based system like the Metric units, they are fractional based. Because the measurement system is based on a fractional scale, people unfamiliar with fractions can find using it difficult. This system is falling out of favor with many industries. As technology becomes more demanding, more precision and standardization is required for both effectiveness and efficiency. Standard bolts can generally be identified by hash marks cast into the head of the bolt, indicating the bolt's grade (**FIGURE 4-9**).

▶ Bolts, Studs, and Nuts

Bolts, studs, and nuts are threaded fasteners designed for jobs requiring a fastener heavier than a screw and tend to be made of metal alloys, making them stronger (**FIGURE 4-10**). Typical **bolts** are cylindrical pieces of metal with a hexagonal head on one end and a thread cut into the shaft at the other end. The thread acts as an inclined plane; as the bolt is turned, it is drawn into or out of the mating thread. Hexagonal **nuts** thread onto the bolt thread. The hexagonal heads for the bolt and nut are designed to be turned by tools such as wrenches and sockets. Note that other bolt head designs are used as well. These require special shaped tools as covered in the Hand Tools chapter.

A **stud** does not have a fixed hexagonal head; rather, it has a thread cut on each end. It is threaded into one part where it stays. The mating part is then slipped over it, and a nut is threaded onto the other end of the stud to secure the part. Studs are commonly used to attach one component to another, such as a throttle body to the intake manifold. Studs can have different threads on each end, to work best with the material they are threaded into. Coarse threads work well in aluminum; in such cases, one end of the stud may use a coarse thread to grip the threads in an aluminum intake manifold and to position the throttle body. On the other end, there may be a fine thread for pulling everything together tightly with a steel nut. Bolts, nuts, and studs have either standard or metric threads. They are designated by their thread diameter, thread pitch, length, and grade. The diameter is measured across the outside of the threads, in fractions of an inch for standard-type fasteners, and millimeters for metric-type fasteners. So a 3/8" (9.5 mm) bolt has a thread diameter of 3/8" (9.5 mm). It is important to note that a 3/8" (9.5 mm) bolt does not have a 3/8" (9.5 mm) bolt head.

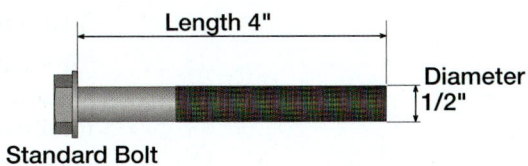
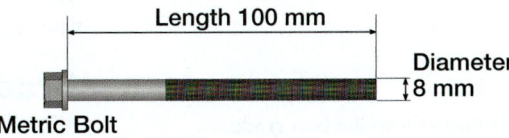

FIGURE 4-11 Bolt diameter and length.

Sizing Bolts

Bolts are sized either by metric or standard measuring systems. A ½" (12.7 mm) bolt does not mean that the bolt has a head that fits into a ½" (12.7 mm) socket. It means that the distance across the outside diameter of the bolt's threads measures ½" (12.7 mm) in diameter. The same goes for the metric equivalent; an 8 mm bolt is the diameter of the outside diameter of the bolt's threads. The length of a bolt is fairly straightforward. It is measured from the end of the bolt to the bottom of the head and is listed in inches or millimeters (**FIGURE 4-11**).

K04003 Describe how bolts are sized.

Thread Pitch

The coarseness of any thread is called its **thread pitch** (**FIGURE 4-12**). In the standard system, bolts, studs, and nuts are measured in threads per inch (TPI). To determine the TPI, simply count the number of threads there are in 1" (25.4 mm). Each bolt diameter in the standard system can typically have one of two thread pitches, **Unified National Coarse Thread (UNC)** or **Unified National Fine Thread (UNF)**. For example, a 3/8–16 is coarse, and a 3/8–24 is fine. In the metric system, the thread pitch is measured in millimeters by the distance between the peaks of the threads. So course threads have larger distances, and fine threads have smaller distances. Each bolt diameter in the metric system can have up to four thread pitches (**FIGURE 4-13**). Consult a metric thread pitch chart, because there is no clear pattern of thread pitches for metric fasteners. These charts can be found in tap and die sets or on the Internet.

K04004 Describe thread pitch and how it is measured.

FIGURE 4-12 UNC and UNF standard bolt thread pitch.

FIGURE 4-13 Metric thread pitches.

Thread Pitch Gauge

Thread pitch gauges make identifying the thread pitch on any bolt quick and easy. Without a thread pitch gauge, technicians are stuck measuring the number of threads per inch, or the distance from peak to peak in millimeters, with a ruler. Each thread pitch gauge matches a particular thread pitch that is stamped or engraved on the gauge. To use a thread pitch gauge, find the gauge that fits perfectly into the threads on the bolt you are measuring. Just make sure you keep the gauge parallel to the shaft of the bolt; otherwise, you will get a wrong match (**FIGURE 4-14**).

Grading of Bolts

K04005 Describe bolt grade.

What is bolt grading? Bolt grade or class means that the fastener meets the strength requirements of that grade or classification. There are a variety of grades, so it is important to use the specified grade of bolt for each component. When bolts are tested for grade, the bolt is put in a testing tool and abused to the point of breakage so that the maximum readings can be determined to properly grade the fastener. We need to know a bolt's grade so that we are able to use the correct fastener for the application and avoid bolt failures down the road. Standard bolts are typically grade 1 to grade 8, with grade 8 being the strongest. The grade can be identified by counting the hash marks on the head of the bolt and adding two to that number. For example, there are three hash marks on the head of a grade 5 bolt and six on the head of a grade 8 bolt (**FIGURE 4-15**).

Metric bolt grades are typically grade 4.6 to grade 12.9, with grade 12.9 being the strongest. In metric bolts, the grade number is cast into the top of the bolt head. These numbers on metric bolts have specific meanings. The number before the decimal indicates its tensile strength in megapascals (MPa), which is found by multiplying the number by 100. For example a metric 12.9-grade bolt would have a tensile strength of 1200 MPa. The number to the right of the decimal, when multiplied by 10 indicates the yield point as a percentage of the tensile strength (**FIGURE 4-16**).

Just because the higher grades are stronger doesn't mean you should use those in all applications. There are times when lower-graded bolts are more appropriate for particular applications, such as with some head bolts that need to have a little bit of give in them to allow for expansion and contraction of the cylinder head when the head gasket is clamped in place. Otherwise, the head gasket seal could fail. In other situations, using the wrong grade of bolt could cause either the bolt to fail or the threads in the threaded hole to fail. So you should always use the specified grade bolt when replacing bolts with new ones.

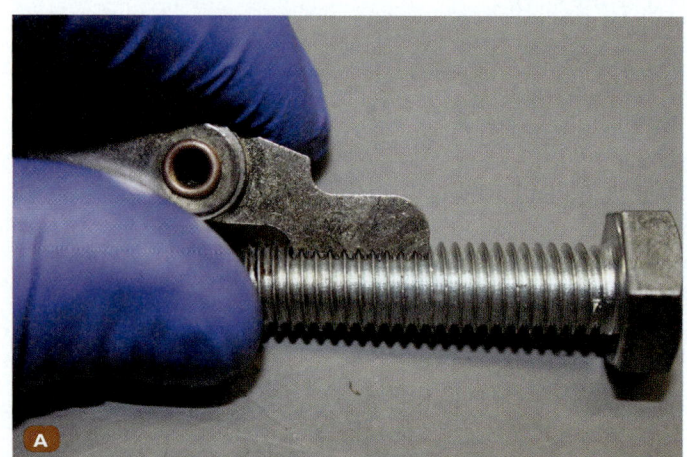

FIGURE 4-14 Thread pitch gauge being used. **A.** Standard. **B.** Metric.

FIGURE 4-15 A. Grade 5 standard bolt. **B.** Grade 8 standard bolt.

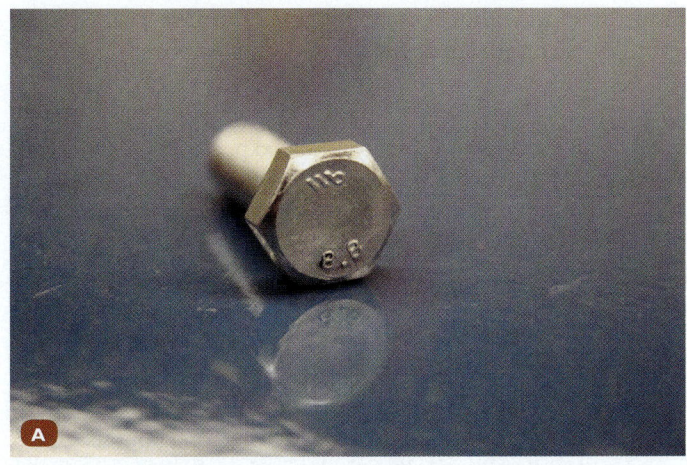

FIGURE 4-16 **A.** Grade 8.8 metric bolt **B.** Grade 10.9 metric bolt.

Strength of Bolts

Tensile Strength

Tensile strength is the maximum tension (applied load) the fastener can withstand without being torn apart or the maximum stress used under tension (lengthwise force) without causing failure. Tensile strength is determined by the strength of the material and the size of the stress area. When a typical threaded fastener fails in pure tension, it usually fractures through the threaded portion, as that is the weakest area because it is also the smallest area of the bolt (**FIGURE 4-17**). Generally speaking, the higher the grade of bolt, the higher the tensile strength.

K04006 Describe the types of bolt strength required.

FIGURE 4-17 Bolt failure (tension).

Shear Strength

Shear strength is defined as the maximum load that can be supported prior to fracture, when applied at a right angle to the fastener's axis. A load occurring in one transverse plane is known as single shear (**FIGURE 4-18**). Double shear is a load applied in two planes, where the fastener could be cut into three pieces. For most standard threaded fasteners, shear strength is not specified, even though the fastener may be commonly used in shear applications. Generally speaking, the higher the grade of bolt, the higher the shear strength.

Proof Load

The proof load represents the usable strength range for certain standard fasteners. By definition, the proof load is an applied

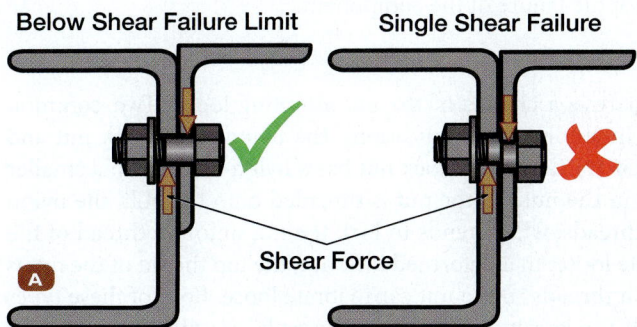
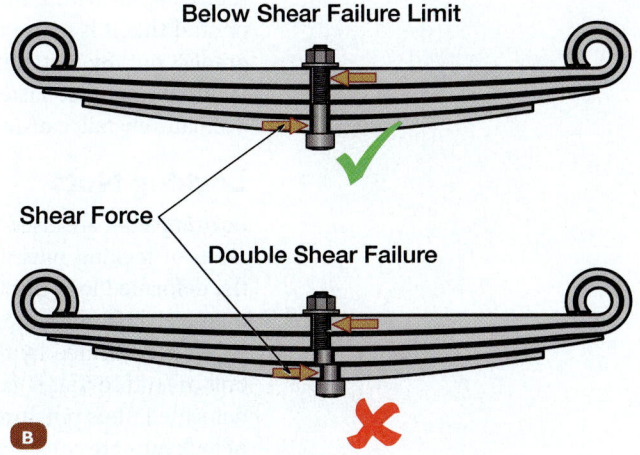

FIGURE 4-18 **A.** Single shear. **B.** Double shear.

tensile load that the fastener must support without exceeding the elastic phase, which is the point up to which the bolt returns to its original point when tension is removed.

Fatigue Strength

A fastener subjected to repeated cyclic loads can suddenly and unexpectedly break, even if the loads are well below the strength of the material. The repeated cyclic loading weakens the fastener over time, and it fails from fatigue. The fatigue strength is the maximum stress a fastener can withstand for a specified number of repeated cycles before failing. Connecting rod bolts are an example of a situation where fatigue strength would need to withstand millions of cyclic loads.

Torsional Strength

Torsional strength is a measure of a material's ability to withstand a twisting load, usually expressed in terms of torque, in which the fastener fails by being twisted off about its axis. A fastener that fails due to low torsional strength typically fails during installation or removal. Generally speaking, the higher the grade, the higher the torsional strength of the fastener.

Ductility

Ductility is the ability of a material to deform before it fractures. A material that experiences very little or no plastic deformation before fracturing is considered brittle. Think of a piece of flat glass: If you try to bend it, it breaks easily. This is an example of a material with low ductility material. Most automotive fasteners need to have some measure of ductility to avoid catastrophic failure. A reasonable indication of a fastener's ductility is the ratio of its specified minimum yield strength to the minimum tensile strength. The lower this ratio, the more ductile the fastener, meaning that the more ductile the bolt, the more it can flex or stretch without breaking. Generally speaking, the higher the grade, the lower the ductility of the fastener. So a balance between tensile strength and ductility is needed that depends on the requirements of the application.

Toughness

Toughness is defined as a material's ability to absorb impact or shock loading. Impact strength toughness is rarely a specified requirement for automotive applications. In addition to its specification for various aerospace industry fasteners, ASTM A320 specification for alloy steel bolting materials for low-temperature service is one of the few specifications that requires impact testing on certain grades.

Nuts

K04007 Describe nuts and their application.

Nuts are screwed onto the threads of a bolt to hold something together. They have internal threads that match the threads on the same size bolt. The many different kinds of nuts are mostly application specific. We cover some of the more popular ones below. When you replace a nut with a new nut on a vehicle, you need to match the nut to the grade of bolt or stud that it is going to be screwed onto. For example, if the bolt is a grade 8, you need a grade 8 nut. By matching the grade of fasteners, you are keeping the integrity of the manufacturer's intended fastening system. If you mix different grades of bolts and nuts, one could prematurely fail, causing catastrophic failure of the component.

Locking Nuts

Locking nuts are used when there is a chance of the nut vibrating loose. Two common types of locking nuts are used in automotive applications: the nylon insert lock nut and the deformed lock nut (**FIGURE 4-19**). The nylon lock nut has a nylon insert with a smaller diameter hole than the threads in the nut. As the nut is threaded onto the bolt, the nylon insert is deformed by the bolt threads, which tends to lock the nut onto the thread of the bolt or stud so that it can't vibrate loose. In a deformed lock nut, the top thread of the nut is deformed, thus pinching the bolt threads so the nut can't vibrate loose. Both of these types of lock nuts are considered single use, meaning they should be replaced with new ones once they have been removed.

FIGURE 4-19 **A.** Nylon lock nut. **B.** Deformed lock nut.

Castle Nuts

Castle nuts, sometimes called castellated nuts, are used with a cotter pin to keep them from moving once they are installed. Protrusions and notches are machined into the top of the nut. The notches then can be lined up with a hole in the threaded bolt or stud (**FIGURE 4-20**). A cotter pin passes through the notch on one side of the nut, then through the hole in the stud and out the notch in the other side of the nut. The cotter pin physically prevents the castle nut from backing off the stud. These nuts are usually used on suspension components or other critical components so that they can't loosen up. It is good practice to always use new cotter pins and visually check to verify if they are installed in the castle nuts before finishing up a job (**FIGURE 4-21**).

Specialty Nuts

In the automotive industry. there are lot of specialty nuts with special characteristics—for example, an extended shoulder, a special head size, or a special material—required for a particular application. You must use the correct nut for the correct application (**FIGURE 4-22**). Sheet metal nuts are a cheap alternative to a regular nut that manufactures are using on different components. Sometimes called J nuts, these nuts are folded

FIGURE 4-20 Castle nut used on a ball joint.

FIGURE 4-21 New cotter pins installed in the castle nuts.

FIGURE 4-22 A variety of specialty nuts.

K04008 Describe washers and their applications.

piece of sheet metal that fits over a hole in a piece of sheet metal; they are threaded to accept a fastener, and they usually do not need to be held as they are clipped to the component. Failure to use the proper fasteners for the proper uses usually results in failure of the component.

Washers

Why are washers used underneath bolt heads and nuts? There are three main purposes of a washer:

- To distribute the force exerted on the component that it is pressing against evenly
- To prevent the surface around the hole from being worn down due to tightening of the head of the bolt or nut against a surface
- To provide a measure of locking force to keep the bolt or nut from loosening due to vibration

Not all applications require washers that provide all three functions, so make sure you know what the application requires. When selecting a washer for a bolt or nut, make sure you use the correct type, size, and grade.

Grades of Washers

Just as with nuts, you want to match the grade of washer to the nut and bolt so that you can maximize the clamping force on the component (**FIGURE 4-23**). Using dissimilar metals that have different hardness characteristics can cause the surface of the softer metal to wear away. This reduces the clamping force over time, which can cause the bolt or component to fail. Not all washers have grade ratings on them, so be careful to select the proper one when replacing a damaged or missing washer with a new one.

Flat Washers

A flat washer is a piece of steel that has been cut in a circular shape with a hole in the middle for the bolt or stud to fit through. Washers are sized and paired with a metric or standard bolt so that the pressure created by torquing the bolt down is spread out on the component that it is being used on (**FIGURE 4-24**). They prevent marring of the surface of the component around the bolt hole. They also act as a bearing surface so that the bolt achieves the expected clamping force at the specified torque. Flat washers are graded just like bolts, which means you need to match the grade on the washer with the fasteners that you are using them on.

Lock Washers

Lock washers are made from spring steel and have a slit cut in them to allow for the spring action to hold tension against the nut so that the nut will not loosen. Also, the slit is cut at an angle, with sharp edges on the top and bottom to cut into the bottom of the bolt head and the surface of the component being clamped (**FIGURE 4-25**). The sharp edges bite into the bolt and component if the bolt tries to loosen. These are used with conventional nuts and are usually used in applications that create a lot of vibration. Lock washers are usually considered to be single use and should be replaced rather than reused.

Star Washers

Star washers, also called toothed lock washers, are used sometimes to stop the rotation of the nut above. They work in a similar way to lock washers in that the teeth are at angles so that they bite into the bottom of the bolt head and the surface being clamped (**FIGURE 4-26**). If these washers are required for your application, they need to be reinstalled before the nut is installed. They also should be replaced with new ones rather than reused.

FIGURE 4-23 Washer with its grade rating marked on it.

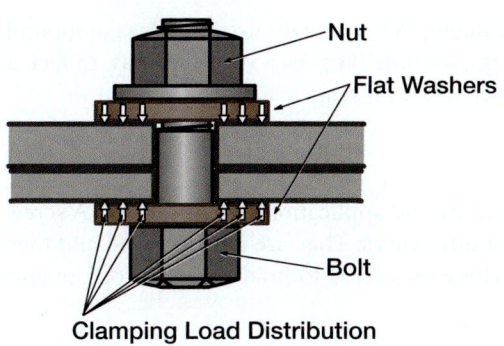

FIGURE 4-24 Flat washer spreading out the clamping force.

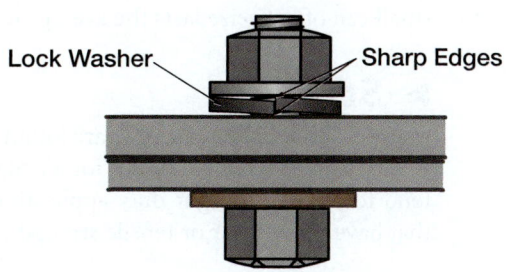

FIGURE 4-25 Lock washers prevent bolts and nuts from loosening. (Note: this nut and bolt is not fully tightened to show the action of the lock washer.)

FIGURE 4-26 Star washers prevent nuts and bolts from loosening.

▶ Threadlocker and Antiseize

Thread-locking compound is a liquid that is put on the threads of a bolt or stud to hold the nut that is threaded over it. The thread-locking compound acts like very strong glue. Once the compound between the nut and bolt cures, it bonds the two together so that they do not move. A popular thread-locking compound (threadlocker) that most automotive technicians prefer is Loctite®. The two main strengths of locking compound that are used in the automotive industry are blue and red (**FIGURE 4-27**). The blue allows for relatively easy removal of the nut or bolt with a socket and ratchet, whereas the red is much stronger. The red version usually requires a fair amount of heat to soften the compound so the nut or bolt can be removed. Thread-locking compound is a good safety item that technicians use on critical parts they do not want to come loose. In fact, the manufacturer may specify its use on certain fasteners, so be aware of situations that require thread-locking compound, and use the strength required.

Antiseize compound is the opposite of thread-locking compound as it keeps threaded fasteners from becoming corroded together or seized due to galling. One common use of antiseize compound is on black steel spark plug threads when the spark plug is installed into an aluminum cylinder head. Antiseize compound is a coating that prevents rust and provides lubrication so that the fastener can be removed in the future. Because of that, it should *not* be used on most fasteners in a vehicle; otherwise, they

K04009 Describe the purpose and application of thread-locking compounds.

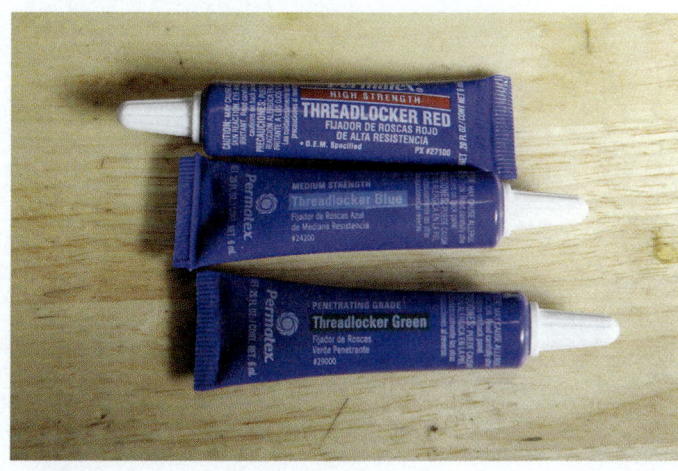

FIGURE 4-27 Thread-locking compound.

could vibrate loose, causing damage or an accident. When you do use antiseize compound on specified fasteners, use a very light coating, as a little bit goes a very long way. In fact, a small can of antiseize lasts the average technician several years.

▶ Screws

K04010 Describe screws and their applications.

Many different types of screws are found in automotive applications (**FIGURE 4-28**). A screw is very similar to a bolt except for a couple of differences: They are not hardened and they tend to be used in light-duty applications where they need to hold together components that have a low shear or tensile strength.

Machine Screws

Similar to a bolt, a machine screw is used to fasten components together, but these are usually driven by a screwdriver with a Philips or slotted head. Machine screws are usually driven into a nut or threaded hole that has been tapped to the thread pitch of the screw. As vehicles become more complex, manufacturers have started to introduce new types of driving tools for these screws. Torx, square, and Allen head screws are becoming prevalent in this fastener group (**FIGURE 4-29**).

Self-Tapping Screws

Self-tapping screws are designed to create their own holes in the material they are being driven into. They have a fluted tip so you can drill a hole into the base material, without needing a pilot hole (**FIGURE 4-30**). The threads on a self-tapping screw are very similar to those on a sheet metal screw as they grip onto the metal as you drill through it. Usually they have a cap screw–type head so they can be driven with a screw gun, but they can have Phillips, square, Allen, or Torx heads as well.

Trim Screws

Trim screws are used in applications where there is a need to hold plastic and metal ornamentation to the vehicle. They are also used to hold door panels, trim pieces, and other small components to the vehicle. They are basic screws that are usually painted either the trim color or black so that they blend in with the material that they are used with (**FIGURE 4-31**).

Sheet Metal Screws

Sheet metal screws are used to attach things to sheet metal. They have a chip on the tip of the screw to push away material as it

FIGURE 4-28 Variety of screws.

FIGURE 4-29 Machine screw heads.

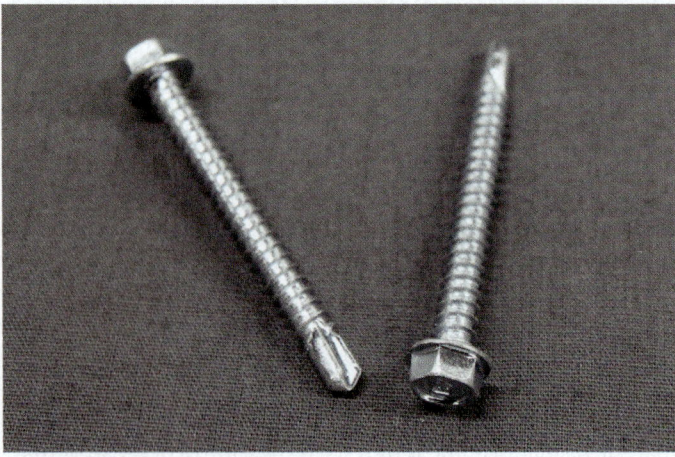

FIGURE 4-30 Self-tapping screws.

FIGURE 4-31 Trim screws.

FIGURE 4-32 Sheet metal screws.

comes off (**FIGURE 4-32**). These screws are usually threaded all the way up to the head so that they can be run down all the way, clamping the pieces together.

▶ Torque-to-Yield and Torque Angle

Torque is not always the best method of ensuring that a bolt is tightened enough as to give the proper amount of clamping force. If the threads are rusty, rough, or damaged in any way, the amount of twisting force required to tighten the fastener increases. Tightening the rusty fastener to a particular torque does not provide as much clamping force as a smooth fastener torqued the same amount. This also brings up the question of whether threads should be lubricated. In most automotive cases, the torque values specified are for dry, non-lubricated threads. But always check the manufacturer's specifications.

When bolts are tightened, they are also stretched. As long as they are not tightened too much, they will return to their original length when loosened. This is called **elasticity**. If they continue to be tightened and stretch beyond their point of elasticity, they will not return to their original length when loosened. This is called the **yield point**. As threaded fasteners are torqued, they go through the following phases:

- **Rundown Phase:** Free running fastener (may or may not have prevailing torque).
- **Alignment Phase:** Fastener and joint mating surfaces are drawn into alignment.
- **Elastic Phase:** *This is the third and final stage for normal bolts!* The slope of the torque/angle curve is constant. The fastener is elongated but will return to original length upon loosening.
- **Plastic or Yield Phase:** *Over-torqued condition for normal bolts. TTY bolts are tightened just into the beginning points of this phase.* Permanent deformation and elongation of the fastener and/or joint occur. Necking of the fastener occurs.

Torque-to-yield means that a fastener is torqued to, or just beyond, its yield point. With the changes in engine metallurgy that manufacturers are using in today's vehicles, bolt technology had to change also. To help prevent bolts from loosening over time and to maintain an adequate clamping force when the engine is both cold and hot, manufacturers have adopted **torque-to-yield (TTY) bolts**. TTY bolts are designed to provide a consistent clamping force when torqued to their yield point or just beyond. The challenge is that the torque does not increase very much, or at all, once yield is reached. So using a torque wrench by itself will not indicate the point at which the manufacturer wants the bolt tightened. TTY bolts generally require a new torquing procedure called torque angle. Also, it is important to note that in virtually all cases, TTY bolts cannot be reused because they have been stretched into their yield zone and would very likely fail if retorqued (**FIGURE 4-33**).

K04011 Describe the torque-to-yield and torque angle.

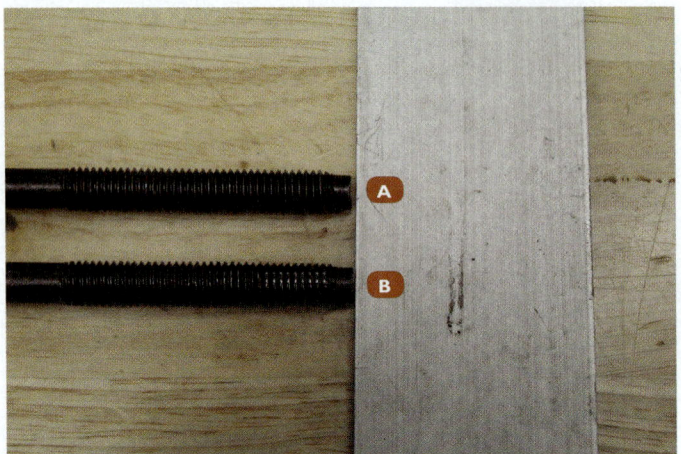

FIGURE 4-33 A. New TTY bolt. **B.** Used TTY bolt.

Torque angle is considered a more precise method to tighten TTY bolts and is essentially a multistep process. Bolts are first torqued in the required pattern, using a standard torque wrench to a required moderate torque setting. They are then further tightened one or more additional specified angles (torque angle) using an angle gauge, thus providing further tightening, which tightens the bolt to, or just beyond, their yield point. In some cases, after the initial torquing, the manufacturer first wants all of the bolts to be turned to an initial angle and then turned an additional angle. And in other cases, the manufacturer wants all of the bolts torqued in a particular sequence, then detorqued in a particular sequence, then retorqued once again in a particular sequence, and finally tightened an additional specified angle. So always check the manufacturer's specifications and procedure before torquing TTY bolts.

To use a torque angle gauge in conjunction with a torque wrench, follow the steps in **SKILL DRILL 4-1**.

SKILL DRILL 4-1 Using a Torque Angle Gauge with a Torque Wrench

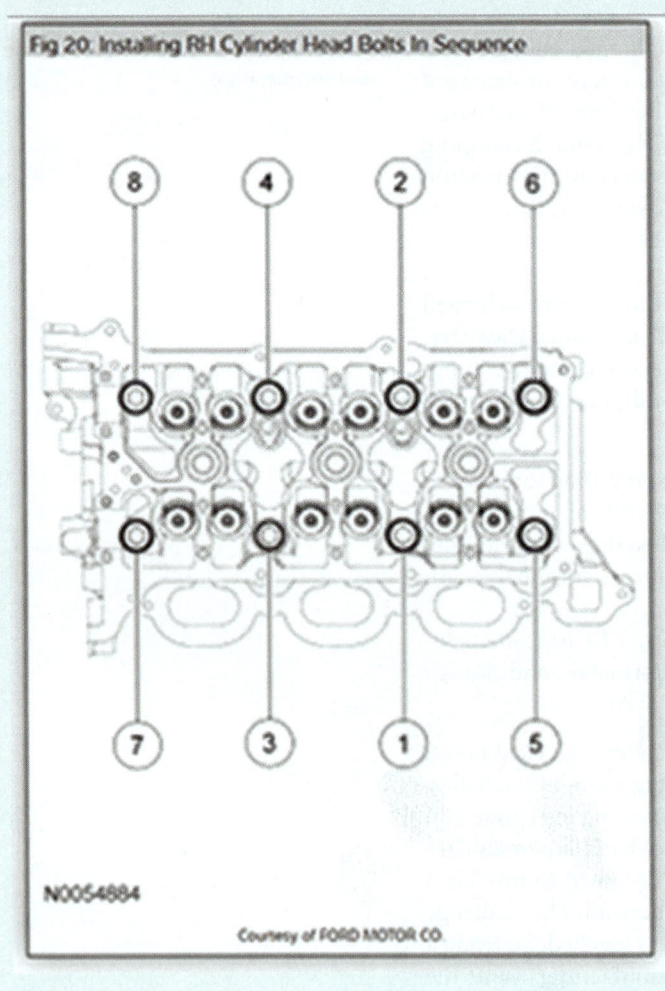

1. Check the specifications. Determine the correct torque value (ft-lb or N·m) and sequence for the bolts or fasteners you are using. Also, check the torque angle specifications for the bolt or fastener, and whether it involves one step or more than one step.

Continued

Torque-to-Yield and Torque Angle   127

2. Tighten the bolt to the specified torque. If the component requires multiple bolts or fasteners, make sure to tighten them all to the same torque value in the sequence and steps that are specified by the manufacturer.

3. Install the torque angle gauge over the head of the bolt, and then put the torque wrench on top of the gauge and zero it, if necessary.

4. Turn the torque wrench the specified number of degrees indicated on the angle gauge.

5. If the component requires multiple bolts or fasteners, make sure to tighten them all to the same torque angle in the sequence that is specified by the manufacturer. Some torquing procedures could call for four or more steps to complete the torquing process properly.

K04012 Describe how to avoid broken fasteners.

How to Avoid Broken Fasteners

Broken bolts can turn a routine job into a nightmare job by adding hours dealing with the broken bolt. In most cases, it is worth taking a bit of extra time doing whatever is needed to avoid breaking a bolt. One of the first things to do is to make sure that you are not dealing with a left-handed bolt (**FIGURE 4-34**). Left-hand bolts loosen by turning them clockwise. So trying to loosen them by turning them counterclockwise will actually tighten them, making it likely you will break them off.

Another cause of broken bolts is fasteners that are rusted or corroded in place. Although it is impossible to prevent all broken bolts in this case, there are several things you can do to minimize the percentage of bolts you break. One of the first things to do is use a good penetrating oil. Penetrating oil actually penetrates the rust between the nut and bolt and breaks it down, making it easier to break the nut or bolt loose. Once the penetrating oil has been given time to soak into the threads of the fastener, you may attempt, with the proper tool, to try to loosen the fastener. Note that you may want to try both loosening and tightening the bolt to help break it loose. It may be good to try using an impact wrench as well, because the hammering effect might assist in breaking it loose. Just try it a little at a time in each direction, and not with too much force.

If the fastener is still not loosening, you may need to heat up the component to help break up the rust or corrosion. The heat not only expands the metal components, it also expands the space between the nut and bolt. This also helps break the nut or bolt loose. The bolt or nut can be heated up in two primary ways. The first is with an oxyacetylene torch, but be careful because the flame from the torch also burns everything in its path. When heating up a bolt on a vehicle, there are almost always other things in the way that could be affected by the heat. The second is with an inductive heater, which uses electrical induction to heat up any ferrous metal that is placed in the inductive coil (**FIGURE 4-35**). The induction heater is preferred by many technicians for many applications, as it doesn't have a flame and it only heats up what is in the induction coil. Most induction heaters come with several sizes of induction coils that make them handy for most applications.

When heating a fastener to remove it, you usually heat it up to a dull orange color and then let it cool a bit before attempting to remove it. Once a heated fastener is removed, it must be replaced with a new bolt because the heat alters the strength of the fastener.

N04001 Perform common fastener and thread repair, to include: remove broken bolt, restore internal and external threads, and repaire internal threads with thread insert.

K04013 Identify situations where thread repair is necessary.

S04001 Perform common fastener and thread repair.

Thread Repair

Thread repair is used in situations where it is not feasible to replace a damaged thread. This may be because the damaged thread is located in a large expensive component, such as the engine block or cylinder head of a vehicle, or because replacement parts are expensive or not available. The aim of thread repair is to restore the thread to a condition that restores

FIGURE 4-34 Left-hand lug nuts are installed on some vehicles.

FIGURE 4-35 Heating a stuck nut with an induction heater.

the fastening integrity (**FIGURE 4-36**). It can be performed on internal threads, such as in a housing, engine block, or cylinder head, or on external threads, such as on a bolt. However, it is usually easier to replace a bolt if the threads become damaged than repairing it.

Types of Thread Repair

Many different tools and methods can be used to repair a thread. The least invasive method is to reshape the threads. If the threads are not too badly damaged, such as when the end thread is slightly damaged from the bolt being started crooked (cross-threaded), then a thread file can be used to clean them up, or a restoring tool can be used to reshape them (**FIGURE 4-37**). Each thread file has eight different sets of file teeth that match various thread pitches. Select the set that matches the bolt you are working on, and file the bolt in line with the threads. The file removes any distorted metal from the threads. Only file until the bad spot is reshaped.

The thread-restoring tool looks like an ordinary tap and die set, but instead of cutting the threads, it reshapes the damaged portion of the thread (**FIGURE 4-38**). Threads that have substantial damage require other methods of repair.

A common method for repairing damaged internal threads is a thread insert. A number of manufacturers make thread inserts, and they all work in a similar fashion. There are two main types of thread inserts: helical thread inserts and sleeve-type thread inserts (**FIGURE 4-39**). A helical thread insert of the same internal thread pitch is used when the threads need to be replaced with the same size thread. The damaged threaded hole is drilled out bigger, tapped with a special tap to a larger size that matches the outside size of the helical insert, and then the threaded coil insert is screwed into place. The insert provides a brand new internal thread that matches the original size. A common name for this type is a HeliCoil®.

The solid sleeve type of insert also replaces the damaged thread with the same size thread as the original. The difference between this type of insert compared to the helical insert is that this one is a slightly thicker insert that resembles a sleeve or bushing, but with threads on both the inside and outside. It may also have a small flange at the top of the insert that prevents it from being installed too deeply in the hole. Like the previous insert, the damaged hole needs to be drilled out to the proper oversize. Then a special tap is used to cut new threads in the hole that match the external thread on the insert. The insert is then installed in the hole and locked into place with either thread-locking compound or keys that get driven down into the base metal. A common name for this type is a Time-Sert.

Self-tapping inserts are a type of insert that is driven in a similar way to a self-tapping screw, making its own threads and at the same time locking the insert into the base material. Bushing inserts are a type of threaded insert that is pressed into a blind hole with an arbor press. These types of inserts can be inserted in any material, so you must be aware of material that is soft as it may break if too much force is put to it. It has internal threads, so bolts and screws can be threaded into them.

The process of repairing a thread should first start with attempting to remove the broken bolt without damaging the threads. If this can be accomplished, then you will likely save time as well as avoid the possibility of making the problem worse, such as breaking off an

FIGURE 4-36 Repaired thread.

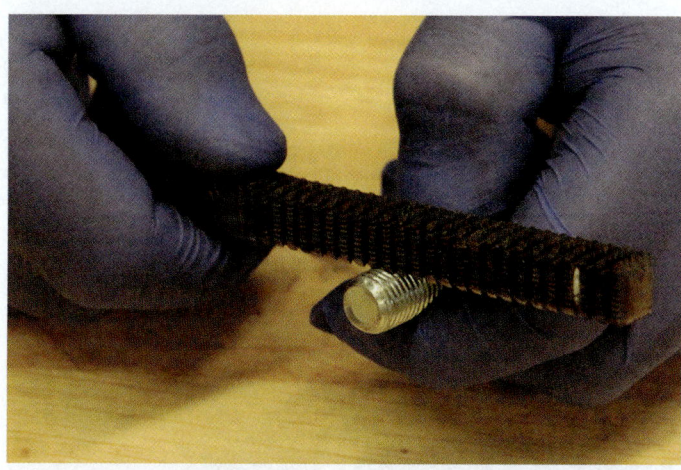
FIGURE 4-37 Threads being cleaned up with a thread file.

FIGURE 4-38 Thread-restoring tool being used.

CHAPTER 4 Fasteners and Thread Repair

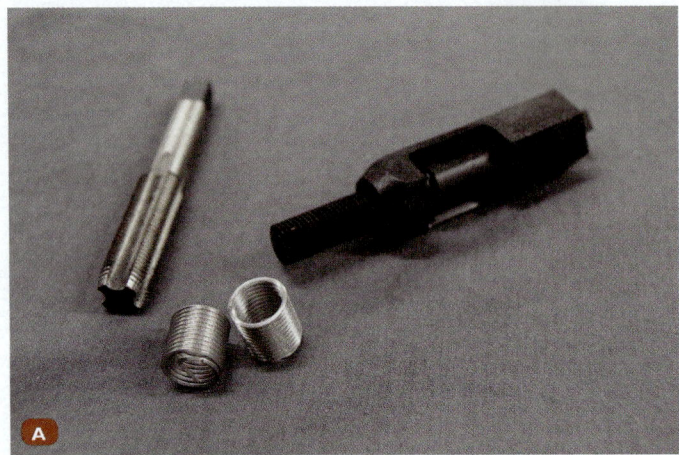

FIGURE 4-39 Thread inserts. **A.** Heli-Coil. **B.** Time-Sert.

easy out in the broken bolt. If the bolt can't be removed without damaging the threads, then the use of a thread insert to repair an internal thread is probably needed.

To remove a broken bolt, inspect the site. If enough of the bolt is sticking out of the surface, then a pair of pliers or locking pliers may be enough to turn and remove the bolt. The author has had quite a bit of luck using a pair of special curved jaw Channellock pliers. The curved jaw tends to get a better grip on the broken-off bolt than regular straight jaw Channellock pliers (**FIGURE 4-40**). Using penetrant or heat may help coax the bolt out. If the bolt is broken off flush with the surface, and the bolt is large enough in diameter, then you may be able to use a small center punch to turn the bolt by tapping on the outside diameter of the bolt, but in the reverse direction (**FIGURE 4-41**). If that doesn't work, or the bolt is too small in diameter, then a screw or bolt extraction tool can be tried.

Use a center punch to mark the center of the bolt to assist in centering the hole to drill. Select the correctly sized **screw extractor**, and drill the designated hole size in the center of the broken bolt to accommodate the extractor. Once the hole is drilled, insert the extractor and turn it counterclockwise (**FIGURE 4-42**).

FIGURE 4-40 Using pliers to remove a broken bolt.

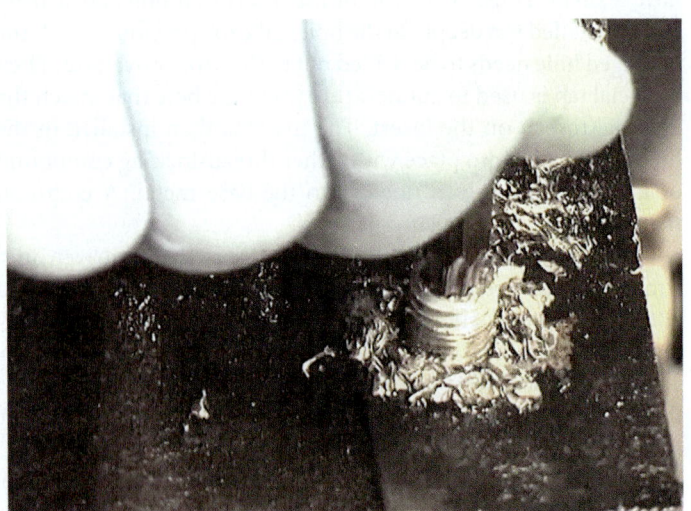

FIGURE 4-41 Removing a broken bolt with a center punch or cold chisel.

FIGURE 4-42 Removing a bolt with a screw extractor.

The flutes on the extractor should grab the inside of the bolt and hopefully enable you to back it out. Be careful not to exert too much force on the extractor if the bolt is extremely stuck in place. If the extractor breaks, it is almost impossible to remove it, as it is made of hardened steel that cannot be cut by most drill bits. Once the broken bolt is removed, run a lubricated tap or thread-restoring tool of the correct size and thread pitch through the hole, to clean up any rust or damage. If the screw extractor can't remove the broken bolt, then it will need to be drilled out. If the internal treads are damaged during the removal process, then the thread will have to be repaired with a thread insert, as described earlier.

To conduct thread repair, follow the steps in **SKILL DRILL 4-2**.

SKILL DRILL 4-2 Conducting Thread Repair

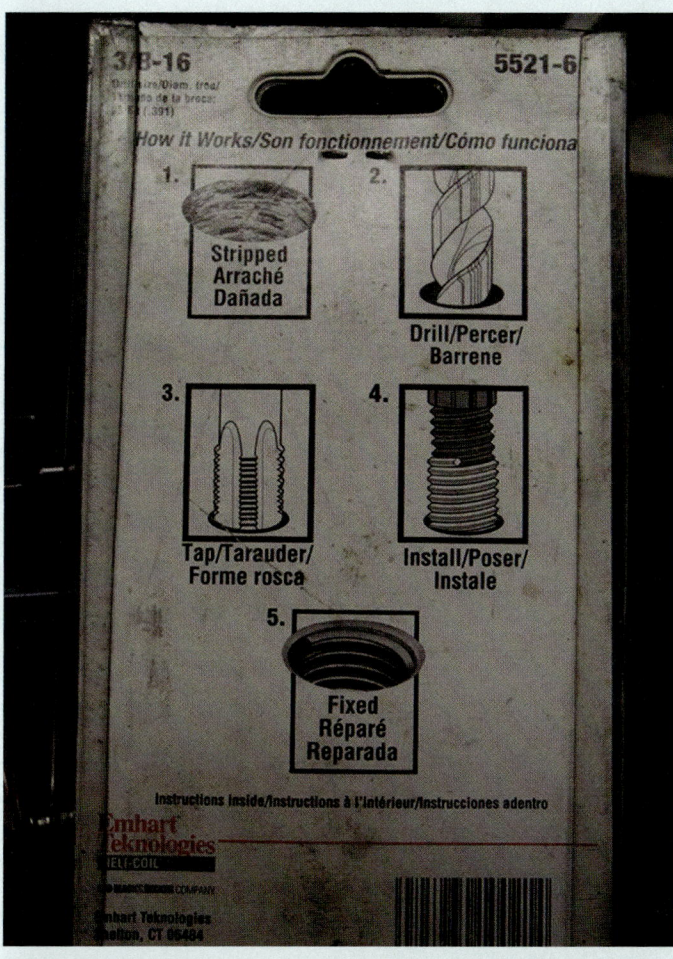

1. Always refer to the manufacturer's manual for specific operating instructions.

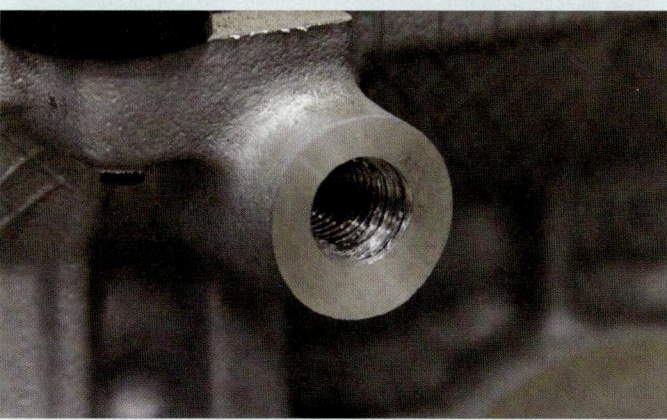

2. Inspect the condition of the threads, and determine the repair method.

Continued

132 CHAPTER 4 Fasteners and Thread Repair

3. Determine the type and size of the thread to be repaired. Thread pitch gauges and vernier calipers may be used to measure the thread.

4. Prepare materials for conducting the repair: dies and taps or a drill bit and drill; cutting oil, if required; and inserts.

5. Select the correctly sized tap or die if conducting a minor repair. Run the die or tap through or over the thread; be sure to use cutting lubricant.

6. If using inserts, select the correctly sized insert, based on the original bolt size. Drill the damaged hole, ensuring the drill is in perfect alignment with the hole.

Continued

7. Cut the new thread, using the proper size tap. Make sure you use cutting lubricant if required.

8. Using the insert-installing tool, install the insert by screwing it into the newly cut threads. Make sure the insert is secure or locked into the hole using the method specified by the manufacturer. Some inserts use a locking tab, whereas others use a liquid thread locker that hardens and holds the insert in place.

9. Test the insert to ensure it is secure and that the bolt will screw all the way in.

Wrap-Up

Ready for Review

- Threaded fasteners include bolts, studs, and nuts, and are designed to secure vehicle parts under stress.
- Torque defines how much a fastener should be tightened.
- Bolts, nuts, and studs use threads to secure each part; these threads can be in standard or metric measures.
- Thread pitch refers to the coarseness of the thread; bolts, nuts, and studs are measured in threads per inch (TPI), classified as Unified National Coarse Thread (UNC) or Unified National Fine Thread (UNF).
- Fasteners are graded by tensile strength (how much tension can be withstood before breakage).
- The SAE rates fasteners from grade 1 to grade 8; always replace a nut or bolt with one of the same grade.
- Torque specification indicates the level of tightness each bolt or nut should be tightened to; torque charts list torque specifications for nuts and bolts.
- Torque (or tension) wrenches tighten fasteners to the correct torque specification.
- Torque value—the amount of twisting force applied to a fastener by the torque wrench—is specified in foot-pounds (ft-lb), inch-pounds (in-lb), or newton meters (N·m).
- Bolts that are tightened beyond their yield point do not return to their original length when loosened.
- Torque-to-yield (TTY) bolts can be torqued just beyond their yield point, but should not be reused.
- Torque angle can be used to tighten TTY bolts and requires both a torque wrench and an angle gauge.
- Thread repair is performed to restore fastening integrity to a damaged fastener.
- Threads can be reshaped with a file, or a thread insert may be used.

Key Terms

bolt A type of threaded fastener with a thread on one end and a hexagonal head on the other.

coarse (UNC) Used to describe thread pitch; stands for Unified National Coarse.

elasticity The amount of stretch or give a material has.

fasteners Devices that securely hold items together, such as screws, cotter pins, rivets, and bolts.

fine (UNF) Used to describe thread pitch; it stands for Unified National Fine.

nut A fastener with a hexagonal head and internal threads for screwing on bolts.

screw extractor A tool for removing broken screws or bolts.

square thread A thread type with square shoulders used to translate rotational to lateral movement.

stud A type of threaded fastener with a thread cut on each end rather than having a bolt head on one end.

tensile strength In reference to fasteners, the amount of force it takes before a fastener breaks.

threaded fasteners Bolts, studs, and nuts designed to secure parts that are under various tension and sheer stresses. These include bolts, studs, and nuts, and are designed to secure vehicle parts under stress.

thread pitch The coarseness or fineness of a thread as measured by either the threads per inch or the distance from the peak of one thread to the next. Metric fasteners are measured in millimeters.

thread repair A generic term to describe a number of processes that can be used to repair threads.

torque Twisting force applied to a shaft, which may or may not result in motion.

torque angle A method of tightening bolts or nuts based on angles of rotation.

torque specifications Supplied by manufacturers and describes the amount of twisting force allowable for a fastener or a specification showing the twisting force from an engine crankshaft.

torque-to-yield A method of tightening bolts close to their yield point or the point at which they will not return to their original length.

torque-to-yield (TTY) bolts Bolts that are tightened using the torque-to-yield method.

torque wrench A tool used to measure the rotational or twisting force applied to fasteners.

yield point The point at which a bolt is stretched so hard that it will not return to its original length when loosened; it is measured in pounds per square inch of bolt cross section.

Review Questions

1. Which of these are threaded fasteners?
   a. C-clips
   b. Trim screws
   c. Cotter pins
   d. Solid rivets
2. All of the following statements are true *except*:
   a. Metric bolts can be identified by the grade number on the top of the bolt heads.
   b. Measuring units of standard bolts use a fractional number–based system.
   c. Standard bolts can be identified by the hash marks on the top of the bolt heads.
   d. Measuring units of metric bolts use a roman number–based system.
3. In an 8 mm bolt, 8 mm refers to the:
   a. length of the bolt.
   b. size of the bolt head.
   c. threads' outer diameter.
   d. number of threads per inch.

4. The easiest way to identify the TPI is to:
   a. use a vernier caliper.
   b. count the number of threads.
   c. measure distance from peak to peak.
   d. use a thread pitch gauge.
5. A bolt has 3 hash marks on its head. What is its grade?
   a. 2
   b. 5
   c. 6
   d. 8
6. Which nut is made with a deformed top thread?
   a. Lock nut
   b. J nut
   c. Castle nut
   d. Specialty nut
7. When you want a nut or bolt to hold tight such that it needs heat to be removed with a socket and ratchet, you should use a(n):
   a. lubricant.
   b. blue thread-locking compound.
   c. antiseize compound.
   d. red thread-locking compound.
8. Which of these has a fluted tip to drill a hole into the base material so that there is no need for a pilot hole?
   a. Machine screw
   b. Self-tapping screw
   c. Trim screw
   d. Sheet metal screw
9. Torque-to-yield bolts:
   a. should be tightened using torque angle.
   b. can be reused multiple times.
   c. cannot be torqued beyond the elastic phase.
   d. do not get stretched.
10. All of the following are ways to avoid breaking a bolt when removing it *except*:
    a. using penetrating oil with bolts that are rusted or corroded in place.
    b. applying maximum force in both directions.
    c. first checking whether it is a left-hand bolt.
    d. as a last resort, you may heat up the bolt that is rusted or corroded in place.

ASE Technician A/Technician B Style Questions

1. Tech A says that most metric fasteners can be identified by the grade number cast into the head of the bolt. Tech B says that standard bolts can generally be identified by hash marks cast into the head of the bolt indicating the bolt's grade. Who is correct?
   a. Tech A
   b. Tech B
   c. Both A and B
   d. Neither A nor B
2. Two technicians are discussing bolt sizes. Tech A says that the diameter of a bolt is measured across the flats on the head. Tech B says that the length of a bolt is measured from the top of the head to the bottom of the bolt. Who is correct?
   a. Tech A
   b. Tech B
   c. Both A and B
   d. Neither A nor B
3. Tech A says that a standard grade 5 bolt is stronger than a standard grade 8 bolt. Tech B says that a metric grade 12.9 bolt is stronger than a metric grade 4.6 bolt. Who is correct?
   a. Tech A
   b. Tech B
   c. Both A and B
   d. Neither A nor B
4. Tech A says that torque-to-yield head bolts are tightened to, or just past, their yield point. Tech B says that the yield point is the torque at which the bolt breaks. Who is correct?
   a. Tech A
   b. Tech B
   c. Both A and B
   d. Neither A nor B
5. Tech A says that some locking nuts have a nylon insert that has a smaller diameter hole than the threads in the nut. Tech B says that castle nuts are used with cotter pins to prevent the nut from loosening. Who is correct?
   a. Tech A
   b. Tech B
   c. Both A and B
   d. Neither A nor B
6. Tech A says that antiseize compound is a type of thread-locking compound. Tech B says that thread-locking compounds acts like very strong glue. Who is correct?
   a. Tech A
   b. Tech B
   c. Both A and B
   d. Neither A nor B
7. Tech A says that flat washers act as a bearing surface, making it so that the bolt achieves the expected clamping force at the specified torque. Tech B says that lock washers are usually considered single use and should be replaced rather than reused. Who is correct?
   a. Tech A
   b. Tech B
   c. Both A and B
   d. Neither A nor B
8. Tech A says that a common method for repairing damaged internal threads is a thread insert. Tech B says that a thread file is used to repair the threads on the inside of a bolt hole. Who is correct?
   a. Tech A
   b. Tech B
   c. Both A and B
   d. Neither A nor B

9. Tech A says that if a bolt breaks off above the surface, you might be able to remove it using locking pliers or curved jaw Channellock pliers. Tech B says that a hammer and punch can sometimes be used to remove a broken bolt. Who is correct?
   a. Tech A
   b. Tech B
   c. Both A and B
   d. Neither A nor B

10. Tech A says that when using a Heli-Coil to repair a damaged thread, the Heli-Coil is welded into the bolt hole. Tech B says that when using a solid sleeve type of insert, the damaged threads need to be drilled out with the correct size drill bit. Who is correct?
    a. Tech A
    b. Tech B
    c. Both A and B
    d. Neither A nor B

CHAPTER 5
Principles of Electrical Systems

NATEF Tasks

There are no NATEF tasks for this chapter.

Knowledge Objectives

After reading this chapter, you will be able to:

- K05001 Describe the basic principles of electricity and the units of voltage resistance and current flow.
- K05002 Describe common electrical terms and their application.
- K05003 Describe basic operation of electrical circuits.
- K05004 Describe common semiconductors, how they work, and their use.
- K05005 Identify the difference between AC and DC.
- K05006 Identify power, and ground sources.
- K05007 Diagnosis open circuits, short circuits, and high resistance.

Skills Objectives

There are no Skills Objectives for this chapter.

You Are the Technician

The customer has complained that the fog lights no longer illuminate when turned on. Checking for a blown fuse is one of the first steps in diagnosing an electrical problem on any system in a car. The fuse box on the vehicle you are working on today is located on the passenger's side cabin. The inside cover of the fuse box identifies the location of the fog light fuse. You test it with a digital volt-ohm meter (DVOM) and find battery power on only one side of the fuse, which indicates that it is blown. You then disconnect the wiring harness connectors from both fog lights and measure the resistance at the harness connector, between the input wire and the frame of the vehicle. The DVOM reads 0.2 ohms, indicating a short circuit to ground. You see that the vehicle has had a winch installed recently. As you visually inspect the area, you see a bundle of wires pinched between the winch bracket and the vehicle's frame. Loosening the winch bracket and removing the wires causes the ohmmeter to read an expected OL. To solve the problem permanently you will need to perform wire repairs on each of the damaged wires.

1. What caused the fuse to blow?
2. What other types of circuit protection devices do manufacturers use?
3. What is the process for repairing wires in a harness?

Introduction

The application of electrical principles in the repair of modern vehicles has become increasingly important for vehicle technicians as the electrical and electronic complexity of vehicles has increased. As hydrocarbons become scarcer and more expensive, the increased use of sophisticated electrical and electronic systems in vehicles to improve efficiency and economy will continue into the future. This trend is supported by the increasing popularity of hybrid vehicles and the investment by manufacturers in future technology, such as electric vehicles and fuel cell technology. To work on current and future vehicles, it is increasingly important for the technician to have a sound understanding of electrical terminology, the behavior of electricity, and electrical component and circuit theory. The technician also needs to be able to read wiring diagrams, measure electrical quantities in the shop, calculate electrical values, and understand the quantities used. Your success as a technician will depend upon your ability to apply these electrical principles and understand how they relate to the operation of virtually every vehicle system.

Electrical Fundamentals

K05001 Describe the basic principles of electricity and the units of voltage resistance and current flow.

Understanding the behavior of electricity can be more difficult than understanding other concepts such as four-stroke engine theory or the operation of disc brakes. For all intents and purposes, electricity cannot be seen, so you will have to use your imagination and visualize what it is doing. At the same time, electricity is governed by the laws of physics, so learning how electricity behaves can be approached in a logical manner, as with any other science. If you apply yourself, over time it will make more and more sense.

In this chapter, we will explore the various types of circuits and how electricity behaves within each type. We will also explore electrical components and see how they either use electricity to perform various types of work or control electricity so that it can be applied to various devices in an appropriate manner.

To get started, it is helpful to know that electricity is made up of tangible objects. Even if we cannot normally see these objects with our eyes, we can imagine them in our minds. In fact, it may be helpful to think of electricity as nothing more than the movement of specific particles from one point to another. Imagine a line of marbles rolling through a tube or drops of water through a pipe. The moving marbles or drops of water can perform work if they are directed against another object with force. In the same way, electricity can perform work if it is directed at objects that can extract energy from the moving particles, such as lights and electric motors. This is one way in which electricity works its magic. The moving electrical particles carry a negative charge and are attracted to a positive charge or repelled by a negative charge. These attractive and repelling forces are what cause the particles to move and perform work. As we continue, just remember that electricity is the movement of particles from one place to another. The fascinating way it does this will be explained next.

Basic Electricity

All questions about the nature of electricity lead to the composition of matter. All matter is made up of atoms (**FIGURE 5-1**). Every atom has a nucleus, with at least one positively charged particle called a proton and, in most atoms, at least one neutron, or uncharged particle. Moving around the nucleus are one or more negatively charged particles called electrons. Electrons travel in different rings (called valence rings) or shells around the nucleus. Each ring or shell can contain a specific maximum number of electrons. Any additional electrons must fit into the next higher ring or shell. With equal numbers of protons and electrons, the charges within an atom cancel each other out, leaving the atom with no overall charge. In this state, the electrons and protons are content to stay in the atom as they are.

An atom with more electrons than protons has an overall negative charge and is called a negative ion (**FIGURE 5-2**). Ion simply means the

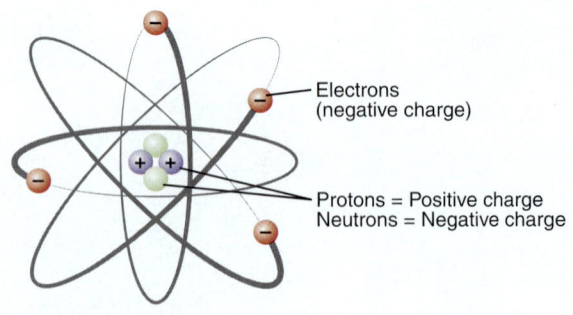

FIGURE 5-1 Parts of an atom.

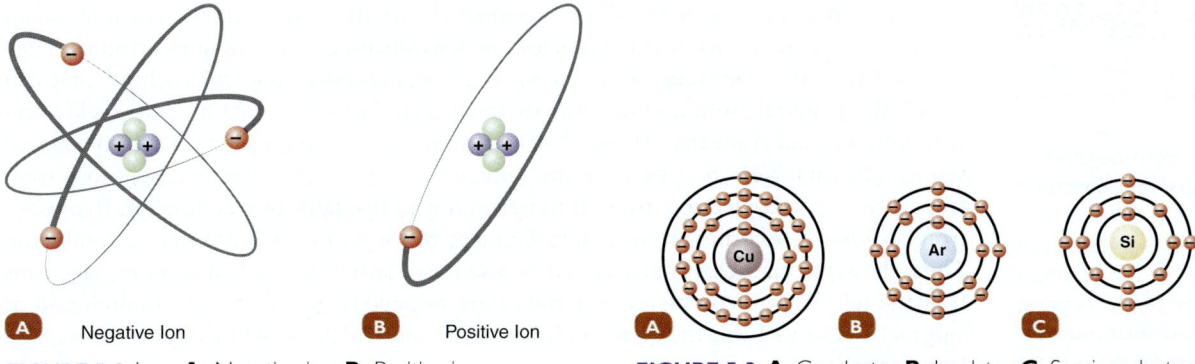

FIGURE 5-2 Ions. **A.** Negative ion. **B.** Positive ion.

FIGURE 5-3 **A.** Conductor. **B.** Insulator. **C.** Semiconductor.

atom has an imbalance of electrical charges due to the gain or loss of electrons. This negative ion is not balanced, so it is looking for a change. Since the electrons have the same negative charge, they repel one another, and the repelling force wants to push one of the electrons away from the atom.

A deficiency of electrons gives the atom an overall positive charge. This atom is called a positive ion. It is also not balanced and is looking for a change. In this case, it is exerting an attracting force on electrons to try to pull one into its atom from another atom. If a negative ion and positive ion are close enough, the negative charge of the negative ion exerts a repelling force on the extra electron, causing it to be pushed away from its atom; at the same time, the positive ion exerts an attracting force on the extra electron. These repelling and attracting forces cause the electron to be pushed and pulled from the negative atom to the positive atom, balancing both atoms out. The flow of electrons from atom to atom is called current flow and is the basic concept of electricity.

Not all atoms can give up or accept electrons easily. Materials that can do so easily are called conductors, while those that cannot do so easily are called insulators (**FIGURE 5-3**). When discussing electrical conductors and insulators, it is safe to say that in some materials there are electrons, called **free electrons,** located on the outer ring, called the valence ring. These electrons are only loosely held by the nucleus and are free to move from one atom to another when an electrical potential (pressure, or voltage) is applied. In fact, atoms with fewer electrons in the valence ring are the best **conductors**. Materials with one electron are the best conductive materials. This is because the single electron by itself in the valence ring is held the most loosely by the nucleus. Materials made up of atoms with one to three valance ring electrons are considered conductors. The more free electrons a particular material has, the better it can conduct electrons. Metals typically have many free electrons because of the atoms' structure and are therefore good conductors (**FIGURE 5-4**).

Every substance, even air, will conduct an electrical current if enough electrical pressure (voltage) is applied to it, but the word conductor normally is used for materials that allow current flow with little resistance. Most metals are good conductors. The most common conductor used in automobiles is copper. It is used in virtually all of the wiring that connects automotive components together. The more electrons a conductor must carry, the heavier the gauge or thickness the wire needs to be.

Materials that do not conduct electrons easily are called **insulators**. Most plastics are good insulators. The plastic covering on a wire is an example of this. The ceramic portion of a spark plug is also an example of a good insulator. In insulators, electrons in the valence ring are bound much more tightly to the nucleus. A good insulator does not support current flow because it has no or very few free electrons, and the electrons it does have cannot move freely; therefore, an insulator prevents the movement of electrons when an electrical potential is applied. Insulators are made up of atoms that have five to eight valence ring electrons. The greater the number of valence ring electrons, the better the insulator.

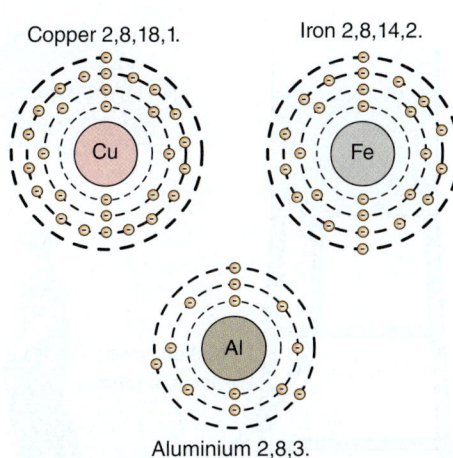

FIGURE 5-4 Different types of conductors used in automobiles and their respective free electrons.

▶ TECHNICIAN TIP

Since electrical current is the flow of electrons, it is natural to say that the direction of current is the direction in which the free electrons move—from negative to positive—which is called the **electron theory**. However, before the discovery that they are negatively charged, it was thought the natural way for electrons to flow was from positive to negative, which is called the **conventional theory**. Most wiring diagrams are written from the conventional theory perspective, while electronic circuits are typically designed and operate on the electron theory perspective. Thus, both concepts are still in use. In fact, a third theory exists that closely mirrors the conventional theory, called the **hole theory**. It states that while negative electrons do move from negative to positive, holes move from positive to negative as electrons move from atom to atom; in this case, holes (current) flow from positive to negative, and are sometimes called "positive holes."

What's most important to remember is that voltage causes current to flow through conductive paths (resistance). We will be using the conventional and hole theories when explaining the electrical concepts throughout this text unless otherwise noted.

Semiconductors are materials that conduct electricity more easily than insulators but not as well as conductors. Semiconductors such as silicon are crucial in electronics. They are used to make electronic components, such as transistors and microchips, that can switch the material from a conductor to an insulator and back again very quickly and without mechanical means. Atoms that have four valance ring electrons are considered semiconductors. Note that because the semiconductor material has precisely four electrons, it is only one electron away from becoming an insulator or a conductor. If an electron is added, it becomes an insulator; if an electron is removed, it becomes a conductor. Thus, the semiconductor material can be used as a switch to control whether electrons flow though the semiconductor material or are stopped by it. All we have to do is add or subtract electrons from the semiconductor material, which we will explore further in a later section.

Movement of Free Electrons

Free electrons are necessary for electrical current, but for the free electrons to move easily, they need two things—a complete pathway, or circuit, and a force that makes them move. The force from a battery can cause electrons to move. Like charges repel, so the negative electrons repel each other and are forced from the negative terminal of the battery. Unlike charges attract, so the electrons are attracted toward the positive protons in the positive side of the battery. In this case, free electrons flow in one direction only. This is called direct current. Most, but not all, circuits in passenger vehicles operate on direct current. The larger the charge between the negative and the positive terminal, the more strongly the positive terminal attracts and the negative terminal repels the free electrons. This attraction/repelling acts as a force driving the electrons along. The greater the force, the stronger the electrical current. The force is called **electromotive force**, and it is also referred to as voltage. So you could say that voltage is the force that motivates electrons to move. Voltage is measured in volts.

Insulators, Conductors, and Electrical Resistance

Also affecting the current flow in a circuit is **electrical resistance**, measured in **ohms**. All materials have resistance—even good conductors. There are four factors that determine the level of electrical resistance:

1. Type of material: This refers to how many free electrons a material has.
2. Length of the conductor: As length increases, so does resistance.
3. Diameter of the conductor: The larger the conductor, the greater the amount of current it can carry.
4. Temperature of the conductor: The higher the temperature, the harder it is for free electrons to pass through and the higher the electrical resistance.

While all materials have some resistance to current flow based on the number of free electrons the material has, a **resistor** is a component designed to extract energy from the current flow by forcing it through a restriction in the circuit (**FIGURE 5-5**). This act of extraction generates heat; the amount of heat generated by a resistance is directly related to the amount of current that passes through the resistance (**FIGURE 5-6**). This will be discussed at length in the following section. A typical resistor has a set resistance, usually marked or coded on its surface. Electrical resistance is somewhat like the electrical equivalent of friction in the mechanical world: It is the degree to which a material opposes, or resists, the passage of an electrical current. Good conductors have low resistance, while insulators have high resistance. Electrical energy lost through resistance is converted into heat (**FIGURE 5-7**).

Resistance is measured in ohms. Under most conditions, except temperature change, the resistance of an object is a constant and does not depend on the amount of the voltage or the amount of current

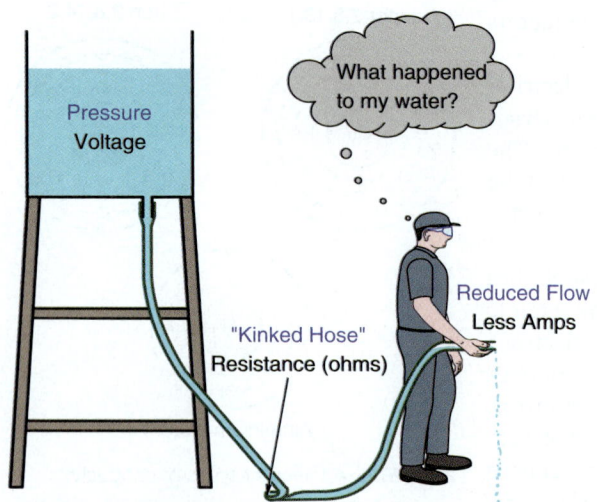

FIGURE 5-5 Resistance (ohms) in an electrical circuit is like a restriction in a water hose.

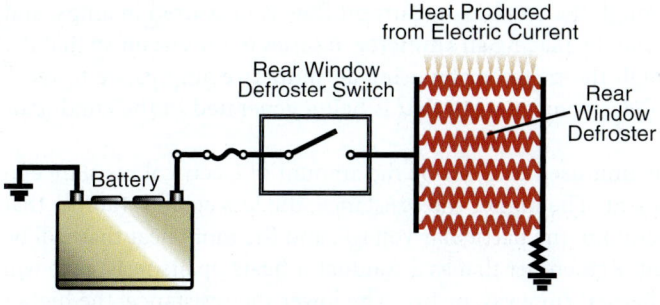

FIGURE 5-6 Heating effect of electricity.

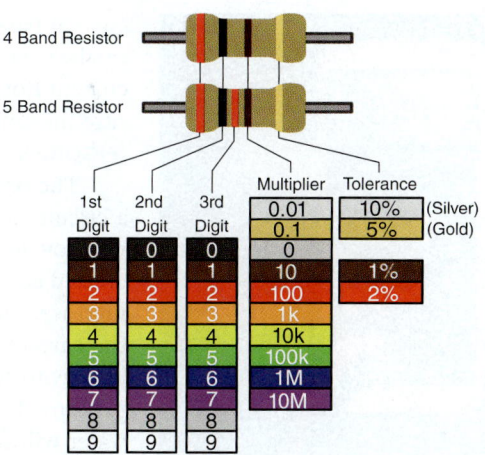

FIGURE 5-7 The process for reading the color bands on resistors.

passing through it. At the same time, **Ohm's law** tells us that if we increase current flow through a resistance, the voltage used by that resistance will increase. So in that sense, the resistance did not change, but the voltage drop did (**FIGURE 5-8**). When diagnosing circuits, we usually use a voltmeter to look for excessive voltage drops that reduce the available voltage at the load device, which is a term used to describe any component that uses electrical energy to perform work (i.e., headlamps using current to produce light, a starter motor that uses electric current to turn over the internal combustion engine). An ohmmeter would not normally work in this case, since it generally cannot read less than 0.1 ohms. This amount of resistance in a 10-amp circuit would cause a 1.0-volt drop, which is way beyond the allowable maximum voltage drop for a wire, which will be discussed further in the Digital Volt, Ohm Meter and Circuit Testing Procedures chapter. The relationships between current, voltage, and resistance are calculated using Ohm's law. This concept will be discussed in detail later in this chapter.

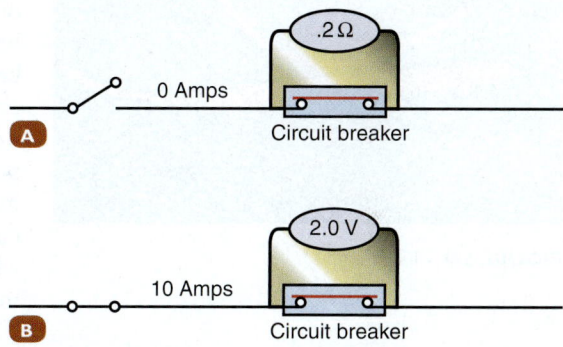

FIGURE 5-8 A. Measuring resistance with an ohmmeter. **B.** Measuring voltage drop of the same resistance.

▶ Volts, Amps, and Ohms

Volts, amps, and ohms are three basic units of electrical measurement. Voltage is the potential or electrical pressure difference between two points in an electrical circuit and is measured in **volts**. For example, the voltage of a typical automotive cranking battery is 12 volts, while the voltage contained by the high voltage battery-packs in hybrid vehicles can range from 36 volts to over 200 volts. These values represent the potential difference or electrical pressure between the positive and the negative battery terminal. This potential difference can be measured with a voltmeter or **multimeter** set to read voltage (**FIGURE 5-9**). Voltage can be measured by hooking a voltmeter across two parts of a circuit where you want to measure the difference in volts. It might be easier to understand if you think of voltage as the electrical force or pressure in a circuit or battery, just like the water pressure that exists in the bottom of a full tank of water or a home plumbing service.

K05002 Describe common electrical terms and their application.

The ampere, or **amp**, is the unit used to describe how much current or how many electrons are flowing at a given point in a given amount of time when work is being performed—for example, when a lamp is operating. An amp is a measure of the number of electrons flowing past a given point in 1 second. An amp is equal to 6.28 billion billion electrons past a given point in 1 second. Yes, billion billion is correct. To say it another way, think of a pile of 1 billion electrons. One amp would equal the number of elections in 6.28 billion of those piles travelling past a given point in a circuit in 1 second. That is impressive! No wonder we can't see them! And just to get you thinking, a starter motor may draw about 200 amps. Amperage, or current flow, can be thought of as a faucet being turned on and water flowing. Each drop of water is like an electron. The size of a conductor directly translates into the amount of current that can be carried. As current flows through a conductor, it generates an electromagnetic force that can be directly linked to the amount of

FIGURE 5-9 Multimeter.

K05003 Describe basic operation of electrical circuits.

current passing through the conductor. Current flow is measured in amps, and can be measured either by placing an **ammeter** in series to the circuit so that the current flows through the meter, or by using an inductive amp probe to measure the amount of electromagnetism that is being generated in the conductor (**FIGURE 5-10**).

The ohm is the unit used to describe the amount of electrical resistance in a circuit or component. The higher the resistance, the less current (amps) that will flow in the circuit for any particular voltage, and the more heat that will be created at that point. Remember that as a conductor heats up, its resistance will increase, causing thermal runaway, or fire. The lower the resistance, the higher the current that will flow in the circuit. Using the water analogy, if you kink a hose with the water running, less water will come out of the hose for a given pressure. If you kink the hose more, more resistance will be added and even less water will flow through the hose. Lessening the kink will lower the resistance and allow more water to flow. Resistance in a simple electrical circuit works the same way.

An ohmmeter is used to measure the amount of resistance in a component or circuit. The ohmmeter pushes a small amount of current through the part being tested, so an ohmmeter is usually used on a component or wire that has been disconnected from the rest of the circuit. The amount of resistance in the component changes the amount of current that the ohmmeter can push through the component. The more current that the ohmmeter can push through the component, the lower the resistance will read on the ohmmeter. This allows us to measure the resistance of just the single component, not the entire connected circuit. You must never attempt to check resistance of a component or circuit while it is connected to power; damage to the circuit, your meter, and your health may result! (**FIGURE 5-11**).

▶ Electrical Circuits

Electrical circuits are designed to perform electrical work in a controlled manner. They can be compared to a small city; the roads are like wires, the stoplights are like switches, and businesses are like electrical devices where work happens. Cars are like electrons that deliver the workers to the workplace. Electrical circuits can be very basic, consisting of a power supply, a fuse, a switch, a component that performs work, and wires connecting them all together (**FIGURE 5-12**). The power source—for example, the

FIGURE 5-10 A current clamp being used with a DMM. Note that the current clamp fits around the wire and is not connected in series.

FIGURE 5-11 When measuring resistance, always isolate the component from the circuit, and place the meter leads across the input and output terminals.

battery—creates a potential difference across its terminals, measured in volts. It pushes a flow of electrons (current flow), measured in amps, when the switch is in the closed position completing the path. The current flows through the fuse into the circuit wires to the lamp where the resistance of the lamp filament causes it to heat up and glow, producing light. As the filament heats up, its resistance increases, thereby limiting the amount of current passing through the filament. The current continues to flow through the lamp filament and the return pathway through the wires back to the battery to complete the circuit. When the switch is moved to the open position, the current path is broken and current flow stops, turning the lamp filament off.

A circuit can be much more complex than the one just described. But even then, most circuits contain a power source, circuit protection device, control mechanism, electrical load device, and connecting wires. As you understand the principles of electricity and how it behaves, you will be able to understand more complicated circuits.

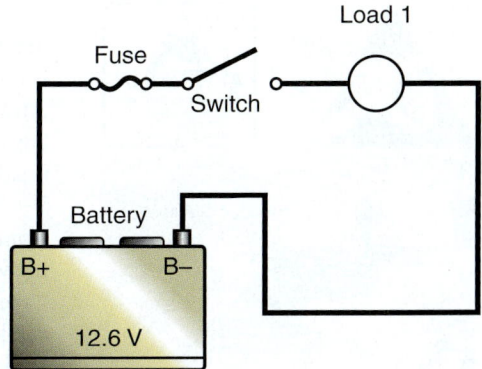

FIGURE 5-12 Simple circuit.

▶ Semiconductors

The term electronics usually refers to devices in which *electrons* are conducted through a vacuum, gas, or **semiconductor**. Automotive applications such as electronic control units mostly use semiconductors such as diodes, transistors, and power transistors. A semiconductor's electrical resistance is higher than that of most conductors, but lower than that of most insulators. A semiconductor's conducting ability depends on two kinds of **charge carriers**:

K05004 Describe common semiconductors, how they work, and their use.

- The first type of charge carrier is the negative electron. This kind of semiconductor contains an excess of free electrons. They can be made to flow and carry charge.
- The second type of charge carrier is the hole. Holes occur when an electron moves from its existing place to a new place, leaving a void where it was. Because the holes are positive, a voltage can make them move toward the negative pole. When connected in a circuit, both the electrons and the holes move, but in opposite directions.

The number of charge carriers, either an excess of electrons or deficiency of electrons (causing holes in a material), can be altered by **doping**, or adding very small quantities of impurities to a semiconductor material. A doped semiconductor always has an excess of one type of charge carrier. For example, electrons in excess make it an **N-type semiconductor** (**FIGURE 5-13**). N stands for negative. Holes in excess make it a **P-type semiconductor** (**FIGURE 5-14**). P stands for positive. Each of these materials responds to positive or negative current flow oppositely.

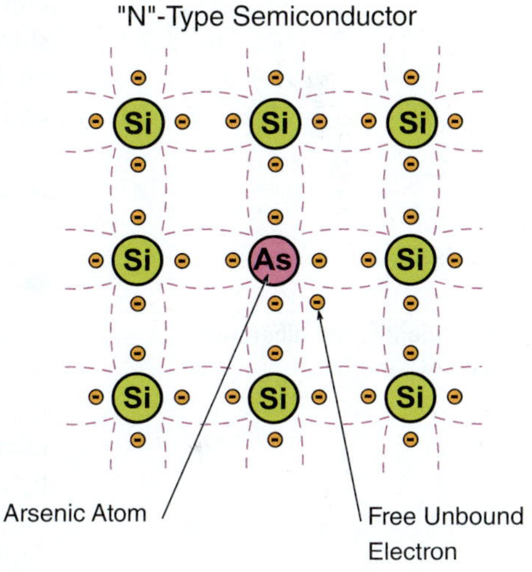

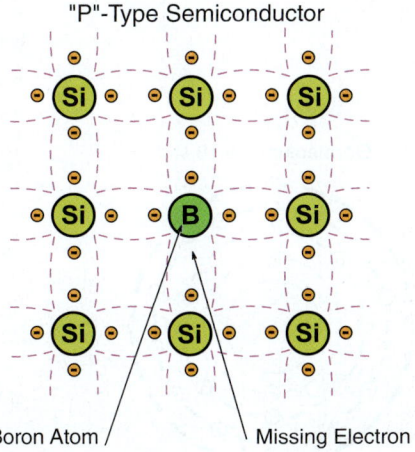

FIGURE 5-13 N-type semiconductor material.

FIGURE 5-14 P-type semiconductor material.

Semiconductor Operation

Most electronic components combine P-type and N-type semiconductors. The point where they join is called the **PN junction**. In this area, the **depletion layer** occurs: some electrons and holes cancel each other out and few charge carriers are present. The depletion layer is very thin and acts like an insulator (**FIGURE 5-15**).

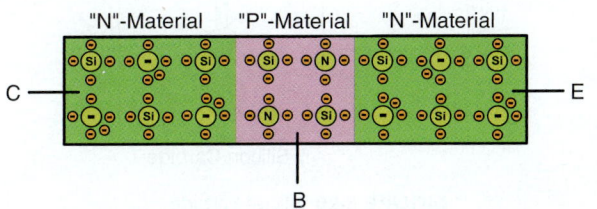

FIGURE 5-15 Transistors. The NP–N transistor has a P-type semiconductor between two N-type semiconductors.

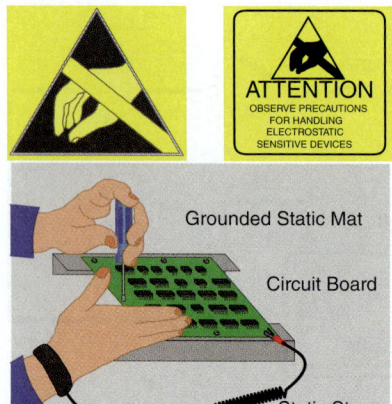

FIGURE 5-16 An example of static electricity.

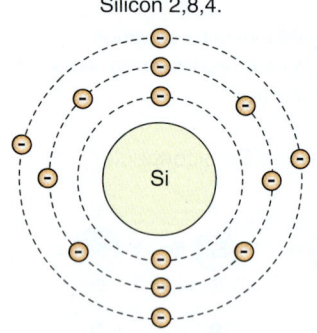

FIGURE 5-17 Silicon.

K05005 Identify the difference between AC and DC.

A **diode** is one P-type semiconductor material joined to one N-type semiconductor material with a single PN junction. If it is connected to a voltage source so that the P region is connected to a negative pole and the N region to a positive pole, then the negative pole attracts the holes, and the positive pole attracts the electrons. This enlarges the depletion layer and the insulated space. As a result, current cannot flow across the junction. Diodes are discussed further later on in the chapter.

Semiconductors are very versatile substances and are widely used to make various electronic components. Their conductivity can be manipulated and precisely controlled by doping with impurities to make transistors. They can also be designed to vary their conductance in magnetic fields to make Hall-effect devices and to vary their light to make photoelectric diodes and transistors. Semiconductor devices are replacing many other kinds of switching devices such as mechanical switches and relays. They are small, light, and use low operating voltages. They are reliable, require no maintenance, and are relatively easy to manufacture, but they are sensitive to heat and voltage spikes. Care must be taken when handling semiconductors and various electronic devices; if you are not careful to ground yourself before handling these components, static electricity can permanently damage them (**FIGURE 5-16**).

Semiconductor Materials

Many different materials have an atomic structure that can be used to make semiconductors. **Silicon** is the most widely used semiconductor material because it has a useful temperature range and is abundant, cheap, and easy to manufacture (**FIGURE 5-17**). **Germanium** was among the first semiconductor materials to be developed. While it is less widely used than silicon, germanium is useful in high-speed devices when alloyed with silicon (**FIGURE 5-18**). High-speed semiconductors widely use **gallium-arsenide**; however, it is more expensive and more difficult to manufacture (**FIGURE 5-19**). **Silicon carbide** has been used to create blue **light-emitting diodes (LEDs)**, and it can withstand high operating temperatures (**FIGURE 5-20**).

▶ Direct Current and Alternating Current

Electrons need to flow in a circuit for work or action to occur. For example, the action of a lamp glowing brightly is caused by the flow of electrons through the filament, heating it up and causing it to glow. There are two fundamental types of **current flow**: **direct current (DC)** (**FIGURE 5-21**) and **alternating current (AC)** (**FIGURE 5-22**). DC is produced by a battery. The battery maintains the same positive and negative **polarity**; therefore, the current flows in one direction only. The characteristics of DC are the fixed polarity of the applied voltage and the flow of charges in only one direction. It is possible to have varying DC; however, the charges always flow in the one direction and the applied voltage polarity remains the same.

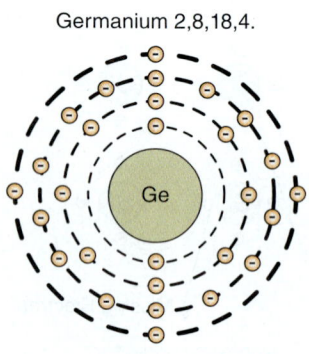

FIGURE 5-18 Germanium.

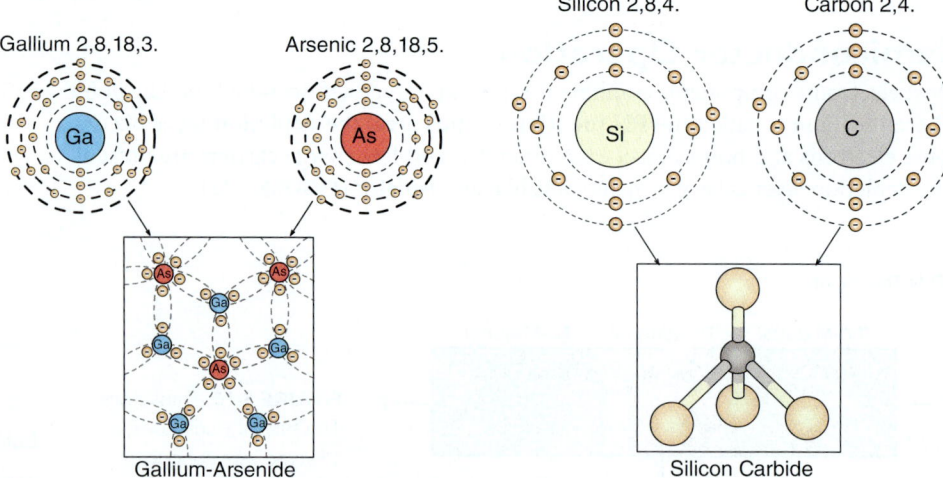

FIGURE 5-19 Gallium-arsenide.

FIGURE 5-20 Silicon carbide.

AC is the type of current in your home. It continuously changes its direction of current flow, and the alternating voltage repeatedly reverses or alternates its polarity. Thus, the current flow moves back and forth within a circuit. AC is produced in what is called a **sine wave**. It operates on a cycle, gradually building to a maximum current flow in one direction (positive value), then gradually reducing to zero, then gradually building to a maximum current flow in the other direction (negative value), and finally gradually reducing back to zero current. In most cases, this cycle can occur many times a second. For example, the AC flow in the house supply changes direction 60 times per second. **Hertz** is the measurement of frequency and indicates the number of cycles per second. So, since the AC supply in the home typically changes polarity 60 times a second, its frequency is 60 hertz, or 60 Hz. Hertz simply means "cycles per second." While frequency describes how rapidly the AC current reverses polarity, amplitude describes how high the voltage level goes as it cycles. The higher the amplitude, the more amperage that can be pushed through the circuit, and the more work that can be done.

AC is used in vehicles to a lesser extent than DC. Alternators use it to create current flow to charge the battery and run the electrical accessories. The AC is first transformed to DC before it leaves the alternator so it can be effectively put to use in the DC electrical system. This conversion is accomplished in the alternator's rectifier bridge, which will be discussed further in the chapter on Charging Systems.

AC is used in the electric motors on most hybrid vehicles. Since these motors generally require high amounts of electrical current, and because AC is more efficient than DC, AC is more advantageous for that application. The AC is created by a sophisticated electronic inverter. Most of these inverters use a semiconductor device called an Insulated-Gate Bi-Polar-Transistor (IGBT). Think of this as a large electronic switch that can turn large amounts of current on and off at a very fast rate. Since we know that DC current can only travel in one direction, IGBTs are used to cycle the DC current's polarity forward and backward, mimicking the AC voltage's sine wave, and powering the traction motors in the driveline (**FIGURE 5-23**). Hybrid Vehicle Operation will be covered further in the Advanced Electrical textbook.

In general, electrical components are designed to work on either AC or DC, but not both. For example, a DC motor will not work on AC, and vice versa. Devices can be made to convert or change AC to DC and DC to AC. For example, a battery charger that has an input AC of 110 volts at 60 Hz power can change this to 14 volts DC through a transformer and rectifier to charge a battery. Devices are also available to change DC into AC, and are called **inverters**. For example, an inverter could have an input of 12 volts DC and convert that to 110 volts AC at 60 Hz. AC and DC power sources can be of any voltage and are not limited to the relatively low 12 volts DC found in the batteries on vehicles or the 110 volts AC found in the home power supply. For example, some hybrid vehicles use high-voltage DC invertors drawing 200 to 300 volts DC to power the 500-volt three-**phase** AC to power the main electric traction motors or accessory motors (**FIGURE 5-24**). Phase refers to the number of separate staggered **power windings** in the motor or alternator. Generally, three-phase motors or alternators produce more output than single-phase motors or alternators because they can maintain a much more even voltage compared to single-phase AC.

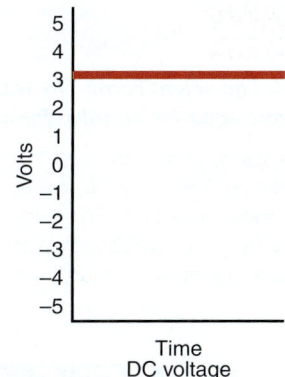

FIGURE 5-21 Direct current (DC).

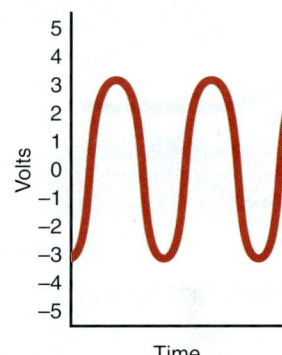

FIGURE 5-22 Alternating current (AC).

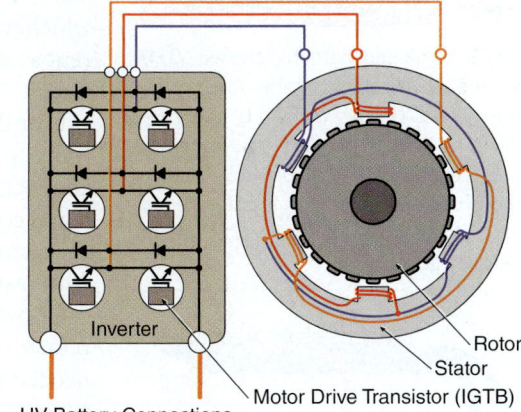

FIGURE 5-23 Schematic of inverter which uses insulated gate bipolar transistors to convert high-voltage DC from the battery to the high-voltage AC that is used to drive the traction motors.

Applied Science

AS-58: AC/DC: The technician can explain the difference between direct and alternating current.

Direct current (DC) and alternating current (AC) are two forms of electricity that are produced differently and have different uses. DC electricity is the simpler form. It starts in one place and then flows in the same direction to its destination. AC electricity flows in one direction for a period of time, then changes direction over and over again continuously.

In modern automotive applications, AC electricity is generated by the alternator using electromagnets. The AC electricity is converted to DC, or rectified, by diodes in the alternator before being supplied to the vehicle's electrical systems or being stored in the battery.

| Applied | Math |

AM-46: Equivalent Form: The technician can write or rewrite an algebraic equation to solve for any unknown variables.

Ohm's law is commonly used to calculate the value of an unknown variable from known values in electrical circuits. The equation is commonly stated as voltage (V) equals current flow (A) multiplied by resistance (R), or $V = AR$. To calculate current flow or resistance, instead of voltage, the equation must be rewritten. To apply the rules of basic algebra, anything we do to one side of the equation we must also do to the other. To calculate current flow, we need the A by itself on one side of the equation, so we divide both sides by R to arrive at $A = V/R$. To calculate resistance, the R must stand alone, so we divide both sides of the equation by A to arrive at $R = V/A$. Ohm's law is commonly represented in a triangle or circle format, making the correct algebraic equation very simple to find.

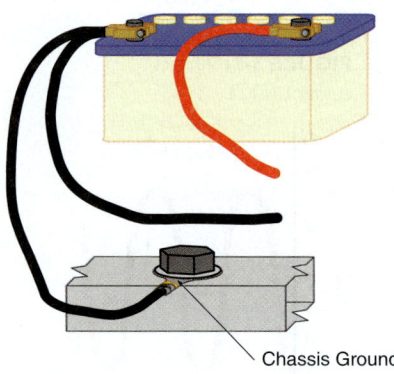

FIGURE 5-24 Body frame ground connections.

K05006 Identify power, and ground sources.

▶ **TECHNICIAN TIP**

While the United States has chosen 60 Hz for the electrical grid, many other countries use 220 volts at 50 Hz and have different-shaped plugs. If you travel to these countries, make sure to purchase the proper converter or transformer so your electrical devices will work and not be damaged.

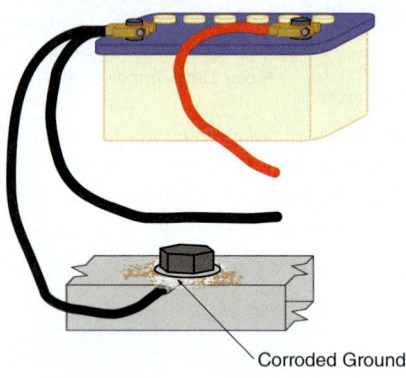

FIGURE 5-25 Corroded ground connection at body or frame.

K05007 Diagnosis open circuits, short circuits, and high resistance.

Some manufacturers use sensors that create an AC signal that varies in frequency. This varying signal is sent to the vehicle's computer as an indication of changes within the system being monitored. The computer can use that signal to make adjustments based on its programmed software.

▶ Power (Source or Feed) and Ground

Power and ground are terms to describe the beginning and end of a circuit. Power, source, and feed signify the supply side (beginning) of the circuit, where the electricity originates. Ground signifies the return side of the circuit (Figure 5-21). Recall that in this text we will use the conventional theory, in which the supply side is the positive side of the circuit and the return side is the negative side of the circuit. In a vehicle, the positive battery post is considered the source. The power or feed side of a circuit refers to the wires and components that originate at the positive post of the battery and end at either a switch or an electrical load device, whichever occurs first.

Ground is a term used by technicians to indicate the portion of the circuit that returns current flow to the negative side of the battery. The ground-side of the circuit starts at the negative post of the battery and ends at either a load device or a switch, whichever occurs first. Ground circuits must be rated to carry the same amount of current as the power side of the circuit, as the electrons move in a loop through the entire circuit. The term ground is also used to describe a direct connection with the negative side of the circuit, as in "the switch is grounded." Also, if you are told to ground something, you are being told to connect to the negative side of the circuit. Many vehicles connect the chassis and body to the negative battery terminal, which means most of the metal components on the vehicle are grounded (Figure 5-24). Frame and body grounds are commonly affected by rust and corrosion. If rust and corrosion are present in a circuit, resistance will increase, generating heat, and lowering the amount of current that can flow through the circuit (**FIGURE 5-25**). Many manufacturers use the chassis as the return path to the negative battery terminal, as this cuts down on the amount of wire needed in the vehicle. At the same time, many computer circuits use dedicated ground wires from their sensors back to the computer so that the electrical signal is accurate and not affected by any stray electrical signals on the ground circuit.

▶ Continuity, Open, Short, and High Resistance (Voltage Drop)

The terms *continuity, open, short*, and *high resistance* are often used to describe a circuit's or component's condition. For example, the wiring harness could be described as having a short or the connector as having an open circuit. **Continuity** is achieved when an electrical circuit has a continuous and uninterrupted electrical connection and is thereby capable of conducting current and working as designed. No continuity means there is a break in the circuit and current cannot flow past the break. It can be measured with a multimeter set to measure in ohms that has been connected between two points in a circuit (**FIGURE 5-26**). For example, if a technician suspects that a circuit has a break or bad connections in the wiring, the continuity could be measured between two points.

Continuity, Open, Short, and High Resistance (Voltage Drop) 147

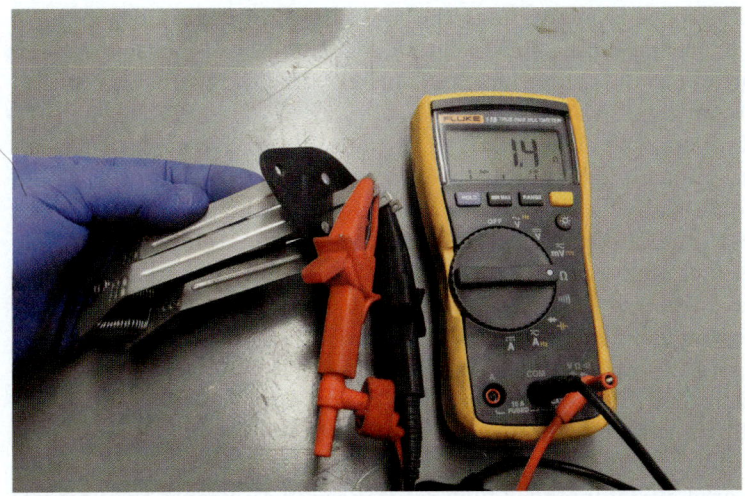

FIGURE 5-26 Multimeter used to test continuity in a portion of a circuit.

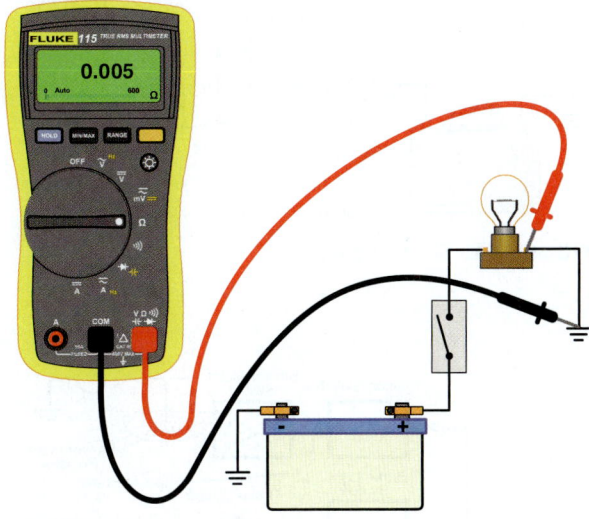

FIGURE 5-27 Circuit fault being tested for continuity and resistance.

A faulty circuit has high resistance or no continuity (**FIGURE 5-27**). A circuit with continuity has very low resistance. But be careful, checking the circuit for continuity cannot determine all electrical faults. If the continuity test determines that there is a break in the circuit, you can bank on that and track it down to determine the fault. However, if the continuity test determines that there is continuity, you still don't know for sure that the circuit is good. It is possible that while the circuit has continuity, it has excessive resistance that could affect the operation of the electrical device under normal current flow. Thus, continuity testing is limited in what it can indicate. It is a great test for determining whether a wire or component is open or whether two circuits are shorted together.

The term **open** describes a circuit that does not have a complete circuit and therefore cannot conduct current (**FIGURE 5-28**). In other words, the circuit does not have continuity. For example, when a switch is turned off, the circuit is open and no current can flow. Open can also be used to describe a fault in a circuit. For example, if the fuse is open-circuited, which describes a blown fuse, no current will be able to flow. A multimeter or test lamp can be used to test for an open circuit. An open circuit has infinite resistance—so much resistance that it is not measurable.

The term **short**, in its purest definition, describes a circuit fault in which current takes a shorter path, resistance-wise, through an accidental or unintended route (**FIGURE 5-29**). The low-resistance fault causes abnormally high current flow in the circuit and may cause the circuit protection devices, such as fuses or circuit breakers, to open the circuit. An example of a pure short would be insulation on the windings within a relay coil that has worn through and is allowing current to bypass many of the windings. In this case, resistance decreases, amperage increases, and the magnetic field created by the winding becomes weaker, potentially causing the relay to not operate.

A pure short circuit is not the only type of short circuit. There are three additional types of short circuits that you need to understand. The first is a short to ground. In this case, the circuit has an unintended path directly to ground (**FIGURE 5-30**). For example, if the wire from the brake switch to the brake lights rubs through the wire insulation

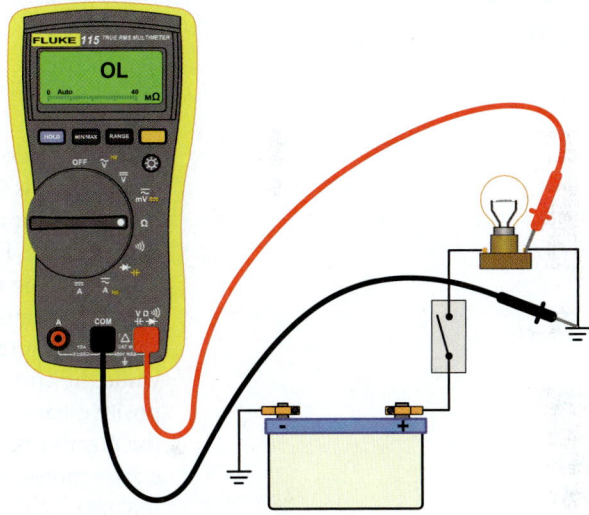

FIGURE 5-28 An open in a circuit.

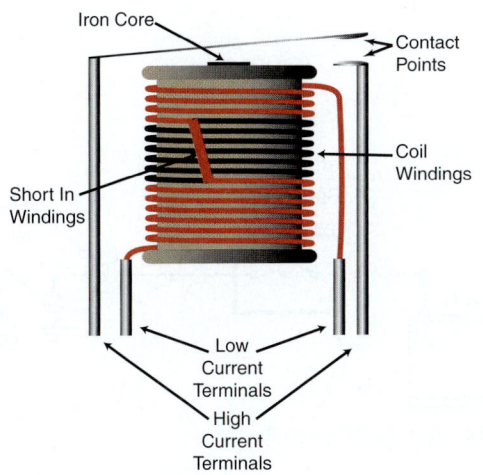

FIGURE 5-29 Short circuit.

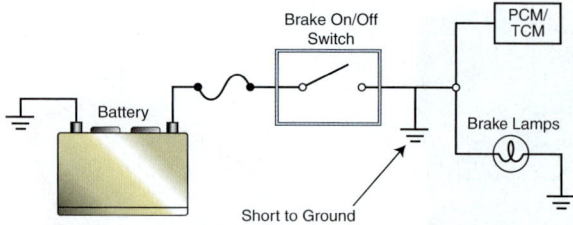

FIGURE 5-30 Short to ground.

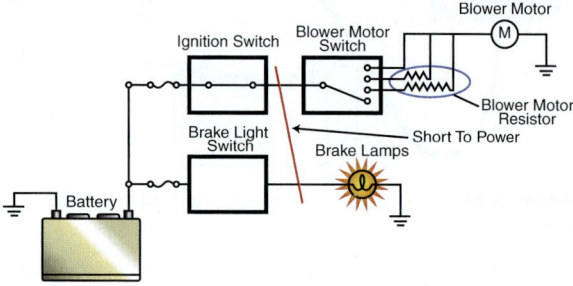

FIGURE 5-31 Short to power.

on a sharp edge of a body panel, the bare wire may make contact with the metal panel and cause a short to ground when the brake pedal is pressed. A short to ground causes increased current flow and will typically blow the circuit fuse.

Another type of short is a short to power. In this case, the circuit has an unintended path directly to a power source. An example would be two wires in a harness that have melted together, one that supplies power to the blower motor and one that feeds power from the brake light switch to the brake lights (**FIGURE 5-31**). In this example, turning on the ignition switch would cause the brake lights to come on because power would be sent to the blower fuse and then, due to the short in the wiring, to the brake lights.

Unintended **high resistance** in a circuit causes a reduction in current flow in the circuit as well as a drop in voltage at the resistance. Both of these cause the intended circuit device to not operate effectively or at all. Unintended high resistance can also cause an overheating condition at the area of resistance, which can melt wire insulation or plastic **connectors**. This condition can be caused by a number of faults, including corroded or loose harness connectors, wire that is too thin for the circuit current flow, incorrectly connected terminals, and poorly soldered joints (**FIGURE 5-32**).

Voltage drop is another name for high resistance, and voltage drop testing is the best way of finding high resistance in the feed side or ground side of the circuit. For a voltage drop to occur, two conditions must be present: resistance and current flow. Thus, a voltage drop in a circuit indicates that there is resistance present and current is flowing, or trying to flow, through the resistance. If the high-resistance fault is excessive, it will reduce both the voltage and the current flow in the circuit, affecting its performance. For example, a voltage drop in the headlight circuit will cause the lights to be dim, reducing their performance. A high resistance in the main battery lead to the starter motor will cause the starter to crank the engine over slowly or not at all. The voltage drop can be measured in a circuit by placing a voltmeter across two different points in a circuit while the circuit is being operated. For example, to measure the voltage drop in the main positive battery lead to the starter motor, the voltmeter's black lead would be placed on the positive battery lead and the red lead would be placed on the main battery lead of the starter motor, and the voltage would then be read on the meter while the engine is cranked (**FIGURE 5-33**).

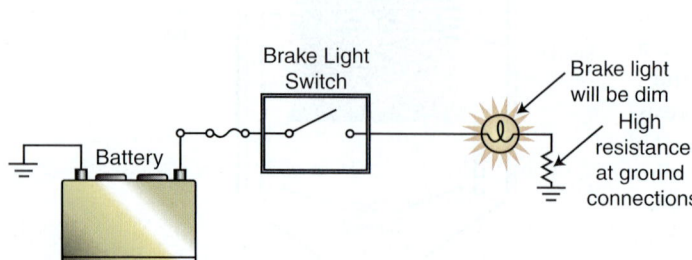

FIGURE 5-32 High-resistance fault.

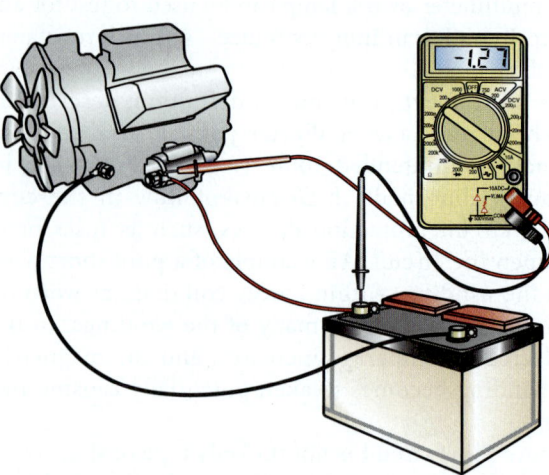

FIGURE 5-33 Voltage drop testing the positive battery cable.

Wrap-Up

Ready for Review

- All questions about the nature of electricity lead to the composition of matter.
- All matter is made up of atoms.
- Every atom has a nucleus, with at least one positively charged particle called a proton and, in most atoms, at least one neutron, or uncharged particle.
- Moving around the nucleus are one or more negatively charged particles called electrons.
- Electrons travel in different rings or shells around the nucleus. Each ring or shell can contain a specific maximum number of electrons.
- With equal numbers of protons and electrons, the charges within an atom cancel each other out, leaving the atom with no overall charge. In this state, the electrons and protons are content to stay in the atom.
- An atom with more electrons than protons has an overall negative charge and is called a negative ion.
- An ion is an atom with an imbalance of electrical charges due to the gain or loss of electrons.
- A deficiency of electrons gives the atom an overall positive charge. This atom is called a positive ion. It is also not balanced and is looking for a change.
- If a negative ion and positive ion are close enough, the negative charge of the negative ion exerts a repelling force on the extra electron, causing it to be pushed away from its atom; at the same time, the positive ion exerts an attracting force on the extra electron.
- The flow of electrons from atom to atom is called current flow and is the basic concept of electricity.
- Not all atoms can give up or accept electrons easily; those that can do so easily are called conductors, while those that cannot do so easily are called insulators.
- Free electrons, located on the outer ring or valence ring are loosely held by the nucleus and are free to move from one atom to another when an electrical potential (pressure or voltage) is applied.
- Atoms with fewer electrons in the valence ring are the best conductors, like copper.
- Materials with one electron are the best conductive materials.
- Materials made up of atoms with one to three valance ring electrons are considered conductors.
- The most common conductor used in automobiles is copper.
- Materials that do not conduct electrons easily are called insulators.
- Most plastics are good insulators.
- Semiconductors are materials that conduct electricity more easily than insulators but not as well as conductors.
- Atoms that have four valance ring electrons are considered semiconductors.
- Like charges repel, so the negative electrons repel each other and are forced from the negative terminal of the battery.
- Free electrons flow in one direction only. This is called direct current.
- Most circuits in passenger vehicles operate on direct current.
- The larger the charge between the negative and the positive terminal, the more strongly the positive terminal attracts and the negative terminal repels the free electrons.
- The greater the force, the stronger the electrical current. The force is called electromotive force, and it is also referred to as voltage.
- Voltage is the force that motivates electrons to move and is measured in volts.
- There are four factors that determine the level of electrical resistance:
  - Type of material
  - Length of the conductor
  - Diameter of the conductor
  - Temperature of the conductor
- A resistor is a component designed to extract energy from the current flow by forcing it through a restriction in the circuit.
- A typical resistor has a set resistance, usually marked or coded on its surface.
- Electrical resistance is the degree to which a material opposes, or resists, the passage of an electrical current.
- Resistance is measured in ohms.
- Ohm's law says if we increase current flow through a resistance, the voltage used by that resistance will increase (voltage = current × resistance). So, the resistance did not change, but the voltage drop did.
- When diagnosing circuits, we usually use a voltmeter to look for excessive voltage drops that reduce the available voltage at the load device.
- Volts, amps, and ohms are three basic units of electrical measurement.
- Voltage is the potential or electrical pressure difference between two points in an electrical circuit and is measured in volts.
- This potential difference can be measured with a voltmeter or multimeter set to read voltage.
- Voltage can be measured by hooking a voltmeter across two parts of a circuit where you want to measure the difference in volts.
- The ampere, or amp, is the unit used to describe how much current or how many electrons are flowing at a given point in a given amount of time when work is being done.
- The ohm is the unit used to describe the amount of electrical resistance in a circuit or component.

- The higher the resistance, the less current (amps) that will flow in the circuit for any particular voltage, and the more heat that will be created at that point.
- An ohmmeter is used to measure the amount of resistance in a component or circuit.
- Electrical circuits can be very basic, consisting of a power supply, a fuse, a switch, a component that performs work, and wires connecting them all together.
- The power source—the battery—creates a potential difference across its terminals, measured in volts. It pushes a flow of electrons (current flow), measured in amps, when the switch is in the closed position completing the path.
- The term electronics usually refers to devices in which electrons are conducted through a vacuum, gas, or semiconductor.
- A semiconductor's conducting ability depends on two kinds of charge carriers:
  - Negative electron, which contains an excess of free electrons.
  - Holes occur when an electron moves from its existing place to a new place, leaving a void where it was. Because the holes are positive, a voltage can make them move toward the negative pole.
- The number of charge carriers, either an excess of electrons or deficiency of electrons (causing holes in a material), can be altered by *doping*, or adding very small quantities of impurities to a semiconductor material.
- Electrons in excess make it an N-type semiconductor; N stands for negative.
- Holes in excess make it a P-type semiconductor.
- Most electronic components combine P-type and N-type semiconductors. The point where they join is called the PN junction.
- A diode is one P-type semiconductor material joined to one N-type semiconductor material with a single PN junction.
- If a diode is connected to a voltage source so that the P region is connected to a negative pole and the N region to a positive pole, then the negative pole attracts the holes, and the positive pole attracts the electrons.
- Silicon is the most widely used semiconductor material because it has a useful temperature range and is abundant, cheap, and easy to manufacture.
- Germanium was among the first semiconductor materials to be developed.
- There are two fundamental types of current flow: direct current (DC) and alternating current (AC).
- DC is produced by a battery. The battery maintains the same positive and negative polarity; therefore, the current flows in one direction only.
- The characteristics of DC are the fixed polarity of the applied voltage and the flow of charges in only one direction. It is possible to have varying DC; however, the charges always flow in the one direction and the applied voltage polarity remains the same.
- AC continuously changes its direction of current flow, and the alternating voltage repeatedly reverses or alternates its polarity.
- AC current flow moves back and forth within a circuit.
- AC is produced in what is called a sine wave.
- AC is used in the electric motors on most hybrid vehicles. Since these motors generally require high amounts of electrical current, and because AC is more efficient than DC.
- Hertz is the measurement of frequency and indicates the number of cycles per second.
- Phase refers to the number of separate staggered power windings in the motor or alternator.
- Power and ground are terms to describe the beginning and end of a circuit.
- Power, source, and feed signify the supply side (beginning) of the circuit, where the electricity originates.
- Ground signifies the return side of the circuit.
- Ground is a term used to indicate the portion of the circuit that returns current flow to the negative side of the battery.
- The ground-side of the circuit starts at the negative post of the battery and ends at either a load device or a switch, whichever occurs first.
- Ground circuits must be rated to carry the same amount of current as the power side of the circuit, as the electrons move in a loop through the entire circuit.
- The terms continuity, open, short, and high resistance are often used to describe a circuit's or component's condition.
- Continuity is achieved when an electrical circuit has a continuous and uninterrupted electrical connection and is thereby capable of conducting current and working as designed.
- No continuity means there is a break in the circuit and current cannot flow past the break and is measured with a DVOM.
- The term "open" describes a circuit that does not have a complete circuit and therefore cannot conduct current; the circuit does not have continuity.
- The term short describes a circuit fault in which current takes a shorter path, resistance-wise, through an accidental or unintended route.
- The low-resistance fault causes abnormally high current flow in the circuit and may cause the circuit protection devices, such as fuses or circuit breakers, to open the circuit.
- A short to ground has an unintended path directly to ground like the wire insulation on a sharp edge of a body panel. The bare wire may make contact with the metal panel and cause a short to ground.
- A short to ground causes increased current flow and will typically blow the circuit fuse.
- A short to power has an unintended path directly to a power source. An example would be two wires in a harness that have melted together, one that supplies power to the blower motor and one that feeds power from the brake light switch to the brake lights.
- Unintended high resistance in a circuit causes a reduction in current flow in the circuit as well as a drop in voltage at the resistance.
- Voltage drop is another name for high resistance, and voltage drop testing is the best way of finding high resistance in the feed side or ground side of the circuit.

Key Terms

alternating current (AC) A type of current flow that flows back and forth.

ammeter A device used to measure current flow.

amp An abbreviation for amperes, the unit for current measurement.

charge carrier A mobile particle that has a positive or negative electrical charge.

conductor A material that allows electricity to flow through it easily. It is made up of atoms with one to three valance ring electrons.

connector The plastic housing on the end of a wiring harness that holds the wire terminals in place. It can also refer to a type of wire terminal that connects wires together or to a common point such as a bolt.

continuity A conductive path between two points.

conventional theory The theory that electrons flow from positive to negative.

current flow The flow of electrons, typically within a circuit or component.

depletion layer An area of neutral charge in semiconductors.

diode A two-lead electronic component that allows current flow in one direction only.

direct current (DC) Movement of current that flows in one direction only.

doping The introduction of impurities to pure semiconductor materials to provide N- and P-type semiconductors.

electrical resistance A material's property that slows down the flow of electrical current.

electromotive force An electrical pressure or voltage.

electron theory The theory that electrons, being negatively charged, repel other electrons and are attracted to positively charged objects; thus electrons flow from negative to positive.

free electron An electron located on the outer ring, called the valence ring, that is only loosely held by the nucleus and that is free to move from one atom to another when an electrical potential (pressure) is applied.

gallium-arsenide A semiconductor used in high-frequency circuits.

germanium A type of semiconductor.

ground The return path for electrical current in a vehicle chassis, other metal of the vehicle, or dedicated wire.

hertz The unit for electrical frequency measurement.

high resistance The resistance of a component or circuit relative to a low resistance. It can also refer to a faulty circuit where a section or component has excess unwanted resistance.

hole theory The theory that as electrons flow from negative to positive, holes flow from positive to negative.

insulator A material that has properties that prevent the easy flow of electricity. These materials are made up of atoms with five to eight electrons in the valance ring.

invertor A device that changes direct current into alternating current.

light-emitting diode (LED) A diode that emits light when current flows through it.

multimeter A test instrument used to measure volts, ohms, and amps. A digital multimeter may also be called a digital volt-ohmmeter (DVOM).

N-type semiconductor Semiconductor material with a small amount of extra electrons.

ohm The unit for measuring electrical resistance.

Ohm's law A law that defines the relationship between current, resistance, and voltage.

open A term used to describe a circuit that does not have a complete path for current to flow.

P-type semiconductor Semiconductor material with holes where electrons are missing.

phase A term used to describe one set of windings from an alternator or alternating current electric motor.

PN junction The junction between N- and P-type semiconductor materials.

power windings The current-carrying windings in an alternator or motor.

resistor A component designed to have a fixed resistance.

semiconductor A material used to make microchips, transistors, and diodes.

short Also called a short circuit, the flow of current along an unintended route.

silicon A material commonly used to make semiconductors.

silicon carbide A type of material used to make semiconductors.

sine wave A mathematical function that describes a repetitive waveform such as an alternating current signal.

volt The unit used to measure potential difference or electrical pressure.

Review Questions

1. All of the following statements describing the basic principles of electricity are true *except*:
   a. An atom with less deficiency of electrons than protons has an overall positive charge and is called a positive ion.
   b. An atom with more electrons than protons has an overall negative charge and is called a negative ion.
   c. A positive ion exerts a repelling force on the extra electron.
   d. The flow of electrons from atom to atom is called current flow.

2. Which of the following materials is typically used in the construction of semiconductors?
   a. Silicon
   b. Copper
   c. Plastic
   d. Ceramic

3. Electromotive force can be best described as the:
   a. force of electrons repelling each other.
   b. repelling force of the negative terminal.

c. attracting force of the positive terminal.
d. force of attraction/repelling which drives the electrons to move along.

4. The degree to which a material opposes the passage of electrical current through it is called:
   a. voltage.
   b. resistance.
   c. discharge.
   d. insulation.

5. Which of the following is a measure of the number of electrons flowing past a given point in 1 second?
   a. Volt
   b. Watt
   c. Ohm
   d. Amp

6. The device which is used to measure the amount of resistance in a component or a circuit is a(n):
   a. ohmmeter.
   b. ammeter.
   c. voltmeter.
   d. multimeter.

7. All of the following statements describing an electrical circuit are true except:
   a. Electrical circuits are designed to perform electrical work in a controlled manner.
   b. The current flows through the fuse into the circuit wires.
   c. Electrical circuits consist of a power source, a fuse, a switch, a component that performs work, and wires connecting them all together.
   d. When the switch is moved to the closed position, the current path is broken and current flow stops.

8. Choose the correct statement regarding semiconductors.
   a. Semiconductor devices use high operating voltages.
   b. Semiconductor devices are sensitive to heat and voltage spikes.
   c. Semiconductor devices are not reliable.
   d. Semiconductor devices need regular maintenance.

9. Which of the following materials is widely used to build a high-speed semiconductor?
   a. Silicon
   b. Germanium
   c. Gallium-arsenide
   d. Silicon carbide

10. All of the following statements describing AC and DC are true *except*:
    a. In DC, the charges always flow in one direction but the applied voltage polarity changes.
    b. Frequency describes how rapidly the AC current reverses polarity.
    c. An alternator converts AC into DC.
    d. An inverter converts DC into AC.

ASE Technician A/Technician B Style Questions

1. Tech A says that a thorough understanding of electrical circuits is critical to accurately diagnose today's vehicles. Tech B says that you only need a small amount of electrical knowledge to repair modern automobiles. Who is correct?
   a. Tech A
   b. Tech B
   c. Both A and B
   d. Neither A nor B

2. Tech A says that electricity is governed by the laws of physics. Tech B says that electricity can be forced to disobey the laws of physics if enough voltage is applied. Who is correct?
   a. Tech A
   b. Tech B
   c. Both A and B
   d. Neither A nor B

3. Tech A says that every atom has a core, which is called the nucleus. Tech B says that every atom must have at least three protons in its structure. Who is correct?
   a. Tech A
   b. Tech B
   c. Both A and B
   d. Neither A nor B

4. Tech A says that protons have a negative charge. Tech B says that electrons have a positive charge. Who is correct?
   a. Tech A
   b. Tech B
   c. Both A and B
   d. Neither A nor B

5. Tech A says more free electrons in the valence ring mean that the material is a better conductor of electricity. Tech B says that the materials that are the best conductors of electricity have the fewest amounts of free electrons in their valence ring. Who is correct?
   a. Tech A
   b. Tech B
   c. Both A and B
   d. Neither A nor B

6. Tech A says that rubber and plastic are effective electrical insulators. Tech B says that certain materials can be both insulators and conductors. Who is correct?
   a. Tech A
   b. Tech B
   c. Both A and B
   d. Neither A nor B

7. Tech A says that voltage is the amount of pressure that is applied to move electrons through a circuit. Tech B says that voltage and amperage is essentially the same thing, and can therefore be used interchangeably. Who is correct?
   a. Tech A
   b. Tech B
   c. Both A and B
   d. Neither A nor B

8. Tech A says that amperage is the amount of voltage that is dropped in a circuit. Tech B says that amperage is a measure of the amount of electrons that are flowing through a circuit. Who is correct?
   a. Tech A
   b. Tech B

c. Both A and B
   d. Neither A nor B
9. Tech A says that resistance should be increased to maximize current flow through a circuit. Tech B says that when resistance goes up, than current flow will go down if the source voltage remains the same. Who is correct?
   a. Tech A
   b. Tech B
   c. Both A and B
   d. Neither A nor B

10. Tech A says that a short is a generic term used to describe a myriad of electrical problems. Tech B says that a short to ground or power can cause excessive current in a circuit. Who is correct?
    a. Tech A
    b. Tech B
    c. Both A and B
    d. Neither A nor B

CHAPTER 6
Sources and Effects of Electricity

NATEF Tasks

- **N06001** Demonstrate knowledge of the causes and effects from shorts, grounds, opens, and resistance problems in electrical/electronic circuits.

Knowledge Objectives

After reading this chapter, you will be able to:

- **K06001** Identify common sources of electricity.
- **K06002** Understand and identify the effects of electricity.

Skills Objectives

There are no Skills Objectives for this chapter.

You Are the Technician

Whenever current passes through a conductor, magnetic lines of flux are created. The higher the current that passes through a conductor, the stronger the magnetic flux that is created. When aftermarket accessories are installed improperly, and power cables are routed too close to existing wire harnesses, the magnetic lines of flux created during component operation can cut through the nearby harnesses and induce voltages in these wires. This can create interference with module communication and wreak havoc with the vehicles systems. When attempting to diagnose strange electrical problems, always take a close look at aftermarket accessories, as they can often be the root cause of your problem.

1. True or False: When more current passes through a conductor the magnetic field surrounding the conductor becomes stronger.
2. True or False: Improperly routed amplifier power cables cannot be a source of electrical interference.
3. True or False: When diagnosing strange electrical problems, it is good practice to disable any aftermarket components to rule them out as a cause.

▶ Introduction

Energy is present in everything that we come into contact with. Energy can be manifested in many forms and may be transferred between objects. While energy cannot be created or destroyed, it can be converted into different forms and states that can be harnessed to accomplish many types of work. Of all the types of energy that exist, electricity is unique. Not only can it be transformed into thermal energy, light energy, chemical energy, and mechanical energy, it can be created easily, especially when compared to other types of energy such as coal, oil, ethanol, and biodiesel. These compounds are much harder to create and can require long periods of time to reach a useable state.

▶ Sources of Electricity

K06001 Identify common sources of electricity.

The most common and easily observable source of electricity is the static electricity produced in thunderstorms. This electricity is formed when warm air and cold air fronts meet, forming thunderclouds. When the two fronts collide, the warm air, which contains water droplets and usually a positive charge, rises, and the cold front, which contains ice crystals and a negative charge, sinks. As the water droplets and ice crystals collide and rub together in the thundercloud, static electricity is created. As the static electricity builds in the cloud, positive charges accumulate at the top of the cloud, and negative charges accumulate in the bottom of the cloud. When the charge is strong enough, it is released in the form of lightning, which contains hundreds of thousands of volts (**FIGURE 6-1**). Static electricity can also be created as you get into a vehicle, walk across carpet, or wipe your hands with a rag. The amount of static electricity created during these ordinary activities can severely damage electronic components; it is critical that you ALWAYS ground yourself by touching metal somewhere on the chassis before handling electronics to ensure the static electricity has passed from your body BEFORE attempting to service any type of electronic component. Electricity can also be produced by moving a conductor through a magnetic field, applying pressure to a special type of crystal, converting sunlight into electricity through photovoltaic solar cells, and chemical reactions that take place inside of batteries. One fact remains constant: electricity is always the movement of electrons in a conducting circuit. In this chapter, you will explore the many different ways in which electricity can be produced, and how electricity affects the materials that it passes through.

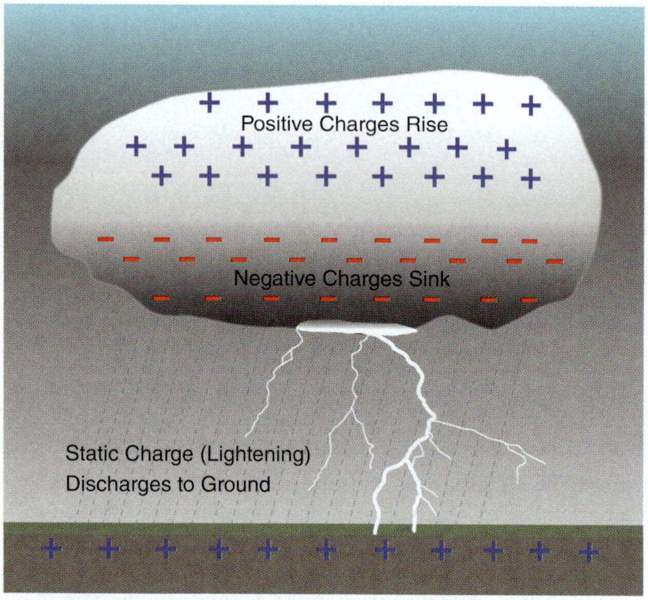

FIGURE 6-1 Demonstrating electrical charges in thundercloud.

FIGURE 6-2 Positive and negative charges creating static.

Electrostatic Electricity

Static electricity can be induced by rubbing two insulators together. You may be thinking, "why would I purposely rub two insulators together?" What you may not realize is a great majority of the materials that you come into contact with on a daily basis are insulators, and as a result, merely going about your daily activities will usually generate a certain amount of static electricity on your person. As you move, these insulators rub against each other, causing one material to lose electrons to the other. The insulator losing electrons becomes positively charged. The other insulator gains electrons to become negatively charged. For example, when you walk across a wool carpet with leather shoes, your shoes pick up free electrons from the carpet with each step you take (**FIGURE 6-2**). This process happens very quickly, and by the time you lift your shoe from the carpet, the extra electrons picked up from the carpet distribute themselves around your entire body. This collection of electrons happens over and over again, until you have no more room on and around your body for extra electrons. By this time, you can have 15,000 to 25,000 volts built up in a negative charge of static electricity! The surplus of

FIGURE 6-3 Static electricity being released in form of a spark.

FIGURE 6-4 Thermocouple.

a negative charge on your body creates an imbalance between positive and negative charges. Since electrons always seek to be in perfect balance, when a positively charged surface, such as a door knob, another person that has fewer electrons than you, or an electronic control unit that you touched without first grounding yourself, comes into close proximity of your body, the electrons run from you, and towards whatever you touched, in the form of a spark. You will likely feel a pretty good shock when this transfer occurs (**FIGURE 6-3**). It is important to know that static electricity is capable of creating large and powerful sparks, which are sufficient to ignite flammable vapors. As a safety precaution, you should always ground yourself before attempting to pump gasoline or handle any flammable materials. Failure to do so can ignite any flammable vapors present and start a fire or explosion. These high voltages can also cause severe damage to electronic components if they are introduced into the device through any wires or connectors pins. For this reason, NEVER touch an electronic device on its pins or connectors, and always ground yourself before servicing these components.

Thermoelectric Energy

In some cases, it is desirable to use heat to create an electrical voltage signal. This is accomplished when two dissimilar metals are joined and heated; the point that is heated is called the **hot junction**, and the whole device is called a **thermocouple** (**FIGURE 6-4**). When heat is applied to the hot junction, a small electrical current is generated. A good rule of thumb is that for a temperature rise of around 392°F (200°C), the potential difference created ranges from 9 to 15mV on a loaded circuit, and up to 30mV on an unloaded circuit. In some cases, it is desirable to generate a larger voltage in the presence of heat. In these cases, several thermocouples are connected in series, effectively boosting the output voltage. These stronger thermoelectric devices are called **thermopiles**, and they are capable of generating 450mV to 750mV (**FIGURE 6-5**). These devices are very simple, and can easily be tested by heating the hot junction, and testing the output with a voltmeter (**FIGURE 6-6**). These thermoelectric devices are often used in gas-fired appliances as a means to prove the presence of a flame, and prevent raw gas from being vented into the home. Thermoelectric devices are also used in the development of engine designs; manufacturers use thermocouples to measure the high temperatures of components, such as spark plugs and exhaust systems (**FIGURE 6-7**). Thermocouples are also used to measure exhaust gas temperature on race cars to determine the best air–fuel ratio so that the engine parts will not be overheated. Pilots in general aviation also use thermocouples to measure the exhaust gas temperature to adjust the fuel mixture as the airplane changes altitude.

SAFETY TIP

There have been many fires caused by the operator starting the refueling process, getting back into the vehicle to check on a child or put a credit card away, sliding back out across the seat of the vehicle (without touching the metal of the vehicle), and then grabbing the refueling nozzle, at which time a spark ignites the fumes escaping from the filling tank. Always touch the metal of the vehicle or fuel pump before touching the refueling nozzle. Also, fires have reportedly resulted from customers sliding their plastic gas cans out of the back of a plastic-lined pickup bed, creating a spark when the nozzle touched the can or when the pour spout touched the gas tank.

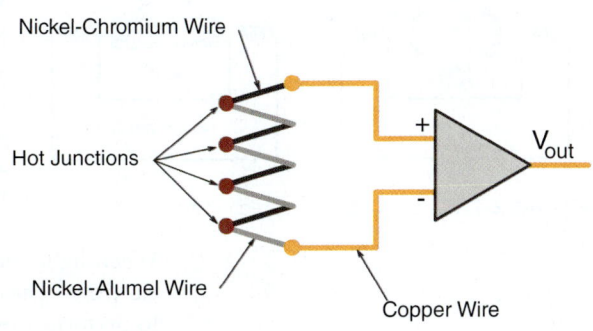

FIGURE 6-5 Thermopile.

FIGURE 6-6 Thermocouple being tested with a voltmeter.

FIGURE 6-7 Thermocouple in an exhaust system.

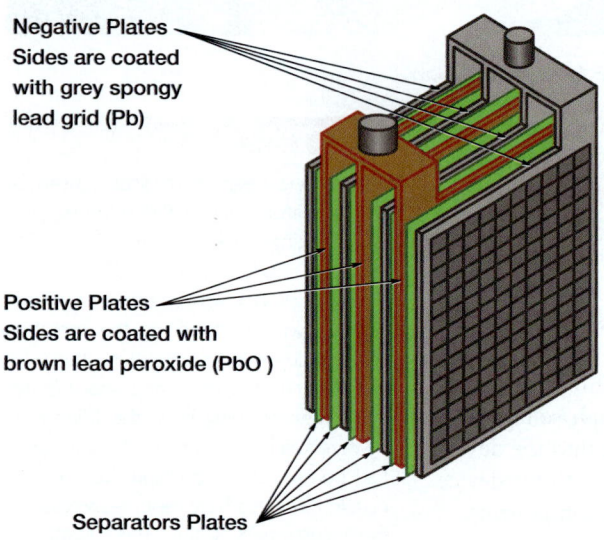

FIGURE 6-8 Automotive battery construction.

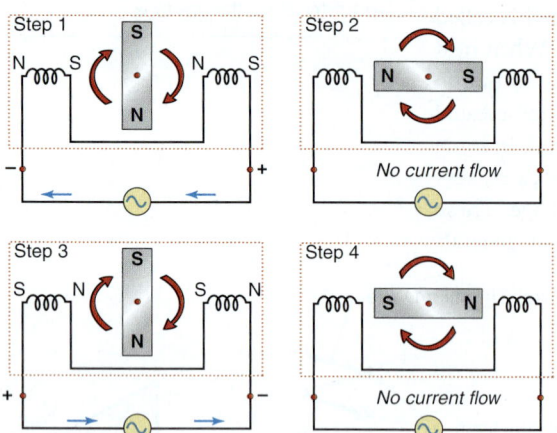

FIGURE 6-9 Battery cells connected in series.

Electrochemical Energy

When two dissimilar metals are immersed in an acidic liquid called an **electrolyte**, the breakdown of chemicals into charged particles, called **ions**, results in a flow of electricity. The process is called **electrolysis**. This principle is applied in the standard lead-acid battery used in most vehicles, of which there are two prevalent types, the standard flooded-cell lead acid battery that contains a liquid electrolyte, and an absorbed glass mat battery, which uses an electrolyte paste. It is important to make the distinction that batteries do not generate voltage on their own; they merely store voltage from another source, which in automotive applications would be either a battery charger or a vehicle's alternator. This is why automotive batteries are referred to as storage batteries. Automotive batteries consist of a group of cells connected in series. These cells are made up of two lead plates; a positive plate coated with lead dioxide paste, a negative plate consisting of sponge lead, and a separator between the plates to prevent them from making contact and shorting out (**FIGURE 6-8**). A typical 12-volt automotive battery contains six cells that generate 2.1 volts each. These cells are connected in series to achieve a fully charged battery voltage of 12.6 volts (**FIGURE 6-9**). This voltage is very specific, and for good reason. When a battery measures 12.4 volts, it is at approximately 75% state of charge (SOC), and when it reads 12.2 volts, it is at approximately 50% SOC. Many electrical concerns can be traced to a faulty battery, and as such, you must inspect this component of the electrical system carefully when diagnosing electrical concerns. This is a basic introduction to automotive batteries, which are covered in greater detail in the Batteries, Starting, and Charging chapter.

Photovoltaic Energy

Solar cells convert sunlight directly into electricity and are used in a wide variety of applications to supply electricity. They are made of semiconducting materials similar to those used in computer chips. When sunlight is absorbed by these materials, the photons knock electrons loose from their atoms, causing the electrons to flow through the material to produce electricity. This process of converting light (photons) to electricity (voltage, which moves electrons) is called the **photovoltaic (PV) effect** (**FIGURE 6-10**). When light energy strikes the surface of PV semiconductor materials, the materials emit electrons. These freed electrons can then be made to flow in a circuit where they are used to perform work. The principle is applied in some ignition systems and in vehicle speed

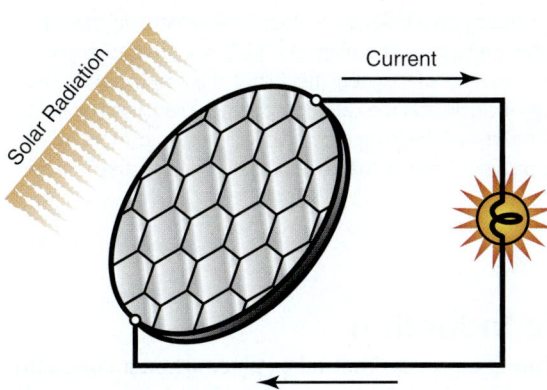

FIGURE 6-10 Photovoltaic effect.

FIGURE 6-11 Knock sensor.

sensors by creating signals used to time the ignition sparks or by telling the computer how fast the vehicle is moving. It is also used to sense the amount of daylight to operate automatic headlights and is used in heating, ventilation, and air-conditioning (HVAC) systems to determine the sun load on the vehicle. In addition, it is the principle of the solar cell.

Piezoelectric Energy

When crystals of certain materials, such as quartz, are subjected to mechanical stress, an electrical potential is created. The best-known application of **piezoelectric** energy are the sparking lighters that ignite flammable vapors in gas grills, hot water heaters, and residential furnaces. The phenomenon of piezoelectricity is often used to measure vibration, and the amount of stress that is put on an object. Automotive knock sensors use piezoelectric crystals to generate a small voltage when pre-ignition detonation is present, which prompts the Engine Control Module to adjust ignition timing and/or fuel delivery (**FIGURE 6-11**). Piezoelectric devices are also used in strain gauges, which are designed to detect stress in a component. These devices are often used to detect the weight of passengers to determine airbag deployment (**FIGURE 6-12**). In addition to generating voltage when subjected to stress, piezoelectric crystals deform when an electric current is passed through them. This phenomenon is used to actuate fuel injectors in diesel and gasoline direct-injection applications (**FIGURE 6-13**).

> **▶ TECHNICIAN TIP**
>
> The term photovoltaic comes from the Greek *photos*, meaning light, and the name of the Italian physicist Volta, after whom the volt (and consequently voltage) is named. Photovoltaic means literally "of light and electricity."

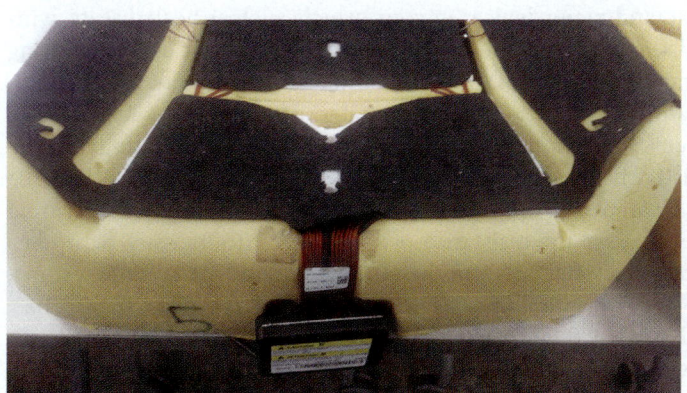

FIGURE 6-12 Automotive strain gauge for airbag systems.

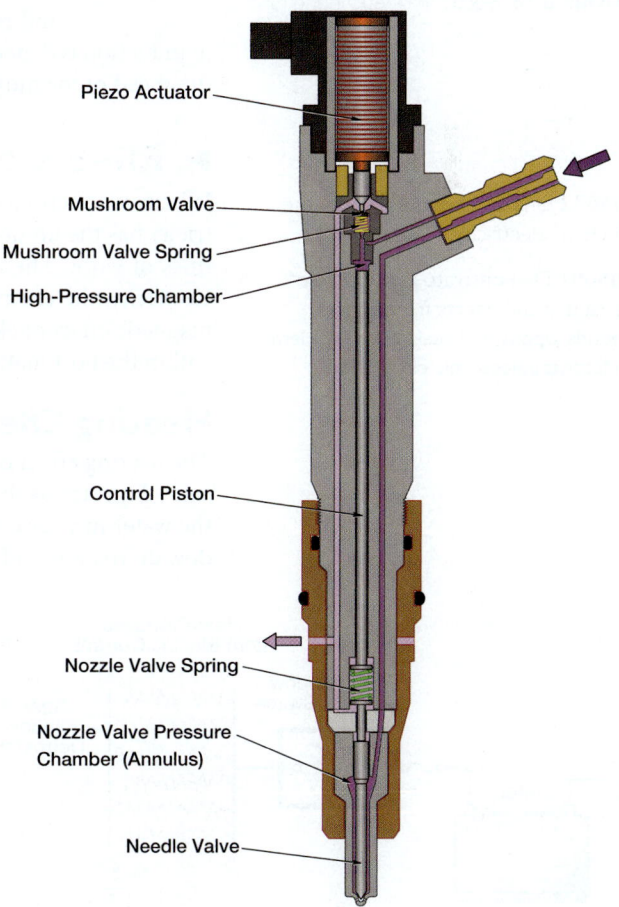

FIGURE 6-13 Piezo fuel injector.

Applied Science

AS-75: Electromagnetism: The technician can explain the relationship between current in a conductor and strength of the magnetic field.

Faraday's law of induction describes the ways that a voltage can be generated within a conductor moved through a magnetic field. Simply stated, whenever there is a change in the number of magnetic field lines passing through a conductor, whether the conductor is moved or changes occur in the magnetic field, a voltage is generated in the conductor. When the strength of the magnetic field is increased, there is an increase in the number of magnetic field lines. If a conductor moves through the magnetic field and the rate of movement remains constant, the increased intensity of the magnetic field will result in more magnetic field lines passing through the conductor, increasing the voltage generated.

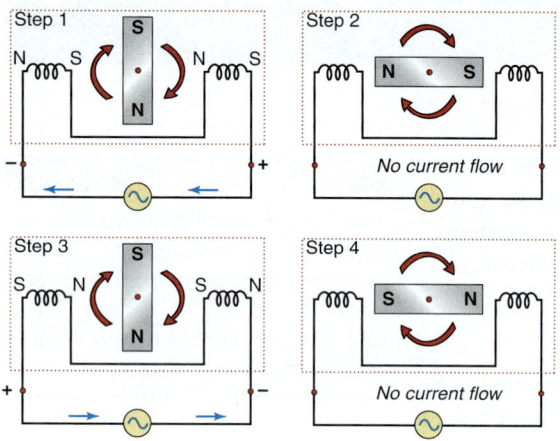

FIGURE 6-14 Electromagnetic induction.

K06002 Understand and identify the effects of electricity.

N06001 Demonstrate knowledge of the causes and effects from shorts, grounds, opens, and resistance problems in electrical/electronic circuits.

Electromagnetic Induction

Electromagnetic induction is a process that takes place when a conductor is placed in a changing magnetic field. This causes an electric current to be created across the conductor, generating voltage (**FIGURE 6-14**). This current flows one way when the conductor cuts the field in one direction, then reverses as it cuts the field in the opposite direction. This reversing action creates an alternating current, or AC voltage. Moving a magnet inside a stationary coil of wire produces the same effect. For every half-revolution, current flow reverses. Increasing the speed of the magnet increases the amount of electrical energy produced. The electromagnetic induction principle is applied in alternators, ignition coils, and some sensors on the vehicle. In the case of alternators, we need to convert the AC voltage produced by the alternator to the DC voltage used in automotive applications. This is accomplished by the use of a group of diodes, called a rectifier bridge. In order for induction to take place, three elements must be present: a winding, a magnet, and relative movement (i.e., movement of one past the other). The amount of induction is dependent upon the strength of the magnetic field, the number of windings, the speed of the movement, and the relative distance between the field and the winding.

▶ Effects of Electricity

Whenever electricity flows, it produces effects in the material through which it flows. Electricity has the unique ability to alter materials, create change, and carry out massive quantities of work. The effects of electricity can take several forms, such as the heat effect of electricity, the chemical effect of electricity, the light effect of electricity, and the electromagnetic effect of electricity. Each of these effects are unique and create a specific reaction within the host material.

Heating Effects of Electricity

The heating effect of electricity is used in many of the appliances and devices that we use everyday, such as the toaster that you use in the morning, the heating element that warms the water in your coffee pot, and many devices in your vehicle such as heated seats, window defrosters, and heated rear-view mirrors. These devices use a device called a heating element (**FIGURE 6-15**), which can be constructed of metal alloys, ceramics, or other composite materials, and introduces a resistance to current flow into the circuit. This type of heating is called resistive heating; when electric current is passed through the heating element, the element material becomes warmer. This warming effect causes the resistance of the element to increase with temperature; a phenomenon that affects all conductors. After a time, the temperature and resistance of the heating element stabilize, and current flow through the circuit is maintained at a manageable level. This effect is not always desirable, and can be magnified if the diameter of the conductor is too small for the level of current flow, the length of the conductor is excessive, rust or corrosion are

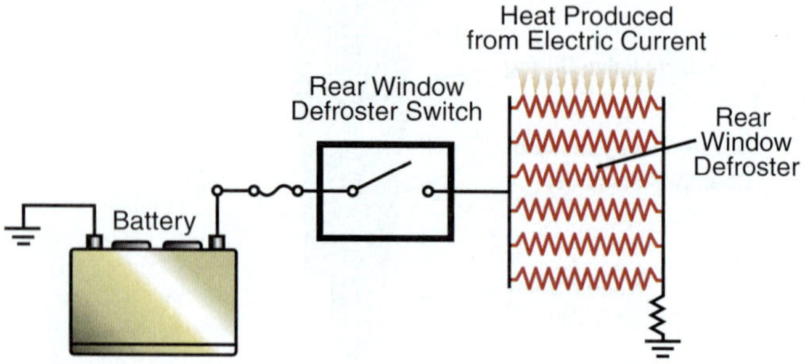

FIGURE 6-15 Heating elements.

present at connections within the circuit, or there are loose connections at any point of the circuit. In these cases, when current flows, the portion of the circuit containing resistance heats up, and the resistance climbs. This action keeps occurring as long as current is being pushed through the circuit, and after a point, thermal runaway takes place, and the circuit will melt down, possibly catching fire (**FIGURE 6-16**).

Chemical Effects of Electricity

Chemical effects of electricity depend on ions. Recall that ions are electrically charged atoms or groups of atoms. Atoms that gain electrons become negatively charged. Atoms that lose electrons become positively charged. The chemical effect of electricity occurs when an electric current passes through a liquid or through conductive materials. When an electric current is passed through water, water dissociates into hydrogen and oxygen. Hydrogen is deposited over the negative pole and oxygen is deposited over the positive pole. In a lead-acid battery, the electrical and chemical differences between the sets of plates create a potential difference or voltage, which makes the current flow in a circuit. It flows in one direction only, so it is called direct current, or DC. Deposition of hydrogen and oxygen at different poles is visible in the form of bubbles that reach the surface of the water. This occurs when an automotive storage battery is charged; great care should be taken so as not to create sparks on or around a charging battery. When you pass an electric current through a solution of metal and salt, the metal gets deposited on the negative pole of the circuit. This action is used in metal-plating operations.

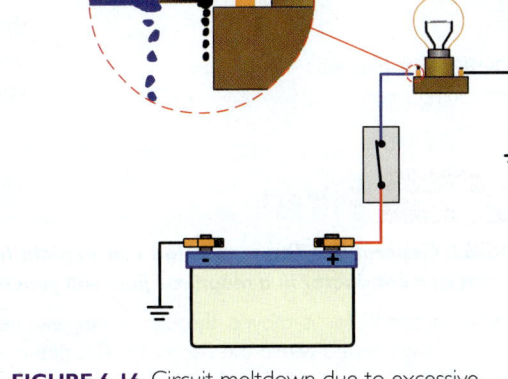

FIGURE 6-16 Circuit meltdown due to excessive resistance.

Light Effects of Electricity

The production of light can also be an effect of electricity. The light effects of electricity can be achieved in many ways. In standard incandescent lamps, the mechanism to create light is similar to a heating element, in that a filament gets glowing hot and as a result emits light. The main difference between an incandescent lamp and a heating element is the lamp filament is contained in a vacuum within the glass bulb. This is an inefficient method of creating light, as only 10% of the energy creates useable illumination, while the other 90% of the energy is wasted to heat. Heat production is a by-product of current flowing through a conductor, and it is a sure sign of inefficiency; the more heat present, the less efficient the conversion process. LEDs are increasingly popular because they are much more efficient than incandescent lamps. An LED is a semiconductor diode that creates light by emitting photons from its PN (**FIGURE 6-17**). At one time, LEDs were available only in the red, green, and yellow tints. More recently, they have become available in many different colors and have grown in light intensity. They are now replacing traditional bulbs as lighting sources and produce as much or more light with less heat loss, using much less energy. LEDs are now used as headlights on many high-end cars; it is likely that as LED technology improves and prices come down, LED headlights will be standard on more and more new vehicles.

Electromagnetic Effects of Electricity

The magnetic effect of electricity, or electromagnetism, occurs when an electric current is passed through a conductor. When current passes through a straight conductor, magnetic lines of flux occurs in a circular pattern perpendicular to the conductor (**FIGURE 6-18**). The more current that passes through the conductor, the stronger the magnetic field that is produced. This effect is magnified further if we wrap a conductor around an iron core. This effectively increases the strength of the magnetic field created, and makes for a stronger electromagnet.

These magnetic effects are used to generate electricity and induce movement. Sometimes they are merely the by-products of current flow. The size of the electromagnet is determined by its application. A fuel injector will contain a small electromagnetic coil using very fine wire, whereas a starter motor will use heavier wire in larger coils. Many components

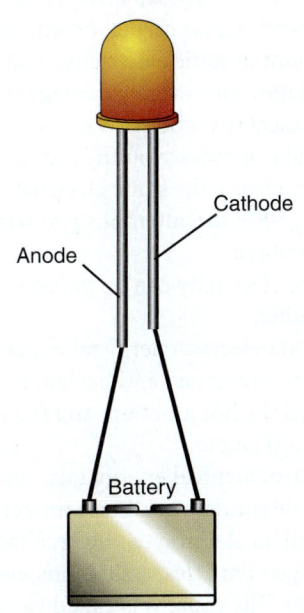

FIGURE 6-17 Light created by an LED.

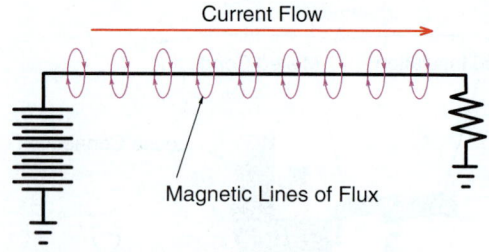

FIGURE 6-18 Magnetic lines of flux surrounding a conductor.

use electromagnetism to operate. If a component has an electrical connection and movement is created by or within the component, then that component will use an electromagnet to facilitate this movement. Devices such as relays, solenoids, and motors use electromagnets to create movement.

When magnetic lines of flux pass through a conductor at a right angle, voltage is induced in the conductor. The amount of voltage induced depends on the strength of the magnetic field, as well as the frequency at which the magnetic field is passed through the conductor. Increases in either of these parameters will cause more voltage to be induced in the conductor. Decreases in these parameters will result in less voltage being induced.

Applied Science

AS-63: Generators: The technician can explain how the movement of a conductor in a magnetic field will generate electricity.

When a conductor is moved through a magnetic field, an electrical current is generated within the conductor. This phenomenon is known as electromagnetic induction. Alternatively, the conductor can remain stationary while the magnetic field is moved in relation to it. The generation of a voltage depends directly on the movement of either the conductor or the magnetic field. Without movement, there is no voltage generated. Electromagnetic induction works by exposing the electrons within the conductor to a magnetic force acting perpendicular to both their motion and the magnetic field, which forces them to move through the conductor, creating a charge.

Wrap-Up

Ready for Review

- Energy is present in everything that we come into contact with. Energy can be manifested in many forms and may be transferred between objects. It cannot be created or destroyed.
- The most common and easily observable source of electricity is the static electricity produced in thunderstorms.
- Static electricity can also be created as you get into a vehicle, walk across carpet, or wipe your hands with a rag. The amount of static electricity created during these ordinary activities can severely damage electronic components; it is critical that you ALWAYS ground yourself by touching metal somewhere on the chassis before handling electronics to ensure the static electricity has passed from your body BEFORE attempting to service any type of electronic component.
- Static electricity can be induced by rubbing two insulators together.
- Thermoelectric energy takes place when two dissimilar metals are joined and heated; the point that is heated is called the hot junction, and the whole device is called a thermocouple.
- Electrochemical energy takes place for example, when two dissimilar metals are immersed in an acidic liquid called an electrolyte, the breakdown of chemicals into charged particles, called ions, results in a flow of electricity. The process is called electrolysis. This principle is applied in the standard lead-acid battery used in most vehicles.
- Photovoltaic energy takes place when solar cells convert sunlight directly into electricity and are used in a wide variety of applications to supply electricity. They are made of semiconducting materials similar to those used in computer chips. When sunlight is absorbed by these materials, the photons knock electrons loose from their atoms, causing the electrons to flow through the material to produce electricity.
- The process of converting light (photons) to electricity (voltage, which moves electrons) is called the photovoltaic (PV) effect.
- Piezoelectric energy takes place when crystals of certain materials, such as quartz, are subjected to mechanical stress and an electrical potential is created.
- The best-known application of piezoelectric energy is the sparking lighters that ignite flammable vapors in gas grills, hot water heaters, and residential furnaces.
- The phenomenon of piezoelectricity is often used to measure vibration and the amount of stress that is put on an object.
- Automotive knock sensors use piezoelectric crystals to generate a small voltage when pre-ignition detonation is present, which prompts the engine control module to adjust ignition timing, and/or fuel delivery.
- Electromagnetic induction is a process that takes place when a conductor is placed in a changing magnetic field.

This causes an electric current to be created across the conductor, generating voltage.

- Whenever electricity flows, it produces effects in the material through which it flows. Electricity has the unique ability to alter materials, create change, and carry out massive quantities of work.
- The heating effect of electricity is used in many devices in your vehicle such as heated seats, window defrosters, and heated rear-view mirrors. These devices use a device called a heating element, which introduces a resistance to current flow into the circuit and creates the heat from the resistance.
- Chemical effects of electricity depend on ions. Recall that ions are electrically charged atoms or groups of atoms. Atoms that gain electrons become negatively charged. Atoms that lose electrons become positively charged. The chemical effect of electricity occurs when an electric current passes through a liquid or through conductive materials.
- The production of light is an effect of electricity. In standard incandescent lamps, the mechanism to create light is similar to a heating element, in that a filament gets glowing hot and as a result emits light. The main difference between an incandescent lamp and a heating element is the lamp filament is contained in a vacuum within the glass bulb. Heat production is a by-product of current flowing through a conductor.
- An LED is a semiconductor diode that creates light by emitting photons from its PN junction.
- The magnetic effect of electricity, or electromagnetism, occurs when an electric current is passed through a conductor. When current passes through a straight conductor, magnetic lines of flux occur in a circular pattern perpendicular to the conductor.
- The more current that passes through the conductor, the stronger the magnetic field that is produced. This effect is magnified further if we wrap a conductor around an iron core and create a stronger electromagnet.
- Magnetic effects are used to generate electricity and induce movement. Sometimes they are merely the by-products of current flow.
- Many components use electromagnetism to operate. If a component has an electrical connection and movement is created by or within the component, then that component will use an electromagnet to cause this movement.
- Relays, solenoids, and motors use electromagnets to create movement.
- When magnetic lines of flux pass through a conductor at a right angle, voltage is induced in the conductor. The amount of voltage induced depends on the strength of the magnetic field and frequency at which the magnetic field is passed through the conductor.

Key Terms

electrolysis A method of using electrical current to create a chemical reaction.

electrolyte A mixture of water and acid that contains free ions that make it electrically conductive.

electromagnetic induction The production of an electrical current in a conductor when it moves through a magnetic field or a magnetic field moves past it.

hot junction The heating point of a thermocouple.

ion An atom that has fewer electrons than protons (positive) or that has more electrons than protons (negative).

photovoltaic (PV) effect The conversion of sunlight into electricity.

piezoelectric A type of electricity in which a material such as a quartz crystal produces voltage when mechanical pressure distorts it.

thermocouple A temperature-sensing component that consists of two dissimilar metals that produce voltage proportional to temperature.

thermopile Several thermocouples connected in series to boost output voltage.

Review Questions

1. Which of the following requires the shortest period of time to reach a useable state?
   a. Electricity
   b. Oil
   c. Ethanol
   d. Biodiesel
2. When pre-ignition detonation is present, which of these prompts the Engine Control Module to adjust ignition timing and/or fuel delivery?
   a. Thermoelectric energy
   b. Photovoltaic energy
   c. Piezoelectric energy
   d. Electrochemical energy
3. When two dissimilar metals are immersed in an acidic liquid, the breakdown of chemicals into charged particles results in a flow of electricity. This principle is used in a(n):
   a. HVAC system.
   b. battery.
   c. sensor.
   d. LED.
4. All of the following statements are true with respect to the use of thermoelectric devices *except*:
   a. They are used in gas-fired appliances as a means to prove the presence of a flame.
   b. They are used in the development of engine designs.
   c. They are used to measure the high temperatures of components, such as spark plugs.
   d. They are used to store voltage from another source.
5. Which of the following is the process of converting light energy to electrical energy?
   a. Thermoelectric effect
   b. Photovoltaic effect

c. Electrochemical effect
   d. Piezoelectric effect
6. A temperature-sensing component that consists of two dissimilar metals that produce voltage proportional to temperature is a(n):
   a. thermopile.
   b. thermocouple.
   c. electrolyte.
   d. piezoelectric material.
7. All of the following devices use electromagnets *except*:
   a. solenoids.
   b. motors.
   c. headlights.
   d. relays.
8. Choose the correct statement.
   a. The incandescent lamp filament is present in a vacuum within the glass bulb.
   b. A fuel injector contains a large electromagnetic coil of very fine wire.
   c. Electromagnetism occurs when an electric current is passed through a conductor.
   d. Devices such as relays, solenoids, and motors use piezoelectricity to create movement.
9. The AC voltage produced by the alternator is converted to the DC voltage used in automobile applications using a:
   a. rectifier bridge.
   b. resistor.
   c. magnet.
   d. sensor.
10. Electricity can produce all of the following effects *except*:
    a. heat.
    b. light.
    c. chemical.
    d. mechanical.

ASE Technician A/Technician B Style Questions

1. Tech A says that energy is created within the earth's core, and cannot be destroyed. Tech B says that energy is present in all objects, and cannot be created or destroyed; it can merely be transformed. Who is correct?
   a. Tech A
   b. Tech B
   c. Both A and B
   d. Neither A nor B
2. Tech A says if aftermarket stereo amplifier cables are improperly over a vehicle's wiring harness, the magnetic lines of flux generated by the power cables as current flows through them can induce voltage in the underlying wiring harness, and in turn, cause the related electrical systems to malfunction. Tech B says that it is a good practice to disable any aftermarket accessories before attempting to diagnose strange electrical problems. Who is correct?
   a. Tech A
   b. Tech B
   c. Both A and B
   d. Neither A nor B
3. Tech A says that static electricity can damage an electronic module. Tech B says that you should always ground yourself before attempting to service any electronic module to prevent static electricity from damaging the component. Who is correct?
   a. Tech A
   b. Tech B
   c. Both A and B
   d. Neither A nor B
4. Tech A says that lightning is an example of static electricity. Tech B says that static electricity is generated within the vehicle's body as it drives down the road. Who is correct?
   a. Tech A
   b. Tech B
   c. Both A and B
   d. Neither A nor B
5. Tech A says if you twist to pieces of dissimilar metal together, and then heat the point where the two metals meet, a small voltage will be created across that two metals. Tech B says that one device that generates electricity when it is heated is called a thermocouple. Who is correct?
   a. Tech A
   b. Tech B
   c. Both A and B
   d. Neither A nor B
6. Tech A says if you submerge two dissimilar metals in an electrolyte solution, a small voltage will be created between the two metals. Tech B says that most automotive batteries use sulfuric acid as an electrolyte. Who is correct?
   a. Tech A
   b. Tech B
   c. Both A and B
   d. Neither A nor B
7. Tech A says that the photovoltaic effect is the process of converting light energy into electricity. Tech B says that photovoltaic components are never used in automotive applications. Who is correct?
   a. Tech A
   b. Tech B
   c. Both A and B
   d. Neither A nor B
8. Tech A says that certain materials, such as quartz, emit a small voltage when they are subjected to mechanical stress. Tech B says that this is called the piezoelectric effect. Who is correct?
   a. Tech A
   b. Tech B

c. Both A and B
   d. Neither A nor B
9. Tech A says that anytime magnetic lines of flux pass through a conductor, voltage is induced in the conductor. Tech B says that the amount of voltage induced in a conductor is partly determined by the strength of the magnetic field that passes through it. Who is correct?
   a. Tech A
   b. Tech B
   c. Both A and B
   d. Neither A nor B
10. Tech A says when you heat certain materials, their electrical resistance increases. Tech B says that excessive electrical resistance in a circuit can cause heat to build up at the point of resistance. Who is correct?
    a. Tech A
    b. Tech B
    c. Both A and B
    d. Neither A nor B

CHAPTER 7
Ohm's Law and Circuits

NATEF Tasks

- **N07001** Demonstrate knowledge of the causes and effects from shorts, grounds, opens, and resistance problems in electrical/electronic circuits.
- **N07002** Demonstrate proper use of a test light on an electrical circuit.
- **N07003** Demonstrate proper use of a digital multimeter (DMM) when measuring source voltage, voltage drop (including grounds), current flow, and resistance.

Knowledge Objectives

After reading this chapter, you will be able to:
- **K07001** Explain Ohm's law.
- **K07002** Calculate basic Ohm's law calculations.
- **K07003** Describe series and parallel circuits.
- **K07004** Explain Kirchhoff's current and voltage drop laws.

Skills Objectives

After reading this chapter, you will be able to:
- **S07001** Demonstrate a working understanding of Ohm's law.
- **S07002** Use Ohm's law to solve for missing circuit variables.
- **S07003** Apply Ohm's law to series circuits.
- **S07004** Apply Ohm's law to series-parallel circuits.

You Are the Automotive Technician

Ohm's law applies to any circuit in which electricity flows. It is one of the most valuable pieces of information to be considered when diagnosing electrical circuits. When properly applied, it allows the technician to calculate the resistance of difficult to reach component(s) by measuring their current draw upstream in the circuit; current flow and voltage can also be calculated as the need arises. Applying Ohm's law to the electrical diagnostic procedure will go a long way to increase diagnostic efficiency and accuracy.

1. True or False: Ohm's law applies to all electrical circuits?
2. True or False: Ohm's law is very difficult to use, and has little practical value.

Introduction

Electricity can be one of the most difficult aspects of diagnostics and repair to grasp. This is largely due to the fact that the flow of electricity cannot, under most circumstances, be observed with the naked eye. One thing to keep in mind is that electricity always works the same way, and must always follow the laws of physics. Armed with this knowledge, the technician is able to study the laws of electricity, and use this information to visualize what is happening within an electrical circuit in order to "see" how it operates.

Ohm's Law

Ohm's Law and Circuits

K07001 Explain Ohm's law.

Ohm's law helps us to understand the relationship between volts, amps, and ohms. As with all things in nature, these electrical properties must always balance out. If they do not, you made an error solving the Ohm's law equation. Ohm's law tells us that it takes 1 volt to push 1 amp through 1 ohm of resistance. That gives us the relationship among the three units. Volts and resistance are physical things. Volts exist when a surplus of electrons accumulate, thereby creating electrical pressure. Resistance is the physical restriction of the conductor. Amps are the amount of electrons moved through the conductor. That means that amps are the result of both the volts and resistance. For example, if resistance doubles and voltage stays the same, current flow must be cut in half, due to the added restriction that the electrons must overcome.

To illustrate this relationship, consider the following: If you steadily push (voltage) a loaded wheelbarrow up a hill (ohms), you will move a constant amount of dirt in a given amount of time. If you push a loaded wheelbarrow up a hill twice as long at the same speed, you will effectively move half of the dirt in the same amount of time (**FIGURE 7-1**). It is the same with the electrical circuit: If the pressure (volts) stays the same but the restriction (ohms) doubles, then the electrons (amps) will be reduced by half. Conversely, if the voltage is doubled and the resistance stays the same, then amperage will double. If you can clearly understand this concept, then you are well on your way to diagnosing electrical problems. Consider the following question: If the amps in a circuit are lower than they should be, what are the two possible causes? First, the source voltage could be low. In that case, you would use a voltmeter to measure the source voltage at the battery. The second possibility is that the resistance in the circuit is too high. In that case, you would use a voltmeter to see if there is an excessive voltage drop on both sides of the circuit that would be caused by high resistance (**FIGURE 7-2**). If the voltage drop is within specifications, you would use an ohmmeter to measure the resistance of the load device (**FIGURE 7-3**).

FIGURE 7-1 Wheelbarrow being pushed up two different hills.

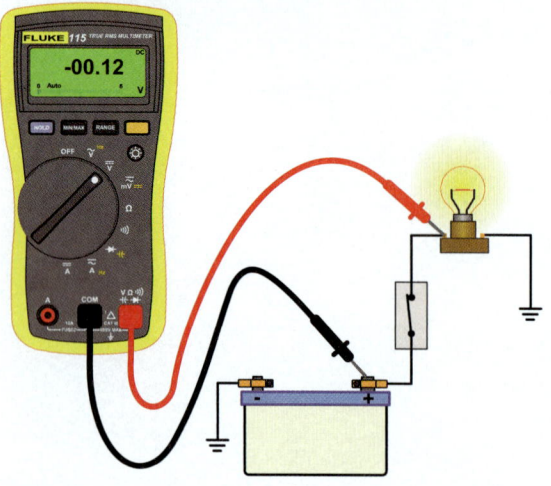

FIGURE 7-2 Meter being used to perform voltage drop test.

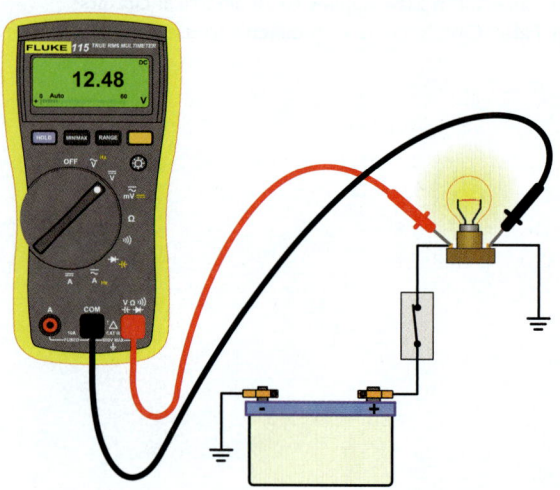

FIGURE 7-3 Load device.

Most of the time when circuits fail it is because the current flow is too low or nonexistent. Although it is possible that the battery is dead or discharged, it is much more likely there is high resistance creating a voltage drop either in the feed side or the ground side of the circuit (**FIGURE 7-4**). Another possibility is the electrical device has too much resistance or is open. Performing a voltage drop test on both sides of the circuit will quickly identify if there is an excessive voltage drop present. Furthermore, a resistance check of the electrical device will indicate a high resistance or open condition in the component. Knowing the specifications that go along with these steps will enable you to repair more than half of the electrical problems you will encounter on a vehicle.

Circuits are made up of components and interconnecting conductors, such as wires, arranged to manage and control the flow of electrons to perform specific electrical tasks. Understanding circuits and how to take electrical measurements is essential to performing electrical repair activities. Circuits come in two basic configurations: series circuits and parallel circuits. The two types can also be combined into what is called a series-parallel circuit. Understanding how electricity behaves within each of these circuits will help you know how the circuit operates and how to approach diagnosis. It will also allow you to apply Ohm's law correctly to each type of circuit. We will look at the types of circuits further after finishing our discussion of Ohm's law.

Ohm's Law and Ohm's Law Calculations

Because Ohm's law is a relationship between volts, amps, and ohms, and because they must always balance out, if we know any two of the values, then we can calculate the third. If resistance stays the same but voltage rises, then the greater force pushes more current through the circuit. If resistance stays the same but voltage decreases, then less current will flow through the circuit. This means the total current flow of a circuit in amps always equals the voltage divided by the resistance. In calculating Ohm's law, R stands for resistance, V for voltage, and A for current (A is for amps).* Depending on which value you wish to solve for, you will apply one of the following three formulas:

- $A = V \div R$
- $V = A \times R$
- $R = V \div A$

Voltage can vary across different points in a circuit, but determining the current at any point can be found without an ammeter by using Ohm's law. Using the Ohm's law circle will help you remember the correct math operation to use. All you have to do is place your finger over the value you are looking for (**FIGURE 7-5**). If you place your finger on the top value (volts), then you would multiply amps by resistance. If you place your finger on one of the side values, then you would divide volts by the other value. This means two things. First, the values always have to balance. And second, as long as you know any two values, the third can be calculated. This is especially helpful when determining current flow, because instead of breaking into a circuit to measure current with a meter, if the voltage and resistance are known, Ohm's law may be used to calculate amps. For example, we can measure battery voltage. Let's assume that 12 volts are present in the battery. The value of the resistor, 4 ohms, is on its casing. Current then equals voltage (12 volts) divided by resistance (4 ohms). We can quickly calculate that there are 3 amps of current flowing through every point in the circuit (**FIGURE 7-6**).

Solving Ohm's Law

Using the rules of Ohm's law gives an accurate method of determining values in an electrical circuit. To assist with these problems, use the Ohm's law circle (Figure 7-5). If the value of V and R are known, then to find A, V is divided by R. Place your thumb over A and the circle tells you this

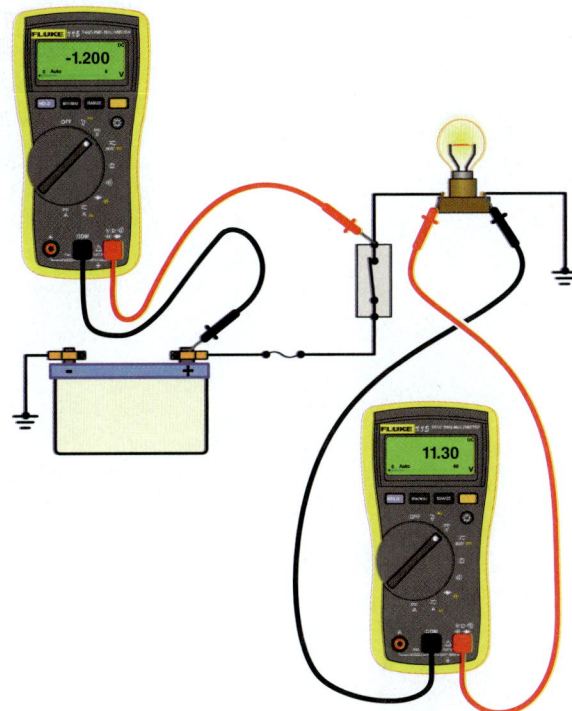

FIGURE 7-4 Excessive voltage drops in an improperly functioning circuit.

K07002 Calculate basic Ohm's law calculations.

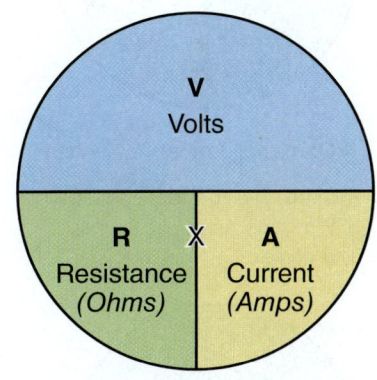

FIGURE 7-5 Ohm's law circle.

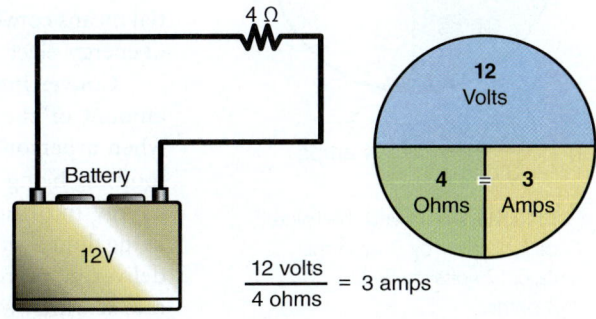

FIGURE 7-6 Calculating current flow in a circuit.

Applied Science

Ohm's Law: The technician can demonstrate an understanding of and explain the use of Ohm's law in verifying circuit parameters (resistance, voltage, amperage).

Ohm's law quantifies the relationship between amperage, voltage, and resistance in any given circuit. It provides a means to calculate any one of these parameters from the values of the other two. Ohm's law is easiest applied through the use of a circle diagram like the one shown here:

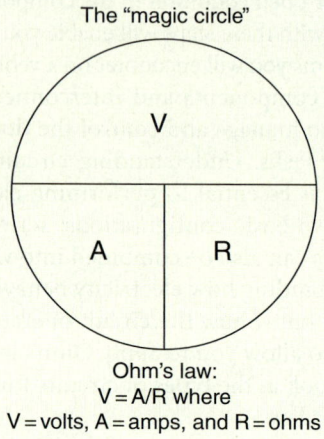

The "magic circle"

Ohm's law:
V = A/R where
V = volts, A = amps, and R = ohms

These diagrams are used by placing your finger over the parameter you are trying to calculate, then reading the remainder of the diagram to identify the calculation you need to perform. For example, if you need to work out the amperage flowing in the circuit, block off the A, and your calculation is V ÷ R, or voltage divided by resistance. For a 12-volt automotive circuit with 20 ohms of resistance, your calculation is 12 ÷ 20, which is 0.6 amps.

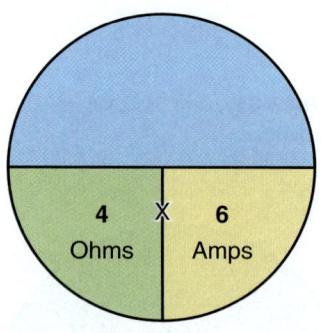

4 Ohms × 6 Amps = 24 Volts

FIGURE 7-7 To find the value of V, multiply A by R, or 6 amps by 4 ohms.

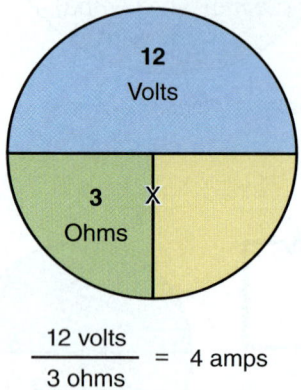

$$\frac{12 \text{ volts}}{3 \text{ ohms}} = 4 \text{ amps}$$

FIGURE 7-8 To find the value of A, divide V by R, or in this case, 12 volts divided by 3 ohms.

formula. Similarly, if V and A are known, then R can be found by dividing V by A. If A and R, are known, then V is found by multiplying A by R.

In the circuit diagram in **FIGURE 7-8**, the value of the amperage is not known. The applied voltage is 12 volts (V) and the resistance is 3 ohms (R). To find the value of A, simply divide V by R, or 12 by 3. The value of A is 4 amps (A).

In the circuit diagram in **FIGURE 7-7**, the value of the applied voltage is not known but the amperage and resistance are. To find the value of V, multiply A by R, or 6 amps by 4 ohms. The result is 24 volts.

Electrical Power and the Power Equation

Energy is the potential to do work. However, work is done only when the energy is released. A disconnected battery is not doing work, but it has the potential to do work and is therefore a source of energy. The difference in electron supply at the battery terminals creates electrical force and is sometimes called the potential difference. In the case of a standard charged automotive battery, it has a potential of approximately 12 volts. Tapping this potential means converting one form of energy, the battery's chemical energy, into another form of energy, electrical energy.

Converting one form of energy into another is called energy transformation. The amount of energy transformed is the amount of work done. Consider the following: when a person's legs turn the pedals of a bicycle, chemical energy (from oxygen and food) is being turned into mechanical energy. A motorcycle engine turns chemical energy into thermal energy, and then into mechanical energy. In each case, **work** is being done, but there is a difference with the motorcycle—it does the work more quickly, delivering more mechanical energy faster. That difference is called **power**. Power is the rate at which work is performed. It is also known as the rate of transforming energy. In

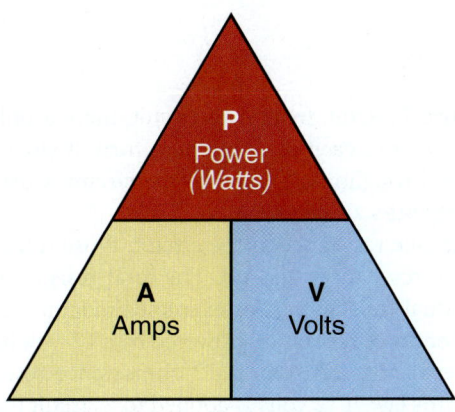

FIGURE 7-9 Watt's law triangle.

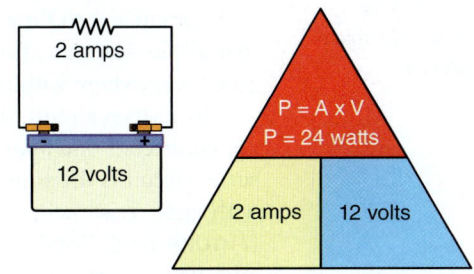

FIGURE 7-10 Circuit flowing 12 volts at 2 amps.

an electrical circuit, power refers to the rate at which electrical energy is transformed into another kind of energy.

The unit of electrical power is the **watt**. One watt is produced when 1 volt causes 1 amp of current to flow. From this comes the power equation: P, the power in watts, equals V, the voltage in volts, multiplied by A, the current in amps. This calculation is applied similarly to Ohm's law and is typically represented as a triangle (**FIGURE 7-9**). When current flows in a circuit with a resistor in it, the resistor becomes hotter as it converts electrical energy into heat energy. If this circuit is powered by a 12-volt battery with a current of 2 amps, using the power equation ($P = V \times A$) we can determine that the resistor is using 24 watts of power (**FIGURE 7-10**). It is also possible to simplify and transpose the power equation. If power equals voltage times current, then:

- Voltage equals power divided by current: $V = P \div A$
- Current equals power divided by voltage: $A = P \div V$

Solving Power Equations

Electrical power is a measurement of the rate at which electricity is consumed or created. When used in relation to loads, it is a measure of electricity consumed. A light bulb uses a certain amount of electrical power, but the power used is not an indication of brightness; it is a measure of power consumption. When used in relation to generators, it is a measure of electrical power produced. One volt pushing 1 amp equals 1 watt of electrical power. To calculate power (P), the current flow in the circuit is multiplied by the voltage used. So, $P = A \times V$, or Watts = Amps × Volts.

This is demonstrated by the diagram (**FIGURE 7-11**). For reference purposes, 746 watts equal 1 horsepower. If a 12-volt circuit with a single light has a current flow of 5 amps, then applying the formula will yield:

- $P = A \times V$
- $P = 5 A \times 12 V$
- $P = 60 W$

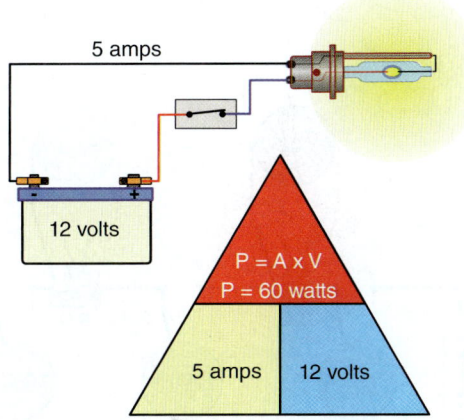

FIGURE 7-11 Circuit containing a blower motor operating at 12 volts, and consuming 15 amps.

The power consumed by the circuit is 60 watts, and the bulb will carry a rating of 60 watts. This rule can be applied to any circuit where the voltage and current flow are known. However, if the values of voltage or current flow are not known, then Ohm's law can be used to determine the missing value. As an example: $V = A \times R$. By expanding $P = A \times V$, it can be said that

$P = A \times (A \times R)$, or $P = A^2 \times R$.

Similarly, by applying $A = V \div R$, we have

$P = (V \div R) \times V$, or $P = V^2 \div R$.

K07003 Describe series and parallel circuits.

N07001 Demonstrate knowledge of the causes and effects from shorts, grounds, opens, and resistance problems in electrical/electronic circuits.

▶ Circuits

Series Circuits

A series circuit is the simplest type of electrical circuit. In a series circuit, there is only one path for current to flow. All of the current flows to each component in turn. It also means that all the electrons flow at the same rate throughout all parts of the circuit. Current is equal everywhere within a series circuit (**FIGURES 7-12** and **7-13**).

In a series circuit, if there is more than one resistance in the circuit, those resistances are connected one after the other; thus the resistances add up. The total resistance in a series circuit is the sum of all of the individual resistances. For example, imagine a circuit with three resistors in series, each having 4 ohms of resistance, powered by a 12-volt battery (**FIGURE 7-14**). Total resistance of the three 4-ohm resistors is 12 ohms, since resistance adds up in a series circuit. According to Ohm's law, if 12 volts is applied to a circuit that has 12 ohms of resistance, the resulting current flow will be 1 amp.

The voltage at different points within the circuit changes: The electromotive force, or pressure, drops from a potential difference of 12 volts as it leaves the battery to virtually no difference, no voltage at all, as it returns to the battery. At each point in the circuit where current flows through a resistance, a drop in voltage occurs, which is called a **voltage drop**.

Voltage drops are good when they occur inside of an intended load, but they cause undesirable effects when they occur where they are not wanted.

In the pictured circuit, after the first resistor, voltage has dropped from 12 to 8 volts. After the second, it is down to 4 volts. After the third, it is 0 volts. The voltage drop across each resistor can be found by subtracting the voltage after a resistor from the voltage before it, or the difference can be measured. The voltmeter will read 4 volts in each case because that is the difference between the two points, the potential difference, or voltage.

Ohm's law can be used in series circuits to calculate voltage, resistance, and current. Any one of these can be calculated as long as the values of the other two are known. The series circuit laws listed here provide a summary of how electricity behaves in a series circuit, which is defined as a circuit with multiple loads but only one path for current to flow:

1. Current flow stays the same in a series circuit. Current flow is the same in all parts of the circuit.
2. Voltage drops as current goes through resistance(s) in series. The applied voltage is equal to the sum of the individual voltage drops (Kirchhoff's voltage law).
3. Resistance adds up in series. Total circuit resistance (R_T) is equal to the sum of the individual resistances; for example, $R_T = R_1 + R_2 + R_3$, and so on.

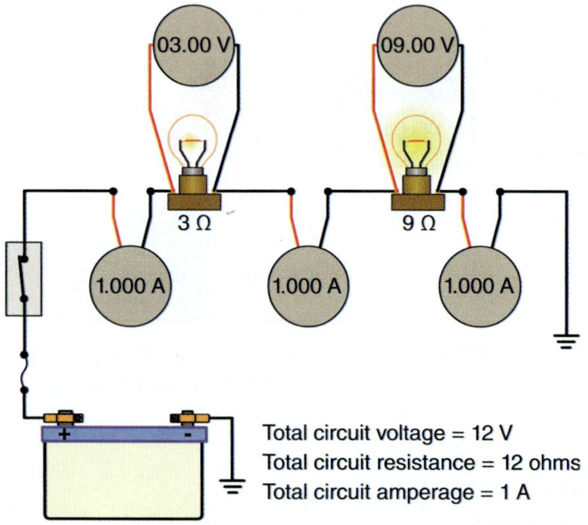

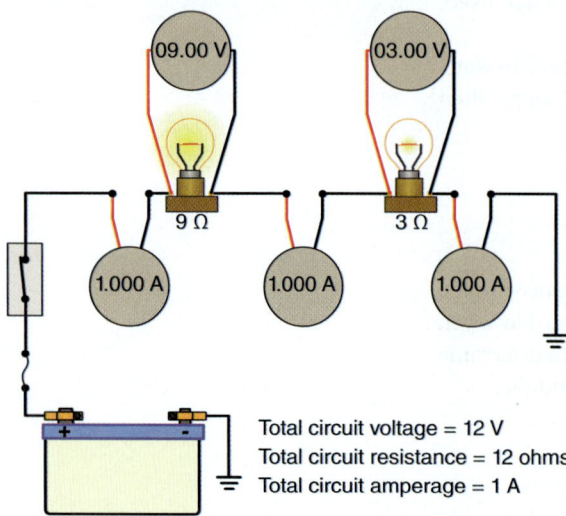

FIGURE 7-12 Electrical behavior in a series circuit with the two unequal-resistance lightbulbs reversed.

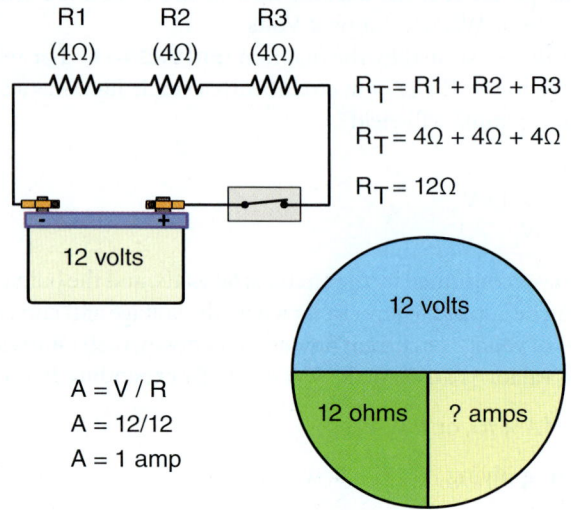

FIGURE 7-13 Electrical behavior in a circuit with two unequal lightbulbs connected in series.

FIGURE 7-14 Circuit with three 4 ohm resistors connected in series.

Parallel Circuits

In a series circuit, components are connected like links in a chain. If any link opens, current to all of the components is cut off. In a parallel circuit, all components are connected directly to the voltage supply (**FIGURE 7-15**). If any connection or component fails in a parallel circuit, current continues to flow normally through the remaining circuits. This is one reason why parallel circuits are used in automotive applications, such as headlight and taillight systems. If one lamp fails, current continues to flow through the other lamps in parallel. In a series circuit, all would go out. You can imagine how this could be disastrous.

Since all components connect directly to the battery terminals, the metal of the vehicle's chassis can become one of the conductors. One terminal of the battery and one terminal of each component can be connected anywhere on the body or chassis to complete the circuit. This is called a common, or ground, connection.

While electricity always behaves according to the laws of physics, when it operates in a parallel circuit, there are some additional laws you need to know. The parallel circuit laws listed here should be applied when working with parallel circuits:

1. The voltage across all branches of a parallel circuit are equal.
2. The total current in a parallel circuit equals the sum of the current flowing in each branch of the circuit.
3. The total resistance of a parallel circuit decreases as more branches are added. Total parallel circuit resistance will always be less than the lowest branch resistance.

Let's look at those laws a bit more closely. A feature of a properly working parallel circuit is that the voltage across each branch is the same as the other branches. No matter how many branches are added, or removed, as long as they are in parallel, the voltage across them will be the same as across each of the other branches, including the battery. Another feature of a parallel circuit is that the current flowing in each branch is determined by the resistance of that branch along with the voltage used by that branch.

In a parallel circuit where the resistors in each branch are the same, the current flowing in each branch is also the same. The sum of their individual currents is equal to the total current flowing in the entire parallel circuit. When the resistances are not equal, the current divides in accordance with the resistance of each branch, but the total current flow is still the sum of the currents flowing in each branch (**FIGURE 7-16**).

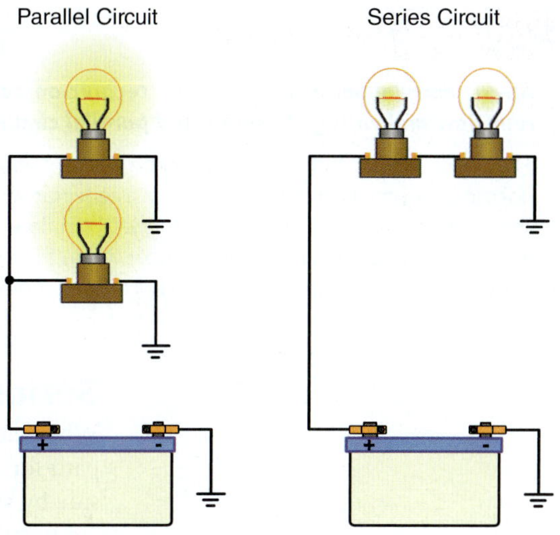

FIGURE 7-15 Typical parallel circuit compared to a series circuit.

FIGURE 7-16 Parallel circuit with unequal resistance.

Parallel Circuit Resistance

Resistance in a parallel circuit is not as easily calculated as it is in a series circuit, because as branches are added, another path for current to flow to ground is added. Adding extra paths reduces the circuit's total resistance to current flow. For example, if you have a 12-volt parallel circuit with three branches, each branch having a 12-ohm resistor that allows 1 amp of current flow, and you then add another parallel branch with a 12-ohm resistor to the circuit, the result will be the opposite of what might be expected. Current increases from 3 amps to 4 amps. This happens because, in a parallel circuit, adding more branches provides more pathways, but decreases the overall circuit resistance; thus, current flow increases. Imagine a freeway with three lanes and bumper-to-bumper traffic. If you add a fourth lane, the resistance to flow decreases and cars can move more easily. Or imagine that you had a bucket full of water and the bucket had three holes in the bottom. If another hole were added, the total flow out of the bucket would speed up, which means the resistance to flow decreases. To calculate resistance, use the formula in (**FIGURE 7-17**) and substitute the value of each resistor in the place of R_1, R_2, etc.

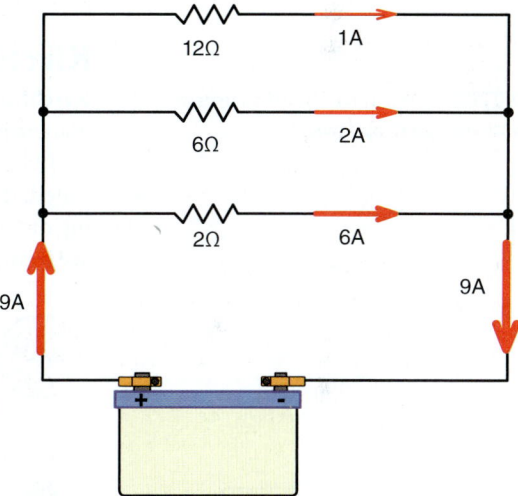

$$R_T = \frac{1}{\frac{1}{R1} + \frac{1}{R1} + \frac{1}{R1}}$$

2 resistors in parallel

$$R_T = \frac{R1 \times R2}{R1 + R2}$$

FIGURE 7-17 Calculating parallel resistance with uneven resistances.

Applied Science

AS-61: Parallel/Series Circuits: The technician can explain current flow and voltage in series and parallel circuits.

The terms "series" and "parallel" describe different ways in which components in a circuit can be connected. In a series circuit, all components are connected along a single path. Electricity can flow only one way, so the flow of electrons, or the current flow, within the circuit is the same at all points. The voltage in the system will change at different points due to voltage drops at various resistors. In a parallel circuit, components are connected so that the circuit divides into two or more paths before recombining. All components are connected to the same voltage supply, so the same voltage is applied to all components. Current flow in different branches of a parallel circuit changes depending on the resistance in each branch; therefore, different branches of the circuit may experience differing amperage if their resistance values are different.

Series-Parallel Circuits

When electrical components are wired together one after another so that there is only one path for current flow, they are said to be wired "in series." When they are wired together side by side, so there is more than one path for current to flow, they are said to be wired "in parallel." A **series-parallel circuit** is made of both a series circuit and a parallel circuit (**FIGURE 7-18**). The series circuit can be before or after the parallel portion of the circuit. Series-parallel circuits can be analyzed using the same electrical laws that apply to separate series or parallel circuits; you just have to apply the series laws to the series portion and the parallel laws to the parallel portion, and then add the two totals together.

Kirchhoff's Current Law

K07004 Explain Kirchhoff's current and voltage drop laws.

Kirchhoff's current law describes a fundamental electrical principle and is used by technicians in understanding how all electrical circuits work. Many technicians would understand the principle behind the law without necessarily associating it with Kirchhoff. Simply stated, the law is that current entering any junction is equal to the sum of the current flowing out of the junction. For example, in a junction of three conductors, current is flowing in at 10 amps from conductor A and out by the other two conductors, B and C (**FIGURE 7-19**).

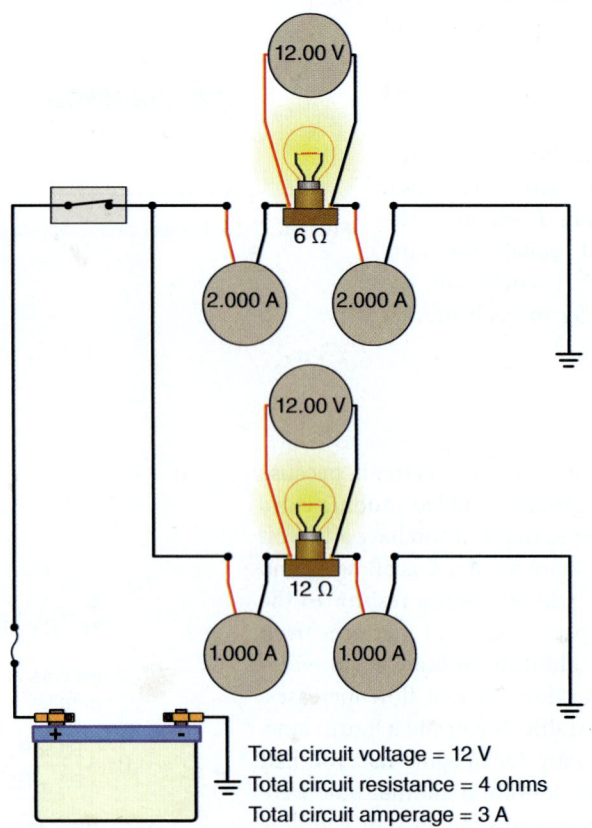

FIGURE 7-18 Electrical behavior in a parallel circuit with unequal-resistance bulbs.

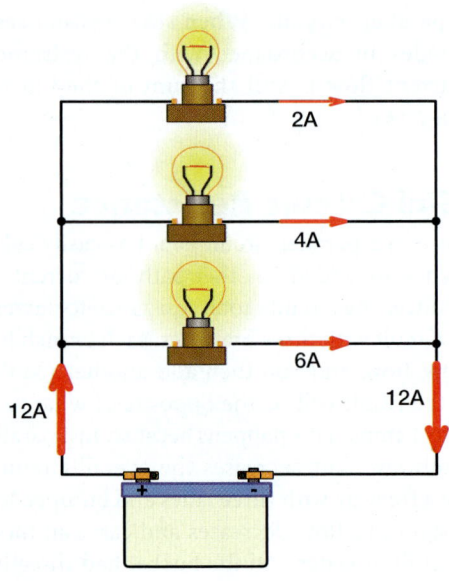

FIGURE 7-19 Kirchhoff's current law: Current entering a junction is equal to the sum of the current flowing out of the junction.

According to Kirchhoff's current law, the sum of the current in conductors B and C will equal the current flowing into the junction at conductor A, or 10 amps.

Kirchhoff's Law of Voltage Drop

Kirchhoff's voltage law tells us that in any closed loop circuit, the total voltage drop around the loop is equal to the sum of all the voltage drops within the same circuit. Remember that whenever current passes through any part of the circuit, the inherent resistance present in all conductors and components causes a voltage drop to occur. For instance, a length of 16-gauge copper wire will induce a small voltage drop when current passes through it due to the inherent resistance of the copper wire. The size of the voltage drop will be very small, around .02 volts, but this does contribute to the total amount of voltage drop in the circuit.

Consider the simple circuit in (**FIGURE 7-20**). To understand Kirchhoff's voltage law, follow the steps in **SKILL DRILL 7-1**.

S07001 Demonstrate a working understanding of Ohm's law.

S07002 Use Ohm's law to solve for missing circuit variables.

S07003 Apply Ohm's law to series circuits.

S07004 Apply Ohm's law to series-parallel circuits.

N07002 Demonstrate proper use of a test light on an electrical circuit.

N07003 Demonstrate proper use of a digital multimeter (DMM) when measuring source voltage, voltage drop (including grounds), current flow, and resistance.

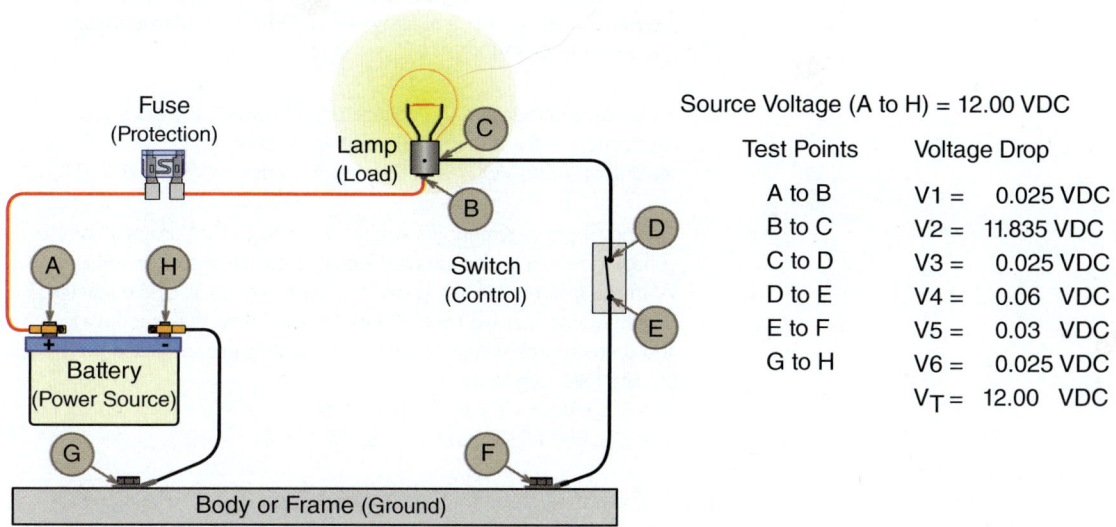

FIGURE 7-20 Circuit representing Kirchhoff's voltage law.

SKILL DRILL 7-1 Understanding Kirchhoff's Voltage Law

1. Battery voltage is 12 volts DC. Using a multimeter set to measure voltage, you proceed to measure the voltage drop that is present in each segment of the circuit. (Remember that in order to measure voltage drop, the circuit must be operating, you must have your multimeter set up to measure DC voltage, and you must connect your multimeter in parallel to each segment of the circuit and record the amount of voltage drop present.)

Continued

2. Starting from the battery you measure the voltage drop between the positive battery post (A) and the positive connection at the lamp (B). A reading of "0.025 VDC" is recorded. You record this measurement as: V1 = 0.025 VDC.

3. Next you measure the voltage drop across the lamp (B-C); a voltage drop of "11.835 VDC" is recorded. You record this measurement as: V2 = 11.835 VDC.

4. Moving forward from the lamp, you measure the voltage drop between the negative terminal of the lamp (C) and the leading terminal of the switch (D). A voltage drop of "0.025 VDC" is observed. You record this value as: V3 = 0.025 VDC.

5. Voltage drop is tested across the switch (D-E), and a reading of "0.06 VDC" is measured. This value is recorded as: V4 = 0.06 VDC.

6. Testing between the ground side of the switch (E) and the body ground connection (F), a reading of "0.03 VDC" is recorded. Record this as: V5 = 0.03 VDC.

7. Finally, you measure the voltage drop between the body ground connection of the battery (G) and the negative battery post (H). A reading of "0.025 VDC" is noted, and recorded as: V6 = 0.025 VDC.

8. Knowing that Kirchhoff's voltage law tells us the sum of all of the voltage drops in a loop circuit is equal to the total source voltage. With this information, we know that if we add up all of the voltage drop readings that we took in the previous steps, they should all add up to source voltage if we did our testing properly. To do this use the following formula:
V_t = V1 + V2 + V3 + V4 + V5 + V6 or
V_t = .025 + 11.835 + .025 + .06 + .03 + .025 = 12 Vt

9. From this we see that Total Voltage Drop (V_T) is 12 VDC, which is equal to source voltage. This is a useful test that can help to locate the source of excessive voltage drops within any electrical circuit.

Wrap-Up

Ready for Review

- Ohm's law helps us to understand the relationship between volts, amps, and ohms.
- Ohm's law tells us that it takes 1 volt to push 1 amp through 1 ohm of resistance. That gives us the relationship among the three units.
- Volts and resistance are physical things. Volts exist when a surplus of electrons accumulate, thereby creating electrical pressure.
- Resistance is the physical restriction of the conductor.
- Amps or amperes are the amount of electrons moved through the conductor.
- Amps or amperes are the result of both the volts and resistance.
- If the pressure (volts) stays the same but the restriction (ohms) doubles, then the electrons (amps) will be reduced by half. Conversely, if the voltage is doubled and the resistance stays the same, then amperage will double.
- Performing a voltage drop test on both sides of the circuit is a dynamic way of identifying if there is an excessive voltage drop present.
- A resistance check of the electrical device will indicate a high resistance or open condition in the component.

- Circuits are made up of components and interconnecting conductors, such as wires, arranged to manage and control the flow of electrons to perform specific electrical tasks.
- Circuits come in two basic configurations: series circuits and parallel circuits. The two types can also be combined into what is called a series-parallel circuit.
- Understanding how electricity works within series and parallel circuits will help you know how the circuit operates and how to approach diagnosis.
- Ohm's law is a relationship between volts, amps, and ohms, and they must always balance out. If we know any two of the values, then we can calculate the third.
- If resistance stays the same but voltage rises, then the greater force pushes more current through the circuit. If resistance stays the same but voltage decreases, then less current will flow through the circuit. This means the total current flow of a circuit in amps always equals the voltage divided by the resistance.
- In calculating Ohm's law, R stands for resistance, V for voltage (some use the letter "E" for volts meaning electromotive force), and A for current (A is for amps, some use the letter "I" for current).
- Using the Ohm's law circle will help you remember the correct math operation to use.
- Energy is the potential to do work.
- Converting one form of energy into another is called energy transformation.
- A motorcycle engine turns chemical energy into thermal energy, and then into mechanical energy. In each case, work is being done, but there is a difference with the motorcycle—it does the work more quickly, delivering more mechanical energy faster. That difference is called power.
- Power is the rate at which work is performed. It is also known as the rate of transforming energy. In an electrical circuit, power refers to the rate at which electrical energy is transformed into another kind of energy.
- The unit of electrical power is the watt. One watt is produced when 1 volt causes 1 amp of current to flow. From this comes the power equation: P, the power in watts, equals V, the voltage in volts, multiplied by A, current in amps. This calculation is applied similarly to Ohm's law and is typically represented as a triangle.
- Electrical power is a measurement of the rate at which electricity is consumed or created.
- In a series circuit, there is only one path for current to flow. All of the current flows to each component in turn. It also means that all the electrons flow at the same rate throughout all parts of the circuit. Current is equal everywhere within a series circuit.
- In a series circuit, if there is more than one resistance in the circuit, those resistances are connected one after the other; thus the resistances add up.
- The total resistance in a series circuit is the sum of all of the individual resistances.
- Current flow stays the same in a series circuit. Current flow is the same in all parts of the circuit.
- Voltage drops as current goes through resistance(s) in series. The applied voltage is equal to the sum of the individual voltage drops (Kirchhoff's voltage law).
- Resistance adds up in series. Total circuit resistance (RT) is equal to the sum of the individual resistances; for example, RT = R1 + R2 + R3.
- In a series circuit, components are connected like links in a chain. If any link opens, current to all of the components is cut off.
- In a parallel circuit, all components are connected directly to the voltage supply. If any connection or component fails in a parallel circuit, current continues to flow normally through the remaining circuits.
- The voltage across all branches of a parallel circuit are equal.
- The total current in a parallel circuit equals the sum of the current flowing in each branch of the circuit (Kirchhoff's current law).
- The total resistance of a parallel circuit decreases as more branches are added. Total parallel circuit resistance will always be less than the lowest branch resistance.
- In a parallel circuit where the resistors in each branch are the same, the current flowing in each branch is also the same. The sum of their individual currents is equal to the total current flowing in the entire parallel circuit. When the resistances are not equal, the current divides in accordance with the resistance of each branch, but the total current flow is still the sum of the currents flowing in each branch.
- Resistance in a parallel circuit is fractional.
- A series-parallel circuit is made of both a series circuit and a parallel circuit. The series circuit can be before or after the parallel portion of the circuit. Series-parallel circuits can be analyzed using the same electrical laws that apply to separate series or parallel circuits; you just have to apply the series laws to the series portion and the parallel laws to the parallel portion, and then add the two totals together.
- Kirchhoff's current law describes a fundamental electrical principle and is used by technicians in understanding how all electrical circuits work. The total current in a parallel circuit equals the sum of the current flowing in each branch of the circuit.
- Kirchhoff's voltage law is the total voltage drop around the loop is equal to the sum of all the voltage drops within the same circuit.

Key Terms

electrical power A measurement of the rate at which electricity is consumed or created.

energy The ability to do work.

Kirchhoff's current law An electrical law stating that the sum of the current flowing into a junction is the same as the current flowing out of the junction.

Kirchhoff's voltage law The sum of all voltages drops in a circuit are equal to source voltage.

power The rate at which work is done; electrical power is measured in watts.

series-parallel circuit A circuit that has both a series and a parallel circuit combined into one circuit.

voltage drop The amount of potential difference between two points in a circuit.

watt The unit for measuring electrical power.

work The process by which one type of energy is transformed into another type of energy.

Review Questions

1. Ohms law helps us to understand the relationship between:
   a. volts, amps, and ohms.
   b. meter, amps, and watts.
   c. volts, amps, and watts.
   d. amps, watts, and ohms.
2. The physical restriction of the conductor is represented as:
   a. current.
   b. resistance.
   c. voltage.
   d. potential.
3. If the voltage is doubled and the resistance stays the same, then amperage will:
   a. double.
   b. triple.
   c. be halved.
   d. remain the same.
4. All of the following mathematical equations are true with respect to Ohm's law *except*:
   a. $A = V \div R$
   b. $V = A \times R$
   c. $R = V + 2A$
   d. $R = V \div A$
5. The value of current in a circuit with 12 volts and resistance 2 ohms is:
   a. 3 amps.
   b. 6 amps.
   c. 2 amps.
   d. 10 amps.
6. The difference in electron supply at the battery terminals creates electrical force and is sometimes called:
   a. potential difference.
   b. voltage drop.
   c. resistance.
   d. power.
7. The unit of electrical power is:
   a. amp.
   b. ohm.
   c. volt.
   d. watt.
8. All of the following statements describing a series electric circuit are true *except*:
   a. Current flow is the same in all parts of the circuit.
   b. Total circuit resistance is equal to the sum of the individual resistances.
   c. Applied voltage is equal to the sum of the individual voltage drops.
   d. Total resistance in the circuit is always less than the individual resistances.
9. Which of the following electrical laws states that the sum of the current flowing into a junction is the same as the current flowing out of the junction?
   a. Ohm's law
   b. Kirchhoff's voltage law
   c. Kirchhoff's current law
   d. Parallel circuit law
10. Choose the correct statement describing the principle of Kirchhoff's voltage drop law.
    a. It takes 1 volt to push 1 amp through 1 ohm of resistance.
    b. $V = A \times R$
    c. The sum of all of the voltage drops in a closed loop circuit is equal to the total source voltage drop across the circuit.
    d. Current entering a junction is greater than the sum of the current flowing out of the junction.

ASE Technician A/Technician B Style Questions

1. Tech A says that Ohm's law is used to calculate the total circuit wattage. Tech B says that Ohm's law allows you to calculate resistance, current, and amps. Who is correct?
   a. Tech A
   b. Tech B
   c. Both A and B
   d. Neither A nor B
2. Tech A says when voltage decreases, current flow must increase. Tech B says when voltage decreases, current flow must also decrease. Who is correct?
   a. Tech A
   b. Tech B
   c. Both A and B
   d. Neither A nor B
3. Tech A says that energy is the potential to do work. Tech B says that energy can only exist in the forms of heat and light. Who is correct?
   a. Tech A
   b. Tech B
   c. Both A and B
   d. Neither A nor B
4. Tech A says that converting energy from one form to another is called transcendence. Tech B says that converting energy from one form to another is called transformation. Who is correct?
   a. Tech A
   b. Tech B
   c. Both A and B
   d. Neither A nor B

5. Tech A says that electrical potential means that the possibility for a circuit to do work is real. Tech B says that electrical potential is a theoretical measurement that is not encountered outside of the classroom. Who is correct?
   a. Tech A
   b. Tech B
   c. Both A and B
   d. Neither A nor B
6. Tech A says that electrical power is a measurement of the rate at which electricity is generated. Tech B says that electrical power is a measurement of the rate at which electricity is consumed. Who is correct?
   a. Tech A
   b. Tech B
   c. Both A and B
   d. Neither A nor B
7. Tech A says that a series circuit is the most complicated type of electrical circuit. Tech B says that series circuits are the most widely used circuits in automotive electrical systems. Who is correct?
   a. Tech A
   b. Tech B
   c. Both A and B
   d. Neither A nor B
8. Tech A says that voltage is divided up across all of the loads that exist in the series. Tech B says that source voltage is supplied to each load in a series circuit. Who is correct?
   a. Tech A
   b. Tech B
   c. Both A and B
   d. Neither A nor B
9. Tech A says that source voltage is applied to each leg of a parallel circuit. Tech B says that total resistance of a parallel circuit is always more that the largest resistance value in the circuit. Who is correct?
   a. Tech A
   b. Tech B
   c. Both A and B
   d. Neither A nor B
10. Tech A says that the series parallel circuit is widely used in automotive electrical systems. Tech B says that electromotive force is another term used to describe voltage. Who is correct?
    a. Tech A
    b. Tech B
    c. Both A and B
    d. Neither A nor B

CHAPTER 8
Electrical Components

NATEF Tasks

- **N08001** Demonstrate knowledge of the causes and effects from shorts, grounds, opens, and resistance problems in electrical/electronic circuits.
- **N08002** Inspect and test fusible links, circuit breakers, and fuses; determine needed action.

Knowledge Objectives

After reading this chapter, you will be able to:

- **K08001** Explain the operation of switches.
- **K08002** Explain the operation of fuses and, fusible links, and circuit breakers.
- **K08003** Explain the operation of flash cans/controllers.
- **K08004** Explain the operation of relays and relay control circuits.
- **K08005** Explain the operation of solenoids and their uses.
- **K08006** Explain the operation of electric motors.
- **K08007** Explain the operation of ignition coils and transformers.
- **K08008** Explain the operation of resistors.
- **K08009** Explain the operation of capacitors.

Skills Objectives

After reading this chapter, you will be able to:

- **S08001** Demonstrate the procedures required to test switches, relays, and variable resistors.

You Are the Technician

Electrical circuit malfunctions can often be traced to a failed relay, but there are many modes of failure that will stop a relay from doing its job. Remember, a relay consists of the control side (terminals 85 and 86), which contains the electromagnet, and the load side (terminals 30, 87, and 87a), which controls the high current load to the device being controlled. It is true that relays fail, but failure to verify the electrical integrity of the control and load sides of the circuit can result in a misdiagnosis. To test the control side of the relay, connect a test light across terminals 85 and 86 on the female portion of the connector, which is located on the vehicle, and visible when the relay is removed. The test light should illuminate when the control side of the relay is engaged. To test the vehicle side of the load portion of the relay, which are terminals 30, 87, and 87a, there should be battery voltage present at terminal 30. To verify that the components downstream of the relay are in working order, a fused jumper wire can be installed between terminal 30 and terminal 87 or 87a. Refer to the vehicle service information to discover the exact location of the relay that you are looking for.

1. True or False: Terminal 85 is on the load side of the relay.
2. True or False: Terminals 85 and 86 should have power and ground when the relay is commanded on.

CHAPTER 8 Electrical Components

N08001 Demonstrate knowledge of the causes and effects from shorts, grounds, opens, and resistance problems in electrical/electronic circuits.

▶ Introduction

Electrical components such as switches, fuses, circuit breakers, resistors, capacitors, and relays are all used in a circuit to modify or manage the flow of current (**FIGURE 8-1**). Each component performs a specific task within the circuit and is connected to its particular circuit (**FIGURE 8-2**). To ensure the components' correct operation, terminals are often numbered or marked with the connections and referenced in wiring diagrams or schematics. Some components are **polarity sensitive**. This means they must be connected into the circuit with the correct polarity to each lead. For example, some capacitors, and most semiconductor components, are polarity sensitive. Polarity-sensitive components have the polarity marked on at least one connection to ensure that the user connects them into the circuit correctly (**FIGURE 8-3**).

▶ Switches

K08001 Explain the operation of switches.

A **switch** is an electrical device used to turn the current on and off in a circuit. When turned off, switches open the circuit, stopping current flow; when turned on, they close the circuit, allowing current to flow (**FIGURE 8-4**). For example, switch off—light goes out; switch on—light comes on. There are many different types and configurations of switches, including toggle switches, push-button switches, and specialty switches for turn signal indicators and

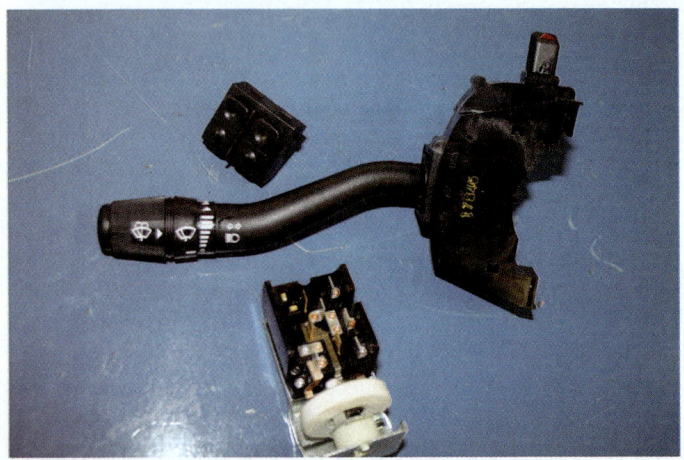

FIGURE 8-1 Circuit breakers and switches used in automotive applications.

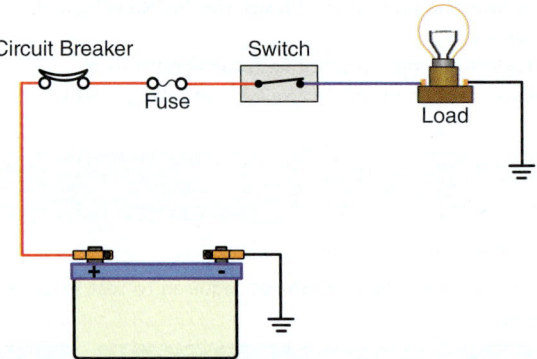

FIGURE 8-2 Simple circuits containing switch, circuit breaker, and fuse.

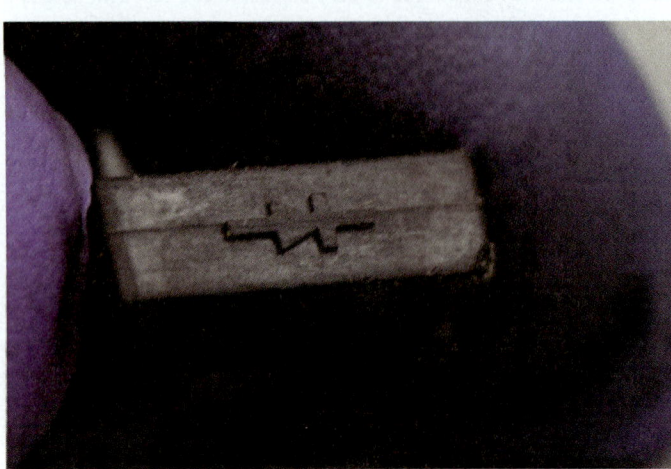

FIGURE 8-3 Polarity-sensitive components marked to ensure proper installation.

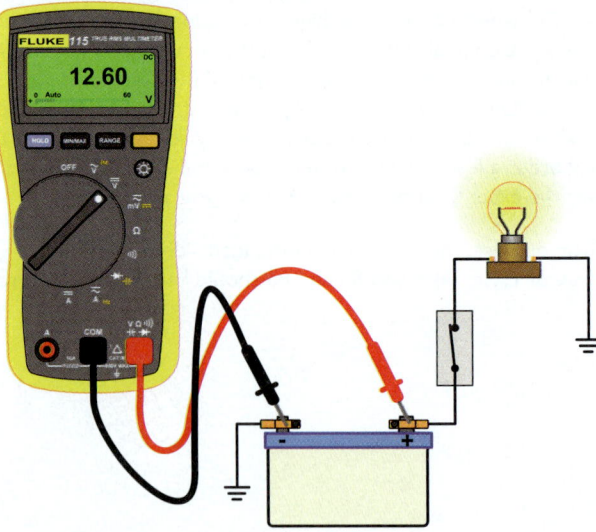

FIGURE 8-4 Simple circuit with a switch in series to the load.

windshield wipers (**FIGURE 8-5**). The most basic switches have only two terminals and are simply on or off.

More complex switches have many terminals and contacts inside them to switch a number of circuits at the same time. The **turn signal switch** is an example of a more complex switch (**FIGURE 8-6**). It has three positions: center for off, pushed in one direction for the left turn signals, pushed in the other direction for the right turn signals. **Circuit or schematic diagrams** show switches, their contacts, and the surrounding circuits so that the technician can identify how they operate in a circuit (**FIGURE 8-7**). The terminals on the switch are often numbered or lettered to indicate the correct way of connecting them into their circuits and to show the mating connectors in the wiring diagrams.

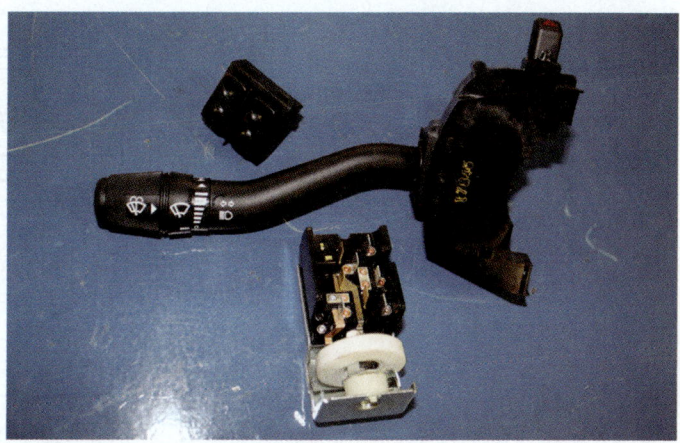

FIGURE 8-5 Typical automotive switches.

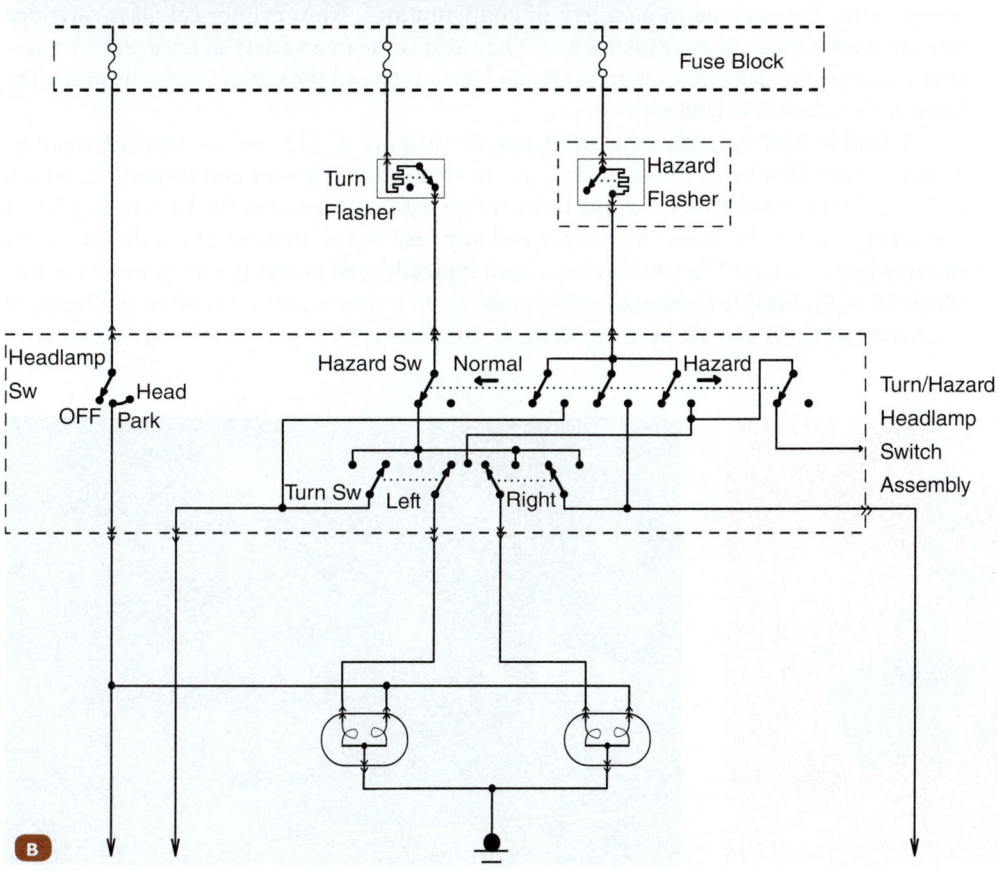

FIGURE 8-6 **A.** Turn signal switch. **B.** Accompanying wiring diagram.

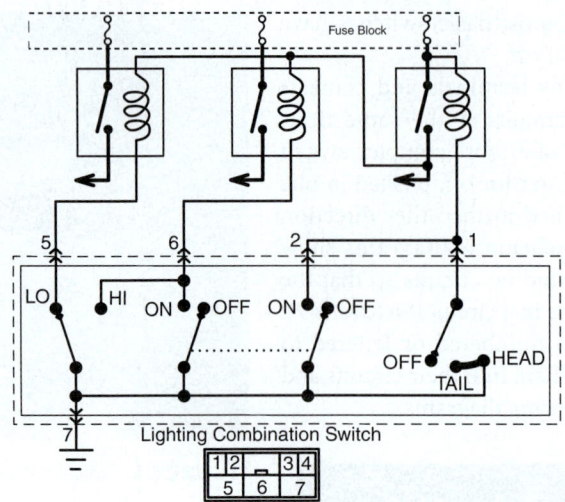

FIGURE 8-7 Wiring diagram containing the operation and construction of automotive switches.

Fuses, Fusible Links, and Circuit Breakers

K08002 Explain the operation of fuses, fusible links, and circuit breakers.

N08002 Inspect and test fusible links, circuit breakers, and fuses; determine needed action.

Fuses and **circuit breakers** are designed to protect electrical circuits by opening the circuit if the current flow is excessive. The most common kinds of circuit protection devices are fuses, fusible links, circuit breakers, and positive temperature coefficient (PTC) thermistor protection devices. Fuses and circuit breakers are rated in amps, and their ratings are usually marked on them. Fusible links are typically rated by their wire size.

Fuses are typically used in lighting and accessory circuits where current flow is usually moderate. Usually, a fuse contains a metal strip that is designed to overheat and melt when subjected to a specified excessive level of current flow, breaking the circuit and stopping the excessive current flow from potentially damaging the wiring harness and more valuable components. Fuses come in a variety of configurations, from cylindrical glass cartridge fuses to plastic blade fuses (**FIGURE 8-8**). They also come in a variety of sizes and amp ratings. Fuses are typically housed in fuse boxes located around the vehicle, typically under the hood and/or dash (**FIGURE 8-9**).

A fusible link is made of a short length (usually 9" [15 cm] or less) of smaller-diameter wire that has a lower melting point than standard wire and insulation, which is fire resistant (**TABLE 8-1**). Fusible links are typically placed near the battery to protect the wiring harness between the battery and any fuse boxes. In most cases, they are used to carry higher current flows than fuses, and typically feed power to one or more circuits (**FIGURE 8-10**). Fusible links are fairly durable; when they do fail, it is often the result of a substantial short circuit in the system, or the fusible link wire is abused by excessive

FIGURE 8-8 Various fuses.

FIGURE 8-9 Automotive fuse box.

TABLE 8-1 Fusible Link Current Rating

Allowable amperage of conductors under 50 Volts with 105°C insulation							
AWG Wire Size	Metric (Sq mm)	AWG CM Area	SAE CM Area	Ohms /1000ft	Ampacity Engine Space Outside	Ampacity Engine Space Inside
18	0.8	1,600	1,537	6.385	20	17
16	1	2,600	2,336	4.016	25	21.3
14	2	4,100	3,702	2.525	35	29.8
12	3	6,500	5,833	1.588	45	38.3
10	5	10,500	9,343	0.9989	60	51
8	8	16,800	14,810	0.6282	80	68
6	13	26,600	24,538	0.3951	120	102
4	19	42,000	37,360	0.2485	160	136
2	32	66,500	62,450	0.1563	210	178.5
1	40	83,690	77,790	0.1239	245	208
0	50	105,600	98,980	0.09827	285	242.3
2/0	62	133,100	125,100	0.07793	330	280.5
3/0	81	167,800	158,600	0.06180	385	327.3
4/0	103	211,600	205,500	0.04901	445	378.3

FIGURE 8-10 A fusible link is typically placed near the battery and carries the current needed to power an individual circuit or a range of circuits.

Applied Science

AS-62: Short Circuit: The technician can demonstrate an understanding of the processes used to locate a short circuit in an electrical/electronic system.

Two types of short circuits can occur in electrical systems. In a short to ground, electrical current finds its way to ground before it was intended, usually due to compromised wiring insulation that allows wiring to touch the metal vehicle body. In a short to power, a circuit is exposed to voltage flowing in another circuit, generally also due to broken wiring insulation. The effect of either type of short will depend on the layout of the circuit and where the short occurs in relation to the load.

The most common indicator of a short circuit is a fuse blowing due to excessive current flow. Traditionally, diagnosing the location of the short has required the segmentation of the circuit through the disconnection of fuses, components, switches, or harness plugs. Individual sections of the circuit are then checked for integrity. Electronic short circuit finders are now available that send an electrical signal along the circuit. A receiver is run along the circuit until it stops receiving a signal, at which point the location of the short has been found.

flexing or pulling on it. Some newer vehicles use maxi-fuses, which are large blade-type fuses, instead of fusible links.

Circuit breakers are different from fuses and fusible links in two ways. First, they are not destroyed by excess current. And secondly, they can be reset, either automatically or manually. In a circuit breaker, a bimetallic strip heats up and bends, opening a set of contacts and breaking the circuit when current flow becomes excessive (**FIGURE 8-11**). In most types, as the strip cools, it resumes its original shape. The contacts then close, completing the circuit once more. These are called self-resetting circuit breakers. Manual breakers must be reset by hand, which could involve flipping a lever or inserting a small rod to reset the bimetal spring once it cools down.

PTC Devices, Virtual Fuses, and FETs

PTC thermistors are also used as circuit protection devices. They have very low resistance at room temperature, but increase in resistance as the temperature increases (**FIGURE 8-12**). If too much current starts to flow through a PTC, the current flow creates heat in the PTC. The increase in temperature causes the resistance of the PTC to increase, which increases the voltage drop that occurs in the PTC. This cycle continues quickly until the PTC reaches its maximum resistance. This heightened resistance effectively shuts off most of the current

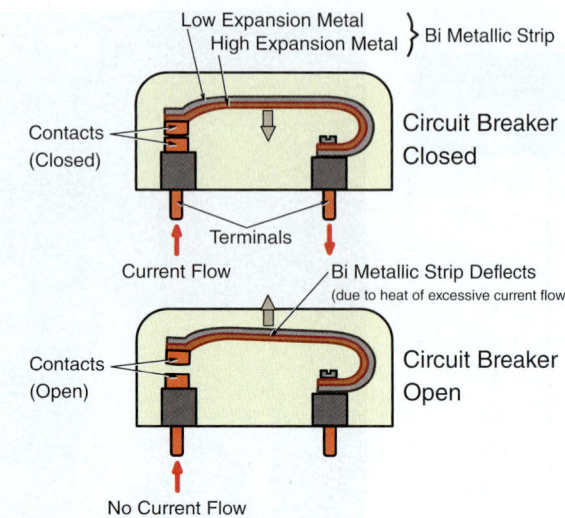

FIGURE 8-11 Circuit breaker in the normal and tripped position demonstrating the bimetallic strip operation.

FIGURE 8-12 PTC, installed and uninstalled.

flow to the protected device. As a rule, PTCs generally reset once power is removed and they are allowed to cool. They are typically integrated into components such as power window motors and door locks.

▶ Flash Can/Control

K08003 Explain the operation of flash cans/controllers.

The **flasher unit** is the control mechanism for the turn signal lights on the vehicle. As the name suggests, it flashes or turns the turn signal lights on and off at a regular rate to indicate a change in the vehicle's direction. **Flasher cans** are mechanical devices, while **flasher controls** are generally electronic devices (**FIGURE 8-13**). Both types perform the same job, but the electronic version is more reliable and consistent. Flashers operate like an automatically resetting circuit breaker, meaning they use a bimetallic strip to open and close the switch contacts. This opening and closing gives them a distinctive clicking sound that, along with the turn signal indicator lights on the dash, tells the driver when the turn signals are on so that they remember to turn them off when needed. Electronic flasher controls control the on/off function electronically, which means they can be designed to operate over a wider range of current flow, which makes them ideal for trailer towing. On some vehicles, when a trailer is wired into the lighting system, the excess current required to power the trailer lights overloads the flasher can, causing the turn signals and flashers to cycle rapidly, mimicking the rapid flashing that occurs when a turn signal bulb burns out. In many cases, the hazard lamps are operated by a separate flasher can or flasher control.

▶ Relays and Relay Control Circuits

K08004 Explain the operation of relays and relay control circuits.

Relays are switches that are turned on and off by a small electrical current. They are ideal for using a small current to control a larger current. An example would be the horn circuit. A small current can be turned on and off by the horn switch in the steering wheel, using a light duty switch and wires. The small current is then used to activate the relay, which sends the larger current to the horn. Thus, larger wires are needed only up front where they can be shorter, and smaller wires can be used to run up the steering column. Most electrical components found in a motor vehicle are controlled by relays. ECUs use relays to control components such as fuel pumps, headlights, the starter motor, and the cooling fan. All of these circuits carry large electrical loads.

The relay is made up of an electromagnet, a set of switch contacts, terminals, and the case (**FIGURE 8-14**). The electromagnet is a winding of fine metal-insulated wire wrapped around an iron core. Each end of the winding is connected to one of the relay terminals, which are usually marked "85 or 1" and "86 or 2." The contacts usually consist of three contacts—two fixed, which are usually marked with an "87a or 4" and "87 or 5," and one movable contact, usually marked "30 or 3." The movable contact is fixed to a spring-loaded armature blade and is held against one of the fixed contacts (87a or 4), which is called the **normally closed (NC)** contact. When the relay coil is activated, the electromagnet pulls the movable armature blade contact away from the NC contact and against the **normally open (NO)** contact (87 or 5), sending power out to the controlled device. When the relay is energized, it opens the NC contact points (87a or 4). When the contact points are closed, current flows across the contact points and out to the electrical device to the rest of the circuit. As long as the small control current flows through the relay windings, the much larger current for the load will flow through the relay's contact points.

Relays are fairly robust components, but they do occasionally fail. Relays can fail in a couple of ways: First, the control side of the relay, or the part that contains the electromagnet, can either short out or become electrically open. A short will cause the amperage draw of the electromagnet to increase as its resistance will be lower, while an open coil will prevent current from flowing through the magnetic coil altogether. The control side of the relay (pins 85 and 86) can be checked with an ohmmeter for an open or shorted condition. Second, the switching contacts on the load side of the relay can become burned, or in some cases, become welded together. In the case of burned contacts, even though the relay contacts may move when energized, the charring on the burned contacts will often create resistance that inhibits current flow. When the contacts are welded together, no movement of the load-carrying contacts can occur. In the case of any of these failures, replacement of the relay is required. In some cases, the relay cover can be removed, and the load contacts can be physically examined. A **solid-state relay** acts like a mechanical relay but does not have any moving parts. This means that electronic

Type 1 Cycling—Automatic Reset

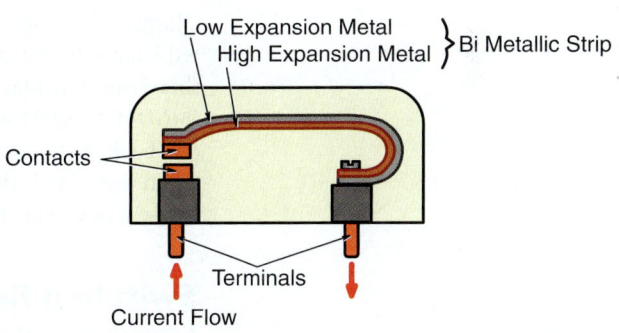

Type 2 Non-Cycling Reset

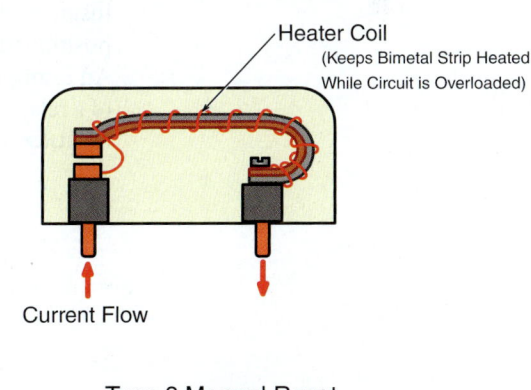

Type 3 Manual Reset

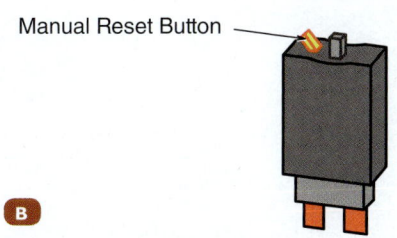

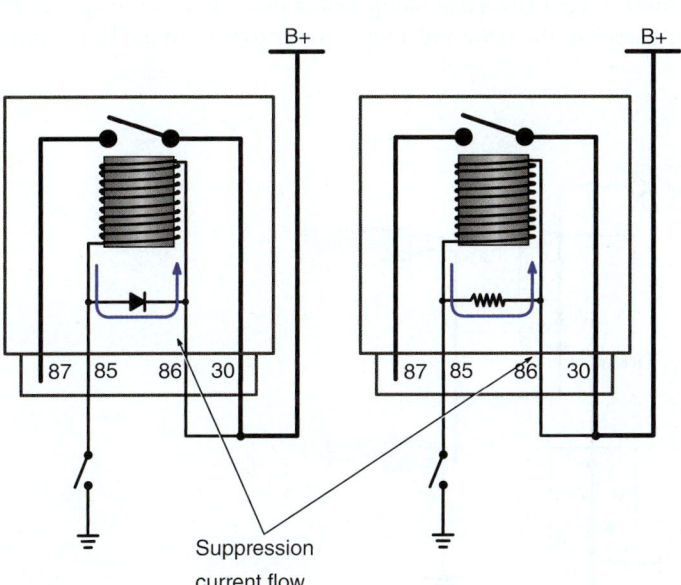

FIGURE 8-14 Spike-protected relays.

FIGURE 8-13 **A.** Flasher control and flasher can. **B.** Mechanical drawings of each.

TABLE 8-2		Standard Labeling Systems
85	1	One end of the relay winding
86	2	The other end of the relay winding
30	3	Common—movable switch contact
87a	4	NC fixed contact to the common
87	5	NO fixed contact to the common

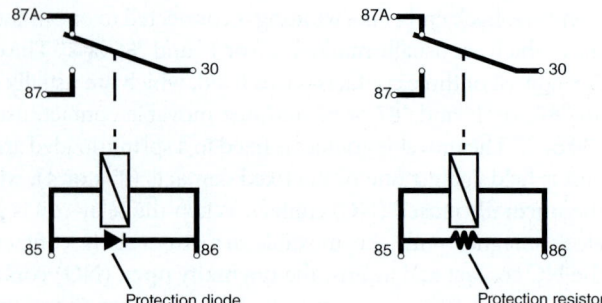

FIGURE 8-15 Spike-protected relays.

relays do not use mechanical switch contacts, but instead use transistorized circuitry to turn the circuit on and off. They also do not make any sound, unless they were specifically designed with components to emulate that function.

Automotive relay terminals use one of two standard labeling systems: 85, 86, 30, 87a, and 87 or 1, 2, 3, 4, 5 (**TABLE 8-2**).

When electromagnetic relays are deenergized, the collapse of the magnetic field induces a large voltage spike in the relay coil windings. If unchecked, this voltage can be transmitted back into the circuit where it can damage electronic components. Some relays manage this danger by placing either a suppression diode or a resistor in parallel with the winding (**FIGURE 8-15**). Doing so reduces the voltage spike by shunting it from the output side of the coil back to the input side of the coil. This design prevents the voltage from spiking so high when the relay is deenergized. When replacing a relay that has spike protection, make sure to use a new relay that is specified for the application.

Switched Relay Controls

A **switch-controlled relay** may be a desirable attribute in certain instances. The main purpose of a relay is that it allows a small current that energizes the relay coil to control the large current of the load device. This allows for the use of small-gauge wires and switches inside of the vehicle to control the relay; the larger-gauge wires to control the load can be positioned outside of the passenger compartment, where space constraints are less critical. An example of this type of control would be a headlamp switch that controls the operation of a relay coil, which in turn energizes the relay and allows power to flow to the headlamps (**FIGURE 8-16**).

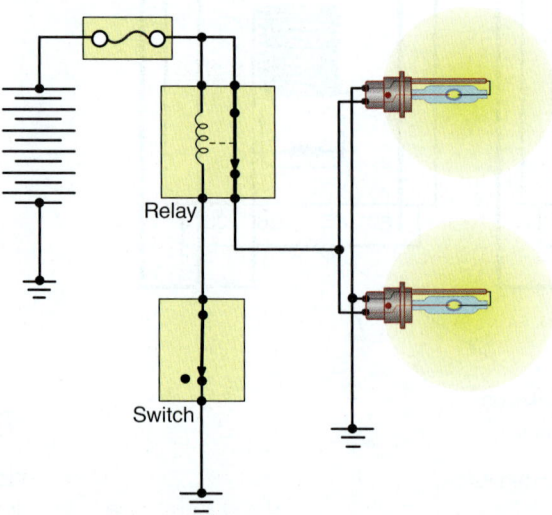

FIGURE 8-16 Wiring diagram of a switch-controlled relay circuit.

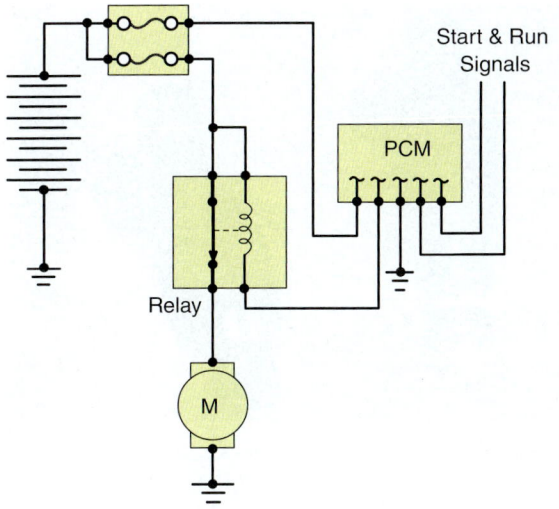

FIGURE 8-17 Wiring diagram of a logic-controlled relay circuit.

FIGURE 8-18 Timer relay.

Logic-Controlled Relays

A **logic-controlled relay** operates in a similar fashion to a switched relay control, except for the fact that the switching of the relay coil is accomplished with an electronic control module. In these cases, the relay coil is usually supplied with battery voltage and the control module employs a low-side driver to control the ground of the circuit. When operation of the relay is desired, the control module completes the path to ground through a transistor. An example of this type of circuit would be the fuel pump relay control via the engine control module (**FIGURE 8-17**).

Timer Delay Relays

Standard relays close their contacts when voltage is applied to the coil. On the other hand, a **timer delay relay (TDR)** is capable of delaying the closing of the contacts once a power input is received. The duration of this delay procedure can be adjusted over a wide variety of parameters, and adjustment is often as easy as turning a dial. TDRs can either be delay-on-device, which delay the contacts from closing for a set period of time, or delay-off-device, in which the relay contacts remain closed for a set period of time after the coil input has been stopped (**FIGURE 8-18**).

▶ Solenoids

A **solenoid** is an electromechanical device that converts electrical energy into mechanical linear (back-and-forth) movement. Solenoids can be used to pull or push. In a simple solenoid, insulated wire is wound many times around a hollow cylindrical form. A sliding mild steel core is made to fit inside the hollow form. When the winding is energized, the resulting magnetic field attracts the mild steel, drawing it into the form and producing linear motion. Solenoids can be very strong and produce a lot of mechanical energy. Solenoid design may incorporate return springs, multiple windings, electrical contacts, and mechanical connections (**FIGURE 8-19**).

Fuel injectors and starter motor solenoids are two of the many solenoid-type components used in a motor vehicle (**FIGURE 8-20**). The operation of a solenoid is similar to a relay, but where a relay uses a magnetic field to close an electrical circuit, a solenoid uses a magnetic field to create lateral movement and, in the case of a starter solenoid, to also close heavy electrical contacts. The metal core used by the electromagnet to strengthen the magnetic field is referred to as an **armature**. It is spring loaded so that it is positioned partially outside the electromagnetic coil and is free to move in and out. When the coil is energized, the magnetic field draws the armature into the center of the coil. If the armature is attached to a lever or plunger, that

K08005 Explain the operation of solenoids and their uses.

CHAPTER 8 Electrical Components

FIGURE 8-20 Fuel injectors.

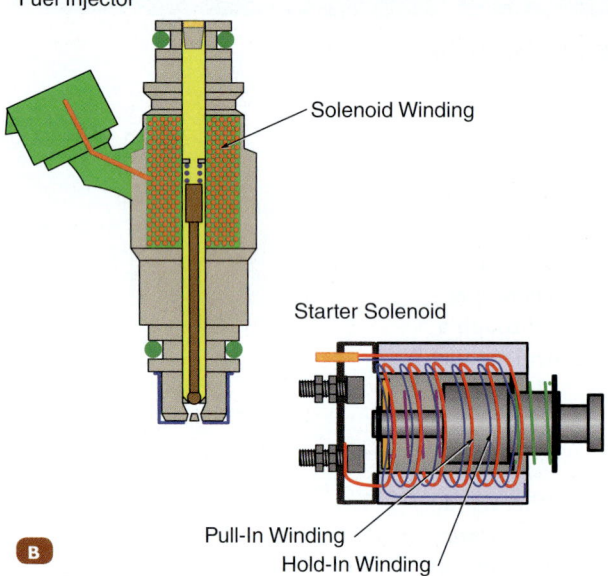

FIGURE 8-19 A. Automotive solenoids. **B.** Examples of solenoids.

K08006 Explain the operation of electric motors.

device will be forced to move as well. Stopping current flow causes the electromagnet to deenergize and the spring pushes the armature out again. Another device that uses a solenoid is a vehicle horn. When the armature is drawn in by the electromagnetic coil, it opens a set of electrical contacts so that current flow through the coil is stopped. This stoppage causes the armature to move out again, closing the contacts, drawing the armature back in. This process occurs at very high speeds. The vibration caused by the rapid movement is transferred to a diaphragm, and the familiar horn sound is produced (**FIGURE 8-21**).

▶ Motors

While solenoids use magnetic fields to create lateral movement, electric motors use magnetic fields to create rotary movement. Motors consist of two main components: the armature and the field. The armature contains electromagnetic coils. The field contains either electromagnets or permanent magnets. The interaction between the magnetic fields of the stationary field coils and the magnetic fields of the moveable armature coils causes the armature to rotate. Since the armature coils can be energized and create electromagnetic fields, those fields can be turned on and off so that they attract and repel the fields of the stationary magnetic fields. The design and orientation of the electromagnetic coils ensure that the armature continues to turn as it rotates through various positions. The armature is connected to the electrical supply by a set of carbon brushes that contact a **commutator**, which is a segmented component of the armature. The commutator and brushes act as switches to control the current flow through the windings of the armature (**FIGURE 8-22**). The brushes allow the electrical connection to occur even when the armature is spinning.

DC Motors

A DC motor converts direct current into mechanical energy. DC motors are widely used in automotive applications: they operate devices such as blower motors, power window motors, electric fuel pumps, and many more. DC motors operate on the magnetic principles that encompass the following rules: (1) opposite magnetic poles attract and (2) like magnetic poles repel one another.

There are four types of DC motors: permanent magnet motors, series motors, shunt motors, and compound motors. With permanent magnet motors, two magnets of opposing magnetic poles are attached to the motor case forming the stator, and the armature spins inside of these magnets (**FIGURE 8-23**). In series motors, the stator field is connected in series with the armature. The field is wired with a few turns of large

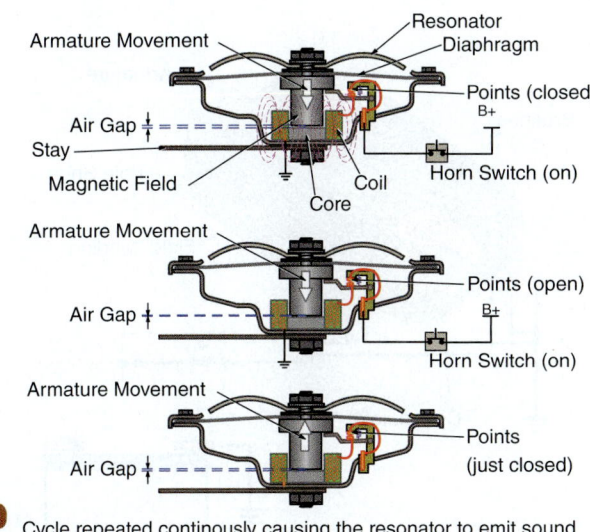

FIGURE 8-21 **A.** Automotive horn. **B.** Cutaway drawing of automotive horn.

wire because it must carry the full armature current (**FIGURE 8-24**). Shunt motors have the field connected in parallel (shunt) with the armature windings. These winding can either be separately excited or common source designs (**FIGURE 8-25**). Compound motors have a field connected in series with the armature and a separately excited shunt field (**FIGURE 8-26**).

Stepper Motors

There are certain applications that make use of stepper motors. A stepper motor is a brushless, synchronous motor that converts a digital pulse width signal into mechanical rotation (**FIGURE 8-27**). Stepper motors differ from standard DC motors in that they contain no brushes to transfer electricity to the rotor, and they can be moved in precise increments. Each complete rotation of the stepper motor is divided into as many as 200 steps (**FIGURE 8-28**). To operate, the stepper motor receives one pulse at a time, which moves the motor one step. Since all the steps are the same, stepper motors can be controlled with precision by any type of position-sensing device.

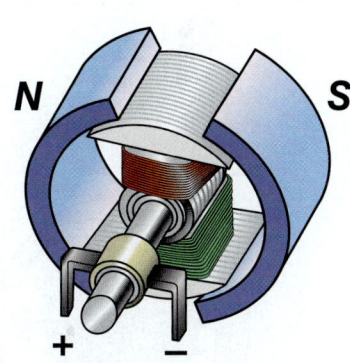

FIGURE 8-22 Simplified electric motor diagram.

AC Motors

An AC motor is an electric motor that is operated by an alternating current (AC). AC motors have a fixed stator that contains coils of wire to generate a rotating magnetic field

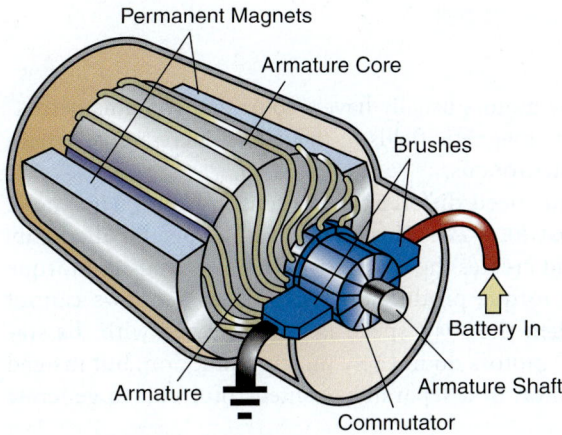

FIGURE 8-23 Permanent magnet motor.

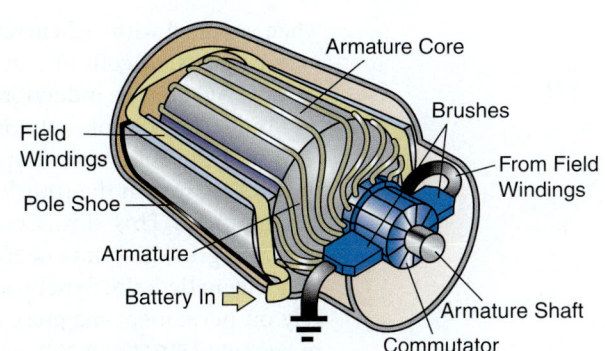

FIGURE 8-24 Series wound motor.

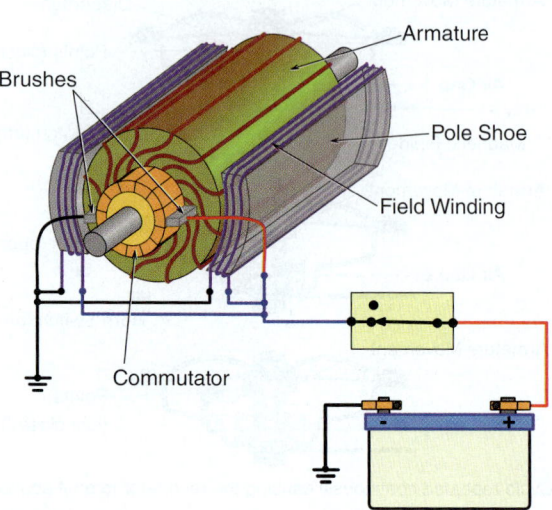

FIGURE 8-25 Separately excited and common source shunt motors.

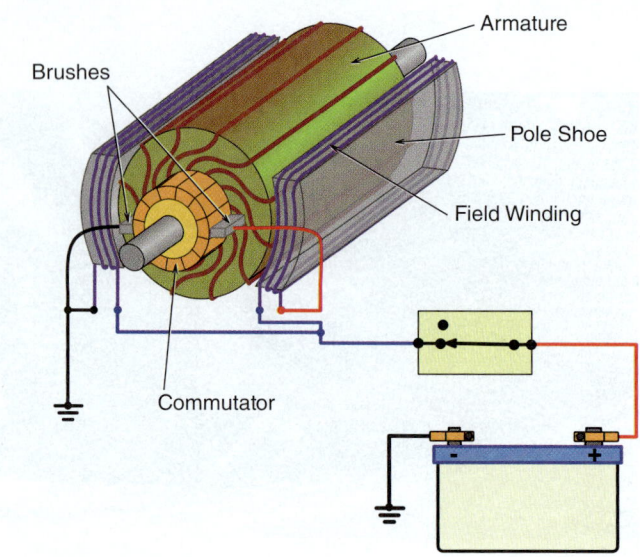

FIGURE 8-26 Compound motor.

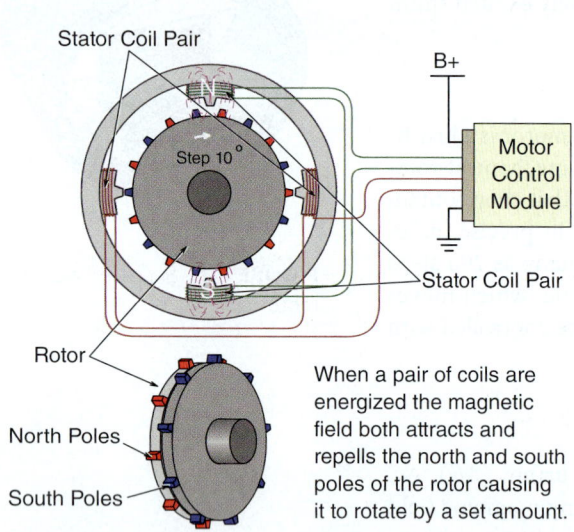

FIGURE 8-27 Stepper motor.

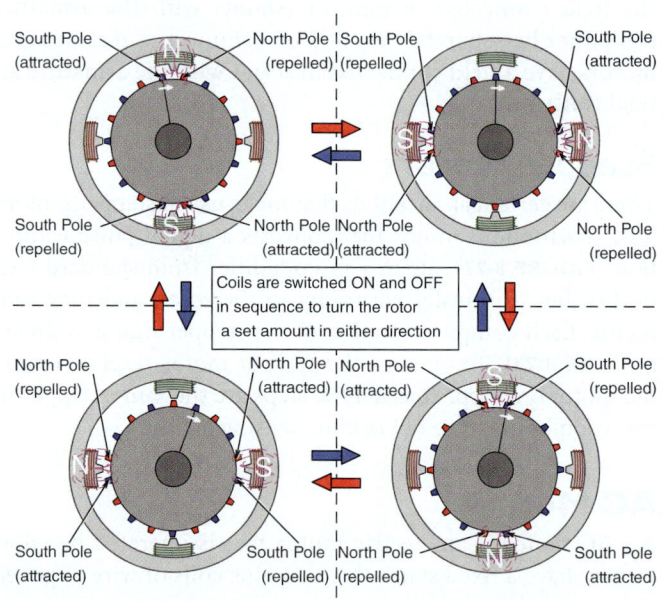

FIGURE 8-28 Internal construction of a stepper motor.

when supplied with AC current. These motors usually have a rotor that contains permanent magnets, or coils of wire to create magnetic fields in the rotor. There are two main types of AC motors: induction and synchronous.

Induction motors rely on a small speed difference between the stator's magnetic field and the rotor's shaft speed (**FIGURE 8-29**). This difference in speed, called slip, induces current in the rotor's coils and creates the magnetism needed to sustain torque development. This slip is crucial to torque production, as induction motors cannot generate great amounts of torque when the shaft speed is synchronized with the stator's magnetic field. Synchronous AC motors do not rely on slip induction, but instead rely on permanent magnets in the rotor, or a separately excited rotor coil to generate movement (**FIGURE 8-30**).

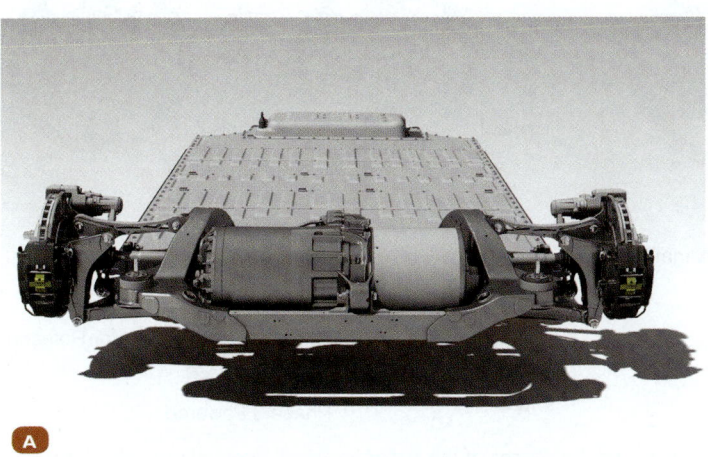
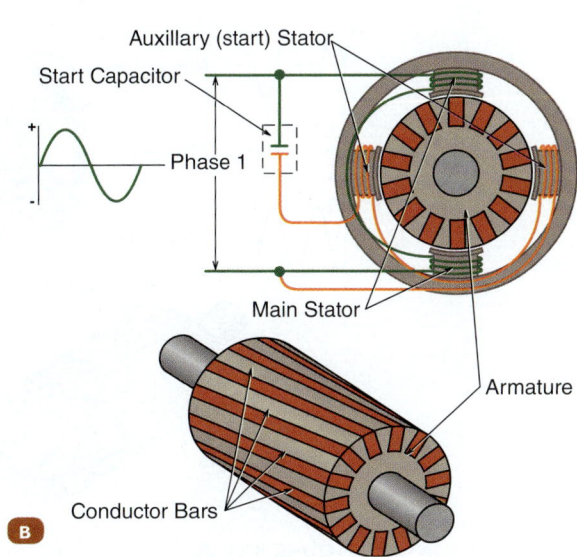

FIGURE 8-29 **A.** Induction motor. **B.** Cutaway of induction motor.

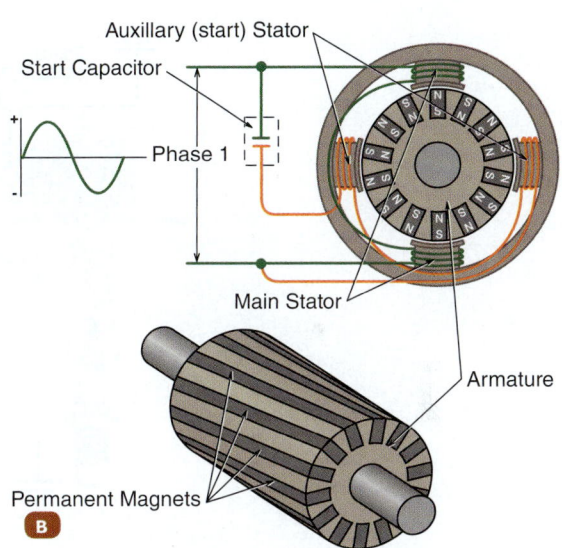

FIGURE 8-30 **A.** Synchronous AC motor. **B.** Cutaway drawing of synchronous motor.

Three-Phase AC Motors

The three-phase AC motor is the most commonly used AC motor, and is frequently seen in hybrid vehicle, electric vehicles, and high-torque applications. The benefit of the three-phase AC motor is that the stator winding is divided into three sections, or phases, that are placed 120 degrees apart (**FIGURE 8-31**). This arrangement allows the motor to smoothly start from any position and also allows it to generate a great deal of torque and efficiency. The rotor of the three-phase AC motor is excited by slip induction, which is described in the previous paragraph.

▶ Ignition Coils and Transformers

Ignition coils and transformers both operate under the principles of electromagnetic induction (**FIGURE 8-32**). They use electromagnetism, rather than mechanical movement, to produce electricity. An ignition coil can be described as a **step-up transformer**. This is because the output can be 60 kilovolts (or more), which is higher than the input, nominally 12 volts.

K08007 Explain the operation of ignition coils and transformers.

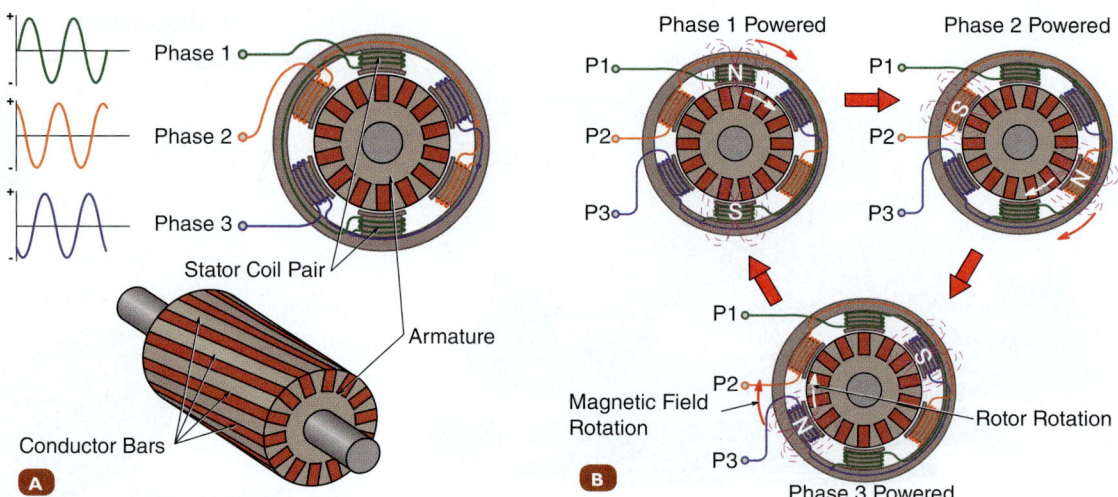

FIGURE 8-31 A. Three-phase AC motor representing the phases of the stator, and their orientation to the rotor. **B.** Cutaway drawing.

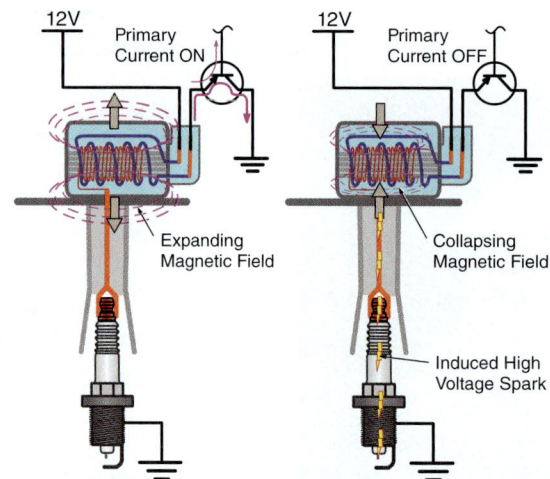

FIGURE 8-32 Internal construction of an ignition coil and brief description of its operation.

▶ TECHNICIAN TIP

You might expect a transformer to be a great way to boost the amount of electrical power that is transmitted, but it isn't. The amount of power after the transformation is relatively the same as before the transformation. For example, if we raise the voltage from 12 volts to 120 volts, the amperage will decrease from, say, 10 amps to 1 amp. Thus, the wattage stays the same:

12 volts × 10 amps = 120 watts

120 volts × 1 amp = 120 watts

Also, remember that transformers are not 100% efficient, so some of the power is lost as heat.

Two sets of coil windings are used. One coil, referred to as a **primary winding**, is wound around a second, the **secondary winding**. The primary coil typically has 200 to 300 turns of light-gauge wire while the secondary has approximately 30,000 to 60,000 turns of very fine wire. When current is passed through the primary winding, the magnetic field builds, surrounding both the windings. When the current is turned off, the magnetic field collapses with enough speed to induce high voltage in the primary winding (self-induction) and very high voltage in the secondary winding (mutual induction) by the rapidly moving (collapsing) magnetic field. Voltage is induced into each of the thousands of windings of the secondary coil. This voltage is strong enough to overcome the infinite resistance of the spark plug gap and push current across the gap, causing a spark with enough heat to ignite the air–fuel mixture in the cylinder.

The **transformer action** causes heat to be produced. In the past, the internal coils were immersed in cooling oil, allowing the heat to be conducted to the case. Modern ignition coils do not use oil. They are usually constructed using a heat-conducting hard resin and are cooled by their location on a heat sink or in a stream of air. The advent of computer controls has allowed the time that current flows through the primary windings to be minimized to reduce heat and electrical loads, while still providing enough spark to ignite the air-fuel mixture.

Step-Down Transformers

Step-down transformers function under the same operating principles. The only difference is that the secondary coil has fewer turns than the primary, providing a lower induced output. These transformers are used on power poles to lower the voltage to your house or school and on low-voltage devices in your home that are plugged into your 110-volt outlets, such as phone chargers.

▶ Resistors

Resistors found on circuit boards are normally fixed in value. Some resistors found in motor vehicles are variable. **Variable resistors** can have their value altered by movement of a slide or by temperature change. The three types of variable resistors are rheostats, potentiometers, and thermistors. Variable resistors can be linear, meaning their resistance value varies proportionally with movement or temperature change, or nonlinear, meaning the resistance change is not proportional with movement (**FIGURE 8-33**).

K08008 Explain the operation of resistors.

Resistor Ratings

Resistance is measured in ohms, represented by the Greek letter omega (Ω). Resistors are rated in ohms as well, to indicate how strongly they will oppose any current flowing through them. Because resistors work by converting some of the electrical energy passing through them into heat, they also have a power rating. Only the resistance value is marked. The resistor's power rating is determined by its size.

Regardless of their power rating, resistors are small, so identification by numbers is impractical. To identify their value, many resistors are marked with four or five colored bands. Each color represents a numeric resistance value. The color bands are set close to each other and read from left to right. The last band, or tolerance band, is spaced farther apart (**FIGURE 8-34**).

Fixed Resistors

Fixed resistors are generally cylinders with connecting metal leads projecting along the axis of the cylinder at each end. Most axial resistors are marked with a series of colored stripes to indicate their resistance and tolerance levels. Fixed resistors can be manufactured as very tiny devices without leads and can be built into integrated circuits with many other miniaturized components.

Rheostats

A **rheostat** is a mechanical variable resistor with two connections. It consists of a resistance wire wrapped in a loose coil connected to the supply at one end only. A moveable wiper is connected to the other circuit connection and is made to move over the wire manually.

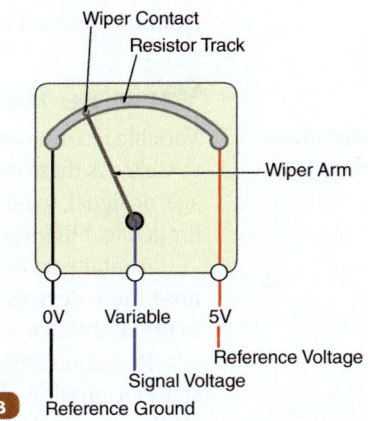

FIGURE 8-33 **A.** Variable resistor. **B.** Cutaway drawing of a variable resistor.

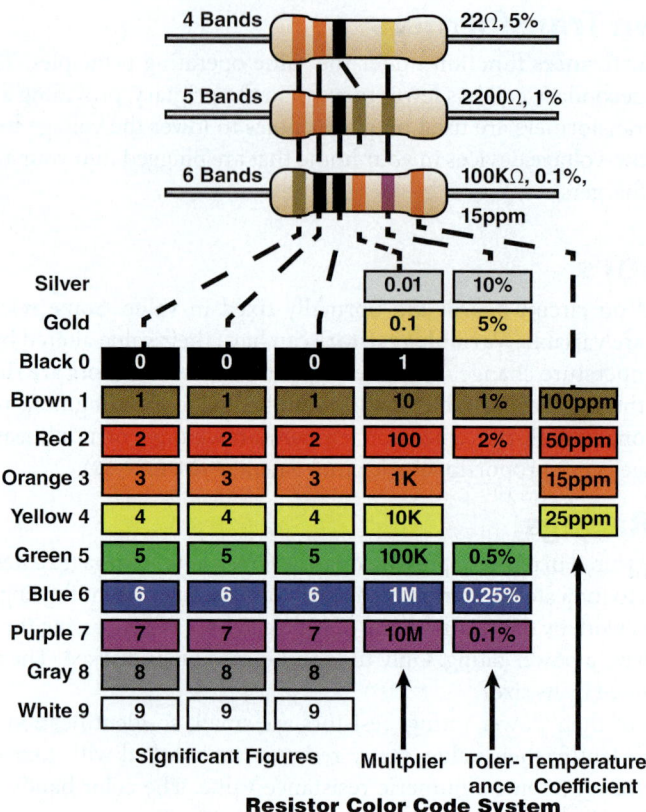

FIGURE 8-34 Resistor colors.

When the wiper is close to the beginning of the coil, the total resistance value is very small. As the wiper is positioned closer to the end of the coil, the resistance value increases. Rheostats are commonly used in dash light dimmer circuits and some fuel gauge sender units. They alter the current flow and voltage in a circuit. Rheostats are simple devices; they usually contain only two wires.

Potentiometers

Potentiometers are mechanical variable resistors with three connections, two fixed and one moveable. They act as voltage dividers and, as such, alter the voltage in a circuit. A resistance wire is wrapped between two fixed connections. One fixed connection is attached to the electrical supply, the other to a ground. The third moveable connection is moved across the coil by a wiper in a similar fashion to a rheostat. The variable voltage output is taken from this point. Throttle position sensors are potentiometers.

Variable Resistor/Potentiometer Exercise

S08001 Demonstrate the procedures required to test switches, relays, and variable resistors.

Variable resistors and potentiometers are mechanical devices, and as such, they are susceptible to wear. As these components age, their performance can degrade, causing their output voltage, or signal, to become incorrect, or erratic. This can lead to many issues, including, but not limited to, blinking dash or interior lights, power window malfunction, or engine hesitation.

A vital part of being an effective automotive technician is the ability to properly diagnose these devices. Let's start with the rheostat. To diagnose the rheostat, follow the steps in **SKILL DRILL 8-1**.

Potentiometers are used for applications that require a bit more accuracy than a rheostat can provide. The three terminals of a potentiometer are its reference voltage input (REF), its signal voltage output (SIG), and its sensor ground (GND). To diagnose a potentiometer, follow the steps in **SKILL DRILL 8-2**.

SKILL DRILL 8-1 Diagnosing the Rheostat

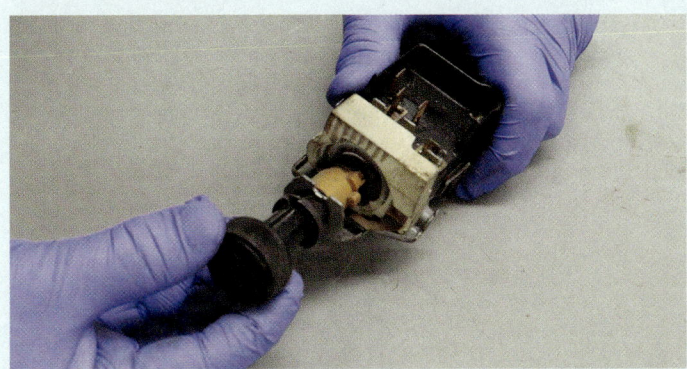

1. To test, isolate the resistor from the circuit.

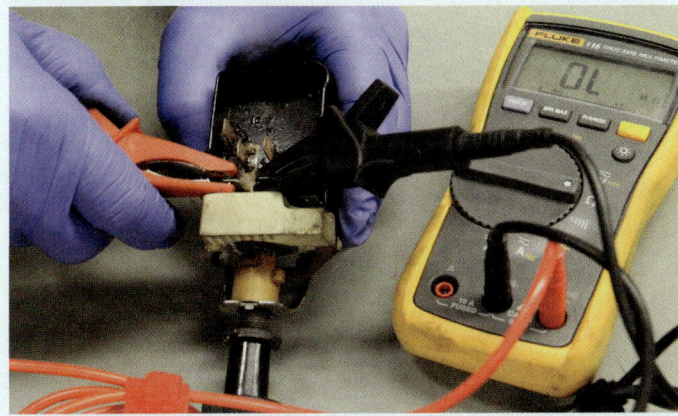

2. Set up your ohmmeter to measure resistance, then connect a lead to each terminal of the rheostat.

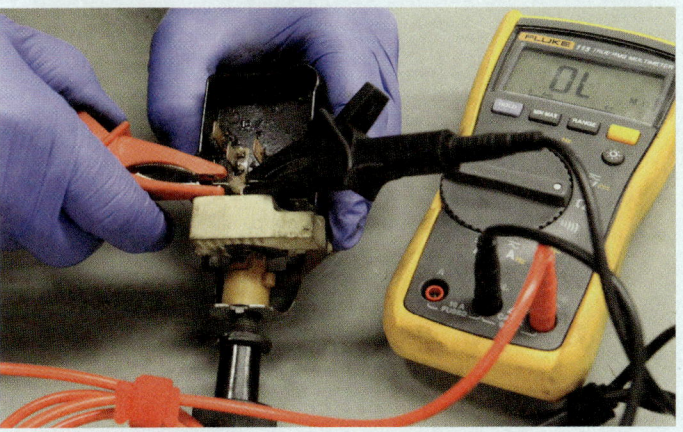

3. Once you have your ohmmeter connected, rotate the rheostat all the way to one direction, and slowly rotate the device until you get to the opposite rotational stop. While you are actuating the rheostat, keep an eye on your ohmmeter display. If the rheostat is good, the ohm reading should steadily change as you actuate the rheostat. If the rheostat is faulty, you will observe places in the rheostat actuation that produce erratic readings.

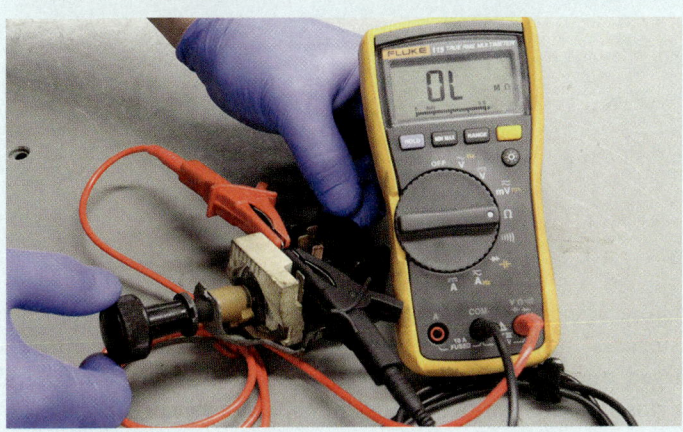

4. In some cases, when a dead spot in the rheostat is found, the meter will read O/L, indicating an open circuit. This indicates that that particular spot in the rheostat has worked to the point that no electrical contact can be achieved.

SKILL DRILL 8-2 Diagnosing the Potentiometer

1. The first step to testing a potentiometer is to test GND and REF. These two inputs are vital to the proper operation of the potentiometer. The GND can be checked with a test light whose alligator clip is connected to battery positive, and the probe can be touched to the GND terminal. If the test light illuminates, you have a ground connection.

2. The REF, or reference voltage, should be checked with your voltmeter. Typical reference voltages seen in automotive applications range from 5 to 7.5 volts.

3. Once GND and REF are verified, it is time to test the signal output (SIG) of the potentiometer. To accomplish this, the potentiometer must be connected to its circuit. Using back probe pins, connect your voltmeter to GND and SIG, and then actuate the potentiometer slowly. As with the rheostat, you should see a steady and equal change in voltage as the potentiometer goes through its range of motion. Any dead spots indicates a worn potentiometer.

K08009 Explain the operation of capacitors.

▶ Capacitors

A **capacitor** can quickly store a small amount of electrical energy, at which point it is charged. There are two surfaces inside the capacitor, separated by insulating material. When the capacitor is charged, one surface is positively charged and the other is negatively charged. When a circuit is closed between its terminals, the capacitor releases its charge. At this point, it is discharged.

A typical capacitor stores the charge on thin sheets of foil, with sheets of insulation between them. These are rolled together to form a protective canister. A capacitor's **capacitance (C)** is a measure of the amount of **charge** (Q) stored on each plate for a given potential difference or **voltage (V)** that appears between the plates. In general, as the capacitance and voltage rating of the capacitor increase, the physical size of the capacitor increases. Some DVOMs can measure the capacitance of a capacitor. That way they can be tested if they are suspected of being faulty.

Capacitor Exercise

Ask your instructor for a capacitor. First, verify that the capacitor has been discharged by placing your voltmeter across the capacitor's terminals. Once it has been verified that the capacitor contains no charge, use your voltmeter to measure the resistance between the positive and negative terminals of the capacitor. There should be no electrical connection between the two sides of the capacitor. Finally, use the capacitor test function on your meter (CAP) to determine the amount of energy that the capacitor can store; this is accomplished by setting your meter to CAP and connecting the leads across the terminals of the capacitor.

> **SAFETY TIP**
>
> Since capacitors can store tremendous amounts of energy, they can be very dangerous. Your skin only has to touch both terminals of a capacitor for a split second for all of the energy stored in the capacitor to be released into your body. Many components used in automotive systems contain capacitors, such as air bag modules, hybrid vehicle inverters, and stereo systems. Before attempting to repair or diagnose any electrical system, refer to the vehicle service information to determine if the system which you are attempting to service contains capacitors, and if so, the proper way to ensure that they are discharged before you proceed with your work.

▶ Wrap-Up

Ready for Review

- Electrical components such as switches, fuses, circuit breakers, resistors, capacitors, and relays are all used in a circuit to modify or manage the flow of current.
- A switch is an electrical device used to turn the current on and off in a circuit. When turned off, switches open the circuit, stopping current flow; when turned on, they close the circuit, allowing current to flow; switch off—light goes out; switch on—light comes on.
- A turn signal switch is an example of a more complex switch because it has three positions: center for off, pushed in one direction for the left turn signals, pushed in the other direction for the right turn signals.
- Circuit or schematic diagrams show switches, their contacts, and the surrounding circuits so that the technician can identify how they operate in a circuit.
- Fuses and circuit breakers are designed to protect electrical circuits by opening the circuit if the current flow is excessive.
- The most common kinds of circuit protection devices are fuses, fusible links, circuit breakers, and positive temperature coefficient (PTC) thermistor protection devices. Fuses and circuit breakers are rated in amps, and their ratings are usually marked on them.
- Fusible links are typically rated by their wire size.
- A fusible link is made of a short length (usually 9" [15 cm] or less) of smaller-diameter wire that has a lower melting point than standard wire and insulation, which is fire resistant.
- Fusible links are typically placed near the battery to protect the wiring harness between the battery and any fuse boxes. In most cases, they are used to carry higher current flows than fuses, and typically feed power to one or more circuits.
- PTC thermistors are also used as circuit protection devices. They have very low resistance at room temperature, but increase in resistance as the temperature increases.
- The flasher unit is the control mechanism for the turn signal lights on the vehicle. As the name suggests, it flashes or turns the turn signal lights on and off at a regular rate to indicate a change in the vehicle's direction.
- Flasher cans are mechanical devices, while flasher controls are generally electronic devices.
- Flashers operate like an automatically resetting circuit breaker, meaning they use a bimetallic strip to open and close the switch contacts.
- Relays are switches that are turned on and off by a small electrical current.
- The relay is made up of an electromagnet, a set of switch contacts, terminals, and the case. The electromagnet is a winding of fine metal-insulated wire wrapped around an iron core.
- A solid-state relay acts like a mechanical relay but does not have any moving parts. This means that electronic relays do not use mechanical switch contacts, but instead use transistorized circuitry to turn the circuit on and off.
- A switch-controlled relay may be a desirable attribute in certain instances. The main purpose of a relay is that it allows a small current that energizes the relay coil to control the large current of the load device. This allows for the use of small-gauge wires and switches inside of the vehicle to control the relay; the larger-gauge wires to control the load can be positioned outside of the passenger compartment.
- A logic-controlled relay operates like a switched relay control, except that the switching of the relay coil is accomplished with an electronic control module. The

- relay coil is usually supplied with battery voltage and the control module uses a low-side driver to control the ground of the circuit.
- A timer delay relay (TDR) is capable of delaying the closing of the contacts once a power input is received.
- A solenoid is an electromechanical device that converts electrical energy into mechanical linear (back-and-forth) movement. Solenoids can be used to pull or push.
- Electric motors use magnetic fields to create rotary movement. Motors consist of two main components: the armature and the field. The armature contains electromagnetic coils. The field contains either electromagnets or permanent magnets. The interaction between the magnetic fields of the stationary field coils and the magnetic fields of the moveable armature coils causes the armature to rotate. Since the armature coils can be energized and create electromagnetic fields, those fields can be turned on and off so that they attract and repel the fields of the stationary magnetic fields.
- A DC motor converts direct current into mechanical energy.
- There are four types of DC motors: permanent magnet motors, series motors, shunt motors, and compound motors.
- A stepper motor is a brushless, synchronous motor that converts a digital pulse width signal into mechanical rotation.
- Stepper motors differ from standard DC motors in that they contain no brushes to transfer electricity to the rotor, and they can be moved in precise increments. Each complete rotation of the stepper motor is divided into as many as 200 steps.
- An AC motor is an electric motor that is operated by an alternating current (AC).
- AC motors have a fixed stator that contains coils of wire to generate a rotating magnetic field when supplied with AC current. These motors usually have a rotor that contains permanent magnets, or coils of wire to create magnetic fields in the rotor.
- There are two main types of AC motors: induction and synchronous.
- Induction motors rely on a small speed difference between the stator's magnetic field and the rotor's shaft speed.
- Induction motors cannot generate great amounts of torque when the shaft speed is synchronized with the stator's magnetic field.
- Synchronous AC motors do not rely on slip induction, but instead rely on permanent magnets in the rotor, or a separately excited rotor coil to generate movement.
- The three-phase AC motor is used in hybrid vehicle, electric vehicles, and high-torque applications. The benefit of the three-phase AC motor is that the stator winding is divided into three sections, or phases, that are placed 120 degrees apart. This arrangement allows the motor to smoothly start from any position and also allows it to generate a great deal of torque and efficiency.
- Ignition coils and transformers both operate under the principles of electromagnetic induction, which is electromagnetism to produce electricity.
- An ignition coil can be described as a step-up transformer. This is because the output can be 60 kilovolts (or more), which is higher than the input, nominally 12 volts.
- Two sets of coil windings are used in an ignition coil. One coil, referred to as a primary winding, is wound around a second, the secondary winding. The primary coil typically has 200 to 300 turns of light-gauge wire while the secondary has approximately 30,000 to 60,000 turns of very fine wire.
- When current is passed through the primary winding, the magnetic field builds, surrounding both the windings. When the current is turned off, the magnetic field collapses with enough speed to induce high voltage in the primary winding (self-induction) and very high voltage in the secondary winding (mutual induction) by the rapidly moving (collapsing) magnetic field. Voltage is induced into each of the thousands of windings of the secondary coil. This voltage is strong enough to overcome the infinite resistance of the spark plug gap and push current across the gap, causing a spark.
- Step-down transformers function under the same operating principles. The only difference is that the secondary coil has fewer turns than the primary, providing a lower induced output.
- Resistance is measured in ohms, represented by the Greek letter omega (Ω).
- Resistors are rated in ohms as well, to indicate how strongly they will oppose any current flowing through them.
- Resistors have a power rating. Only the resistance value is marked. The resistor's power rating is determined by its size.
- Resistors are marked with four or five colored bands. Each color represents a numeric resistance value. The color bands are set close to each other and read from left to right. The last band, or tolerance band, is spaced farther apart.
- Fixed resistors are generally cylinders with connecting metal leads projecting along the axis of the cylinder at each end.
- A rheostat is a mechanical variable resistor with two connections and consists of a resistance wire wrapped in a loose coil connected to the supply at one end only. A moveable wiper is connected to the other circuit connection and is made to move over the wire manually.
- Potentiometers are mechanical variable resistors with three connections, two fixed and one moveable.
- Potentiometers act as voltage dividers and, as such, alter the voltage in a circuit. A resistance wire is wrapped between two fixed connections.
- A capacitor can quickly store a small amount of electrical energy, at which point it is charged. There are two surfaces inside the capacitor, separated by insulating material.
- When the capacitor is charged, one surface is positively charged and the other is negatively charged. When a circuit is closed between its terminals, the capacitor releases its charge. At this point, it is discharged.
- A typical capacitor stores the charge on thin sheets of foil, with sheets of insulation between them. These are rolled together to form a protective canister.

- A capacitor's capacitance (C) is a measure of the amount of charge (Q) stored on each plate for a given potential difference or voltage (V) that appears between the plates.

Key Terms

armature The rotating wire coils in motors and generators. It is also the moving part of a solenoid or relay, as well as the pole piece in a permanent magnet generator.

capacitance (C) The ability of a capacitor to store an electrical charge.

capacitor A device that can quickly store a small amount of electrical energy, at which point it is charged.

charge (Q) In measuring capacitance, the amount of electrical energy present.

circuit or schematic diagram A pictorial representation or road map of the wiring and electrical components.

circuit breaker A device that trips and opens a circuit, preventing excessive current flow in a circuit. It is resettable to allow for reuse.

commutator A device made on armatures of electric generators and motors to control the direction of current flow in the armature windings.

fixed resistor A resistor that has a fixed value.

flasher can A mechanical device that switches the vehicle's turn signal and hazard flasher bulbs on and off.

flasher control An electronic device that switches the vehicle's turn signals on and off.

flasher unit The name given to the system that is responsible for switching the vehicle's flashers on and off.

fuse A safety device that self-destructs to prevent excessive current flowing in a circuit in the event of a fault.

logic-controlled relay A relay that is turned on and off by an electronic control module.

normally closed (NC) An electrical contact that is closed in the at-rest position.

normally open (NO) An electrical contact that is open in the at-rest position.

polarity sensitive A term used to describe a component that must be connected into a circuit with the correct polarity to its terminals.

potentiometer Also called a pot, a three-terminal resistive device with one terminal connected to the input of the resistor, one terminal connected to the output of the resistor, and the third terminal connected to a movable wiper arm that moves up and down the resistor.

primary winding The coil of wire in the low-voltage circuit that creates the magnetic field in a step-up transformer.

relay An electromechanical switching device whereby the magnetism from a coil winding acts on a lever that switches a set of contacts.

rheostat An adjustable resistor that varies current flow through a circuit.

secondary winding The coil of wire in which high voltage is induced in a step-up transformer.

solenoid An electromagnet with a moving iron core that is used to cause mechanical motion.

solid-state relay A relay that performs the function of a mechanical relay but uses only electronic components.

step-down transformer A transformer used to reduce the voltage, such as to allow a battery charger operated on 120 volts to charge a 12-volt battery.

step-up transformer A transformer used to increase the voltage from a lower input voltage to a higher output, such as an ignition coil.

switch An electrical device with contacts that turns current flow on and off.

switch-controlled relay A relay that is controlled by a mechanical switch.

timer delay relay (TDR) A relay that remains on for a set period of time after power has been removed from the relay coils.

transformer action The transfer of electrical energy from one coil to another through induction in a transformer.

turn signal switch A switch that turns the left and right turn signal lights on and off.

variable resistor A component that has a mechanism for varying resistance.

voltage (V) The electrical pressure that causes current to flow in a circuit.

Review Questions

1. When a two-terminal switch connected to a light bulb is turned off, all of the following will happen *except*:
    a. switches open the circuit.
    b. current flow stops.
    c. light goes out.
    d. the circuit is closed.
2. Which of these stops excessive current flow from potentially damaging the components of a circuit?
    a. Fuse
    b. Transformer
    c. Rheostat
    d. NC contact
3. Choose the correct statement.
    a. PTC has high resistance at room temperature.
    b. PTC is generally reset and is allowed to cool after maximum resistance.
    c. PTC resistance increases as the temperature decreases.
    d. PTC resistance increases as the current flow decreases.
4. All of the following are circuit protection devices *except*:
    a. fusible links.
    b. circuit breakers.
    c. PTC thermistors.
    d. capacitors.
5. Which of the following have an operating mechanism similar to that of circuit breakers?
    a. Fusible links
    b. Push buttons
    c. Flash cans and flash controls
    d. Toggle switches

6. All of the following statements are true regarding relays and relay control circuits *except*:
   a. Relays have two fixed contacts and one movable contact.
   b. The movable contact is fixed to armature spring.
   c. Current flows through relay windings and large current through relay points.
   d. The two fixed contacts are designated as "85 or 1" and "86 or 2."
7. A solenoid creates:
   a. electrical energy.
   b. lateral movement.
   c. high resistance.
   d. rotary movement.
8. Which of the following electrical components converts a digital pulse width signal into mechanical rotation?
   a. A solenoid
   b. A timer delay relay
   c. A stepper motor
   d. A logic-controlled relay
9. Rheostats are commonly used in:
   a. dash light dimmer circuits.
   b. throttle position sensors.
   c. the vehicle horn.
   d. circuit breakers.
10. In general, as the capacitance and voltage rating of the capacitor increases, the:
    a. physical size of the capacitor increases.
    b. physical size of the capacitor decreases.
    c. amount of charge stored decreases.
    d. less dangerous it becomes.

ASE Technician A/Technician B Style Questions

1. Tech A says that most switches in modern automobiles are interchangeable with one another. Tech B says that a switch is used to turn a circuit on and off, and should only be replaced with the appropriate part. Who is correct?
   a. Tech A
   b. Tech B
   c. Both A and B
   d. Neither A nor B
2. Tech A says that as long as a fuse fits in the fuse box, it is an acceptable alternative for replacing a blown fuse, as long as the correct fuse is installed after the repair is made. Tech B says that only the correct size and rating of fuse should be used to replace a blown fuse. Who is correct?
   a. Tech A
   b. Tech B
   c. Both A and B
   d. Neither A nor B
3. Tech A says that some circuit breakers need to be manually reset after they have tripped. Tech B says that some circuit breakers reset themselves after they have cooled. Who is correct?
   a. Tech A
   b. Tech B
   c. Both A and B
   d. Neither A nor B
4. Tech A says that a PTC device increases in resistance as its temperature increases. Tech B says that a PTC device decreases its resistance as its temperature increases. Who is correct?
   a. Tech A
   b. Tech B
   c. Both A and B
   d. Neither A nor B
5. Tech A says that relays induce mechanical movement when they are energized. Tech B says that relays use a small current to control a large current. Who is correct?
   a. Tech A
   b. Tech B
   c. Both A and B
   d. Neither A nor B
6. Tech A says that some relays can be controlled by a switch. Tech B says that some relays can be controlled by an electronic control module. Who is correct?
   a. Tech A
   b. Tech B
   c. Both A and B
   d. Neither A nor B
7. Tech A says that solenoids convert electrical energy into linear movement. Tech B says that solenoids are used in automotive starter circuits. Who is correct?
   a. Tech A
   b. Tech B
   c. Both A and B
   d. Neither A nor B
8. Tech A says that all 12-volt motors will rotate at the same speed, as long as the batter is fully charged. Tech B says that electric motors convert electric energy into mechanical movement. Who is correct?
   a. Tech A
   b. Tech B
   c. Both A and B
   d. Neither A nor B
9. Tech A says that stepper motors do not contain brushes. Tech B says that stepper motors can move in precise increments. Who is correct?
   a. Tech A
   b. Tech B
   c. Both A and B
   d. Neither A nor B
10. Tech A says that ignition coils can be dangerous, as they produce between 1000 and 2000 volts. Tech B says that ignition coils typically produce more than 10,000 volts. Who is correct?
    a. Tech A
    b. Tech B
    c. Both A and B
    d. Neither A nor B

CHAPTER 9
Electronic Components

NATEF Tasks

- **N09001** Inspect and test fusible links, circuit breakers, and fuses; determine needed action.
- **N09002** Inspect, test, repair, and/or replace components, connectors, terminals, harnesses, and wiring in electrical/ electronic systems (including solder repairs); determine needed action.

Knowledge Objectives

After reading this chapter, you will be able to:

- **K09001** Explain the function and purpose of the various types of diodes used in automotive circuits.
- **K09002** Describe the operation of transistors, and describe circuits that use these devices.
- **K09003** Describe integrated circuits, and the role that they play in control units.
- **K09004** Describe the basic operation and construction of microprocessors.
- **K09005** Describe the purpose and function of microcontrollers.
- **K09006** Describe the operation of speed control circuits.

Skills Objectives

There are no Skills Objectives in this chapter.

You Are the Technician

Diodes are designed to operate as a one-way check valve for electrons, but they can and do fail. When a diode is reverse-biased, and a large voltage pushes through the diode in the wrong way, breakdown occurs, often causing catastrophic damage to the diode. Burned-out diodes can cause a wide variety of electrical concerns; testing them is a critical skill for the automotive technician. To test a diode, set your meter to ohms and measure resistance across the diode in one direction, then reverse your meter leads on the diode terminals and retest. A properly functioning diode will have little resistance when the meter is connected in a forward-biased state and high resistance when the meter is connected in a reverse-biased manner. A low or high reading in both forward- and reverse-biased conditions indicate that the diode has failed.

1. True or False: If a diode is installed into the circuit with its polarity reversed, current will be allowed to flow.
2. True or False: When testing diodes, you should have continuity in both directions across a good diode.

Introduction

The once-held notion that mechanics only ended up in their profession due to the fact that they did not have the skills for any other career is misguided, and considering the enormous complexity of today's automobiles, this way of thinking could not be farther from the truth. Today's automobiles are far more complicated than the spacecraft that took us to the moon. It is not uncommon for a vehicle to contain at least six separate control modules, while some vehicles make use of more than one hundred computers to operate all of their accessories and functions. A clear understanding of these modules and their corresponding systems is critical if the technician is to become proficient in diagnosing and repairing these advanced systems. This chapter will cover the basic components and functions of today's electronic devices as they are employed in automotive technology. With this knowledge, you will be well on your way to becoming a more proficient automotive diagnostician.

Electronic Components

K09001 Explain the function and purpose of the various types of diodes used in automotive circuits.

N09001 Inspect and test fusible links, circuit breakers, and fuses; determine needed action.

The term *electronic components* tends to be used for those components that have no moving parts, such as diodes, transistors, and integrated circuits (**FIGURE 9-1**). Each component has a particular function and, when arranged in circuits, interacts with other components to control electrical and electronic functions. Individual components may be mounted in a housing to operate as a sensor: for example, an engine temperature sensor. Or, many components can be arranged in circuits and mounted on circuit boards to perform various functions. Examples of this would be the many electronic components used in control units to manage virtually all of the systems on a vehicle, including ignition, fuel, and emission systems, as well as entertainment, lighting, safety, and security systems.

Diodes

A **diode** can be thought of as the electronic version of a one-way check valve. By restricting the direction of movement of charge carriers, it allows an electrical current to flow in one direction, but essentially blocks it in the opposite direction. The most common diode is constructed of semiconductor material, although other types of diodes exist. When a diode is connected in a circuit, if the diode is positioned in such a way that current is allowed to flow, the diode is forward-biased. If the diode is connected in a circuit in a manner that prevents current from flowing, the diode is reverse-biased. A forward-biased diode will conduct current and allows a small amount of voltage (0.5–0.7 volts) to drop across it, leaving the remaining voltage free to continue on and operate the load device. If the diode's polarity gets reversed, all of the voltage will drop across the diode, and no current will flow through the circuit (**FIGURE 9-2**).

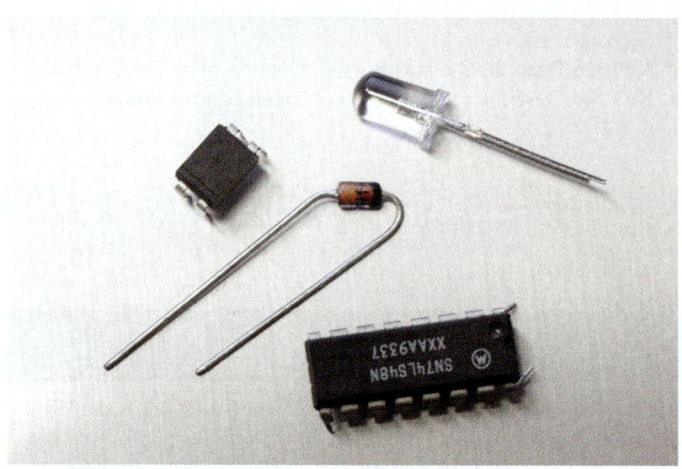

FIGURE 9-1 Diode, transistor, and integrated circuit.

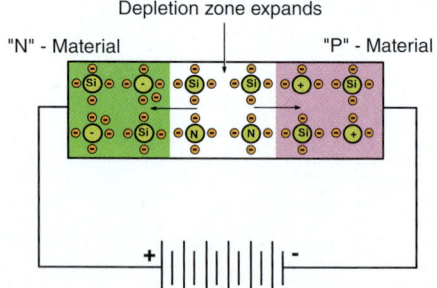

A Charges are drawn to the ends of material by battery polarity creating a depletion zone that has no electrical charge. The depletion zone acts as an insulator.

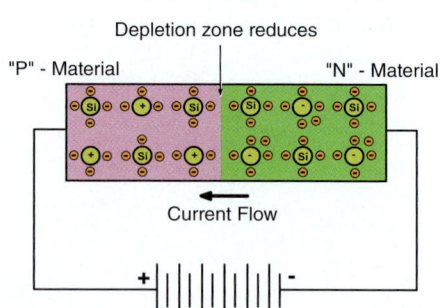

B Charges are drawn into the depletion zone causing it to shrink to almost nothing. The depletion zone acts as a conductor.

FIGURE 9-2 Diode illustrating the PN junction.

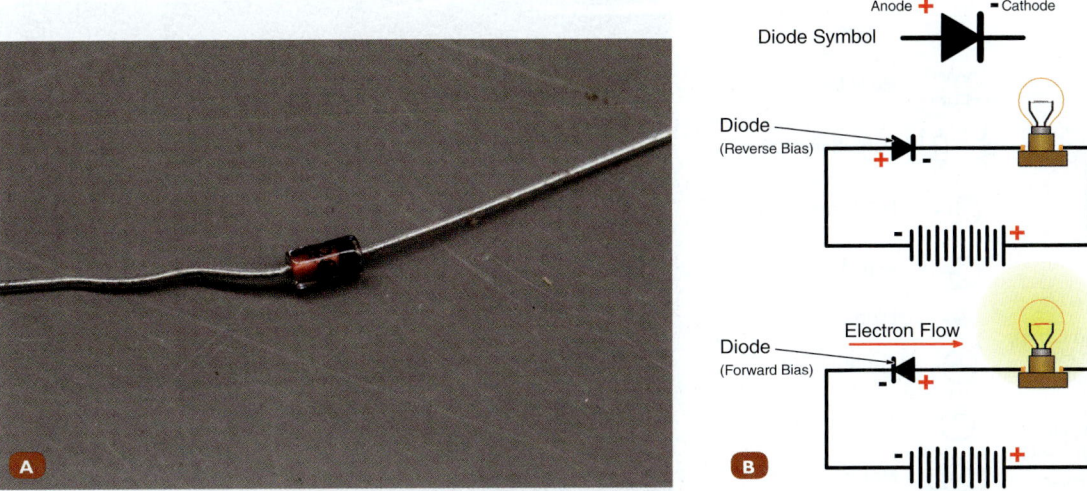

FIGURE 9-3 **A.** Image of diode. **B.** Schematic symbol of diode. Arrows indicate the direction of current flow.

FIGURE 9-4 Diode being tested with an ohmmeter.

Diodes are constructed of two different types of materials, each occupying half of the diode: one half is P material, located on the positive side and commonly referred to as the *anode*, and the other half is N material, located on the negative side, commonly referred to as the *cathode*, on the pointed side of the symbol. Diodes have a single PN junction. If the diode is connected to a current source, with the P region connected to a negative pole and the N region to a positive pole, the holes will be attracted toward the negative pole and the electrons to the positive pole. This movement enlarges the depletion layer, which makes the insulated space larger, stopping current flow across the junction. Think of a diode like you would think of a simple switch: closed when forward-biased, and open when reverse-biased. It is important to note that the "arrow" points in the opposite direction of current flow (**FIGURE 9-3**).

In addition to serving as one-way check valves for electrons, certain diodes emit light when current is passed through them. These diodes are call light emitting diodes, or LEDs; they emit light when they are connected in a forward-biased arrangement within the circuit (**FIGURE 9-5**). In general, LEDs operate at lower voltage than battery voltage, and much lower current draw than their incandescent cousins. For this reason, LEDs will usually have a resistor in series with them to reduce the supplied voltage down to a range where they can safely operate. These resistors are critical to the proper operation and longevity of LEDs. If an LED is connected in a circuit without a resistor in series, and the circuit voltage is higher than the LED is designed for, too much current will pass through the LED, and it will burn up. LEDs emit a characteristically clean, brilliant light and are available in a range of colors. They are often used as indicator lights on electrical and electronic appliances. High-powered white LEDs are increasingly being used to replace traditional incandescent light bulbs.

You will commonly find LEDs in running lights, driving lights, taillights, stoplights, interior lights, button illumination, headlights, and many other applications. The advantage of using LEDs in place of traditional incandescent light bulbs is that they are much more

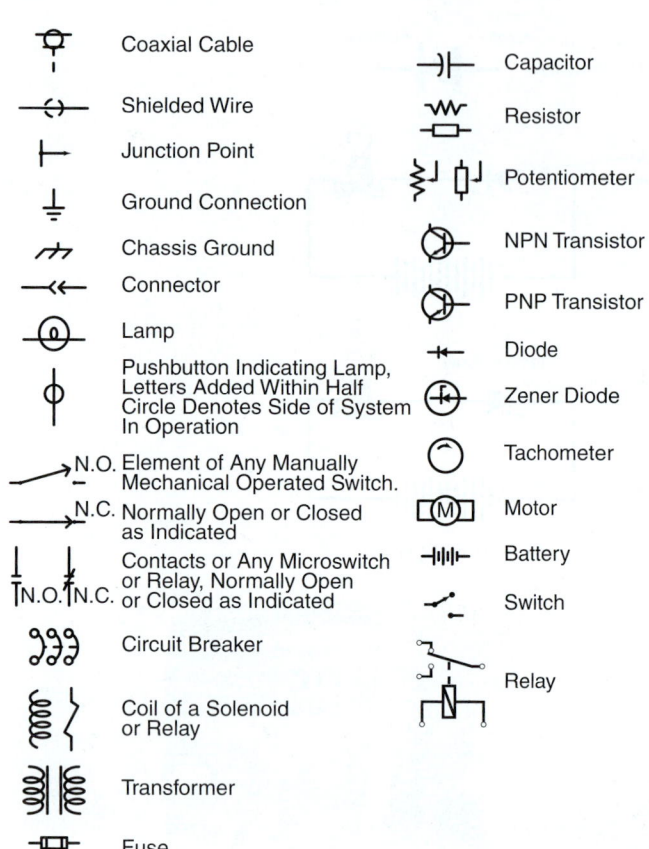

FIGURE 9-5 Every electrical device and component has a corresponding electrical symbol.

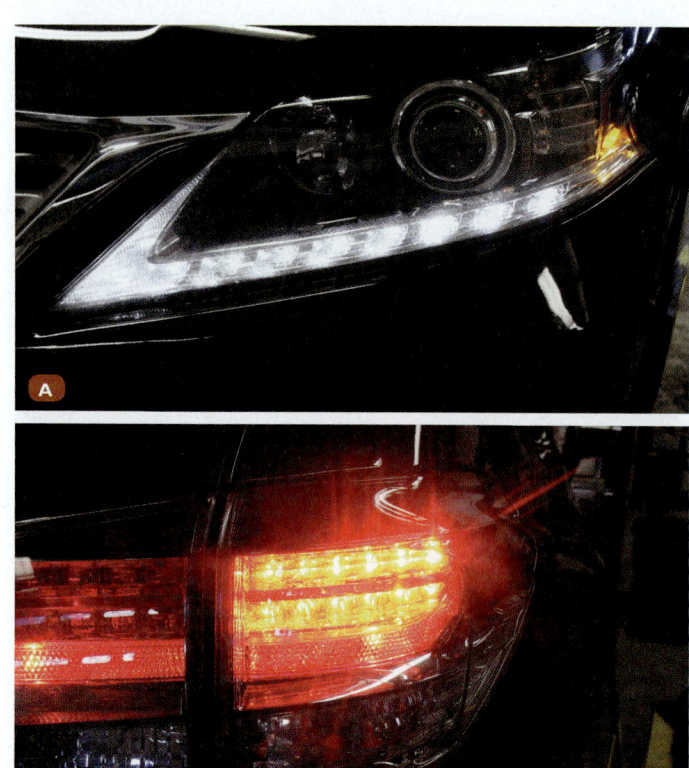

FIGURE 9-6 A. LEDs in marker lights and headlight assemblies. **B.** LED taillight assembly.

efficient: they generate more light while using less current and producing less heat, which allows manufacturers to design smaller, more focused lighting enclosures. The light color of LEDs can be easily modified, which allows them to mimic the color spectrum of sunlight in certain applications (**FIGURE 9-6**).

Zener Diodes

A **Zener diode** is designed to block current flow through it when it is reverse-biased up to a preset threshold. Once this threshold voltage, or *Zener voltage*, is reached, breakdown occurs and current is allowed to pass through in the reverse direction (see Figure 9-5). As breakdown voltage is reached, the Zener diode's resistance suddenly collapses. It lets a large current flow through it, without damage. Once the reverse-biased voltage drops below the threshold, the Zener diode once again acts like a traditional diode, allowing current to flow in only one direction. Because Zener diodes respond to certain voltage changes similar to switches, they are used in voltage regulators commonly found in automotive charging systems.

Rectifier Bridge

A **rectifier bridge** is an arrangement of diodes, typically in sets of three, that converts AC voltage into DC voltage. AC voltage rapidly reverses direction, while DC voltage flows in only one direction. When diodes are arranged to form a rectifier bridge, current flow is routed through the diodes in such a way that cycling AC voltage goes into the rectifier bridge, but only DC voltage is allowed to exit the rectifier bridge. Since diodes generally drop 0.7 volts apiece when current passes through them, they generate considerable amounts of heat. For this reason, rectifier bridges will usually be found mounted to a heat sink of some sort to prevent them from burning up (**FIGURE 9-7**). Rectifier bridges will be discussed in greater detail in the chapter on charging systems.

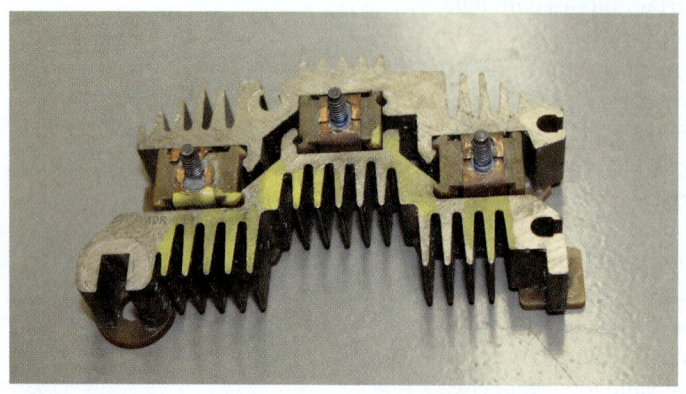

FIGURE 9-7 Rectifier bridge.

Transistors

A **transistor** is a semiconductor device used as a switch and to amplify currents. A transistor is a key component in almost any electronic device. There are two kinds: NPN and PNP. The **NPN transistor** has a P-type semiconductor between two N-type semiconductors. A **PNP transistor** has an N-type semiconductor between two P-types (**FIGURE 9-8**).

Each of the three regions has a terminal. The center region is always called the base. The outer regions are the collector and the emitter. In the symbol for a transistor, the emitter is the terminal with the arrow. It always points toward the negative material (see Figure 9-5)).

In a circuit, a transistor can act as an electronic switch. If an open control switch is connected to the base, the depletion layer at one PN junction will block current from flowing through the transistor and powering the load.

When the control switch is closed, a small current flows through the emitter–base PN junction. The base has only a limited number of charge carriers; so extra ones flow across the emitter–collector PN junction, letting current operate the load. The transistor then operates as a low-resistance conductor. Most transistors have about 0.5 volts of voltage drop across the emitter–collector junction while current is flowing; this drop causes heat to be produced in the transistor. When current flow increases through the transistor, so does heat. For this reason, nearly all transistors have a heat sink of some sort attached to them that allows heat to be conducted away from the transistor. A small current through the base lets larger current flow across the emitter–collector junction, in a similar fashion that a small current through a relay coil closes the contacts, and allows current to pass to the load. When current is applied to the base of the transistor, the transistor is said to be turned on.

One difference between PNP and NPN transistors is whether the base is activated by positive current flow or negative current flow. In the case of an NPN transistor, positive current, through a resistor (to limit current flow to the base), is applied to the base. In a PNP transistor, a ground (through a resistor) is applied to the base. Transistors can be placed in the feed side of a circuit or on the ground side of a circuit. Transistors placed in the feed side of the circuit are commonly referred to as *high-side drivers*; transistors placed in the ground side of the circuit are referred to as *low-side drivers*.

FETs

A **field-effect transistor (FET)** is a type of transistor that uses an electric field to control the shape and conductivity of a channel of one type of charge carrier consisting of a certain type of charge carrier in the semiconductor material (**FIGURE 9-9**). All forms of FETs have high resistance. The conductivity of a non-FET transistor is regulated by the input current (the emitter to base current), which has low resistance. The conductivity of the FET is regulated by how much voltage is applied to the gate, which is insulated from the device; the applied gate voltage imposes an electric field into the device, which in turn attracts or repels charge carriers to or from the region between a source terminal and a drain terminal. The density of charge characters in turn influences the conductivity between the source and drain. The terminals of the FET transistor are as follows:

- Source (S): This is where electrons enter the FET
- Drain (D): This is where the electrons exit the FET when voltage is applied to the gate
- Gate (G): Voltage to this terminal controls how much conductivity exists through the FET

MOSFETs

A **metal–oxide–semiconductor field-effect transistor (MOSFET)** is a type of transistor used for switching and amplifying high current (10–50 amps) electrical signals. MOSFETs

K09002 Describe the operation of transistors, and describe circuits that use these devices.

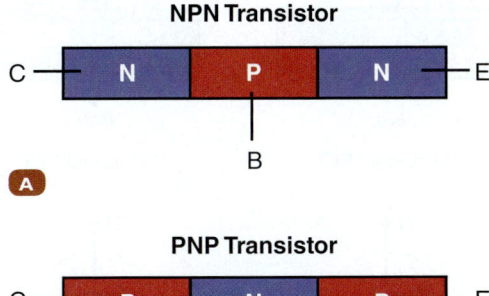
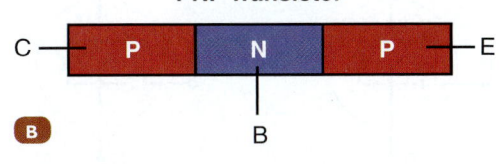

FIGURE 9-8 Transistors. **A.** The NPN transistor has a P-type semiconductor between two N-type semiconductors. **B.** A PNP transistor has an N-type semiconductor between two P-type semiconductors.

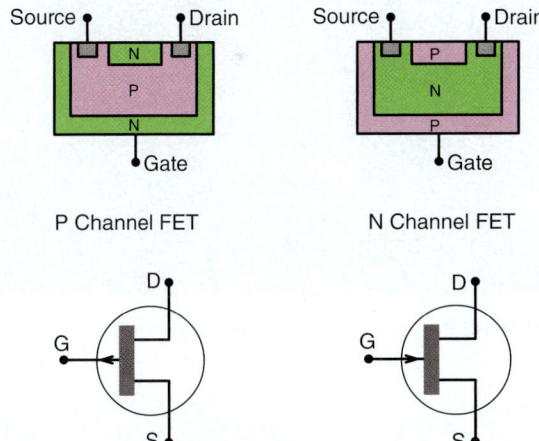

FIGURE 9-9 FET with its terminals marked: source, drain, and gate.

are used extensively in stereo amplifiers, computers, and many other electronic devices that need to rapidly switch high current loads using a very small control current (**FIGURE 9-10**). The MOSFET is a four-terminal device with Source (S), Drain (D), Gate (G), and Body (B). However, the body terminal (B) is commonly connected to the source (S), making MOSFETs similar to other field effect transistors. As stated earlier, the main advantage of the MOSFET over other types of transistors is its ability to control a large current load (10–50 amps) with a very small input current (less than 1 milliamp). As of this writing, MOSFETs are the most popular type of transistor.

BJTs

A **bipolar junction transistor (BJT)** is a type of transistor that relies on the contact of two different types of semiconductor material to operate. BJTs can serve as amplifiers, switches, or oscillators. BJTs can either be found as individual components, or they can be joined together to form the workings of an integrated circuit (IC). Before the advent of MOSFETs, BJTs were the most popular type of transistor in use.

IGBTs

An **insulated-gate bipolar transistor (IGBT)** is a three-terminal power transistor that is capable of switching very high current loads and voltages at high frequencies (**FIGURE 9-11**). IGBTs see wide use in motor controllers and inverters in hybrid and electric vehicles. IGBTs are designed to switch very rapidly. As a result, they are capable of producing sophisticated waveforms by using pulse width modulation techniques. An IGBT combines the gate control characteristics of a MOSFET transistor with the high-amperage capability of a bipolar transistor.

▶ Control Modules

A **control module** or unit is a generic term that identifies an electronic unit that controls one or more electrical systems in the vehicle (**FIGURE 9-12**). The modern control module is a configuration of resistors, transistors, diodes, integrated circuits, and other electronic components. Today's vehicles contain many control units, each with a specific purpose. For example, there are engine, transmission, body and suspension, entertainment, anti-theft,

FIGURE 9-10 MOSFET with the terminals clearly labeled: source (S), drain (D), gate (G), and body (B).

FIGURE 9-11 IGBT in hybrid vehicle inverter.

FIGURE 9-12 Automotive control module, both with the cases closed, and with the cases opened, showing the various electronic components inside.

and many other control units that monitor the function of their respective system. Control modules monitor a number of inputs from sensors and circuits, process those data according to the information programmed into their memory, and send output commands to actuators and electrical devices, which control the operation of various components and systems on the vehicle. On modern vehicles, the control units share information on a data **bus**, which is a common set of wires that connect the modules together so that information or data can be shared across the network. For example, the transmission control unit needs to know road speed, as does the control unit that operates the dash instruments and the anti-lock brake system control unit. The information from the one vehicle speed sensor is shared across the data bus to all three control units.

▶ Integrated Circuits

An **integrated circuit (IC)** is a set of electrical circuits on one small piece of semiconductor material (**FIGURE 9-13**). ICs are also referred to as microchips. The IC chip allows multiple circuits to be combined in a very tiny space. The advent of the IC chip has been a driving force behind the shrinking of electronic devices. This is due to the fact that a single IC chip can perform all of the functions of several dedicated circuits, all the while taking up a fraction of the space and using fewer components. IC chips can be thought of as the nerve center of control units and other computerized devices.

K09003 Describe integrated circuits, and the role that they play in control units.

▶ Microprocessors

A **microprocessor** or central processing unit (CPU) is a programmable device that is found in a vehicle's ECU (**FIGURE 9-14**). It is a computer that monitors information from inputs such as sensors and makes decisions based on its program, then outputs signals to actuators and electrical devices on the vehicle. Microprocessors control and monitor most electrical systems on the modern vehicle. For example, there are microprocessor control units for the engine, transmission, suspension, brakes, steering, and main body control functions (**FIGURE 9-15**). Modern diagnostic equipment connects into the microprocessors on the vehicle via a data-link connector to provide information on system performance to the technician. It is not unusual for the microprocessors on a vehicle to require programming updates. Updating is performed by reflashing or reprogramming the memory in the microprocessors with an update supplied by the manufacturer.

There are many different types of sensors installed into the modern vehicle, and they are used to provide information to the microprocessors on the vehicle. A sensor essentially converts a mechanical action into electrical signals that can be read by the microprocessors in the control units. For example, the sensor installed near the crankshaft converts the rotary movement of the crankshaft into electrical signals that tell the

K09004 Describe the basic operation and construction of microprocessors.

N09002 Inspect, test, repair, and/or replace components, connectors, terminals, harnesses, and wiring in electrical/ electronic systems (including solder repairs); determine needed action.

FIGURE 9-13 IC chips.

FIGURE 9-14 CPU.

FIGURE 9-15 Various automotive control units.

FIGURE 9-16 Crankshaft position sensors, temperature sensor, and throttle position sensor.

microprocessor each time the crankshaft has rotated through a certain number of degrees and where it is in relation to top dead center of a cylinder, typically cylinder 1. The temperature sensor converts the temperature of the engine coolant into an electrical signal that can be read by a microprocessor (**FIGURE 9-16**).

▶ Microcontrollers

Microcontrollers are hidden inside of nearly all electronic devices that you use today. **Microcontrollers** are dedicated computers that perform a specific set of functions. A good example of this would be the Arduino series of microcontrollers, which can perform a variety of functions and are user-programmable (**FIGURE 9-17**). Microcontrollers are also found in many of the components that you interact with in your vehicle, such as the radio, climate-control unit, power windows, power seats, and the like.

FIGURE 9-17 Arduino controller.

K09005 Describe the purpose and function of microcontrollers.

Delay Circuits

Delay circuits are added to vehicles to build in specific time delays in turning on or off electrical devices. For example, most modern vehicles have a time delay built into the interior lights on vehicles. After closing the door, the lamp will often stay on for a predetermined time before slowly turning off. The delay circuit can be constructed from electronic components or it can be programmed into an ECU that controls the circuit. Modern vehicles are increasingly using ECUs with programmed timers to perform time-delay functions.

▶ Speed Control Circuits

K09006 Describe the operation of speed control circuits.

Speed control circuits are used on the vehicle to control the speed of motors or accessories. A common use is the control of the fan speed in the vehicle's HVAC system and the electric radiator fans. Essentially, the speed control circuit manages the speed of a fan motor to control the amount of air movement through the system.

Most speed control systems use pulse width modulation to control motor speed. Pulse width (or duty cycle) refers to the percentage of time a circuit is fully on, versus fully off. A pulse width of 50% means that the circuit is on 50% of the time and off 50% of the time. This provides about 50% of the electrical power to the motor, which allows it to operate at about 50% power. Likewise, a 25% pulse width means that the electrical power is on

25% of the time. Since the pulse width is cycled on and off very quickly (sometimes exceeding 10,000 cycles per second), power to the motor appears seems smooth and is controlled accurately. Like the time-delay circuit, speed control circuits are constructed from electronic components that can be in dedicated modules or integrated into the vehicle's powertrain control module that controls the circuit.

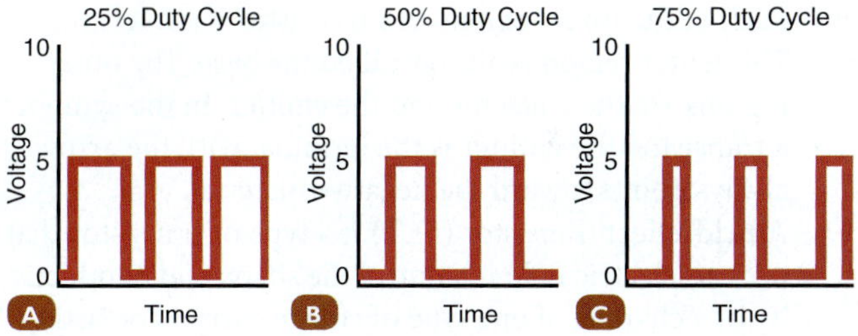

Duty Cycle

Duty cycle describes the amount of time that power flow within a circuit is allowed. Think of a simple switch: If you walk into a room and turn on the light switch for one minute, and then turn off the light switch for 99 minutes, you have created a 1% duty cycle of the lights in that room for the given timeframe of 100 minutes. If you kept the switch on for 5 minutes and off for 95 minutes, you would have a duty cycle of 5%. Duty cycle is very useful for accurately controlling the amount of work that electrical components are doing. As a rule, the higher the duty cycle supplied to a component, the more work that the component will do in a given amount of time, which will result in the component wearing out faster than if it were operated at a lower duty cycle (**FIGURE 9-18**).

FIGURE 9-18 Pulse width modulation, ground side controlled. **A.** 25% duty cycle. **B.** 50% duty cycle, **C.** 75% duty cycle.

Pulse Width Modulation

Pulse width modulation (PWM) refers to the amount of time that power is turned on to a circuit during a given period of time. Pulse width modulation is used to accurately adjust the speed and/or travel of electric motors and devices. PWM can also be used to dim lights. Pulse width functions like this: If a transistor is turned on for 80% of a second, and off for the other 20% of a second, this would be a duty cycle of 80% for the given timeframe. If you apply this 80% pulse width to a motor, it will spin at 80% of its maximum speed. If you apply an 80% pulse width to a lamp, it will operate at 80% of its total brightness. If you apply 80% of pulse width to an actuator, it will only move 80% of its total distance. The principle of pulse width modulation is widely used by control modules to vary the speed, intensity, and travel of the electrical devices controlled.

▶ Wrap-Up

Ready for Review

▸ The term electronic components tends to be used for those components that have no moving parts, such as diodes, transistors, and integrated circuits (also called solid state components).
▸ A diode is an electronic one-way check valve. By restricting the direction of movement of charge carriers, it allows an electrical current to flow in one direction, but essentially blocks it in the opposite direction.
▸ Diodes are constructed of two different types of materials, each occupying half of the diode: one half is P material, located on the positive side and commonly referred to as the anode, and the other half is N material, located on the negative side, commonly referred to as the cathode, on the pointed side of the symbol.
▸ Diodes have a single PN junction.
▸ Certain diodes emit light when current is passed through them. These diodes are call light emitting diodes, or LEDs; they emit light when they are connected in a forward-biased arrangement within the circuit.
▸ A Zener diode is designed to block current flow through it when it is reverse-biased up to a preset threshold. Once this threshold voltage, or Zener voltage, is reached, breakdown occurs and current is allowed to pass through in the reverse direction. As breakdown voltage is reached, the Zener diode's resistance collapses. It lets a large current flow through it, without damage.
▸ A rectifier bridge is an arrangement of diodes, typically in sets of three, that converts AC voltage into DC voltage.
▸ A transistor is a semiconductor device used as a switch and to amplify currents.
▸ There are two kinds of transistors: NPN and PNP. The NPN transistor has a P-type semiconductor between two N-type semiconductors. A PNP transistor has an N-type semiconductor between two P-types.

- Each of the three regions in a transistor has a terminal. The center region is always called the base. The outer regions are the collector and the emitter. In the symbol for a transistor, the emitter is the terminal with the arrow. It always points toward the negative material.
- A field-effect transistor (FET) is a type of transistor that uses an electric field to control the shape and conductivity of a channel of one type of charge carrier consisting of a certain type of charge carrier in the semiconductor material.
- All forms of FETs have high resistance.
- The metal–oxide–semiconductor field-effect transistor (MOSFET) is a type of transistor used for switching and amplifying high current (10–50 amps) electrical signals.
- MOSFETs are used extensively in stereo amplifiers, computers, and many other electronic devices that need to rapidly switch high current loads using a very small control current.
- A bipolar junction transistor (BJT) is a type of transistor that relies on the contact of two different types of semiconductor material to operate. BJTs can serve as amplifiers, switches, or oscillators.
- BJTs can either be found as individual components, or they can be joined together to form the workings of an integrated circuit (IC). Before the advent of MOSFETs, BJTs were the most popular type of transistor in use.
- An insulated-gate bipolar transistor (IGBT) is a three-terminal power transistor that is capable of switching very high current loads and voltages at high frequencies.
- IGBTs see wide use in motor controllers and inverters in hybrid and electric vehicles.
- IGBTs are designed to switch very rapidly.
- A control module or unit is a generic term that identifies an electronic unit that controls one or more electrical systems in the vehicle.
- An integrated circuit (IC) is a set of electrical circuits on one small piece of semiconductor material.
- ICs are also referred to as microchips.
- The IC chip allows multiple circuits to be combined in a very tiny space.
- The advent of the IC chip has been a driving force behind the shrinking of electronic devices.
- A microprocessor or central processing unit (CPU) is a programmable device that is found in a vehicle's ECU (electronic control unit).
- Microcontrollers are hidden inside of nearly all of the electronic devices that you use today.
- Microcontrollers are dedicated computers that perform a specific set of functions.
- Delay circuits are added to vehicles to build in specific time delays in turning on or off electrical devices.
- Speed control circuits are used on the vehicle to control the speed of motors or accessories.
- A common use is the control of the fan speed in the vehicle's HVAC system and the electric radiator fans. Essentially, the speed control circuit manages the speed of a fan motor to control the amount of air movement through the system.
- Duty cycle describes the amount of time that power flow within a circuit is allowed. If you walk into a room and turn on the light switch for 1 minute, and then turn off the light switch for 99 minutes, you have created a 1% duty cycle of the lights in that room for the given timeframe of 100 minutes. If you kept the switch on for 5 minutes and off for 95 minutes, you would have a duty cycle of 5%.
- Pulse width modulation (PWM) refers to the amount of time that power is turned on to a circuit during a given period of time.
- Pulse width modulation is used to accurately adjust the speed and/or travel of electric motors and devices.

Key Terms

bipolar junction transistor (BJT) A semiconductor device constructed with three doped semiconductor regions (base, collector and emitter) separated by two P-N junctions.

control module A generic term that identifies an electronic unit that controls one or more electrical systems in the vehicle; also called a control unit.

delay circuit A combination of electrical and electronic components that provide a time delay for switching an electrical circuit.

diode A two-lead electronic component that allows current flow in one direction only.

duty cycle The percentage of one period of time in which the circuit is powered on.

field-effect transistor (FET) A transistor in which most current is carried along a channel whose effective resistance can be controlled by a transverse electric field.

insulated-gate bipolar transistor (IGBT) A three-terminal power semiconductor device primarily used as an electronic switch.

integrated circuit (IC) An electronic circuit formed on a small piece of semiconducting material.

metal–oxide–semiconductor field-effect transistor (MOSFET) A type of field-effect transistor (FET) which has an insulated gate whose voltage determines the conductivity of the device.

microcontroller A stand-alone electronic control module that can be programmed to perform various functions.

microprocessor An electronic control unit that can process data and control one or more devices.

NPN transistor A transistor in which P-type material is sandwiched between two layers of N-type material.

PNP transistor A transistor in which N-type material is sandwiched between two layers of P-type material. This type of semiconductor material has holes, meaning it is missing electrons.

pulse width modulation (PWM) A digital on/off electrical signal used as a variable control for devices such as solenoids.

rectifier bridge An arrangement of diodes that is used to convert the AC voltage produced in the automotive alternator into DC voltage.

speed control circuit A circuit that controls the speed of a motor.

transistor A semiconductor device that allows a small current in the base lead to control a larger current through the emitter collector leads.

Zener diode A diode that forward biases when a certain voltage is reached.

Review Questions

1. Which of the following statements is true with respect to a common diode in an electrical circuit?
   a. It allows current to pass rapidly through the circuit.
   b. It allows current in one direction and essentially blocks it in the opposite direction.
   c. It functions as an electronic version of a two-way check valve.
   d. If current is flowing through it, then it is reverse-biased.
2. A diode is positioned in such a way that current is allowed to pass through it, causing a small voltage drop across it. The diode is:
   a. forward-biased.
   b. reverse-biased.
   c. a Zener diode.
   d. faulty.
3. All of the following statements are true with respect to LEDs in automobiles *except*:
   a. They emit a clean and brilliant light.
   b. They consume less current than incandescent lights.
   c. They do not need a resistor to operate properly.
   d. They are usually connected with resistors in series.
4. All of the following statements are true with respect to a transistor *except*:
   a. It can be used as a switch.
   b. It can be used to amplify currents.
   c. The central region is called the base.
   d. When current is applied to the base of the transistor, the transistor is said to be turned off.
5. Which type of transistor is capable of switching very high current loads and voltages at high frequencies?
   a. IGBT
   b. MOSFET
   c. FET
   d. BJT
6. All of the following statements with respect to integrated circuits are true *except*:
   a. BJTs can be joined to form the workings of an IC.
   b. An IC can also be referred to as a microchip.
   c. ICs are the main contributors to the failure of components due to overheating.
   d. ICs can perform all of the functions of many circuits.
7. A CPU receives inputs in the form of electric signals from:
   a. sensors.
   b. actuators.
   c. FETs.
   d. MOSFETs.
8. What is the main function of a sensor in an electrical circuit?
   a. Converting rotary motion to linear motion
   b. Converting mechanical action to electrical signals
   c. Converting electrical input to mechanical movement
   d. Conducting current through high resistance
9. Which of the following uses a microprocessor?
   a. Steering wheel
   b. Windshield
   c. Door
   d. Climate control unit
10. What is the most common use of speed control units in vehicles?
    a. To control the fan speed in the HVAC system and the radiator
    b. To control the speed of the windshield
    c. To control the speed of the car
    d. To control the speed of the engine

ASE Technician A/Technician B Style Questions

1. Tech A says that diodes function like a one-way check valve, and only allow electricity to flow through in one direction. Tech B says that standard diodes can allow current to flow in both directions if the voltage is high enough. Who is correct?
   a. Tech A
   b. Tech B
   c. Both A and B
   d. Neither A nor B
2. Tech A says that the cathode of a diode indicates the negative side of the diode. Tech B says that the anode denotes the negative side of the diode. Who is correct?
   a. Tech A
   b. Tech B
   c. Both A and B
   d. Neither A nor B
3. Tech A says that once the threshold voltage of a Zener diode is reached, current will be able to flow in both directions through the diode. Tech B says that once the threshold voltage is reached for a Zener diode, it must be replaced. Who is correct?
   a. Tech A
   b. Tech B
   c. Both A and B
   d. Neither A nor B
4. Tech A says that transistors function like a solid-state relay, wherein a large current controls a smaller current. Tech B says that transistors allow control modules to control relatively large current output devices with a very small input from the control module. Who is correct?
   a. Tech A
   b. Tech B
   c. Both A and B
   d. Neither A nor B

5. Tech A says that control modules are mechanical devices. Tech B says that control modules take inputs from various sources, and determine the required output signal based on the module's programming. Who is correct?
   a. Tech A
   b. Tech B
   c. Both A and B
   d. Neither A nor B
6. Tech A says that BJT stands for bipolar junction transistor. Tech B says that MOSFET stands for metal-oxide-semiconductor-field-effect-transformer. Who is correct?
   a. Tech A
   b. Tech B
   c. Both A and B
   d. Neither A nor B
7. Tech A says that IGBTs are very robust transistors, and they are widely used in the motor-control circuits of hybrid and electric vehicles. Tech B says that MOSFETs are capable of handling much more current than IGBTs. Who is correct?
   a. Tech A
   b. Tech B
   c. Both A and B
   d. Neither A nor B
8. Tech A says that "control module" is a generic term that identifies an electronic unit that controls electrical systems in a vehicle. Tech B says that control modules are very fragile, and require special handling procedures to avoid damaging them. Who is correct?
   a. Tech A
   b. Tech B
   c. Both A and B
   d. Neither A nor B
9. Tech A says that duty cycle refers to the job that an electrical component does when it is turned on. Tech B says that duty cycle refers to the amount of time power is allowed to flow through a circuit. Who is correct?
   a. Tech A
   b. Tech B
   c. Both A and B
   d. Neither A nor B
10. Tech A says that pulse width modulation is the percentage of time that power is turned on to a component over a given period. Tech B says that pulse width modulation is difficult to control, and very seldom used in modern automobiles. Who is correct?
    a. Tech A
    b. Tech B
    c. Both A and B
    d. Neither A nor B

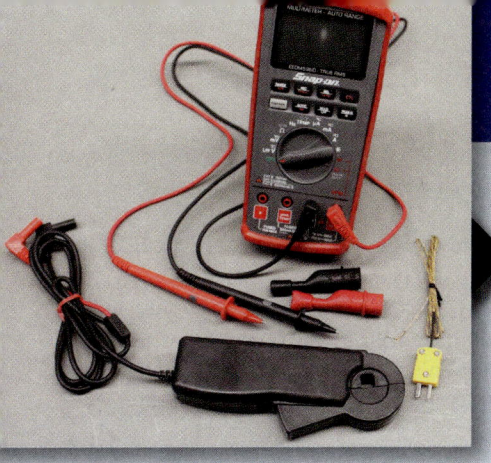

CHAPTER 10
Digital Multimeter Use and Circuit Testing Procedures

NATEF Tasks

- **N10001** Demonstrate the proper use of a digital multimeter (DMM) when measuring source voltage, voltage drop (including grounds), current flow, and resistance.
- **N10002** Inspect and test switches, connectors, and wires of starter control circuits; determine needed action.
- **N10003** Demonstrate knowledge of electrical/electronic series, parallel, and series-parallel circuits using principles of electricity (Ohm's Law).
- **N10004** Use wiring diagrams during the diagnosis (troubleshooting) of electrical/electronic circuit problems.
- **N10005** Demonstrate proper use of a test light on an electrical circuit.
- **N10006** Use fused jumper wires to check operation of electrical circuits.
- **N10007** Demonstrate knowledge of the causes and effects from shorts, grounds, opens, and resistance problems in electrical/electronic circuits.
- **N10008** Inspect and test fusible links, circuit breakers, and fuses; determine needed action.

Knowledge Objectives

After reading this chapter, you will be able to:

- **K10001** Explain the fundamentals of a digital multimeter (DMM).
- **K10002** Describe the uses of a DMM.
- **K10003** Describe how to measure volts, ohms, and amps.
- **K10004** Understand the use of the DMM testing procedures.
- **K10005** Understand how to use a DMM to test current flowing through a circuit.
- **K10006** Understand the use of the DMM in measuring volts, ohms, and amps in a parallel circuit.
- **K10007** Understand circuit types and how electricity behaves within them.
- **K10008** Locate opens, shorts, bad grounds, and high resistance.

Skills Objectives

After reading this chapter, you will be able to:

- **S10001** Test and inspect high-voltage rubber gloves.
- **S10002** Use a DMM to measure voltage.
- **S10003** Switch the current through a resistor.
- **S10004** Use Ohm's law to diagnose circuits.
- **S10005** Use wiring diagrams to diagnose electrical circuits.
- **S10006** Check circuits with a test light.
- **S10007** Check circuits with fused jumper leads.
- **S10008** Inspect and test circuit protection devices.
- **S10009** Inspect and test switches, solenoids, and relays.

You Are the Technician

Your supervisor requested that you train the shop apprentice on how to use a DMM. You will be using the DMM to teach the apprentice how to take simple measurements that will come in handy when doing vehicle inspections such as measuring battery and charging system voltage, testing a fuse to see if it is blown, and performing a voltage drop on a simple circuit. The apprentice explains what volts, ohms, and amps are, so you show him how to set up the meter for each of these tests. Once he is competent in setting up the meter, you show him how to take each type of reading. The following is the list of questions your supervisor has asked you to review with the new employee after the hands-on training:

1. What are the most common measurements taken by a DMM, and how is the meter hooked up in the circuit for each one?
2. Describe the different methods that can be used to measure voltage drop across a component.
3. What two electrical properties mentioned in Ohm's law are needed to have a voltage drop?

▶ Introduction

A **digital multimeter (DMM)**, also known as a **digital volt-ohm meter (DVOM)** or **oscilloscope**, is an electrical measuring tool frequently used to diagnose and repair electrical faults. Like many diagnostic tools, practice is required to understand how the DMMs are used to take electrical measurements. Care must be taken when connecting these tools into electrical circuits to ensure correct readings are obtained. Once a reading is obtained, it needs to be interpreted and applied, in conjunction with knowledge of electrical theory, to diagnose the circuit being tested. This chapter provides an explanation of how to use and set up a DMM for measuring voltage, amperage, and resistance.

This chapter also has a number of exercises that will expand your knowledge on using a DMM, allowing you to practice taking readings and apply the results. It also relates the practical circuit examples and DMM readings to Ohm's law calculations. Basic oscilloscope usage and typical waveforms are also covered. Knowing how to properly use DMMs and oscilloscopes along with interpreting and applying their readings will allow you to diagnose electrical faults, making you very valuable to your employer.

▶ DMM Fundamentals

K10001 Explain the fundamentals of a digital multimeter (DMM).

A digital multimeter (DMM) or digital volt-ohm meter (DVOM) is a versatile and useful piece of test equipment (**FIGURE 10-1**). It is called a digital meter because the meter gives a numerical reading on a digital display. An analog meter, by comparison, uses a needle that hovers over a series of scales, requiring the technician to determine the numerical value of the reading (**FIGURE 10-2**). Digital meters are easier to read, which means that a technician

FIGURE 10-1 Fluke, Snap-On, and various other DMMs.

FIGURE 10-2 Analog volt-ohm meter.

is less likely to get the wrong reading. The DVOM is often the first test tool selected for electrical diagnosis and repairs. Basic DVOMs can measure alternating current (AC) and direct current (DC) voltage, AC and DC amperage, and resistance. Most modern DVOMs can also measure frequency and temperature; many have a dedicated diode test function.

DVOMs come in a variety of layouts and quality. You will want to get used to the meters in your shop so you will know their capabilities and how to use them. Most DVOMs of average quality are "fused," meaning that one or more "fast-blow" fuses are included inside the DVOM (**FIGURE 10-3**). If the amperage passed through the meter is too high, the fuse will blow, protecting the meter. If the meter is unfused, it will not be protected and could be damaged if used incorrectly when measuring amperage.

DVOMs and test leads also should have a CAT rating listed on the front. CAT is short for "category." Each level, or CAT, is designed to work safely on higher-powered electrical systems (**TABLE 10-1**). CAT ratings were not designed initially for automotive meters; most vehicles used low voltage. Today, with more and more hybrid and electric vehicles on the road, which operate at very high voltages, CAT ratings are becoming important for automotive technicians. Hybrid vehicles typically require meters and test leads rated as CAT III (**FIGURE 10-4**). Failure to use a meter with the sufficient CAT rating could result in injury, due to the fact that the meter is not designed for use on high-voltage circuits. Another thing to keep in mind if you are working on high-voltage systems is that you need to wear a pair of certified and tested Class "0" rubber-insulated gloves, with leather protectors over top. Make sure that the leather protectors are at least 4" (101.60 mm) shorter than the rubber gloves, or electrical shock can result! (**FIGURE 10-5**)

It is important to always inspect the rubber high-voltage gloves for damage before each use. NEVER use rubber high-voltage gloves without leather protectors. It is critical that

FIGURE 10-3 DVOM "fast-blow" fuses.

TABLE 10-1 Meter CAT Ratings

Overvoltage Category	Short Description	Examples
CAT I	Electronics	Low-power electronic equipment such as copiers
CAT II	Single-phase plug-in tools and equipment	Portable tools, appliances, etc.
CAT III	Three-phase fixed equipment and single-phase commercial lighting	Equipment in fixed installations (this is the minimum rating required for hybrid vehicles)
CAT IV	Three-phase utility connection, any outdoor wires	Main power wires from the utility company; outside wire for lighting (this rating is for heavier-duty meters and leads for hybrid vehicles)

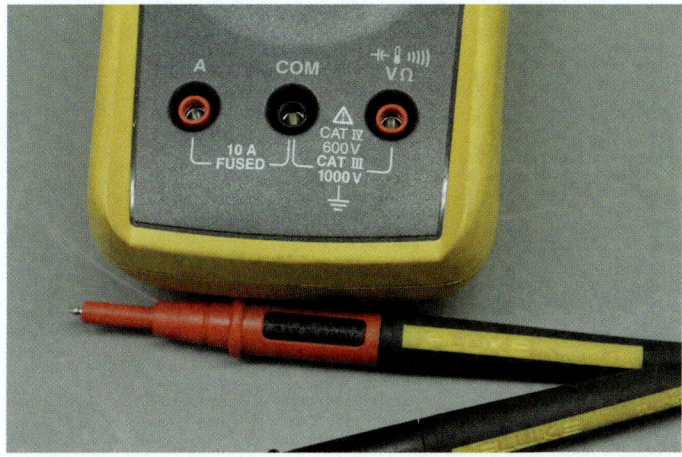
FIGURE 10-4 CAT III meter and leads.

FIGURE 10-5 Class "0" rubber high-voltage gloves and leather protector.

S10001 Test and inspect high-voltage rubber gloves.

you always remember to use the proper CAT-rated meter and leads along with the proper personal protective equipment when working on high-voltage systems. To test and inspect high-voltage rubber gloves, follow the steps in **SKILL DRILL 10-1**.

There are two main components of a DVOM, the main instrument body and the test leads that connect the DVOM to the circuit being tested. The DVOM main instrument body has a function switch to choose the type of electrical measurement to be taken, digital display to report the readings, and sockets to connect test leads (**FIGURE 10-6**). Test leads

SKILL DRILL 10-1 Testing and Inspecting High-Voltage Rubber Gloves

1. Visually inspect gloves for damage.

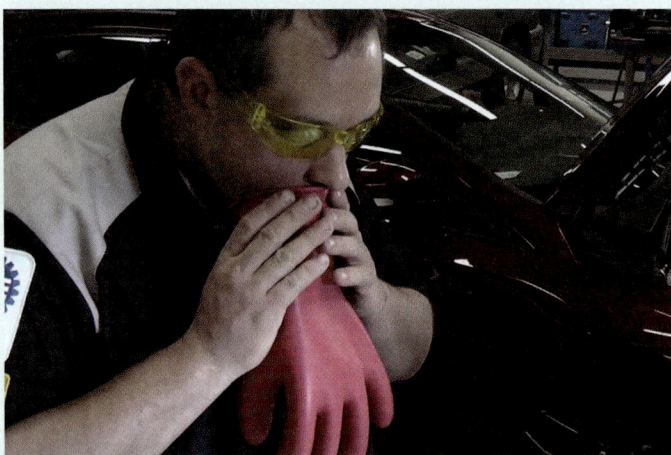

2. Trap air in rubber glove, and roll cuff to increase pressure in the glove.

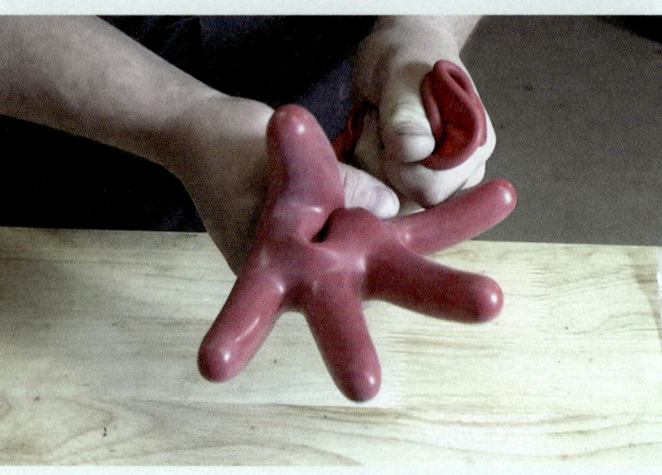

3. Look/feel/listen for air leaks.

FIGURE 10-6 DVOM selector dial.

FIGURE 10-7 DVOM lead adapters and accessories.

are used to connect the DVOM to or into the circuit being tested. They come in pairs: one red, the other black. Basic leads have a probe on one end for making the connection with the electrical circuit being tested and a connector on the other end for plugging into the sockets of the DVOM. A wide variety of test leads and adapters are available to make it easier to use the DVOM; for example, alligator clips enable hands-free connection of the leads, piercing probes allow the technician to obtain voltage signals anywhere on a wire by piercing the wire insulator with a small needle and back-probing adapters (**FIGURE 10-7**). Adapters such as temperature probes and inductive current clamps connect to the input sockets of the DVOM and convert temperature or current flow into a voltage that can be measured by the DVOM.

DVOMs read very small quantities in the one-ten thousandths of a unit range up to very large quantities in the range of millions of units in the case of resistance measurements. It is not possible for DVOMs to effectively and accurately measure such ranges with only a single range or scale; they must have multiple ranges or scales. But before we can talk about those ranges, we need to understand that the DVOM screen can only display four or five digits. This means that symbols must be used to substitute for some of the digits. **TABLE 10-2** shows the common symbols, their prefix, and the factor they represent. You will have to place the appropriate electrical symbol, V, A, or Ω, behind the factor symbol based on what you are measuring. For example, 2168 mV would be the same as 2168 millivolts. It could also be called 2.168 volts, since there are 1000 millivolts in 1 volt. Either designation is correct. The challenge is taking the meter reading and making sense of it, which takes practice.

Once you understand the symbols and what value they represent, you are ready to decide which range to set the meter to. **TABLE 10-3** lists a typical set of DVOM ranges; however, there is no single range or scale value used by DVOM manufacturers. The resolution indicates the accuracy of the count within any given range and is different for each range. To achieve the most accurate reading, always select the lowest range possible for the value being measured. For example, if you are measuring 12 volts, you should select the 60-V range, because 6 volts would be too low and 600 volts would be less accurate (**FIGURE 10-8**).

Most modern DVOMs have an automatic ranging capability, but they maintain the ability to be used in a manually selected range, which means the user can determine the range. When used in the automatic range, the DVOM selects the best

TABLE 10-2 DVOM Values

Factor	Prefix	Symbol
1,000,000	mega	M
1,000	kilo	k
1	No prefix	No symbol
0.001	milli	m
0.000001	micro	μ

TABLE 10-3 DVOM Ranges

Function	Range	Resolution
mV DC	0–600.0 mV	0.1 mV
V DC	0–6.000 V	0.001 V
	0–60.00 V	0.01 V
	0–600.0 V	0.1 V
	0–1000 V	1 V
Ohms	0–600.0 Ω	0.1 Ω
	0–6.000 kΩ	0.0001 kΩ
	0–60.00 kΩ	0.01 kΩ
	0–600.0 kΩ	0.0001 MΩ
	0–40.00 MΩ	0.01 MΩ

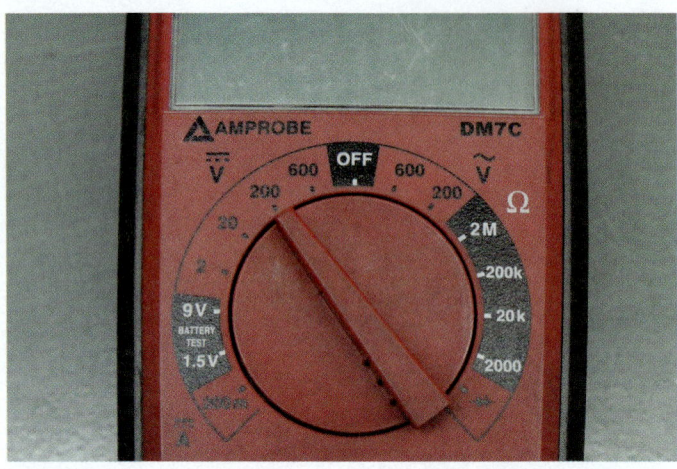

FIGURE 10-8 DVOM that illustrates selecting the different display ranges.

K10002 Describe the uses of a DMM.

range for the value being measured so that the technician does not have to be concerned with manually setting the range. But be careful! The meter does not give you flashing light warnings that it has changed ranges, so it is extremely easy to miss that. Many technicians have been led down the wrong diagnostic path by thinking the 12.6 on the meter was volts when in fact the meter had auto-ranged to millivolts; so instead of having full power, the battery being tested had almost no power. To prevent this mistake, many instructors require their students to use only manual ranging when using their meter.

▶ DMM Uses

DVOMs are used to take many different electrical measurements on electrical circuits and are one of the first tools used when conducting electrical repair or diagnosis work. As a voltmeter, the DVOM can measure electrical voltage within circuits; for example, the available voltage at a fuse, switch, or lamp. It is important to note, however, that DVOMs place an extremely low current load on the circuit, and this can lead to false diagnoses. Imagine that you have a vehicle that does not start. You discover that the electric fuel pump motor is not working, and so you check the fuel pump feed circuit with your voltmeter, which reads battery voltage on the fuel pump power wire. You replace the fuel pump, only to discover that the vehicle still will not start, and the fuel pump is not working. Further diagnosis reveals that the power wire was nearly corroded in two a little ways up the wiring harness. The wire was intact enough to permit a very small current to pass through the meter, but when the fuel pump was introduced into the circuit the higher current draw caused the corroded spot to burn in two, and the wire was severed. Keep this in mind when using a DVOM: it may tell you voltage is present in the circuit, but it will not tell you if the circuit is capable of supporting its load devices. A good stress test of electrical circuits such as these is to use a high-load test lamp composed of a headlamp bulb connected to a pair of test leads. This will introduce a load to the circuit, and ensure it can power the load device. Warning! The average sealed-beam headlamp draws 5–9 amps, so be sure that the circuit you are testing with this device is rated to handle this amount of current (**FIGURE 10-9**). The DVOM can also measure resistance of a component, connector, or cable, such as the resistance of an ignition coil to check against specifications. DVOMs can also measure current flow in circuits, such as when the amount of current flow through a fuse needs to be checked against specifications. However, you should be careful not to exceed the current rating on the DVOM, as this will cause damage to the meter and possibly injure you. Clearly, a DVOM is a very versatile tool, explaining why it is the most commonly used electrical diagnosis tool. In the next several sections, we will further explore DVOMs and how to use them.

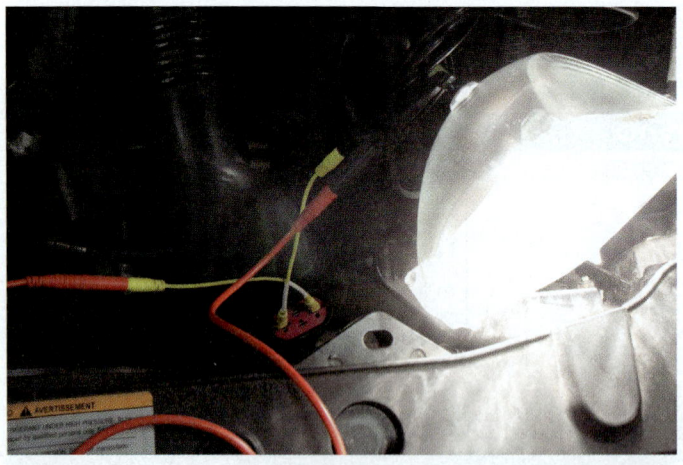

FIGURE 10-9 Sealed-beam headlamp connected to test leads being used to stress-test an electrical circuit.

Applied Math

AM-2: Decimals: The technician can add decimal numbers to determine conformance with the manufacturer's specifications.

A starter has been rebuilt and the technician wants to check the pinion clearance. A feeler gauge will be used to determine this clearance. Manufacturer's specifications for this clearance are from 0.010" to 0.140" with 0.070" considered the midpoint. The technician's feeler gauge set only goes to 0.045" which fits too loosely in the gap. He places a 0.025" blade next to the 0.045" blade, which together, fits in the gap with just the right tension. The selected gauges are 0.025" plus 0.045" to equal a total of 0.070". In this example, we are working with decimal numbers. Decimals are numbers that are expressed using a decimal point.

Setting Up a DVOM

To set up a DVOM to take accurate measurements, you need to know if you will be measuring resistance, voltage, or current. You should also know the reading that you are expecting so you can be sure to set up the meter appropriately. If very high voltages are to be measured, it is important to make sure the DVOM and leads match the appropriate CAT rating for use at the voltages you will be testing (**FIGURE 10-10**). All of this information will determine the way in which you set up the DVOM, including the connections you need to make on the DVOM and the range you select. Resistance measurements should be taken with the circuit disconnected. If measuring the resistance of components, they should be removed from the circuit. Failure to disconnect power from the circuit prior to using a DVOM to measure resistance will lead to erratic and unreliable test results and may damage the meter. Failure to isolate an individual component from a circuit before it is tested will yield an unreliable reading, which may cause the technician to misdiagnose a concern.

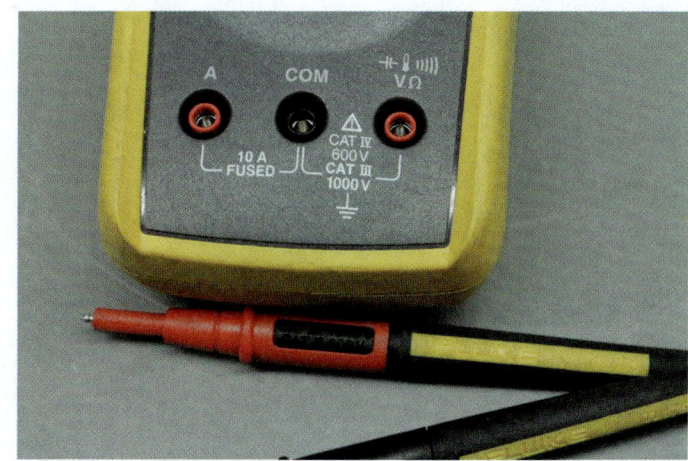

FIGURE 10-10 Meter and leads CAT rating.

The following steps describe how to set up a DMM:

1. Know what you are testing—volts, amps, or ohms.
2. Know the value you expect to be reading (specification).
3. Select leads and probes to suit the measuring task.
4. Connect the leads to the DMM.
5. Use the function switch to select the type of measurement to be undertaken (e.g., resistance, volts, or amps; DC or AC).
6. Select the correct meter range if you are using a manual range meter.
7. Connect the leads to the circuit being tested.
8. Read the meter display.

Min/Max and Hold Setting

Many DVOMs have special settings incorporated into their design to assist you in taking measurements of rapidly changing values or to freeze the display so that an individual reading is not lost. In the **min/max setting**, the DVOM will record in memory the maximum and minimum reading obtained during the time the DVOM is connected to a source to take a reading. The min/max setting is often used to measure vehicle battery voltage while the engine is cranking or the battery is charging. During cranking of the starter motor, current draw from the battery is at its highest for only a fraction of a second when the engine initially starts to crank. Cranking is also when the battery voltage is at its lowest. In min/max mode, a DVOM will capture the minimum and maximum battery voltage. A limitation in the use of the DVOM is the sample rate. The sample rate is the speed at which the DVOM can sample the voltage. The DVOM does not continuously sample the voltage; rather, it checks the voltage at regular intervals or at a sample rate. While this occurs quickly—for example, every 100 milliseconds—it does mean that if a transient voltage occurs between samples, it will not be recorded by the DVOM. Where quicker sample rates are required, other tools such as oscilloscopes can be used.

The **hold function** allows the display to be frozen. When the hold function is activated, the display will hold the value on the display until the function or DVOM is turned off. A variation of the hold function is the "auto hold" function found on some DVOMs. When activated, the auto hold function takes a measurement and freezes or holds the display until the function or DVOM is turned off. This function can be useful when taking measurements in difficult locations, such as underneath a dash where you may not be able to watch the meter display while making the meter connections.

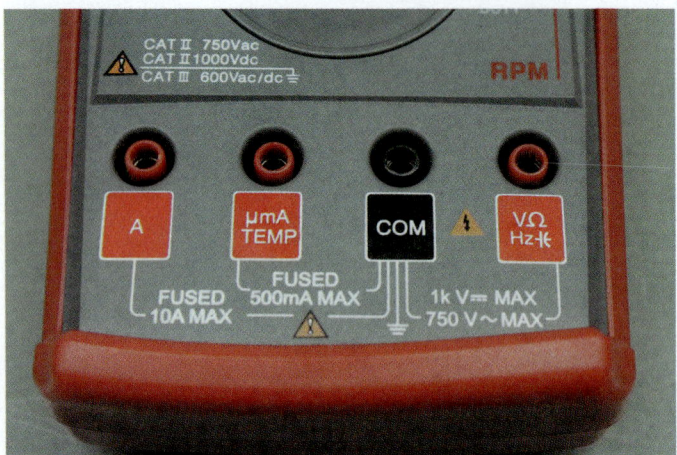

FIGURE 10-11 Labels and ports on a DVOM.

▶ **TECHNICIAN TIP**

Failure to choose the proper meter terminal for the appropriate test is one of the most common causes for blown meter fuses and damaged electronic components. For example, we know from earlier reading that the DVOM is considered a "high-impedance" device, which means that the meter has a very high internal resistance; this allows the meter to draw a minimum of current through a circuit when testing voltage. If, however, the meter is set up to measure current, with the red lead in either the "A" or the "mA" terminal, the meter will have a very low internal resistance because electrons must be allowed to flow without restriction to determine current. If the meter is used to measure voltage at the battery in this configuration, when it is connected across the battery terminals, it will produce a short across the battery and allow high current to flow through the meter until the fuse rating is exceeded and the meter fuse fails. Similarly, if the meter is connected to check the voltage of an input or output of an electronic device, the meter will again act as a short circuit. Excessive current will flow through the circuit until the meter fuse opens or the electronic component connected to the circuit burns up. This can be a very costly mistake! Always be sure to check—and double check—which terminals the leads are connected to BEFORE testing is carried out.

Test Leads: Common and Probing Leads

Many people incorrectly label the red lead as positive and the black lead as negative. However, if you look at your meter near the test lead terminals, you will not see a "+" or a "–" anywhere. What you will see is "A" (typically 10 A), "mA," "common," and "V/Ω" (**FIGURE 10-11**). Common just means that the terminal is "common" to all of the functions of the meter. In other words, this lead does not need to be moved when different functions of the meter are typically accessed. On the other hand, the red lead does have to move, depending on what function of the meter is being used. That is why it is labeled with the V/Ω symbol, and not "+." If you find this distinction confusing, consider the following: When we measure various electrical signals at the same time on an oscilloscope, we need more than just the red lead. In fact, we typically use a yellow, a blue, and a green test lead. In all of these situations, the test lead (no matter the color) acts as a probe into the circuit. So rather than referring to the red lead as the positive lead, it is more accurate to refer to it as the probing lead for the DVOM. Then we can introduce probing leads of other colors when we use an oscilloscope.

It is important to note the meter screen will always read what the probing lead is touching. For example, if the common lead is touching the battery's negative post and the probing lead is touching the positive post, the meter screen will display a "+" before the reading. That means that the probing lead is touching something more positive than the common lead. If we reverse the leads, the meter screen will display a "–" before the number, meaning that the probing lead is touching something more negative than the common lead. When you understand and apply this concept, rather than jump to the conclusion that "the meter leads are hooked up backward," you will be ready to start diagnosing all kinds of strange electrical problems, especially ground issues and charging system issues.

Back-Probing Techniques

A **probing technique** is the way in which the DVOM probes are connected into circuits. There are many different types of probes and probing techniques you can use, depending on the circuit being tested. Some examples are alligator clips, fine-pin probes, and insulation piercing clips (**FIGURE 10-12**). Make sure you know the voltage limits of the probes you use, since high-voltage measurements require special probes that are designed for that purpose.

Never use excessive force when probing; doing so may bend or damage connectors and terminals. Care must be used when probing connectors. As a rule, a connector should never be probed from the front where it engages that other half of the connection. Probing

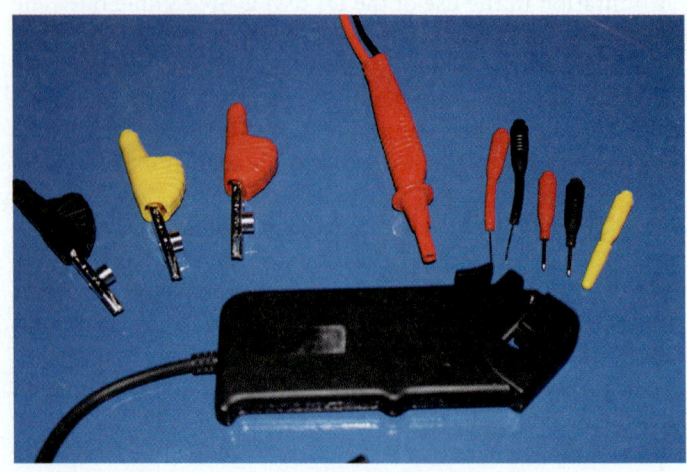

FIGURE 10-12 There are many different leads and probing techniques you can use, depending on the circuit being tested.

a connector from the front can cause the terminals to be spread open too wide, which can cause intermittent circuit operation and high resistance (**FIGURE 10-13**). The standard probe leads that are supplied with a DVOM are basic straight metal probes useful for making quick measurements in circuits, but they do require the use of both hands to hold them in place. Leads with alligator clips, which come in various sizes, allow the DVOM leads to be clipped onto the circuit and held in place, freeing up your hands for other tasks. These clips are particularly useful for connecting to larger terminals, such as battery terminals.

Back-probing occurs when a narrow probe is pushed in from the back of a connector to make a connection. To perform this task, very fine pins are used to reduce the possibility of damage. The pins are designed to slip into the back of connectors and provide contact without causing damage to the connector terminal. In the absence of dedicated back-probing equipment, small paper clips and/or "T" pins can be used in conjunction with alligator clips on the standard meter leads to effectively back-probe a connector. When using this method, care must be taken to ensure that the exposed clips or pins do not touch, thus creating a short circuit condition, and possibly damaging a component downstream (**FIGURE 10-14**).

Piercing Wire Taps

Insulation piercing probes are also available but should be used with caution. They have sharp, fine pins that pierce the insulation on conductors to create a connection. This method of wire-tapping is especially useful when the sensor or device being tested is not easily accessible. Piercing taps allow the technician to obtain information from a sensor from a remote location of the wiring harness, increasing technician speed and productivity. Remember to always reinsulate the hole that the probe makes to prevent any corrosion. A dab of silicone grease applied over the hole is also an effective means of sealing the wire. Use liquid insulation or a similar product to reinsulate; do not use room temperature vulcanizing (RTV) silicone, which attracts moisture as it cures, potentially causing corrosion.

▶ Measuring Volts, Ohms, and Amps

The most common measurements taken with DVOMs are voltage, resistance, and current. To take voltage measurements, the probing lead (red) is connected to the volts/ohms, or V/Ω, terminal, and the common lead (black) is connected to the common, or COM, terminal of the DVOM. An appropriate range or auto range is selected on either AC or DC voltage, depending on the voltage to be measured. The probing lead is typically connected to the positive side of the circuit being tested, and the common lead to the negative side. Watch your screen. If the "+" or "−" is not what you were expecting, check the leads to

▶ TECHNICIAN TIP

A technician recently posted an electrical problem on a technical forum. He said that he had hooked up a voltmeter with the black (common) lead on the negative battery terminal and the red (probing) lead on the vehicle engine ground with the engine running. The meter read a negative number. He asked the forum if he had the meter leads hooked up backward. He received several comments saying yes, he had hooked them up backward. However, those technicians were not correct. His probing lead was registering a reading that was more negative than the common lead. But what could be more negative than the negative post of the battery? When the engine is running, the alternator can be more negative than the negative battery post. So what his voltmeter was trying to tell him was that there was a voltage drop between the negative battery post and the alternator frame. If he would have understood that the probing lead was not lying to him, that it was reporting exactly what it was touching compared to what the common lead was touching, then he could have started down the path to diagnosing what it indicated. In this case, he should have been looking for a dirty ground connection between the negative battery post and the engine block.

K10003 Describe how to measure volts, ohms, and amps.

S10002 Use a DMM to measure voltage.

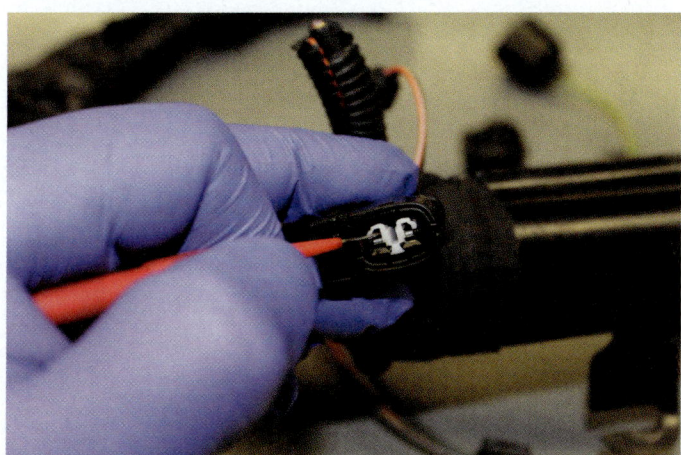

FIGURE 10-13 Terminals should never be probed directly on the contact points.

FIGURE 10-14 Paper clip and "T" pins being used to back-probe a connector.

verify they are connected the way you intended. If you still get an unexpected reading, stop and analyze the situation. Ask yourself, what could cause the meter to read that way? Then brainstorm the options.

Most DVOMs can measure milliamps or 10 to 20 amps directly through the meter. The correct range needs to be selected, along with AC or DC. The red probe is connected to the "A" terminal, and the black lead is connected to the COM terminals. On some DVOMs, there may be a separate "mA" terminal that the red probe plugs into to measure milliamps. To measure current, the DVOM is connected in series with the circuit, with the probing lead closest to the positive terminal of the power supply or battery. As noted earlier, most DVOMs have an internal fuse that will blow if excessive current flows through it. This fuse is designed to help prevent damage to the meter.

To accurately measure the resistance of a component, you should remove or isolate the component from the circuit. Doing so removes the possibility of any parallel-circuit resistance affecting the resistance measurement. If you need to measure resistance in a circuit, always make sure the power is disconnected. In order to read resistance, batteries inside the DVOM supply the circuit with power to take the measurement. If power is not removed from the circuit being tested, it disrupts the measurement and can provide a false reading or potentially damage the DVOM. To take resistance measurements, the red (probing) lead is connected to the V/Ω terminal, and the black (common) lead is connected to the COM terminal of the DVOM. You will need to select an appropriate range or auto range to measure resistance. The red probing lead is connected to one side of the component being tested, and the black common lead is connected to the other side.

To use a DVOM to measure voltage, follow the steps in **SKILL DRILL 10-2**.

SKILL DRILL 10-2 Using a DVOM to Measure Voltage

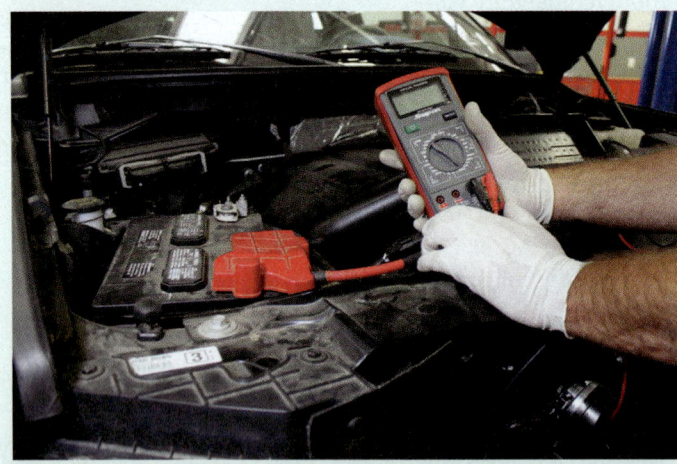

1. Prepare the DVOM for testing voltage by connecting the black lead to the COM terminal and the red lead to the Volt/Ohms (V/Ω) terminal.

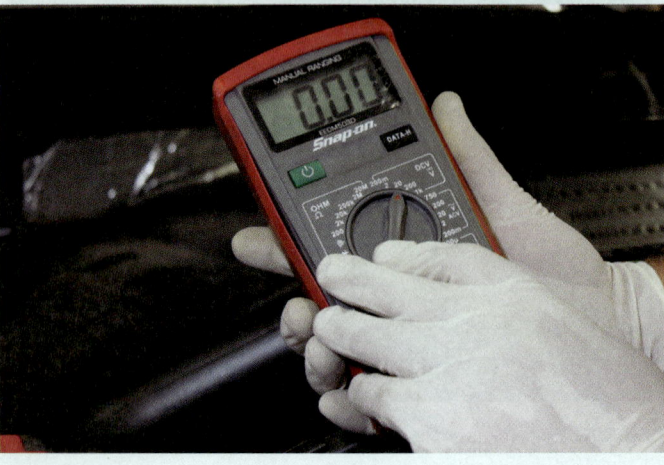

2. Turn the rotary dial until you have selected the mode for volts DC. The reading on the DVOM should now be at zero.

Continued

3. Connect the black lead to the negative battery post and the red lead to the positive post. Measure the voltage and interpret the results.

Interpreting Voltage Ranges

Typically, a DVOM has both an auto range and a manual range capability. The way in which you select auto range and manual range will vary depending on the DVOM. Different DVOMs have different range settings. For example, one DVOM's setting could be 6 V, 60 V, and 600 V, and another's 4 V, 40 V, and 400 V. **FIGURE 10-15** shows a circuit with two resistors in a series with a 12-volt DC supply. Various DVOM ranges can be compared by measuring the voltage drops across each of the resistors. The DVOM has the following ranges: 600.0 mV, 6.000 V, 60.00 V, 600.0 V, and 1000 V. **TABLE 10-4** provides results of voltmeter readings and a DVOM's display to show how different ranges affect the way in which the DVOM readings are displayed.

Manual Ranging Meters

Most DVOMs have a four-digit display, which limits the amount of information that can be displayed at any one time. For this reason, meters are capable of displaying data in multiple ranges. Each range determines where on the display that the decimal point falls. For example, the smallest range that most automotive meters are capable of displaying is "milli," which is displayed with a lower case "m," and is capable of breaking a reading down to one-thousandth of a unit. This would place the decimal point one place to the right of the first digit. Meter ranges can go up from "milli" to whole values. This places the decimal point in the middle of the display, breaking measured units down to the hundredth, or two decimal places. From this point, the range goes up to the "kilo" multiplier, which is denoted with a lowercase "k." One "kilo" is equivalent to one thousand units of measure, so 1 kilo ohm would equal 1,000 ohms. The largest scale that most automotive DVOMs are capable of ranging is "Mega," which is denoted with a capital "M." One Mega is equivalent to one million individual units (**FIGURE 10-16**).

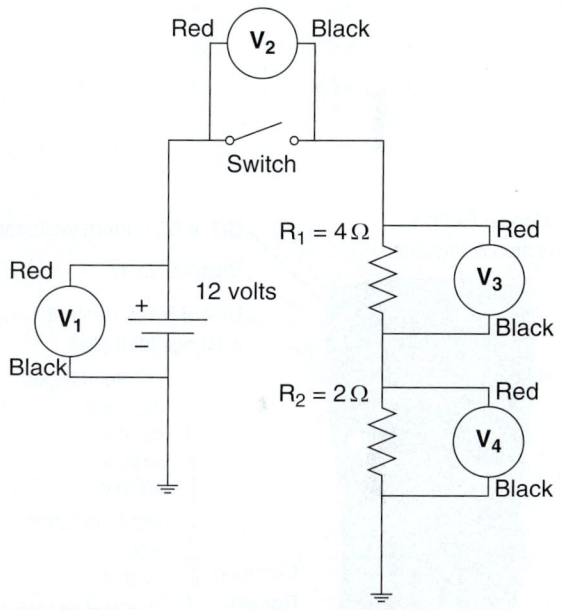

FIGURE 10-15 Measuring voltage in a circuit with two unequal resistors connected in series with a 12-volt battery and the switch open.

Manual ranging meters allow the user to choose what range is the most appropriate for the task at hand. This requires the technician to have an in-depth understanding of both the available ranges and the specification parameters of the component being tested. Using the incorrect range can lead to misdiagnoses and readings that lack the proper resolution to accurately diagnose the component in question. For example, if the meter is set to range kilo "k" ohms, which places the decimal point to the far right of the display and the component specification is .560 ohms +/− 10 ohms, the display will not be able to display the adequate information due to the fact that the decimal is placed to the far right of the display, which is because the meter is manually ranged for "kilo" ohms. For this reason, manual ranging meters must be used with caution, and the meter range should also be checked and double checked by the technician to ensure that the meter is properly set up for the test at hand.

TABLE 10-4 Example of the Effect of Different Ranges on DVOM Readings

Voltmeter	Range	Meter Display	Explanation
V_3	Auto range	8.00	Auto ranging selects the correct range for the voltage being read.
V_4	Auto range	4.000	
V_3	6 V	OL	OL means overload and indicates the voltage being read is higher than the maximum allowed for the range.
V_4	6 V	4.000	The 6-V range is the best range to most accurately measure 4 V. Note how it has three digits after the decimal point.
V_3	60 V	8.00	The 60-V range is the best range to most accurately measure 8 V.
V_4	60 V	4.00	Four volts can also be read on the 60-V range; however, when compared to the 6-V range, there are two digits rather than three after the decimal point. This reading will not be as accurate on this range.
V_3	600 V	8.0	On these ranges, the DVOM still reads but the measurements are not as accurate. To get the most accurate reading, a DVOM range should be selected that is slightly higher than the reading expected.
V_4	600 V	4.0	
V_3	1000 V	8	
V_4	1000 V	4	

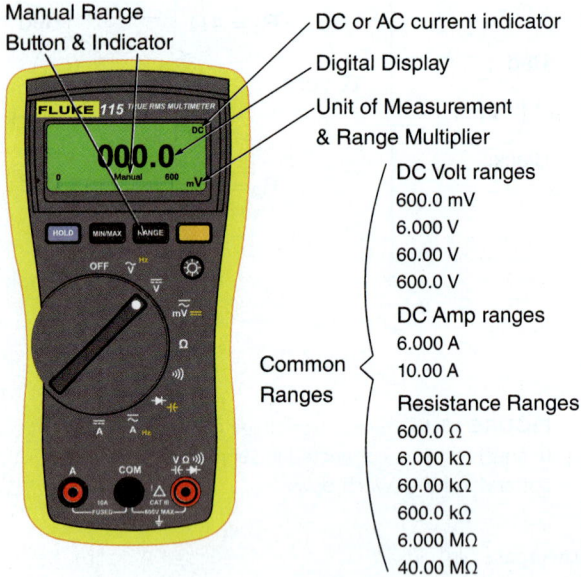

FIGURE 10-16 Diagram displaying the ranges used in automotive meters, and the appropriate value associated with each.

K10004 Understand the use of the DMM testing procedures.

Auto Ranging Meters

Auto ranging meters are capable of selecting the appropriate range needed to obtain the proper resolution for the electrical measurement that is being measured. This type of meter is the most common automotive meter, and the auto-ranging nature affords a great deal of convenience to the technician. However, the technician should always double check what range the meter has selected to display the data in. For example, if the component measured has a specification of 970 ohms +/− 20 ohms, and the meter automatically ranges in "kilo," resulting in a displayed value of 950 "k" ohms, the careless technician will not notice this detail and make an incorrect judgment of the component. Another finicky feature of auto-ranging meters appears when testing a value that fluctuates. This will cause the auto-ranging meter to display erratic information that proves to be of little use. In these instances, it is a good practice to manually range the meter to the appropriate selection in order to obtain a more stable readout.

Low-Amps Probe

When current loads need to be measured that exceed the DVOM's rating, an accessory called a low-amps probe can be used in conjunction with the meter to safely and effectively measure the current that is flowing through the circuit. The low-amps probe will connect to the DVOM using the same ports for the test leads. This device senses the strength of the magnetic field that is generated around a conductor, and converts this into a useable current reading through the DVOM display (**FIGURE 10-17**).

▶ DVOM/DMM Testing Procedures

The DVOM is one of the most versatile tools in the electrical diagnostic toolbox. In fact, most of the diagnostics that a technician will perform on a vehicles' electrical system will include a DVOM. In addition to checking for proper resistance and current flow, DVOMs can also be used to determine how much voltage is dropping in an electrical circuit by performing voltage drop testing on components of the circuit while electrical current is

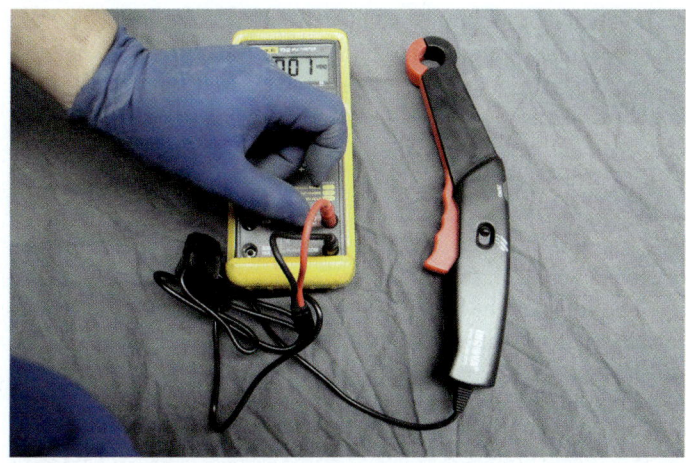

FIGURE 10-17 Low-amps probe attached to a DVOM.

flowing. Voltage drop testing is an accurate and valuable tool to employ when chasing down electrical gremlins.

Testing for Voltage Drops and Circuit Integrity

Voltage drop is measured with a voltmeter while the circuit is operating. It is the potential difference between two points in a circuit. The sum of all the voltage drops in a series circuit equals the supply voltage, while the voltage drop across all parallel circuit branches is the same. Voltage drop does occur in all parts of the circuit, but in a correctly working vehicle circuit, the vast majority of voltage drop is across the component or load we want to do work, such as the headlight bulb.

Unwanted voltage drop becomes a problem if it becomes excessive and occurs in parts of the circuit other than the load. Ideally, the only resistance in a headlight circuit would be the headlight bulb. If this were the case, then all of the battery voltage would be dropped (used up) across the headlight bulb. In practice, however, resistance exists in the cables and connectors within the circuit. In a good circuit, the resistance of the cables, terminals or connectors, and switches is very low, causing small and insignificant amounts of voltage drop. A problem arises when excessive voltage drop occurs in the circuit cables, connectors, and switches, which reduces the efficiency of the circuit. Excessive voltage drop is a fault in the circuit and can cause problems; for example, it will result in yellow or dim headlights, and may also induce excessive heat into the circuit, which can cause components to burn up. To test for unwanted voltage drop of the conductors, switches, and connectors, measure the voltage across each of these parts of the circuit and add the voltage drops together (**FIGURE 10-18**). In a 12-volt system, the total unwanted voltage drop across each side of the whole circuit should not exceed 0.5 volts. In a 24-volt circuit, voltage drop should not exceed 1.0 volts. Voltage drops across an individual wire, connection, or common switch should not exceed 0.2 volts.

To measure voltage drop, the DVOM needs to be used on the voltage range. To perform this measurement process, you will need to set the function switch to "auto range volts DC" on the DVOM (some technicians prefer to use manual range, so they will not be fooled by auto range), and connect the black lead to COM and the red (probing) lead to V/Ω. Voltage drop can be measured across components, connectors, or cables. Remember that the probing lead of the DVOM is normally connected to the point in the circuit where you want to know the voltage.

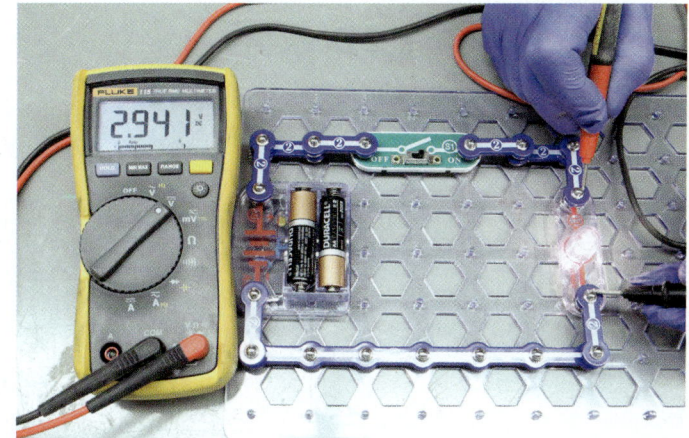

FIGURE 10-18 Voltage drop test being conducted by a DVOM on a simple electrical circuit.

FIGURE 10-19 Voltage drop test being conducted on the feed side of the horn circuit.

If you are performing a voltage drop test, for example, on the feed side of the horn circuit, you could connect the black lead to the positive terminal of the battery and the red lead to the input wire of the horn (the wire connected to the horn) (**FIGURE 10-19**). When you activate the horn, the voltmeter will read the amount of voltage drop in the feed side of the circuit. For example, it might be −4.2 volts. This means that the voltage is 4.2 volts less at the input of the horn (red lead) than it is at the positive battery post (black lead). In this case, the "−" means "less than." So there are 4.2 volts less at the horn than at the positive battery post. Since the voltage drop is more than 0.5 volts, this is an excessive voltage drop in that portion of the circuit, and the voltmeter leads will need to be moved wire by wire closer together until the point of the voltage drop is located.

You could make the same measurement with the DVOM leads reversed. If you place the red lead on the positive battery post and the black lead on the input of the horn, the meter would then read 4.2 volts. In this case it shows positive. This is because the red lead is on the positive post of the battery, which is 4.2 volts higher than the horn input where the black lead is connected. As you can see, voltmeter leads can be hooked up in a couple of ways. Just remember that the meter always reads what the red lead is touching.

FIGURE 10-20 shows a series circuit of two resistors with a 12-volt battery and switch. In this example of how to measure voltage drop, the voltage in various parts of the circuit will be measured with the switch in the open position. **TABLE 10-5** gives an explanation of the voltage measurements and validates the measurements with Ohm's law calculations with the switch in the open position.

FIGURE 10-21 shows a series circuit of two resistors with a 12-volt battery and switch. In this example of how to measure voltage drop, the voltage in various parts of the circuit will be measured with the switch in the closed position. **TABLE 10-6** gives an explanation of the voltage measurements and validates the measurements with Ohm's law calculations with the switch in the closed position.

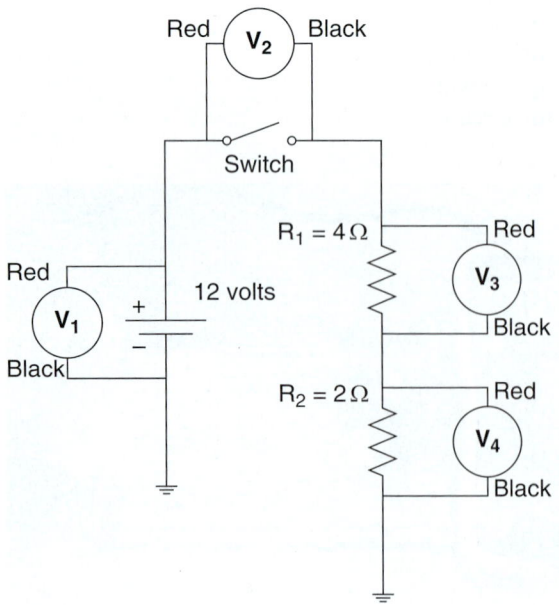

FIGURE 10-20 Measuring voltage in a circuit with two unequal resistors connected in series, with a 12-volt battery and the switch open.

TABLE 10-5 Explanation of Voltage Measurements in Figure 10-20

Measured Voltage	Explanation
$V_1 = 12\,V$	V_1 is measuring the voltage directly across the supply battery, so the voltage reading will be the battery voltage.
$V_2 = 12\,V$	With the switch in the open position, current flows in the circuit. The probing lead is connected directly to the battery via the cable, while the black lead is connected to the negative battery terminal via the two series resistors. Because no current flows in the circuit, there is no voltage drop across the two resistors; therefore, the entire battery voltage drop is across the open circuit switch.
$V_3 = 0\,V$	No current is flowing in the circuit; therefore, resistor R_1 has no voltage drop across it. This can be calculated using Ohm's law: Voltage Drop = Resistance × Current = 4 Ω × 0 A = 0 V drop
$V_4 = 0\,V$	No current is flowing in the circuit; therefore, resistor R_2 has no voltage drop across it. This can be calculated using Ohm's law: Voltage Drop = Resistance × Current = 2 Ω × 0 A = 0 V drop

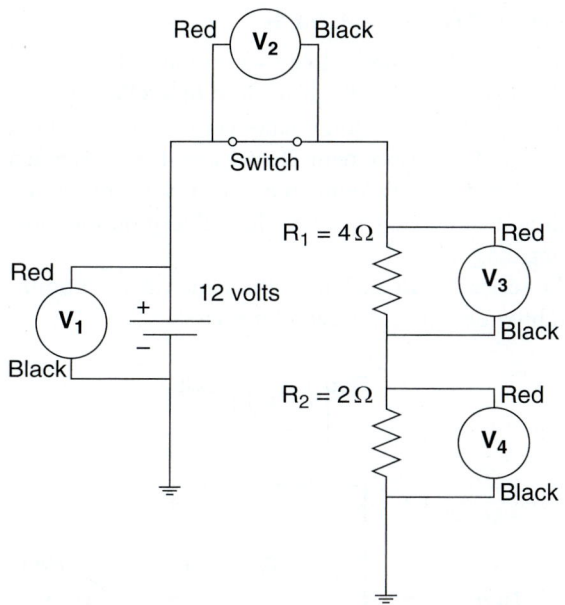

FIGURE 10-21 Measuring voltage in a circuit with two unequal resistors connected in series, with a 12-volt battery and the switch closed.

TABLE 10-6	Explanation of Voltage Measurements in Figure 10-21
Measured Voltage	**Explanation**
$V_1 = 12\,V$	V_1 is measuring the voltage directly across the supply battery, so the voltage reading will be the battery voltage.
$V_2 = 0\,V$	With the switch in the closed position, current flows in the circuit. The switch contacts ideally will have no resistance when closed, so the voltage drop across the contacts will be 0 V. In actual circuit operation, there will always be a slight amount of resistance that will drop a slight amount of voltage in a good switch, but for our examples in this chapter, we will assume there is none.
$V_3 = 8\,V$	With the switch closed, current can flow in the circuit. The amount of voltage drop across R_1 will be 8 V. This can be calculated using Ohm's law: Voltage Drop = Resistance × Current $= 4\,\Omega \times 2\,A$ (see note below) $= 8\,V$ drop Note: To calculate total circuit current, use this formula: $I_T = V_T \div R_T$ $= 12 \div 6\,\Omega$ $= 2\,A$
$V_4 = 4\,V$	With the switch closed, current can flow in the circuit. The amount of voltage drop across R_2 will be 4 V. This can be calculated using Ohm's law: Voltage Drop = Resistance × Current $= 2\,\Omega \times 2\,A$ $= 4\,V$ drop

Unwanted voltage drops in vehicle circuits can cause real problems and faults. For example, a corroded or bad chassis ground can cause a voltage drop that reduces the voltage and current available to components. **FIGURE 10-22** shows a simple circuit with a bulb connected via a switch across a 12-volt circuit. In this circuit, a corroded ground connection has caused a resistance that is dropping 2 volts across it. **TABLE 10-7** analyzes the voltage drops across the corroded ground connection and explains how it reduces the voltage across the bulb, which in turn will cause poor illumination.

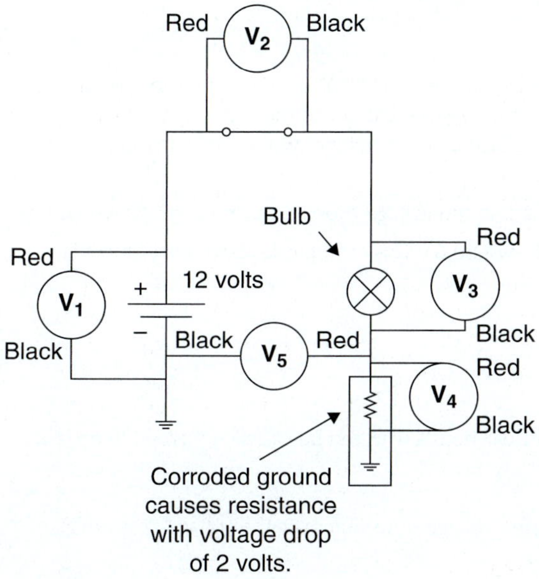

FIGURE 10-22 Measuring voltage in a simple circuit with a switch and one bulb, but with high resistance due to a corroded ground.

TABLE 10-7	Analysis of Voltage Drops in Figure 10-22
Measured Voltage	**Explanation**
$V_1 = 12\,V$	V_1 is measuring the voltage directly across the supply battery, so the voltage reading will be the battery voltage.
$V_2 = 0\,V$	With the switch in the closed position, current flows in the circuit. The switch contacts ideally will have no resistance when closed, so the voltage drop across the contact will be 0 V.
$V_3 = 10\,V$	With the switch closed, current can flow in the circuit. The amount of voltage drop across the bulb is 10 V, because 2 V is being dropped across the corroded ground connection.
$V_4 = 2\,V$	With the switch closed, current can flow in the circuit. The amount of voltage drop across the corroded ground connection is 2 V. This amount of voltage drop robs the bulb of part of the source voltage and causes the illuminated bulb to appear yellow and dim. The box around the resistor and ground indicates the resistance is within the ground connection. High resistance always robs voltage from the designed loads in the circuit and reduces amperage in the circuit.
$V_5 = 2\,V$	V_5 is measuring the same voltage as V_4; the red and black leads of both DVOMs are connected to points with the same potential.

Testing for Voltage Drops Across Multiple Loads

To measure voltage drop, remember that the DVOM needs to be set on the voltage range. Select "auto range volts DC" on the DVOM, and connect the black lead to "COM" and the red lead to "V/Ω." Voltage drop can be measured across components, connectors, or cables, but current has to be flowing to get an accurate measurement. Remember, the leads when checking voltage can be placed in either direction. Just remember which way you placed them so you understand what the reading means. We will show the red lead on the most positive side for purposes of the following diagrams.

In **FIGURE 10-23**, the resistors in the series circuit each have the same value—3 ohms. **TABLE 10-8** lists the circuit voltages and provides an explanation for each.

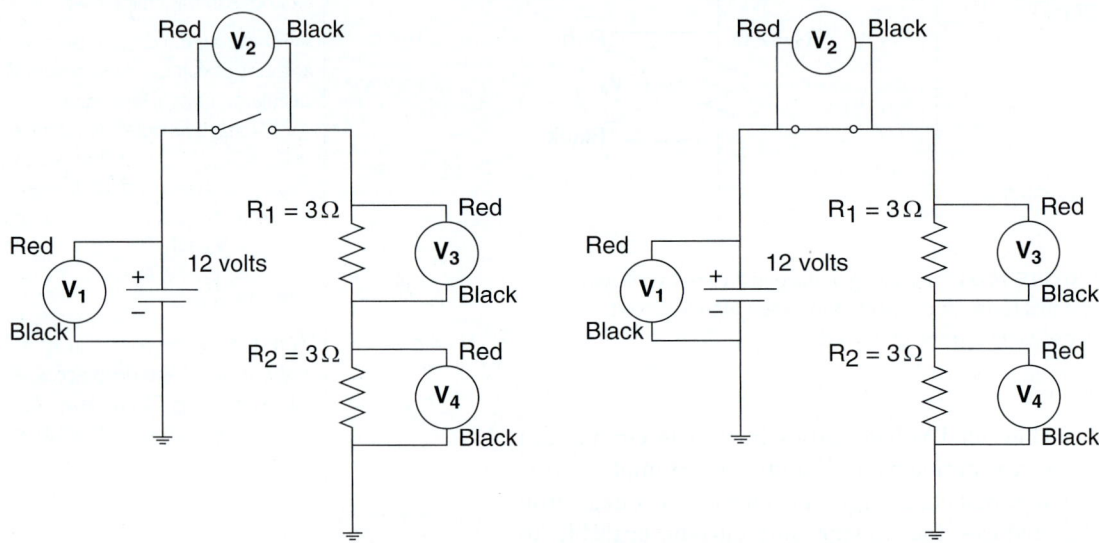

FIGURE 10-23 Measuring voltage in a circuit with two equal resistance bulbs connected in series.

TABLE 10-8 Explanation of Circuit Voltages in Figure 10-23

Figure	Measured Voltage	Explanation
A	$V_1 = 12\,V$	V_1 is measuring the voltage directly across the supply battery, so the voltage reading will be the battery voltage regardless of the position of the switch.
B	$V_1 = 12\,V$	
A	$V_2 = 12\,V$	With the switch in the open position, no current flows in the circuit. The red DVOM lead is connected directly to the battery via the cable, while the black lead is connected to the negative battery terminal through the two series resistors. Because no current flows in the circuit, there is no voltage drop across the two resistors; therefore, the entire battery voltage drop is across the open circuit switch.
B	$V_2 = 0\,V$	The switch contacts ideally will have no resistance when closed, so the voltage drop across the contacts will be 0 V.

B	$V_3 = 6\,V$	With the switch closed, the amount of voltage drop across R_1 will be 6 V. This can be calculated using Ohm's law:
		Voltage Drop = Resistance × Current
		   = 3 Ω × 2 A (see note below)
		   = 6 V drop
		Note: To calculate total circuit current, use this formula:
		$I_T = V_T \div R_T$

B	$V_4 = 6\,V$	With the switch closed, the amount of voltage drop across R_2 will be 6 V. This can be calculated using Ohm's law:
		   = 3 Ω × 2 A
		   = 6 V drop
		R_1 and R_2 are of the same resistance, so they will have the same voltage drop. Summing the voltage drops of R_1 and R_2 will equal the supply voltage:
		V_T = The sum of the voltage drops in a series circuit
		   = VD R_1 + VD R_2
		   = 6 V + 6 V
		   = 12 V

Voltage Drops Across Unequal Loads

Voltage drop can be measured accurately across components, connectors, or cables only when current is flowing. Remember, when checking voltage, the leads can be placed in either direction. Just remember which way you placed them so you understand what the reading means. We will show the red lead on the most positive side for purposes of the following diagrams.

FIGURE 10-24 shows that the resistors in the series circuit have different values; R_1 is 4 ohms and R_2 is 2 ohms. **TABLE 10-9** lists the circuit voltages and provides an explanation for each.

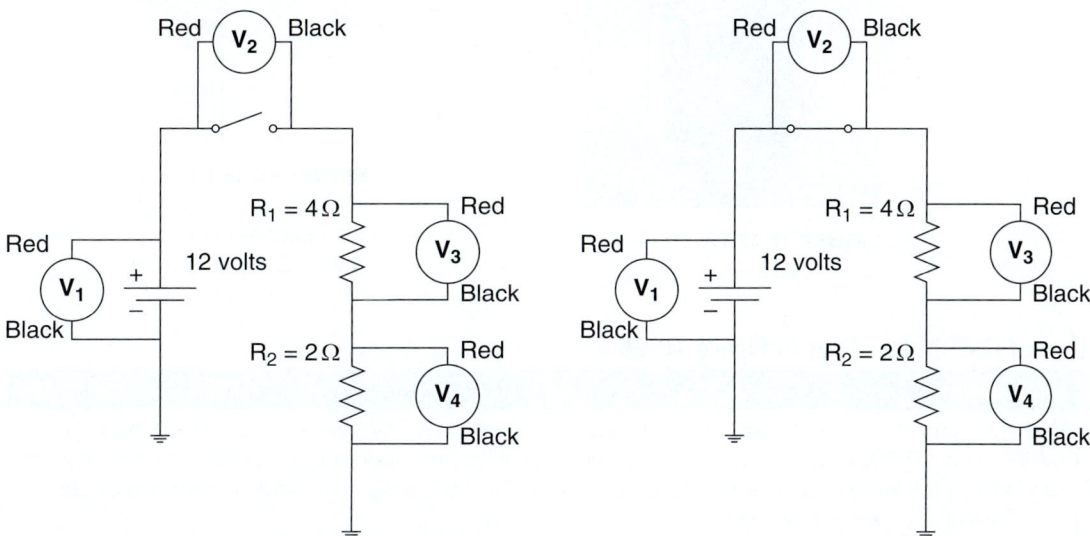

FIGURE 10-24 Measuring voltage in a circuit with two bulbs of unequal resistance connected in series.

TABLE 10-9 Explanation of Circuit Voltages in Figure 10-24

Figure	Measured Voltage	Explanation
A	$V_1 = 12\,V$	V_1 is measuring the voltage directly across the supply battery, so the voltage reading will be the battery voltage regardless of the position of the switch.
B	$V_1 = 12\,V$	
A	$V_2 = 12\,V$	With the switch in the open position, no current flows in the circuit. The red DVOM lead is connected directly to the battery via the cable, while the black lead is connected to the negative battery terminal via the two series resistors. Because no current flows in the circuit, there is no voltage drop across the two resistors; therefore, the entire battery voltage drop is across the open circuit switch.
B	$V_2 = 0\,V$	With the switch in the closed position, current flows in the circuit. The switch contacts ideally will have no resistance when closed, so the voltage drop across the contacts will be 0 V.
B	$V_3 = 8\,V$	With the switch closed, current can flow in the circuit. The amount of voltage drop across R_1 will be 8 V. This can be calculated using Ohm's law: Voltage Drop = Resistance × Current 　　　　　　 = 4 Ω × 2 A (see note below) 　　　　　　 = 8 V drop Note: To calculate total circuit current, use this formula: 　$I_T = V_T \div R_T$
B	$V_4 = 4\,V$	With the switch closed, current can flow in the circuit. The amount of voltage drop across R_2 will be 4 V. This can be calculated using Ohm's law: Voltage Drop = Resistance × Current 　　　　　　 = 2 Ω × 2 A (see note below) 　　　　　　 = 4 V drop R_1 and R_2 are of different resistances, so each will have a voltage drop proportional to its resistance. In a series circuit, the higher resistance will have a greater proportion of the voltage drop. You will notice that R_1 has twice the amount of resistance that R_2 has; thus it takes twice the amount of voltage to push current through R as it does through R_2. Summing the voltage drops of R_1 and R_2 will equal the supply voltage: 　V_T = The sum of the voltage drops in a series circuit 　　 = VD R_1 + VD R_2 　　 = 8 V + 4 V = 12 V

FIGURE 10-25 DVOM set up to measure amps.

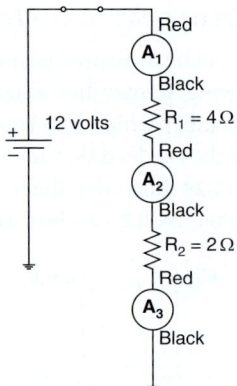

FIGURE 10-26 Measuring amperage in a circuit with two resistors in series with a 12-volt DC supply.

TABLE 10-10 Explanation of DVOM Readings in Figure 10-26

Measured Voltage	Explanation
$A_1 = 2\,A$	With the switch closed, the current will flow through the series circuit. The amount of current flow through R_1 will be 2 A, or 2000 milliamps (mA). This can be calculated using Ohm's law. As this is a series circuit, the current flow in the circuit will be the same in all parts of the circuit. To calculate the current flow in the series circuit: Current = Voltage ÷ Resistance = 12 V ÷ 6 Ω (see note below) = 2 A Note: To calculate the total circuit current, use this formula: $I_T = V_T \div R_T$
$A_2 = 2\,A$	With the switch closed, the current flow will be the same in all parts of the series circuit; therefore, current flow on the ammeter will always be the same regardless of its position in the circuit. The same current flow will be read on A_1, A_2, and A_3.

Measuring Current

To conduct this exercise, the DVOM must be set to read "DC amps." The red lead will be connected to the "A" socket and the black lead connected to the "COM" socket (**FIGURE 10-25**). If using a manual-range DVOM, select an appropriate range. If unsure which range is appropriate, start with the largest range and work down.

FIGURE 10-26 shows a circuit with two resistors in series with a 12-volt DC supply. The DVOM can be connected in various parts of the circuit to measure the current flow. **TABLE 10-10** shows the results and an explanation of the ammeter readings in a series circuit.

K10005 Understand how to use a DMM to test current flowing through a circuit.

S10003 Switch the current through a resistor.

N10001 Demonstrate the proper use of a digital multimeter (DMM) when measuring source voltage, voltage drop (including grounds), current flow, and resistance.

N10002 Inspect and test switches, connectors, and wires of starter control circuits; determine needed action.

▶ Current and Resistance Exercises

At this point, you know that DVOMs are capable of measuring nearly every aspect of an electrical circuit. It is a proven fact, however, that having the knowledge that a test can be performed, and actually knowing how to set up and carry out the test are two very different things. In the following pages, you will learn how to use your DVOM to test current flowing through a circuit.

Current and Magnetic Fields

In this example, a relay controlled by a switch will be used to switch the current through a resistor. The compass is used to demonstrate that a magnetic field is produced around the relay winding when the current flows through it. To conduct this experiment, follow the steps in **SKILL DRILL 10-3**.

SKILL DRILL 10-3 Switching the Current Through a Resistor

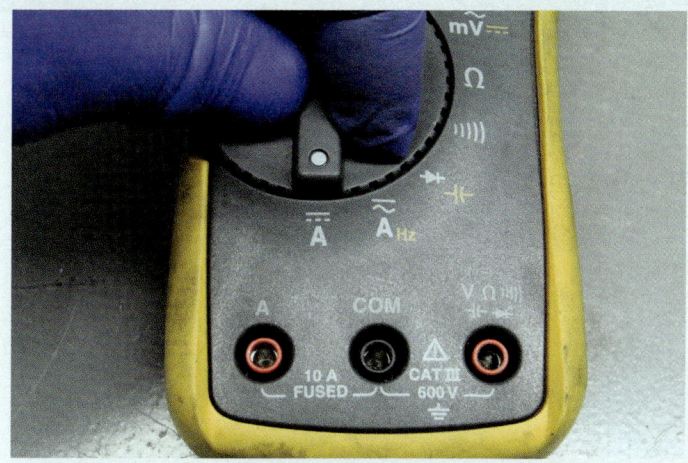

1. To conduct this experiment, set the DVOM to measure DC amps.

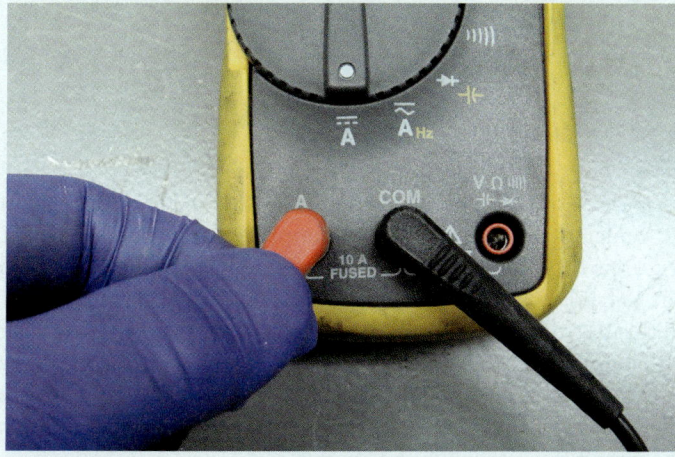

2. Connect the red lead to the A socket and the black lead to the COM socket.

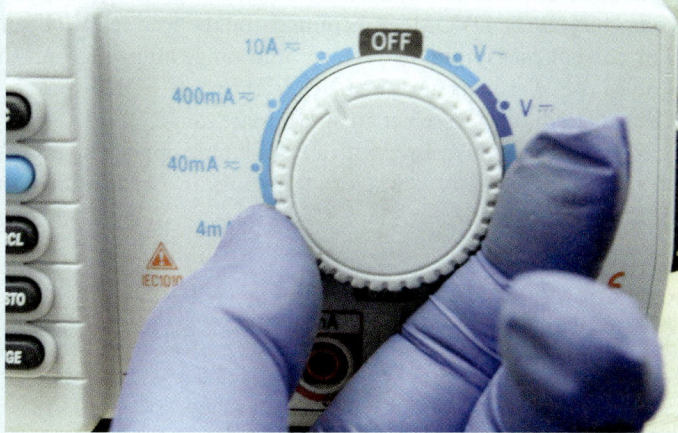

3. If using a manual-range DVOM, select an appropriate range.

FIGURE 10-27 shows a circuit with a relay controlled by a switch and a single resistor with a 12-volt DC supply. The compass is used to show that when energized the relay winding produces a magnetic field. The DVOM will be used to measure current. **TABLE 10-11** shows the current flow through the circuit and provides an explanation of the circuit, current flow, and how Ohm's law calculations can be used.

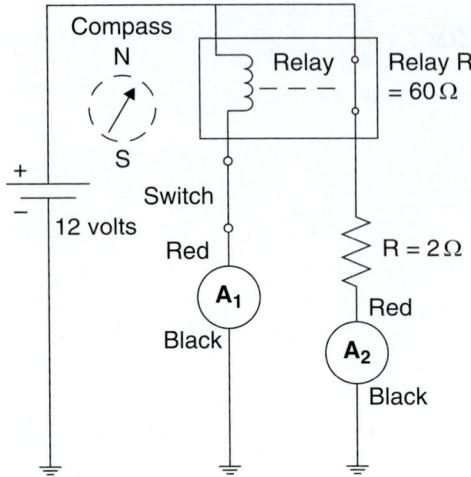

FIGURE 10-27 Measuring amperage in a circuit with a relay controlled by a switch and a single resistor with a 12-volt DC supply.

TABLE 10-11 Explanation of DVOM Readings in Figure 10-27

Measured Current	Explanation
$A_1 = 200$ mA	When the switch is closed, the current will flow through the relay winding. The magnetic field generated by current flowing through the winding will cause the main relay contacts to close, which in turn will supply current to the resistor. The compass positioned near the winding will change its heading indicating the presence of a magnetic field from the winding. The amount of current through the relay winding is 200 mA. The resistance of the winding can also be calculated using Ohm's law if the current flow through the winding resistance is known: $R = V \div I$ $= 12 \div 0.2$ A $= 60\ \Omega$
$A_2 = 6$ A	With the switch closed, the relay contact will close and the current flow through the resistor will be 6 A. As there is only one resistor in the circuit, its current flow will be the same as total current flow for the circuit. To calculate the current flow in the resistor circuit, use this formula: Current = Voltage ÷ Resistance $= 12$ volts $\div 2\ \Omega$ $= 6$ A

Current Exercises

In this section, the exercises are designed to explain the use of the DVOM when taking DC current measurements. Undertaking the exercises will improve your understanding of Ohm's law and current measurements. Examples are given to demonstrate measuring current and to show the magnetic fields produced around a conductor when current flows. It is important to understand that current is the same in all parts of a properly working series circuit. Always remember that an ammeter must be connected in series within the circuit (**FIGURE 10-28**). That means that the circuit must be broken in two and each end of the ammeter connected to one of the two broken ends. This method will ensure that all of the current flowing through the circuit flows through the ammeter.

In this exercise, voltage and current measurements will be taken. For voltage measurements, select "auto range volts DC" on the DVOM, and connect the red lead to "V/Ω" and the black lead to "COM." For current measurements, select "auto range amps DC" on the DVOM. Connect the red lead to the "A" socket and the black lead to the "COM" socket. If using a manual-range DVOM, you will need to select an appropriate range.

FIGURE 10-29 shows a circuit with a single resistor with a 12-volt DC supply. The DVOM will be used to measure both voltage and current. **TABLE 10-12** provides an explanation of the DVOM readings and shows how they relate to Ohm's law.

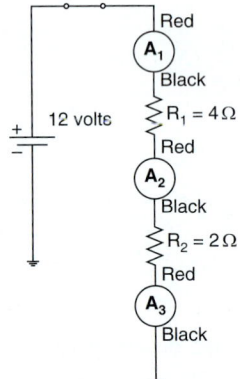

FIGURE 10-28 Measuring amperage across loads in a series circuit with a 12-volt DC supply.

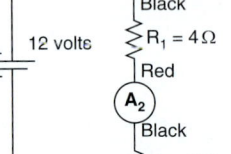

FIGURE 10-29 Measuring voltage and current flow in a circuit with a single resistor and a 12-volt DC supply.

TABLE 10-12 Explanation of DVOM Readings in Figure 10-29

Measurements	Explanation
$V_1 = 12$ V	With the switch closed, current will flow through the resistor. The amount of voltage drop across R_1 will be 12V—the full battery voltage. This can be calculated using Ohm's law: Voltage Drop = Resistance × Current $= 2\ \Omega \times 6$ A (see note below) $= 12$ V drop Note: To calculate total circuit current, use this formula: $I_T = V_T \div R_T$ $= 12 \div 2\ \Omega$ $= 6$ A
$A_1 = 6$ A	With the switch closed, the current flow through the resistor will be 6 amps. Since there is only one resistor in the circuit, its current flow will be the same as total current flow for the circuit.

Resistance Exercises

In this section, the exercises are designed to explain the use of the DVOM in measuring resistance (**FIGURE 10-30**). Resistance measurements are used to check components or circuits against the manufacturer's specifications; for example, the resistance of sensors. Examples are given to demonstrate how to measure resistance and to describe how the resistance affects current flow. It is important to understand that current flow is inversely proportional to resistance: The higher the resistance, the less current that will flow. The reverse is also true: The lower the resistance, the higher the current flow.

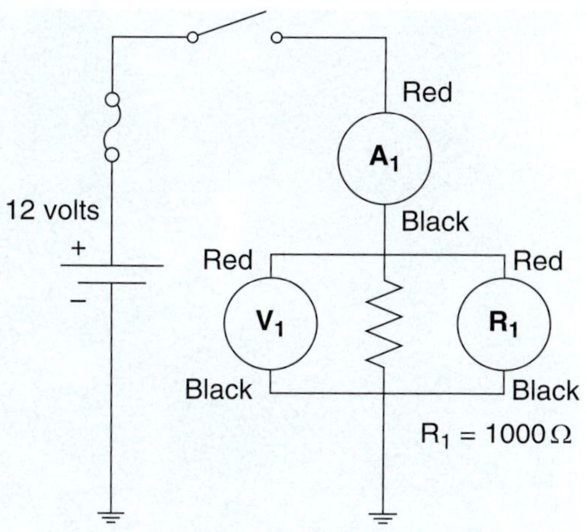

Measuring Resistance

In this exercise, a resistance measurement will be taken. For resistance measurements, you need to select "auto range Ω" on the DVOM, and connect the red lead to "V/Ω" and the black lead to "COM." If using a manual range DVOM, select an appropriate range by starting at the highest range and working your way down. Resistance measurements should only be taken with power disconnected, and ideally, with the component disconnected from the circuit.

FIGURE 10-31 shows a circuit with a lamp in series with a resistor and a 12-volt DC supply. The DVOM will be used to measure resistance. **TABLE 10-13** shows the measurement that can be expected from the circuit.

How Rust, Corrosion, and Debris Effect Resistance

As automobiles have evolved, the electrical systems they contain have become more and more complex. Nearly all automobiles today contain a 12-volt electrical system that operates various lights, control modules, blower motors, power seats, and other systems. All of the systems receive power from dedicated circuits, and many of these systems receive their ground from a connection to the chassis, rather than a direct wire from the battery negative (−) terminal.

A fact of life is that vehicles are made of metal, and metals corrode. Driving vehicles in the winter in the presence of salt and water accelerate this process. Areas where the chassis is drilled or tapped, and a bolt is installed, are often more prone to rust than other areas. This causes a problem for any and all ground circuits that are connected to the chassis. One thing to keep in mind is that rust and corrosion equal electrical resistance. The more rust and/or corrosion that is present in an electrical circuit, the higher the resistance at these points will be. The best way to locate excessive voltage drops caused by resistance is to first perform a voltage drop test from the battery positive to the feed side of the circuit in question. Once it is determined that voltage drops occur in the circuit, you can begin to perform voltage drop tests across each portion of the circuit until the source of the resistance is located. If rust and/or corrosion is found to be the cause, all of the corrosion must be mechanically removed with a wire brush and or contact cleaner, and any corroded wires or switches found to drop excessive voltage should be replaced (**FIGURE 10-32**).

FIGURE 10-30 DVOM set up to measure resistance.

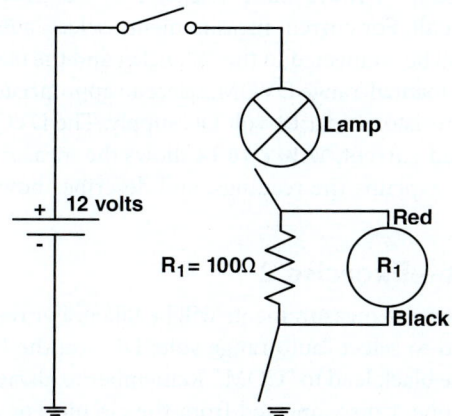

FIGURE 10-31 Measuring resistance in a circuit with a lamp in series with a resistor and a 12-volt DC supply.

TABLE 10-13 Explanation of DVOM Readings in Figure 10-31

Measurement	Explanation
$R_1 = 100\ \Omega$	The switch should be open to measure resistance. If possible, one lead of the resistor should be disconnected from the circuit. The DVOM leads should be connected to the "V/Ω" and "COM" sockets. The DVOM switch should be placed in the "Ω" or resistance position. Reversing the DVOM connections will have no effect on the DVOM reading.

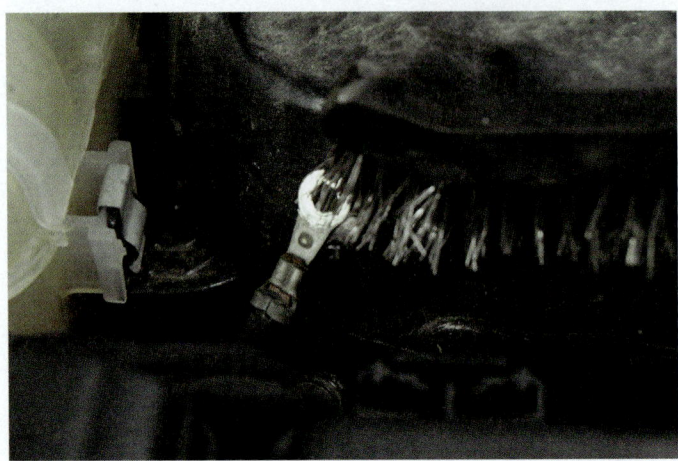

FIGURE 10-32 Technician removing corrosion from a chassis ground connection with a wire brush.

FIGURE 10-33 Wiring harness that has been damaged by being struck with an object.

In addition to causing voltage drops due to high resistance, rust and corrosion can cause intermittent electrical connections, which can be the source of many electrical "gremlins," or scenarios in which the system in question seems to work when it wants to, and only when it chooses. Rust and corrosion can accumulate in electrical connections and then break away with movement, causing the connection to become loose, and thereby operate erratically. In some instances of rust intrusion into a circuit, corrosion can create an open in the circuit, and then "heal" the connection, allowing a very small current to flow. When the load is operated, the corrosion can burn up and cause the open circuit to reappear. For this reason, a voltmeter alone may not be sufficient to diagnose electrical circuits, because it may not put enough of a current load on the circuit to reveal the weak connection. For this reason, a "high-load" test lamp, such as the sealed-beam device described earlier in this chapter, is an effective way to locate any weak connections.

Road debris and other hazards can also wreak havoc on a vehicle's electrical system. The most common mode of failure associated with road debris occurs when a wire, or wiring harness, is struck by an object, thereby damaging strands of wire within a conductor. This can cause high resistance if some of the conductor strands have been severed and there are now fewer paths over which current can flow (**FIGURE 10-33**). These failures can be found with a thorough visual inspection and, if needed, a voltage drop test across all points of the circuit.

Resistance Effects on Current—Exercise 1

In this exercise, resistance, voltage, and current measurements will be taken. For resistance and voltage measurements, select "auto range volts DC" on the DVOM, and connect the red lead to the "V/Ω" and the black lead to "COM." Always make resistance measurements with the component disconnected from the circuit. For current measurements, select "auto range amps DC" on the DVOM. The red lead will be connected to the "A" socket and the black lead connected to the "COM" socket. If using a manual-range DVOM, select an appropriate range.

FIGURE 10-34 shows a circuit with a resistor and a 12-volt DC supply. The DVOM will be used to measure resistance, voltage, and current. **TABLE 10-14** shows the measurements that a DVOM would read in the circuit, explains the readings, and describes how Ohm's law is applied.

Resistance Effects on Current—Exercise 2

In this exercise, resistance, voltage, and current measurements will be taken. For resistance and voltage measurements, you will need to select "auto range volts DC" on the DVOM, and connect the red lead to "V/Ω" and the black lead to "COM." Remember to always make resistance measurements with the component disconnected from the circuit. For current measurements, select "auto range amps DC" on the DVOM. The red lead will be connected to the "A" socket and the black lead connected to the "COM" socket. If using a manual-range DVOM, select an appropriate range.

Current and Resistance Exercises

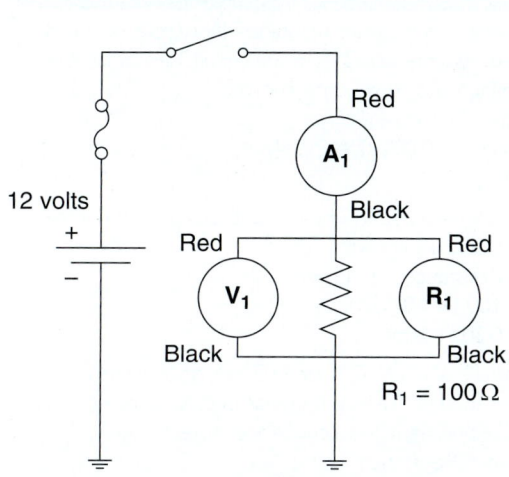

FIGURE 10-34 Measuring volts, amps, and ohm in a circuit with a 100-ohm resistor and a 12-volt DC supply.

TABLE 10-14 Explanation of DVOM Readings in Figure 10-34

Measurements	Explanation
$R_1 = 100\ \Omega$	The switch should be open to measure resistance. If possible, one lead of the resistor should be disconnected from the circuit. Reversing the DVOM connections will have no effect on the DVOM reading. Resistance can be calculated if the voltage and current are known: Resistance = Voltage ÷ Current = 12 V ÷ 0.12 A = 100 Ω
$V_1 = 12\ V$	With the switch closed, V1 = 12 V. This can also be calculated if the current flow and resistance are known: Voltage Drop = Resistance × Current = 100 Ω × 0.12 A = 12 V drop Note: In practice, fuses have a very low resistance and, therefore, will have a very small voltage drop.
$A_1 = 0.12\ A$	With the switch closed, A1 = 0.12 A. To calculate the current flow: Current = Voltage ÷ Resistance = 12 V ÷ 100 Ω = 0.12 A

FIGURE 10-35 shows a circuit with a resistor and a 12-volt DC supply. The DVOM will be used to measure resistance, voltage, and current. **TABLE 10-15** shows the measurements of the circuit by a DVOM, explains the measurement, and describes how Ohm's law is applied.

Resistance Effects on Current—Exercise 3

In this exercise, resistance, voltage, and current measurements will be taken. For resistance and voltage measurements, you will need to select "auto range volts DC" on the DVOM, and connect the red lead to "V/Ω" and the black lead to "COM." Always make resistance measurements with the component disconnected from the circuit. For current measurements, select "auto range amps DC" on the DVOM. Connect the red lead to the "A" socket and the black lead to the "COM" socket. If using a manual-range DVOM, select an appropriate range.

FIGURE 10-36 shows a circuit with a resistor and a 12-volt DC supply in series with a light-emitting diode (LED). The DVOM will be used to measure resistance, voltage, and current through R_1, a 100-ohm resistor in **FIGURE 10-36**. **TABLE 10-16** provides the

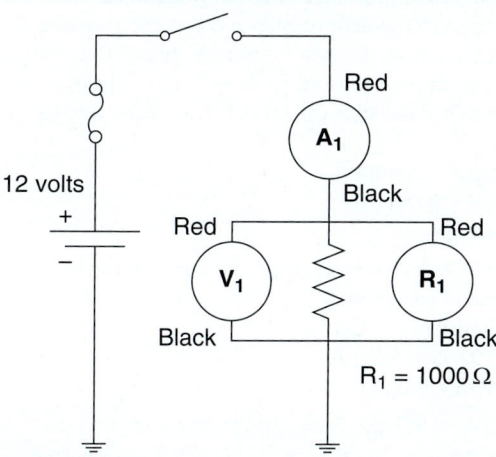

FIGURE 10-35 Measuring volts, amps, and ohm in a circuit with a 1,000 ohm resistor and a 12-volt DC supply.

TABLE 10-15 Explanation of DVOM Readings in Figure 10-35

Measurements	Explanation
$R_1 = 1000\ \Omega$, or 1 kΩ	The switch should be open to measure resistance. If possible, one lead of the resistor should be disconnected from the circuit. Resistance can be calculated if the voltage and current are known: Resistance = Voltage ÷ Current = 12 V ÷ 0.012 A = 1000 Ω
$V_1 = 12\ V$	With the switch closed, V_1 = 12 V. This can also be calculated if the current flow and resistance are known: Voltage Drop = Resistance × Current = 1000 Ω × 0.012 A = 12 V drop
$A_1 = 0.012\ A$	With the switch closed, A_1 = 0.012 A. To calculate the current flow: Current = Voltage ÷ Resistance = 12 V ÷ 1000 Ω = 0.012 A

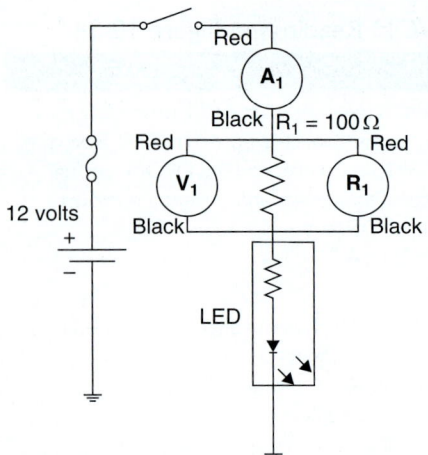

FIGURE 10-36 Measuring volts, amps, and ohm in a circuit with a 100-ohm resistor and a 12-volt DC supply in series with an LED.

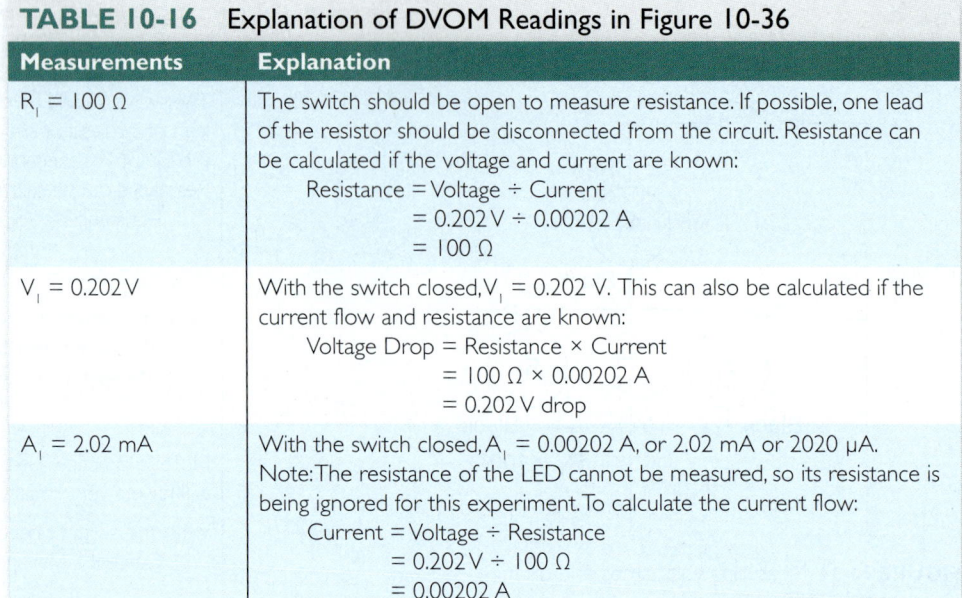

TABLE 10-16	Explanation of DVOM Readings in Figure 10-36
Measurements	**Explanation**
$R_1 = 100\ \Omega$	The switch should be open to measure resistance. If possible, one lead of the resistor should be disconnected from the circuit. Resistance can be calculated if the voltage and current are known: Resistance = Voltage ÷ Current = 0.202 V ÷ 0.00202 A = 100 Ω
$V_1 = 0.202\ V$	With the switch closed, $V_1 = 0.202$ V. This can also be calculated if the current flow and resistance are known: Voltage Drop = Resistance × Current = 100 Ω × 0.00202 A = 0.202 V drop
$A_1 = 2.02\ mA$	With the switch closed, $A_1 = 0.00202$ A, or 2.02 mA or 2020 μA. Note: The resistance of the LED cannot be measured, so its resistance is being ignored for this experiment. To calculate the current flow: Current = Voltage ÷ Resistance = 0.202 V ÷ 100 Ω = 0.00202 A

measurement taken of the circuit by a DVOM, explains the circuit and the measurements, and describes how Ohm's law is applied.

Resistance Effects on Current—Exercise 4

In this exercise, measurements for resistance, voltage, and current will be taken. For resistance and voltage measurements, you will need to select "auto range volts DC" on the DVOM, and connect the red lead to "V/Ω" and the black lead to "COM." Always make resistance measurements with the component disconnected from the circuit. For current measurements, select "auto range amps DC" on the DVOM. Connect the red lead to the "A" socket and the black lead to the "COM" socket. If using a manual-range DVOM, select an appropriate range.

FIGURE 10-37 shows a circuit with a resistor and a 12-volt DC supply in series with an LED. The DVOM will be used to measure resistance, voltage, and current through R_1, a 10 kΩ resistor. **TABLE 10-17** provides the measurement taken of the circuit by a DVOM, explains the circuit and the measurements, and describes how Ohm's law is applied.

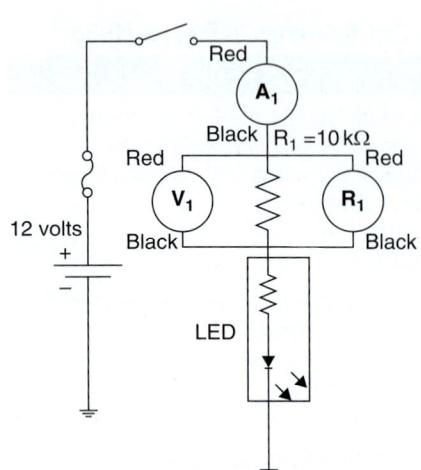

FIGURE 10-37 Measuring volts, amps, and ohms in a circuit with a 10 kΩ resistor and a 12-volt DC supply in series with an LED.

TABLE 10-17	Explanation of DVOM Readings in Figure 10-37
Measurements	**Explanation**
$R_1 = 10{,}000\ \Omega$, or 10 kΩ	The switch should be open to measure resistance. If possible, one lead of the resistor should be disconnected from the circuit. Note: The resistance of the LED cannot be measured, so its resistance is being ignored for this experiment. Resistance can be calculated if the voltage and current are known: Resistance = Voltage ÷ Current = 9.75 V ÷ 975 μA = 10 kΩ
$V_1 = 9.75\ V$	With the switch closed $V_1 = 9.75$ V. This can also be calculated if the current flow and resistance are known: Voltage Drop = Resistance × Current = 10 kΩ × 975 μA = 9.75 V drop
$A_1 = 975\ \mu A$	With the switch closed, $A_1 = 975$ μA. For a given voltage, circuit current decreases as resistance increases. To calculate the current flow through R_1: Current = Voltage ÷ Resistance = 9.75 V ÷ 10 kΩ = 975 μA

▶ Series Circuit Exercises

These series circuit exercises are designed to explain the use of the DVOM for current and voltage in a series circuit. Current and voltage measurements are often used to diagnose faults in electrical circuits. Examples are given to demonstrate the use of measuring voltage and current and to describe how current flows and how voltage drop and current are affected by resistance in a series circuit. It is important to remember that current flow is the same in all parts of a good series circuit, and the sum of the voltage drops across individual resistors in a series circuit is equal to the supply voltage. The exercise also examines how the addition of resistors in a series circuit affects the current flow and voltage drops.

K10005 Understand how to use a DMM to test current flowing through a circuit.

Series Circuit—Exercise 1

In this exercise, voltage measurements will be taken from the series circuit. To measure voltage drop, set the DVOM on the voltage range. Select "auto range volts DC" on the DVOM, and connect the black lead to "COM" and the red lead to "V/Ω." You can measure voltage drop across components, connectors, or cables, but the current has to be flowing to get accurate measurements. Remember, when checking voltage, the leads can be placed in either direction. Just remember which way you placed them so you understand what the reading means. We will show the red lead on the most positive side for purposes of the following diagrams.

FIGURE 10-38 shows a typical circuit with a battery supply, fuse, switch, and resistor. In figure A the switch is open, and in figure B the switch is closed. **TABLE 10-18** provides the measurements taken of the circuit by a DVOM, explains the circuit and the measurements, and describes how Ohm's law is applied.

Series Circuit—Exercise 2

In this exercise, voltage measurements will be taken from the series circuit. To measure voltage drop, the DVOM must be set on the voltage range. Select "auto range volts DC" on the DVOM, and connect the black lead to "COM" and the red lead to "V/Ω." Voltage drop can be measured across components, connectors, or cables only when current is flowing. Remember, the leads when checking voltage can be placed in either direction. Just remember which way you placed them so you understand what the reading means. We will show the red lead on the most positive side for purposes of the following diagrams.

FIGURE 10-39 shows two resistors of equal value connected in series. **TABLE 10-19** provides the measurements taken of the circuit by a DVOM, explains the circuit and the measurements, and describes how Ohm's law is applied.

Series Circuit—Exercise 3

In this exercise, voltage measurements will be taken from the series circuit. To measure voltage drop, the DVOM is used on the voltage range. Select "auto range volts DC" on the DVOM, and connect the black lead to "COM" and the red lead to "V/Ω." Voltage drop can be

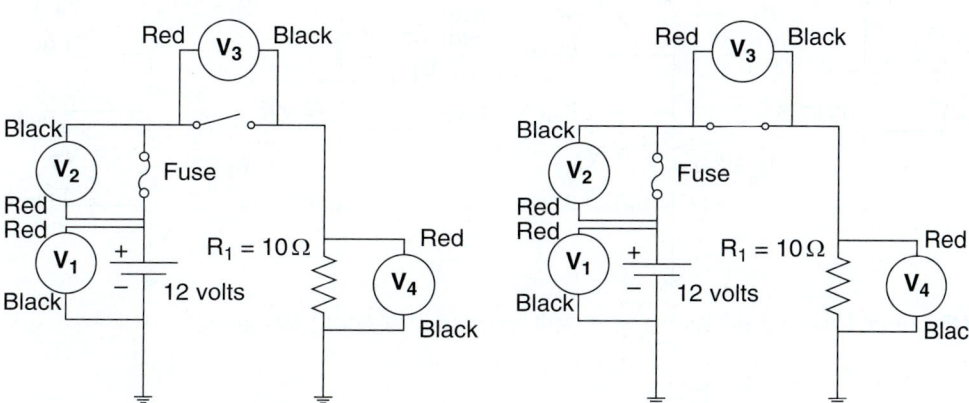

FIGURE 10-38 Measuring voltage in a typical circuit with a battery supply, fuse, switch, and resistor.

TABLE 10-18 Explanation of DVOM Readings in Figure 10-38

Figure	Measured Voltage	Explanation
A	$V_1 = 12\,V$	V_1 is measuring the voltage directly across the supply battery, so the voltage reading will be the battery voltage regardless of the position of the switch.
B	$V_1 = 12\,V$	
A	$V_2 = 0\,V$	With the switch in the open position, no current flows in the circuit. The red DVOM lead is connected directly to the battery via the cable, while the black lead is connected to the switch side of the fuse. Because the switch is open, there is no access to the negative side of the battery; therefore, the DVOM will read 0 V.
B	$V_2 = 0\,V$	With the switch in the closed position, current flows in the circuit. Ideally, the fuse has no resistance, so the voltage drop across the fuse will be 0 V. Note: In practice, fuses will have a very low resistance and will therefore have a very small voltage drop.
A	$V_3 = 12\,V$	With the switch in the open position, no current flows in the circuit. The red DVOM lead is connected directly to the battery via the cable, while the black lead is connected to the negative battery terminal via the resistor. Because no current flows in the circuit, there is no voltage drop across the resistor; therefore, the entire battery voltage drop is across the open circuit switch.
B	$V_3 = 0\,V$	With the switch in the closed position, current flows in the circuit. The switch contacts ideally will have no resistance when closed, so the voltage drop across the contacts will be 0 V.
A	$V_4 = 0\,V$	With the switch in the open position, no current flows in the circuit. Because no current flows in the circuit, there is no voltage drop across the resistor. This can be calculated using Ohm's law: Voltage Drop = Resistance × Current = 10 Ω × 0 A = 0 V drop
B	$V_4 = 12\,V$	With the switch closed, current can flow in the circuit. The amount of voltage drop across R_1 will be 12 volts. This can be calculated using Ohm's law: Voltage Drop = Resistance × Current = 10 Ω × 1.2 A = 12 V drop
B	$V_T = 12\,V$	The sum of all the voltage drops in Figure 10-39B should add up to the supply voltage. This can be calculated using Ohm's law: V_T = VD Fuse + VD Switch + VD Resistor = 0 + 0 + 12 = 12 V

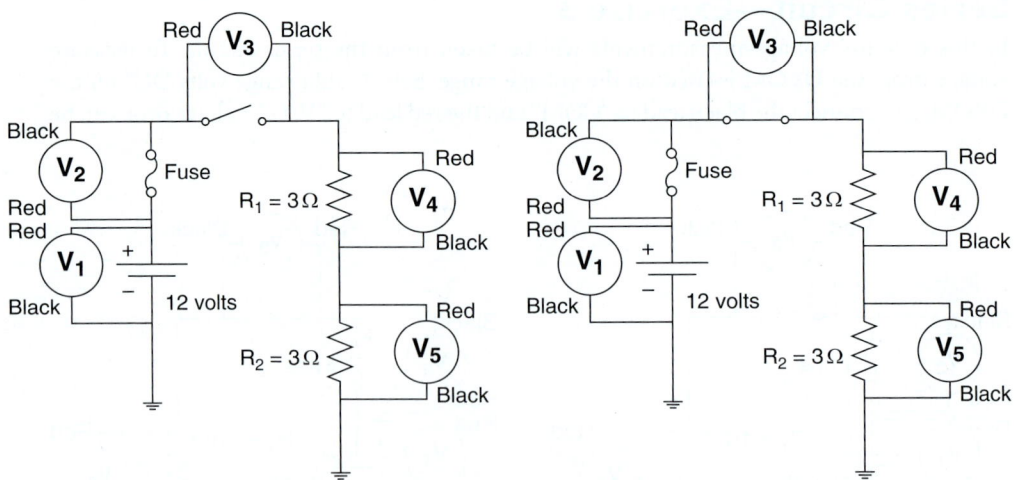

FIGURE 10-39 Measuring voltage in a circuit with two resistors of equal value connected in series.

TABLE 10-19 Explanations of DVOM Readings in Figure 10-39

Figure	Measured Voltage	Explanation
A	$V_1 = 12\,V$	V_1 is measuring the voltage directly across the supply battery, so the voltage reading will be the battery voltage regardless of the position of the switch.
B	$V_1 = 12\,V$	
A	$V_2 = 0\,V$	With the switch in the open position, no current flows in the circuit. The red DVOM lead is connected directly to the battery via the cable, while the black lead is connected to the switch side of the fuse. Because the switch is open, there is no access to the negative side of the battery; therefore, the DVOM will read 0 V.
B	$V_2 = 0\,V$	With the switch in the closed position, current flows in the circuit. The fuse ideally has no resistance, so the voltage drop across the fuse will be 0 V.
A	$V_3 = 12\,V$	With the switch in the open position, no current flows in the circuit. The red DVOM lead is connected directly to the battery via the cable, while the black lead is connected to the negative battery terminal via the series resistors. Because no current flows in the circuit, there is no voltage drop across the resistor; therefore, the entire battery voltage drop is across the open circuit switch.
B	$V_3 = 0\,V$	With the switch in the closed position, current flows in the circuit. The switch contacts ideally will have no resistance when closed, so the voltage drop across the contacts will be 0 V.
A	$V_4 = 0\,V$	With the switch in the open position, no current flows in the circuit. Because no current flows in the circuit, there is no voltage drop across the resistor. This can be calculated using Ohm's law: Voltage Drop = Resistance × Current $= 3\,\Omega \times 0\,A$ $= 0\,V$ drop
B	$V_4 = 6\,V$	With the switch closed, current can flow in the circuit. The amount of voltage drop across R_1 will be 6 V. This can be calculated using Ohm's law: Voltage Drop = Resistance × Current $= 3\,\Omega \times 2\,A$ (see note below) $= 6\,V$ drop Note: To calculate total circuit current, use this formula: $I_T = V_T \div R_T$ $= 12 \div 6$ $= 2\,A$
A	$V_5 = 0\,V$	With the switch in the open position, no current flows in the circuit. Because no current flows in the circuit, there is no voltage drop across the resistor. This can be calculated using Ohm's law: Voltage Drop = Resistance × Current $= 3\,\Omega \times 0\,A$ $= 0\,V$ drop
B	$V_5 = 6\,V$	With the switch closed, current can flow in the circuit. The amount of voltage drop across R_2 will be 6 volts. This can be calculated using Ohm's law: Voltage Drop = Resistance × Current $= 3\,\Omega \times 2\,A$ (see note below) $= 6\,V$ drop Note: To calculate total circuit current, use this formula: $I_T = V_T \div R_T$ $= 12 \div 6$ $= 2\,A$
B	$V_T = 12\,V$	The sum of all the voltage drops in Figure 10-40B should add up to the supply voltage. This can be calculated using Ohm's law: $V_T = $ VD Fuse + VD Switch + VD R_1 + VD R_2 $= 0 + 0 + 6 + 6$ $= 12\,V$

measured across components, connectors, or cables as long as current is flowing in the circuit. The red lead of the DVOM is normally connected to the positive side of the component.

FIGURE 10-40 shows two resistors in a series; this time each resistor is of a different value. **TABLE 10-20** provides the measurement taken of the circuit by a DVOM, explains the circuit and the measurements, and describes how Ohm's law is applied.

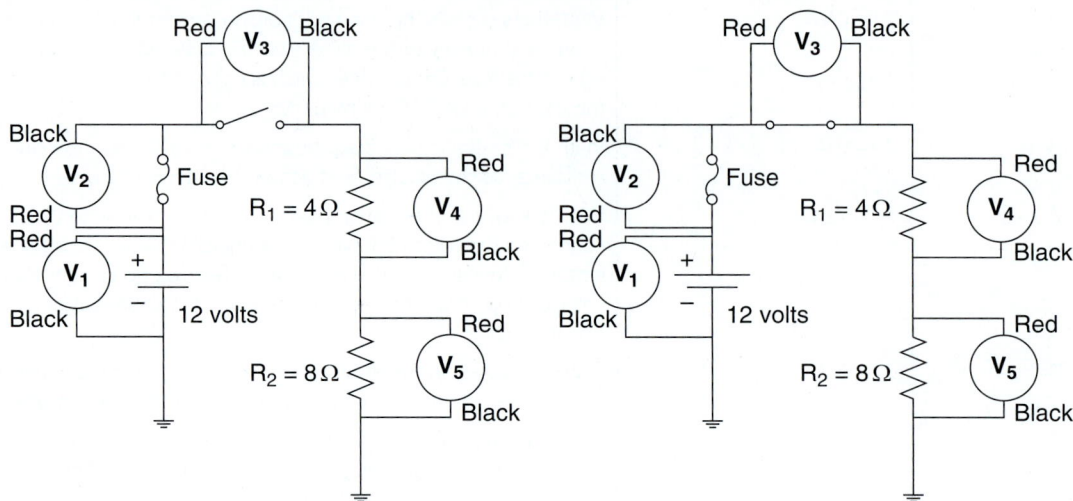

FIGURE 10-40 Measuring voltage in a circuit with two unequal resistors in series.

TABLE 10-20 Explanation of DVOM Readings in Figure 10-40

Figure	Measured Voltage	Explanation
A B	$V_1 = 12\,V$ $V_1 = 12\,V$	V_1 is measuring the voltage directly across the supply battery, so the voltage reading will be the battery voltage regardless of the position of the switch.
A	$V_2 = 0\,V$	With the switch in the open position, no current flows in the circuit. The red DVOM lead is connected directly to the battery via the cable, while the black lead is connected to the switch side of the fuse. Because the switch is open, there is no access to the negative side of the battery; therefore, the DVOM will read 0 V.
B	$V_2 = 0\,V$	With the switch in the closed position, current flows in the circuit. The fuse ideally has no resistance, so the voltage drop across the fuse will be 0 V.
A	$V_3 = 12\,V$	With the switch in the open position, no current flows in the circuit. The red DVOM lead is connected directly to the battery via the cable, while the black lead is connected to the negative battery terminal via the series resistors. Because no current flows in the circuit, there is no voltage drop across the resistor; therefore, the entire battery voltage drop is across the open circuit switch.
B	$V_3 = 0\,V$	With the switch in the closed position, current flows in the circuit. The switch contacts ideally will have no resistance when closed, so the voltage drop across the contacts will be 0 V.

A	$V_4 = 0\,V$	With the switch in the open position, no current flows in the circuit. Because no current flows in the circuit, there is no voltage drop across the resistor. This can be calculated using Ohm's law:
		Voltage Drop = Resistance × Current
		    = 4 Ω × 0 A
		    = 0 V drop

B	$V_4 = 4\,V$	With the switch closed, current can flow in the circuit. The amount of voltage drop across R_2 will be 4 V. This can be calculated using Ohm's law:
		Voltage Drop = Resistance × Current
		    = 4 Ω × 1 A (see note below)
		    = 4 V drop
		Note: To calculate total circuit current, use this formula:
		$I_T = V_T \div R_T$
		    = 12 ÷ 12
		    = 1 A

Continued

TABLE 10-20 Continued

Figure	Measured Voltage	Explanation
A	$V_5 = 0\,V$	With the switch in the open position, no current flows in the circuit. Because no current flows in the circuit, there is no voltage drop across the resistor. This can be calculated using Ohm's law: Voltage Drop = Resistance × Current = 8 Ω × 0 A = 0 V drop
B	$V_5 = 8\,V$	With the switch closed, current can flow in the circuit. The amount of voltage drop across R_1 will be 8 V. This can be calculated using Ohm's law: Voltage Drop = Resistance × Current = 8 Ω × 1 A (see note below) = 8 V drop Note: To calculate total circuit current, use this formula: $I_T = V_T \div R_T$ = 12 ÷ 12 = 1 A
B	$V_T = 12\,V$	The sum of all the voltage drops in Figure 10-41B should add up to the supply voltage. Also, note that the higher value series resistor has the higher voltage drop in the series circuit. This can be calculated using Ohm's law: VT = VD Fuse + VD Switch + VD R_1 + VD R_2 = 0 + 0 + 4 + 8 = 12 V

Series Circuit—Exercise 4

In this exercise, voltage measurements will be taken from the series circuit. To measure voltage drop, the DVOM must be set on the voltage range. Select "auto range volts DC" on the DVOM, and connect the black lead to "COM" and the red lead to "V/Ω." Voltage drop can be measured across components, connectors, or cables as long as current is flowing in the circuit. The red lead of the DVOM is normally connected to the positive side of the component.

FIGURE 10-41 shows diagrams of a circuit with two series resistors; R_1 is now an 8-ohm resistor, and R_2 is a 4-ohm resistor. In **FIGURE 10-41A** the switch is open, and in **FIGURE 10-41B** the switch is closed. **TABLE 10-21** provides the measurement taken of the circuit by a DVOM, explains the circuit and the measurements, and describes how Ohm's law is applied.

Series Circuit—Exercise 5

In this exercise, voltage measurements will be taken from the series circuit. To measure voltage drop, set the DVOM on the voltage range. Select "auto range volts DC" on the DVOM, and connect the black lead to "COM" and the red lead to "V/Ω." Voltage drop can

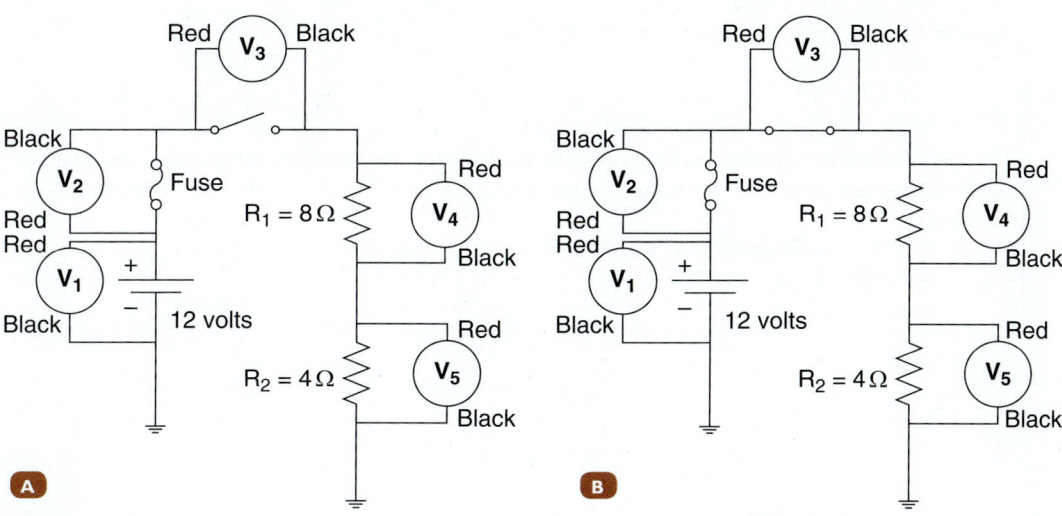

FIGURE 10-41 Measuring voltage in a circuit with two unequal resistors in series (resistors reversed).

TABLE 10-21 Explanation of DVOM Readings in Figure 10-41

Figure	Measured Voltage	Explanation
A B	$V_1 = 12\,V$ $V_1 = 12\,V$	V_1 is measuring the voltage directly across the supply battery, so the voltage reading will be the battery voltage regardless of the position of the switch.
A	$V_2 = 0\,V$	With the switch in the open position, no current flows in the circuit. The red DVOM lead is connected directly to the battery via the cable, while the black lead is connected to the switch side of the fuse. Because the switch is open, there is no access to the negative side of the battery; therefore, the DVOM will read 0 V.
B	$V_2 = 0\,V$	With the switch in the closed position, current flows in the circuit. The fuse ideally has no resistance, so the voltage drop across the fuse will be 0 V.
A	$V_3 = 12\,V$	With the switch in the open position, no current flows in the circuit. The red DVOM lead is connected directly to the battery via the cable, while the black lead is connected to the negative battery terminal via the series resistors. Because no current flows in the circuit, there is no voltage drop across the resistor; therefore, the entire battery voltage drop is across the open circuit switch.
B	$V_3 = 0\,V$	With the switch in the closed position, current flows in the circuit. The switch contacts ideally will have no resistance when closed, so the voltage drop across the contacts will be 0 V.

A	$V_4 = 0\,V$	With the switch in the open position, no current flows in the circuit. Because no current flows in the circuit, there is no voltage drop across the resistor. This can be calculated using Ohm's law:
		  Voltage Drop = Resistance × Current
		  $= 8\,\Omega \times 0\,A$
		  $= 0\,V$ drop

B	$V_4 = 8\,V$	With the switch closed, current can flow in the circuit. The amount of voltage drop across R_1 will be 8 V. This can be calculated using Ohm's law:
		  Voltage Drop = Resistance × Current
		  $= 8\,\Omega \times 1\,A$ (see note below)
		  $= 8\,V$ drop
		Note: To calculate total circuit current, use this formula:
		  $I_T = V_T \div R_T$
		  $= 12 \div 12$
		  $= 1\,A$

A	$V_5 = 0\,V$	With the switch in the open position, no current flows in the circuit. Because no current flows in the circuit, there is no voltage drop across the resistor. This can be calculated using Ohm's law:
		  Voltage Drop = Resistance × Current
		  $= 4\,\Omega \times 0\,A$
		  $= 0\,V$ drop

B	$V_5 = 4\,V$	With the switch closed, current can flow in the circuit. The amount of voltage drop across R_2 will be 4 V. This can be calculated using Ohm's law:
		  Voltage Drop = Resistance × Current
		  $= 4\,\Omega \times 1\,A$ (see note below)
		  $= 4\,V$ drop
		Note: To calculate total circuit current, use this formula:
		  $I_T = V_T \div R_T$
		  $= 12 \div 12$
		  $= 1\,A$

B	$V_T = 12\,V$	The sum of all the voltage drops in Figure 10-42B should add up to the supply voltage. Also, note that the higher value series resistor has the higher voltage drop in the series circuit. This can be calculated using Ohm's law:
		  $V_T =$ VD Fuse $+$ VD Switch $+$ VD R_1 $+$ VD R_2
		  $= 0 + 0 + 8 + 4$
		  $= 12\,V$

be measured across components, connectors, or cables as long as current is flowing in the circuit. The red lead of the DVOM is normally connected to the more positive side of the component.

FIGURE 10-42 shows diagrams of a series circuit with three unequal resistances. **TABLE 10-22** provides the voltage measurements taken of the circuit by a DVOM, explains the circuit and the measurements, and describes how Ohm's law is applied.

Series Circuit—Exercise 6

To conduct this exercise, the DVOM must be set to read "DC amps." The red lead will be connected to the "A" socket and the black lead connected to the "COM" socket. If using a manual-range DVOM, select an appropriate range.

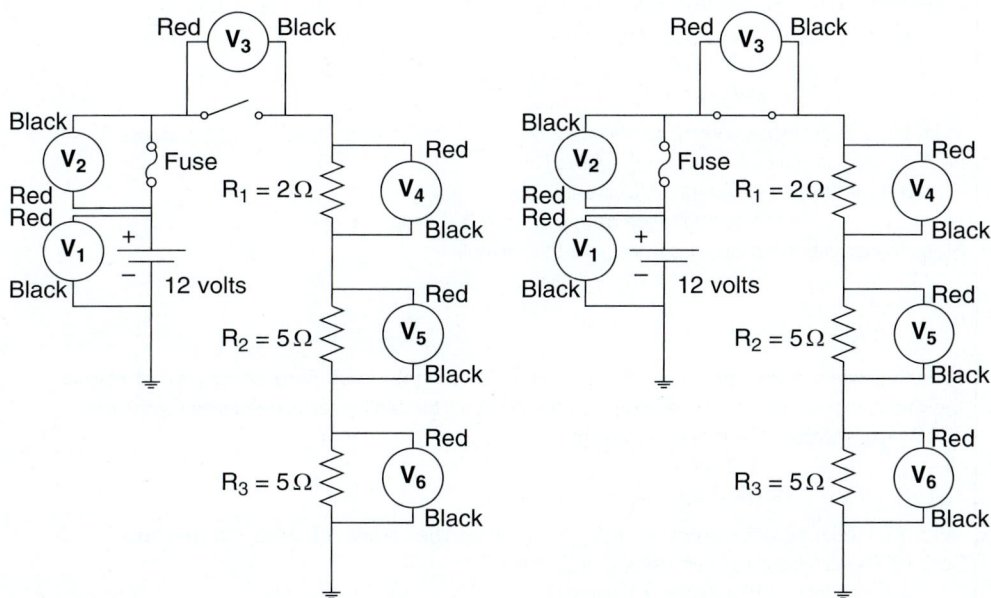

FIGURE 10-42 Voltage measurements in a series circuit with unequal resistances.

TABLE 10-22 Explanation of DVOM Readings in Figure 10-42

Figure	Measured Voltage	Explanation
A B	$V_1 = 12\,V$ $V_1 = 12\,V$	V_1 is measuring the voltage directly across the supply battery, so the voltage reading will be the battery voltage regardless of the position of the switch.
A	$V_2 = 0\,V$	With the switch in the open position, no current flows in the circuit. The red meter lead is connected directly to the battery via the cable, while the black lead is connected to the switch side of the fuse. Because the switch is open, there is no access to the negative side of the battery; therefore, the meter will read 0 V
B	$V_2 = 0\,V$	With the switch in the closed position, current flows in the circuit. The fuse ideally has no resistance, so the voltage drop across the fuse will be 0 V
A	$V_3 = 12\,V$	With the switch in the open position, no current flows in the circuit. The red meter lead is connected directly to the negative battery terminal via the series resistors. Because no current flows in the circuit, there is no voltage drop across the resistor; therefore, the entire battery voltage drop is across the open circuit switch
B	$V_3 = 0\,V$	With the switch in the open position, no current flows in the circuit. The switch contacts ideally will have no resistance when closed so the voltage drop across the contacts will be 0.V
A	$V_4 = 0\,V$	With the switch in the open position, no current flows in the circuit. Because no current flows in the circuit, there is no voltage drop across the resistor. This can be calculated using Ohm's law: Voltage Drop = Resistance × Current = 5 Ω × 0 A = 0 V drop.

Continued

TABLE 10-22 Continued

Figure	Measured Voltage	Explanation

B	$V_4 = 2\,V$	With the switch closed, current can flow in the circuit. The amount of voltage drop across R_1 will be 2.4 V. This can be calculated using Ohm's law:
		    Voltage Drop = Resistance × Current
		                  = 2 Ω × 1 A (see note below) = 2 V drop
		Note: To calculate total circuit current, use this formula:
		    $I_T = V_T \div R_T$
		       = 12 ÷ 12
		       = 1 A

A	$V_5 = 0\,V$	With the switch in the open position, no current flows in the circuit. Because no current flows in the circuit, there is no voltage drop across the resistor. This can be calculated using Ohm's law:
		    Voltage Drop = Resistance × Current
		                  = 5 Ω × 0 A
		                  = 0 V drop.

B	$V_5 = 5\,V$	With the switch closed, current can flow in the circuit. The amount of voltage drop across R_2 will be 4.8 V. This can be calculated using Ohm's law:
		    Voltage Drop = Resistance × Current
		                    = 5 Ω × 1 A (see note below) = 5 V drop
		Note: To calculate total circuit current, use this formula:
		    $I_T = V_T \div R_T$
		       = 12 ÷ 12
		       = 1 A

A	$V_6 = 0\,V$	With the switch in the open position, no current flows in the circuit. Because no current flows in the circuit, there is no voltage drop across the resistor. This can be calculated using Ohm's law:
		    Voltage Drop = Resistance × Current
		                    = 5 Ω × 0 A
		                  = 0 V drop

B	$V_6 = 5\,V$	With the switch closed, current can flow in the circuit. The amount of voltage drop across R_2 will be 4.8 V. This can be calculated using Ohm's law:
		    Voltage Drop = Resistance × Current
		                  = 5 Ω × 1 A (see note below)
		Note: To calculate total circuit current, use this formula:
		    $I_T = V_T \div R_T$
		       = 12 ÷ 12
		       = 1 A

B	$V_T = 12\,V$	The sum of all the voltage drops in Figure 37-21B should add up to the supply voltage. This can be Calculated using Ohm's law:
		    V_T = Fuse + VD Switch + VD R_1 + VD R_2 + VD R_3
		       = 0 + 0 + 2 + 5 + 5
		       = 12 V
		Also, note that the higher value series resistor has the higher voltage drop in the same value have the same voltage drop.

FIGURE 10-43 shows a circuit with three resistors in series with a 12-volt DC supply. The DVOM can be connected in various parts of the circuit to measure the current flow. In **FIGURE 10-43A** the switch is open, and in **FIGURE 10-43B** the switch is closed. **TABLE 10-23** provides the measurement taken of the circuit by a DVOM, explains the circuit and the measurements, and describes how Ohm's law is applied.

▶ Parallel Circuit Exercises

K10006 Understand the use of the DMM in measuring volts, ohms, and amps in a parallel circuit.

In this section, the exercises are designed to explain the use of the DVOM in measuring volts, amps, and ohms in a parallel circuit. Parallel circuits are commonly used in the vehicle's electrical system, especially for lights (**FIGURE 10-44**). Understanding how they work and the relationship between voltage, amperage, and resistance in parallel circuits will help

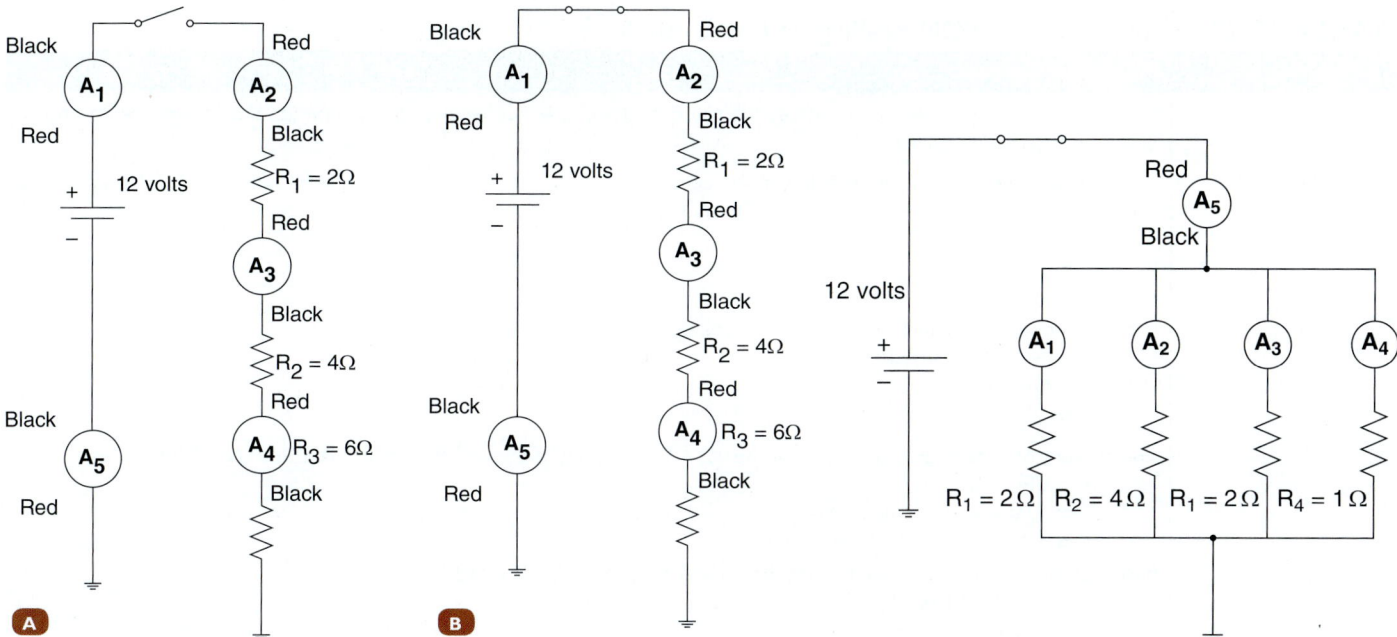

FIGURE 10-43 Current flow in a circuit with three unequal resistors in series.

FIGURE 10-44 Wiring diagram of parallel circuit.

TABLE 10-23 Explanation of DVOM Readings in Figure 10-43

Figure	Measured Current	Explanation
A	$A_1 = 0$ A	With the switch off, the current flow through A_1 will be 0 A.
B	$A_1 = 1$ A	With the switch closed, the current will flow through the series circuit. The amount of current flowing through the resistors will be proportional to the 50 Ω of total resistance and will be 0.24 A. This can be calculated using Ohm's law: Current = Voltage ÷ Resistance = 12 V ÷ 12 Ω (see note below) = 1 A Note To calculate the total circuit current, use this formula: $I_T = V_T ÷ R_T$
B	$A_2 = 1$ A	With the switch closed, the current flow will be the same in all parts of the series circuit; therefore, current flow on the ammeter will always be the same, regardless of its position in the circuit. The same current flow will be read on $A_1, A_2, A_3, A_4,$ and A_5.

you to diagnose electrical faults. Examples are given to demonstrate how to measure volts, amps, and ohms and to show how current flows and voltage drops in a parallel circuit. It is important to remember the laws for a parallel circuit: resistance decreases when more parallel paths are added, current flow from individual legs add up in parallel, and voltage stays the same at all common parallel circuit inputs. The following exercises help to reinforce the understanding of these laws.

Parallel Circuit—Exercise 1

In this exercise, voltage measurements will be taken from the parallel circuit. To measure voltage drop, the DVOM must be set up to measure voltage. Select "auto range volts DC" on the DVOM, and connect the black lead to "COM" and the red lead to "V/Ω." Voltage drop can be measured across components, connectors, or cables as long as current is flowing in the circuit. The red lead of the DVOM is normally connected to the positive side of the component.

FIGURE 10-45 shows three resistors connected in parallel across a 12-volt supply. **TABLE 10-24** provides the measurements taken of the

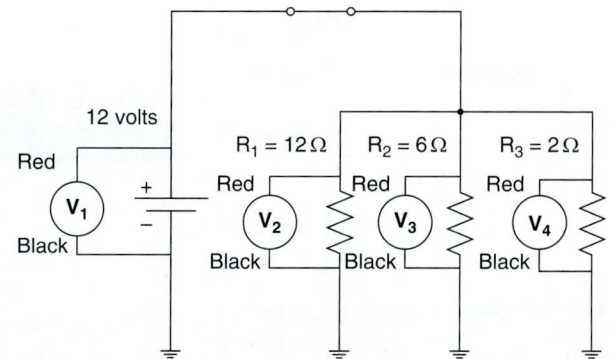

FIGURE 10-45 Measuring voltage across three unequal resistors connected in parallel across a 12-volt supply.

TABLE 10-24 Explanation of DVOM Readings in Figure 10-45

Measured Current	
$V_1 = 12V$	V_1 is measuring the voltage directly across the supply battery, so the voltage reading will be the battery voltage, regardless of the position of the switch.
$V_2 = 12V$	With the switch closed, current can flow in the circuit. The amount of voltage drop across R_1 will be 12 V. The amount of current through R_1 can be calculated using Ohm's law: $IR_1 = VR_1 \div R_1$ $= 12V \div 12\,\Omega$ $= 1A$ If the current and resistance are known, then voltage drop can be calculated. $VR_1 = IR_1 \times \text{Resistance } R_1$ $= 1A \times 12\,\Omega$ $= 12V$
$V_3 = 12V$	The amount of voltage drop across R_2 will be 12 V The amount of through R_2 can be calculated using Ohm's law: $IR_2 = VR_2 \div R_2$ $= 12V \div 6\,\Omega$ $= 2A$ If the current and resistance are known, the voltage drop can be calculated: $VR_2 = IR_2 \times \text{Resistance } R_2$ $= 2A \times 6\,\Omega$ $= 12V$
$V_4 = 12V$	The amount of voltage drop across R_3 will be 12 V. The amount of through R_2 can be calculated using Ohm's law: $IR_3 = VR_3 \div R_3$ $= 12V \div 2\,\Omega$ $= 6A$ If the current and resistance are known, the voltage drop can be calculated: $VR_3 = IR_3 \times \text{Resistance } R_3$ $= 6A \times 2\,\Omega$ $= 12V$

From the above, we can determine that the voltage drop across a parallel branch is the same for all components, regardless of their resistance. The total circuit current flow is equal to the sum of the individual branch currents.

circuit by a DVOM, explains the circuit and the measurements, and describes how Ohm's law is applied.

Parallel Circuit—Exercise 2

To conduct this exercise, the DVOM must be set to read "DC amps." Connect the red lead to the "A" socket and the black lead to the "COM" socket. If using a manual-range DVOM, select an appropriate range.

FIGURE 10-46 shows a circuit with a single resistor and a 12-volt DC supply. The DVOM can be connected in various parts of the circuit to measure the current flow. **TABLE 10-25**

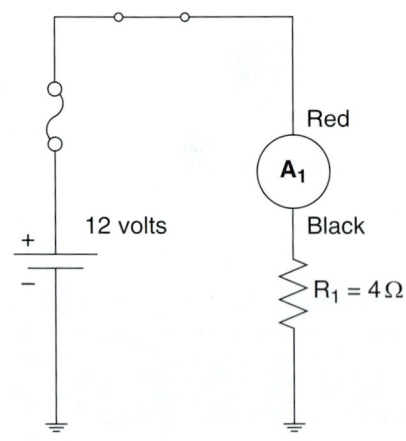

FIGURE 10-46 Measuring amperage in a circuit with a single resistor and a 12-volt DC supply.

TABLE 10-25 Explanation of DVOM Readings in Figure 10-46

Measured Current	Explanation
$A_1 = 3A$	With the switch closed, current will flow through the circuit. The amount of current flow through R_1 will be 3 A. As this is a series circuit, the current flow in the circuit will be the same in all parts of the circuit. This can be calculated using Ohm's law: $\text{Current} = \text{Voltage} \div \text{Resistance}$ $\phantom{\text{Current}}= 12V \div 4\,\Omega$ $\phantom{\text{Current}}= 3A$

provides the measurements taken of the circuit by a DVOM, explains the circuit and the measurements, and describes how Ohm's law is applied.

Parallel Circuit—Exercise 3

To conduct this exercise, the DVOM must be set to read DC amps. Connect the red lead to the "A" socket and the black lead to the "COM" socket. If using a manual-range DVOM, select an appropriate range.

FIGURE 10-47 has two resistors in parallel. The additional resistor in parallel will cause an increase in circuit current flow and a decrease in total circuit resistance. **TABLE 10-26** provides the measurements taken of the circuit by a DVOM, explains the circuit and the measurements, and describes how Ohm's law is applied.

Parallel Circuit—Exercise 4

To conduct this exercise, the DVOM must be set to read DC amps. Connect the red lead to the "A" socket and the black lead to the "COM" socket. If using a manual-range DVOM, select an appropriate range.

FIGURE 10-48 has three resistors in parallel. With the additional resistors, the total circuit current will increase while the total circuit resistance will decrease. **TABLE 10-27** provides the measurements taken of the circuit by a DVOM, explains the circuit and the measurements, and describes how Ohm's law is applied.

TABLE 10-26 Explanation of DVOM Readings in Figure 10-47

Measured Current	Explanation
$A_3 = 9\ A$	With the switch closed, the current will flow through the circuit. Due to its position in the circuit, A_3 will measure total current flow or I_T and will be 9 A.
$A_1 = 6\ A$	A_1 is positioned so that it only measures the current flow through R_1 and its current flow will be 6 A.
$A_2 = 3\ A$	A_2 is positioned so that it only measures the current flow through R_2 and its current flow will be 3 A.
The current flows can be calculated using Ohm's law in two ways	
Calculate current flow through each resistor and add them together: $IR_1 = VR_1 \div R_1$ $\quad = 12V \div 2\ \Omega$ $\quad = 6\ A$ $IR_2 = VR_2 \div R_2$ $\quad = 12V \div 4\ \Omega$ $\quad = 3\ A$ $I_T = IR_1 + IR_2$ $\quad = 6 + 3$ $\quad = 9\ A$	Calculate total resistance R_T and use it to then calculate I_T total current: $R_T = \dfrac{1}{\dfrac{1}{R_1} + \dfrac{1}{R_2}}$ $R_T = \dfrac{1}{\dfrac{1}{2} + \dfrac{1}{4}}$ $R_T = 1.3333\ \Omega$ $I_T = V_T \div R_T$ $\quad = 12V \div 1.3333\ \Omega$ $\quad = 9.0002\ A$ Note: R_T is 1.3 recurring. It has been rounded to four decimal places; this gives the variance of 0.0002 in the I_T calculation. In practice, this small variation can be ignored.

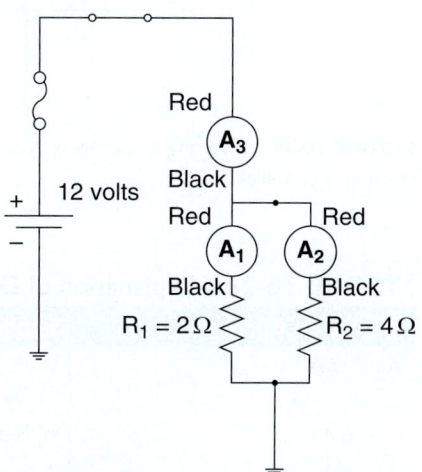

FIGURE 10-47 Measuring amperage in a circuit with two unequal resistors in parallel.

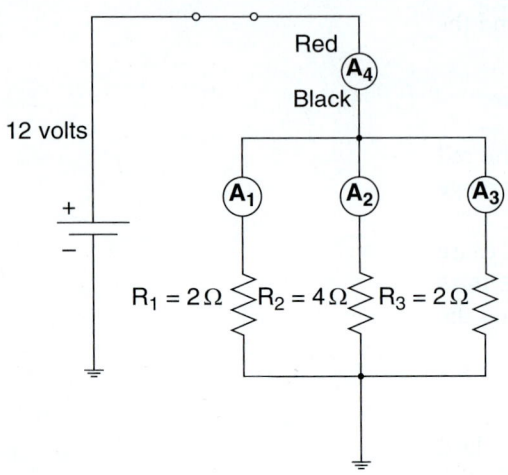

FIGURE 10-48 Measuring amperage in a circuit with three resistors in parallel.

TABLE 10-27 Explanation of DVOM Readings in Figure 10-48

Measured Current	Explanation
$A_4 = 15\ A$	With the switch closed, the current will flow through the circuit. Due to its position in the circuit, A_3 will measure total current flow or I_T and will be 15 A. The total current in the circuit increases as additional, load or resistances are added in parallel.
$A_1 = 6\ A$	A_1 is positioned so that it only measures the current flow through R_1 and its current flow will be 6 A.
$A_2 = 3\ A$	A_2 is positioned so that it only measures the current flow through R_2 and its current flow will be 3 A.
$A_3 = 6\ A$	A_3 is positioned so that it only measures the current flow through R_3 and its current flow will be 6 A.

The current flows can be calculated using Ohm's law in two ways:

Calculate current flow through each resistor and add them together:	Calculate total resistance R_T and use it to then calculate I_T total current:
$IR_1 = VR_1 \div R_1$ $= 12V \div 2\ \Omega$ $= 6\ A$ $IR_2 = VR_2 \div R_2$ $= 12V \div 4\ \Omega$ $= 3\ A$ $IR_3 = IR_1 + IR_2$ $= 12V \div 2\ \Omega$ $= 6\ A$ $I_T = IR_1 + IR_2 + IR_3$ $= 6 + 3 + 6$ $= 15\ A$	$R_T = \dfrac{1}{\dfrac{1}{R_1} + \dfrac{1}{R_2} + \dfrac{1}{R_3}}$ $R_T = \dfrac{1}{\dfrac{1}{2} + \dfrac{1}{4} + \dfrac{1}{2}}$ $R_T = 0.8\ \Omega$ $I_T = V_T \div R_T$ $= 12V \div 0.8\ \Omega$ $= 15\ A$

Parallel Circuit—Exercise 5

To conduct this experiment, the DVOM must be set to read DC amps. Connect the red lead to the "A" socket and the black lead to the "COM" socket. If using a manual-range DVOM, select an appropriate range.

FIGURE 10-49 has four resistors in parallel. With the additional resistors, the total circuit current will increase while the total circuit resistance decreases even more. **TABLE 10-28** provides the measurements taken of the circuit by a DVOM, explains the circuit and the measurements, and describes how Ohm's law is applied.

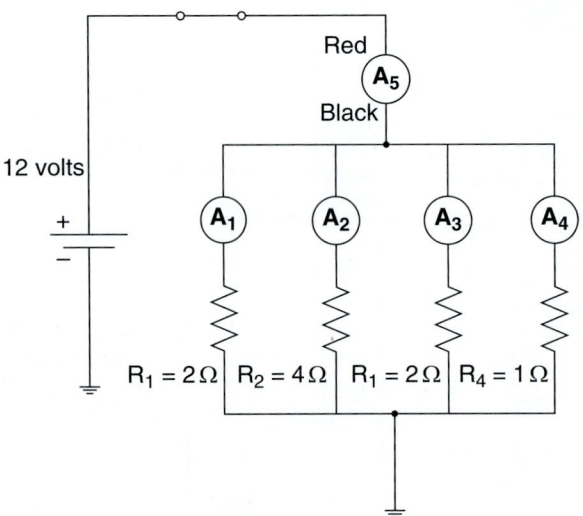

FIGURE 10-49 Measuring amperage in a circuit with four resistors in parallel.

TABLE 10-28 Explanation of DVOM Readings in Figure 10-49

Measured Current	Explanation
$A_5 = 27\ A$	With the Switch closed, the current will flow through the circuit. The amount of current flow through A_3 will be 27 A. The total current in the circuit increases as additional load or resistances are added in parallel.
$A_1 = 6\ A$	A_1 is positioned so that it only measures the current flow through R_1 and its current flow will be 6 A.
$A_2 = 3\ A$	A_2 is positioned so that it only measures the current flow through R_2 and its current flow will be 3 A.
$A_3 = 6\ A$	A_3 is positioned so that it only measures the current flow through R_3 and its current flow will be 6 A.
$A_4 = 12\ A$	A_4 is positioned so that it only measures the current flow through R_4 and its current flow will be 12 A.

TABLE 10-28 Continued

Measured Current	Explanation
The current flows can be calculated using Ohm's law in two ways:	
Calculate current flow through each resistor and add them together: $IR_1 = VR_1 \div R_1$ $\quad = 12V \div 2\,\Omega$ $\quad = 6\,A$ $IR_2 = VR_2 \div R_2$ $\quad = 12V \div 4\,\Omega$ $\quad = 3\,A$ $IR_3 = VR_2 \div R_2$ $\quad = 12V \div 2\,\Omega$ $\quad = 6\,A$ $IR_4 = VR_2 \div R_2$ $\quad = 12V \div 1\,\Omega$ $\quad = 12\,A$ $I_T = IR_1 + IR_2 + IR_3$ $\quad = 6 + 3 + 6 + 12$ $\quad = 27\,A$	Calculate total resistance R_T and use it to then calculate I_T total current: $R_T = \dfrac{1}{\dfrac{1}{R_1} + \dfrac{1}{R_2} + \dfrac{1}{R_3} + \dfrac{1}{R_4}}$ $R_T = \dfrac{1}{\dfrac{1}{2} + \dfrac{1}{4} + \dfrac{1}{2} + \dfrac{1}{1}}$ $R_T = 0.4444\,\Omega$ $I_T = V_T \div R_T$ $\quad = 12\,V \div 0.4444\,\Omega$ $\quad = 27.0027\,A$ Note: R_T is 0.4 recurring. It has been rounded to four decimal places; this gives the variance of 0.0027 in the I_T calculation. In practice, this small variation can be ignored.

▶ Understand Circuit Types

Electrical circuit testing begins with understanding circuit types and how electricity behaves within them. Add to that the ability to use meters and oscilloscopes to measure the values of voltage, amperage, and resistance, along with understanding how to read wiring diagrams so you will know how the circuits are constructed, and you will be well on your way to diagnosing electrical faults successfully. Those are the concepts we will be exploring in this section. Feel free to refer back to the previous sections to help you remember how electricity behaves, as well as how meters are hooked up for specific measurements. Let's kick this off by seeing how Ohm's law can help us predict the behavior of electricity.

K10007 Understand circuit types and how electricity behaves within them.

S10004 Use Ohm's law to diagnose circuits.

N10003 Demonstrate knowledge of electrical/electronic series, parallel, and series-parallel circuits using principles of electricity (Ohm's Law).

Using Ohm's Law to Diagnose Circuits

Ohm's law can be used in two ways to help in diagnosing electrical circuit faults. The first is by using it to perform the math to predict and verify measurements. The second method is by using the relationships it demonstrates to guide you through the diagnosis process.

When Ohm's law is employed in the first manner, it is used to calculate electrical quantities in a circuit and is valuable in cross-checking actual measured results within the circuit. For example, if the resistance and voltage of a circuit are known, then the theoretical current can be calculated using Ohm's law. The calculated result can then be compared to the measured results from an ammeter to determine if the circuit is functioning correctly. Technicians will often do a quick calculation to obtain an approximate value of an electrical quantity before they take actual measurements. Doing so allows them to anticipate what they will be measuring and to set the measuring tool to the correct range. Always remember that a calculation may only yield an approximate value; in actual circuits, variations or tolerances exist in components, causing differences between calculated values and actual measurements.

Using Ohm's law the second way helps you to understand the relationship between volts, amps, and ohms. For example, if voltage stays the same but resistance decreases, amperage must increase. In the case of a short circuit, the resistance decreases and the amperage increases, potentially blowing the fuse, and/or burning up a wire or component. In the opposite scenario, where resistance increases, current flow decreases. This is the case when a corroded or loose connection introduces excessive resistance to the circuit. It also results in less electrical power (volts and amps) to operate the intended load.

What does Ohm's law tell us to expect when the voltage changes? If voltage decreases and the resistance stays the same, then amperage will decrease. This results in less power being available to operate the load. If the voltage increases and the resistance stays the same, then amperage will increase. If amperage and voltage both increase, then the electrical power operating the load will also increase. This condition can shorten the life of, or even burn out, the load device.

So, how do amperage changes affect volts and amps? That is a good question. But if you think about it, amperage is a result of, or product of, the voltage and resistance that exists within an electrical circuit. Amperage cannot exist without both voltage and resistance—a means of pushing the amperage (voltage) and a path for the amperage to flow (resistance). If you ask yourself, "what is the amperage doing in a circuit?" The answer will always be, it is doing whatever the voltage and resistance allow it to do. If the amperage is low, then you know that one of two conditions is present—either the voltage is too low or the resistance is too high. If the amperage is high, then either the voltage is too high or the resistance is too low. Understanding this relationship between volts, amps, and ohms will help you know what test you need to perform next during diagnosis.

Since amperage is a product of voltage and resistance, it is a good idea to keep your eye on the amperage. What this means is that if you have a circuit fault, you can generally see what the amperage is doing; it is either high or low. If you truly cannot see what the current is doing, like in a solenoid, you will have to measure it. But in most cases, you can see it; the fuse blew because the current was too high, the lights are dim because the current is too low, etc. If current is low (the most common scenario), then Ohm's law tells you that it is because either the voltage is low or the resistance is high. On the flip side, if your eye determines that the current is high, then either the voltage is high or the resistance is low.

For example, let's say that the left front headlight is dim. Your eye determines that the current in that circuit is low. Thus, either the voltage is low or the resistance is high. Now you just have to test for those two things. Use a voltmeter to measure the voltage at the battery. If the battery voltage is low, determine why this is so. If battery voltage is normal, check the voltage across both sides of the headlight with the circuit on. It should be within 1.0 volt of the battery voltage. If it is not, look for the high resistance. This is accomplished by first measuring the voltage drop on each side of the headlight. If there is an excessive voltage drop on one side, follow the circuit back toward the battery to identify the cause. If the voltages on both sides of the headlight are within specifications, the problem is most likely the headlight itself. You may be able to check the resistance of the headlight itself and compare it to a known good bulb. Or more likely, you may have to measure the current flow through the bulb and compare that to specifications, since its resistance increases greatly due to heat when it is illuminated.

If the current flow appears too high, then the circuit may have too much voltage or too little resistance. Measuring the battery voltage is an easy way to check for too much voltage. Using an ohmmeter to check the resistance of the load and comparing that to specifications will tell you if it is shorted. If it is not, then use the ohmmeter to check the wire harness for any short circuit conditions

To use Ohm's law to diagnose circuits, follow the steps in **SKILL DRILL 10-4**.

> **SKILL DRILL 10-4** Using Ohm's Law to Diagnose Circuits
>
> **1.** Identify the circuit to be tested and determine the expected voltage, current, and resistance of the component or circuit. Using Ohm's law, calculate the expected voltage, current flow, or resistance of the circuit.
>
> **2.** Set up the DVOM for a continuity or resistance check. Make sure there is no power connected to any circuit that you test for continuity. Next prepare the DVOM for testing just like you did for voltage by inserting the black probe into the "COM" terminal and the red probe into the "V/Ω" terminal.
>
> **3.** Turn the rotary dial of the DVOM to the mode for measuring ohms, which also measures continuity. The digital display should now give you an "Out of Limits" reading indicating that there is not a continuous circuit connection between the two probes (some meters show "OL," and others place "1" on the left of the display). Touch the probe ends together. The display should now give a zero reading, or a reading very near to zero, which indicates no resistance. This means there is a continuous circuit through the probes. Some DVOMs also indicate continuity with an audible tone.
>
> **4.** Check a fuse. One typical use of the test is to determine whether a fuse needs to be replaced. If the fuse has been overloaded and "blown," then it will no longer complete a circuit when a DVOM is used to test it. To check this, place the black probe on one end of the fuse and the red probe on the other. If the fuse is functioning correctly, then the reading will be zero, indicating a complete, or closed, circuit. If the fuse is open, then there will be no reading and no tone, indicating an incomplete, or open, circuit.
>
> **5.** A continuity or resistance test is used to check for a broken circuit caused by a break in a cable or lead or caused by a component becoming disconnected. The same test can also confirm whether there is continuity between components that are not supposed to be connected, a condition known as a short circuit. This test can also be used to check circuits that are suspected to have a high resistance.
>
> **6.** Compare the test results with the calculated results from step 1. Key things to note are any variations between the calculated and measured results. Determine whether the variations can be accounted for within the tolerances of the components or whether a fault exists.

Using Wiring Diagrams to Diagnose Electrical Circuits

Vehicle wiring diagrams or schematics may be available as paper-based manuals, computer programs, or online resources; they are produced by manufacturers and some aftermarket publishing companies. Increasingly, repair information is accessed via the Internet using subscription services that are regularly updated. To use wiring diagrams, an understanding of the symbols, abbreviations, and connector coding used in the diagrams is required. These are usually found on the diagram or in information pages. See the chapter Lighting Systems for examples of some of the common symbols used.

Reading a wiring diagram is like reading a road map. There are a lot of interconnected circuits, wires, and components to decipher. Learning to read wiring diagrams takes a bit of time and experience, but knowing that circuits usually consist of a power source, a switch, a load, and a ground is a good start. Jorge Menchu of AESWave has been promoting a novel approach of using color crayons to help understand how a particular circuit in a wiring diagram operates. The following is a paraphrased version of that process.

Begin by printing out a copy of the wiring diagram for the circuit being diagnosed. Color all of the wires green that are directly connected to "ground." Color all of the wires red that are "hot" at all times. Color all of the wires orange that are "switched to power." Color all of the wires yellow that are "switched to ground." If there are any wires that reverse polarity, such as power window motor wires mark those with side-by-side orange and yellow lines. Finally, color any variable wires, such as signal wires, blue.

Coloring the wires on the wiring diagram in this way does several things. First, it forces you to determine what each wire in the diagram does, which helps you get the total picture. Second, it helps to organize your thoughts so that you can understand how electricity flows through the circuit. Third, it helps to keep you from losing your place or forgetting what a particular wire does. And fourth, it can give you confidence that you have properly diagnosed the problem when you know why the circuit is not working properly and exactly where the problem is located.

To use wiring diagrams to diagnose electrical circuits, follow the steps in **SKILL DRILL 10-5**.

N10004 Use wiring diagrams during the diagnosis (troubleshooting) of electrical/electronic circuit problems.

S10005 Use wiring diagrams to diagnose electrical circuits.

SKILL DRILL 10-5 Using Wiring Diagrams to Diagnose Electrical Circuits

1. Identify the correct wiring diagram for the vehicle and system circuit being repaired and print a copy.

2. Color each wire (using the wiring diagram's color code key) on the wiring diagram for the circuit that requires diagnosis. Note components, wire coding, and harness connectors.

3. Determine circuit test points and their location on the wiring diagram. Find the same test point on the vehicle and perform the appropriate electrical test.

4. Depending on the results of the test, continue to use the wiring diagram to guide you in performing additional tests on the circuit until the fault has been located.

Checking Circuits with a Test Light

N10005 Demonstrate proper use of a test light on an electrical circuit.

S10006 Check circuits with a test light.

Non-powered test lamps are useful in determining if electrical power is present in a part of a circuit. But you should always first test the test light on a known good power and ground before using it to test a circuit, as the bulb in the test light could be burned out, leading to a misdiagnosis of the circuit. If the test light illuminates, the two ends of the test light are touching both a power and a ground. If the light does not illuminate, the circuit is missing one or both of those elements or the test light is faulty. Test lights are great to grab to perform simple tests such as testing fuses. The test light lead can be quickly grounded and the probe end touched to each end of the suspect fuse (**FIGURE 10-50**). If both ends light, the fuse itself is good (but the fuse box terminal could be loose). If only one side of the fuse lights the test lamp, then the fuse is blown.

To avoid damaging the test light, make sure the circuit voltage you are testing does not exceed the test light's rating. Most test lights are rated for 6- or 12-volt systems, and using the light in a 24-volt system will usually blow the bulb. You should not use a test light to test SRS (supplemental restraint systems), as unintended deployment of the airbags could result. This is a costly mistake with the potential to injure you. Also, using a test light on a computer circuit designed for very small amounts of current flow can damage the circuit, due to the fact that the test light will draw an excessive amount of current through the control unit.

To check circuits with a test light, follow the steps in **SKILL DRILL 10-6**.

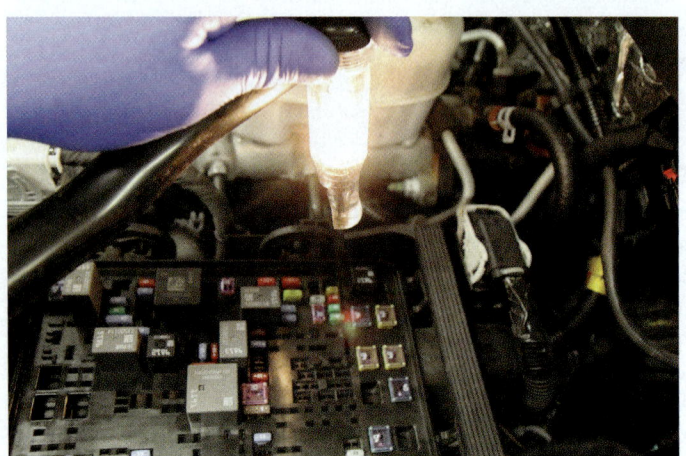

FIGURE 10-50 Testing fuses with a test light.

SKILL DRILL 10-6 Checking a Circuit with a Test Light

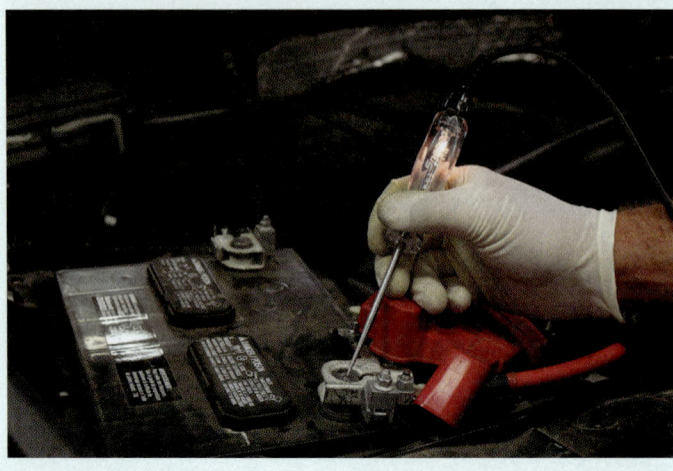

1. Connect the end of the light with the clip on it to the negative battery terminal. Touch the probe end of the test light to the positive battery terminal. The light should come on.

Continued

2. Connect the clip to any known good ground. A typical known good ground is any unpainted metal surface on the vehicle that is directly attached to the battery ground return system.

3. Place the probe on the terminal to be tested. If voltage is present, the light will come on.

Checking Circuits with a Fused Jumper Leads

Jumper leads can be used in a number of ways to assist in checking circuits. They can be created by the technician or purchased in a range of sizes, lengths, and fittings, or connectors (**FIGURE 10-51**). They are used to extend connections to allow circuit readings or tests to be undertaken with a DVOM, an oscilloscope, current clamps on fuses, relays, and connector plugs on components. In some circumstances, jumper leads may provide an alternate current or ground source for components being tested. Regardless of their application, it is

N10006 Use fused jumper wires to check operation of electrical circuits.

S10007 Check circuits with fused jumper leads.

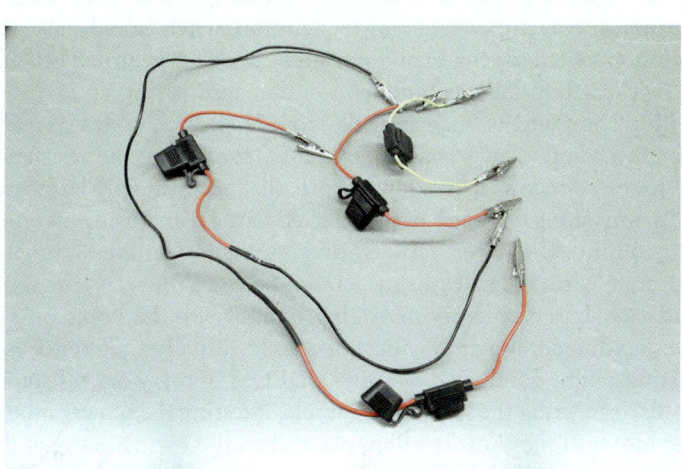

FIGURE 10-51 Fused jumper wires.

SAFETY TIP

Exercise care when you hook up any type of jumper leads, fused or unfused. If you hook them up to the wrong branch of a circuit, especially electronic circuitry, the resulting damage can be extensive. There is the old "magic smoke" saying: "Electrical and electronic components work off of the principle of magic smoke. Once the magic smoke is allowed to escape from the component, the component will never function again." Don't use jumper leads in a way that would let the "magic smoke" out of the circuit.

SKILL DRILL 10-7 Checking Circuits with Fused Jumper Leads

1. Identify the circuit to be checked and determine the fuse rating for the circuit.

2. Select appropriate jumper leads with the correct fuse rating.

3. Install the jumper lead into the circuit. Perform any required circuit checks. Never use a jumper lead to jump across a load; doing so bypasses circuit resistance and will likely cause excessive current to flow in the circuit.

---

important that the circuit remain protected by a fuse of the correct size. To determine the correct size of fuse for any particular application, refer to the manufacturer's information regarding the current rating for the circuit in question.

To check circuits with fused jumper leads, follow the steps in **SKILL DRILL 10-7**.

▶ Locating Opens, Shorts, Bad Grounds, and High Resistance

N10007 Demonstrate knowledge of the causes and effects from shorts, grounds, opens, and resistance problems in electrical/electronic circuits.

K10008 Locate opens, shorts, bad grounds, and high resistance.

DVOMs, test lamps, and simulated loads tend to be the tools used most often for locating opens, shorts, grounds, and high-resistance faults. Refer to the chapter Principles of Electrical Systems for more information on opens, shorts, grounds, and high resistance faults. An **open circuit** is a break in the electrical circuit where either the power or ground circuit has been interrupted. Most open circuits can be located by probing along the circuit at various points to test for power and by checking for an effective ground at the ground point. A systematic check of the circuit is required by first performing a voltage drop test on each side of the affected circuit. An open circuit will cause a voltage drop equal to the source voltage. Once the voltage drop is isolated to one side of the circuit, voltage drop testing can continue on that side, working the leads closer together in steps. Use your understanding of electrical systems to consider the most likely places for the open circuit, such as a blown fuse or a faulty switch. Don't forget that the load could also be open. If the voltage drop test on each side of the circuit is within specifications, use an ohmmeter to check whether the load is open, if possible. Some loads, such as diodes, cannot be tested with a standard ohmmeter. In this case, follow the manufacturer's diagnostic procedure.

Shorts, or **short circuits**, can occur anywhere in the circuit and can be difficult to locate, especially if the short is intermittent. A short is a circuit fault in which current travels along an accidental or unintended route and can be thought of as a shorter path for current to flow. The short may occur within the load, such as shorted relay windings, or it can be in the wiring, where a wire is shorted to ground or to supply voltage. A short will typically cause lower-than-normal circuit resistance. The low-resistance fault would cause an abnormally high current flow in the circuit and may cause the circuit protection devices, such as fuses or circuit breakers, to open the circuit. A short to supply voltage may cause the circuit to remain live even after the switch is turned off. For example, a short between a wire with power on all the time and a wire switched by the ignition switch would cause the circuit controlled by the ignition switch to remain on even after the switch is turned off. Just remember that shorts can be caused by faulty components or damaged wiring.

Shorts that happen within components, such as a relay coil, can usually be tested using a DVOM connected across the component terminals, and then comparing the resistance reading of the component to the service specifications. Shorts that occur in wire harnesses are usually best tested by disconnecting each end of the affected harness and using an ohmmeter to test for unwanted continuity between various wires. A reading on the ohmmeter when connected to two separate wires indicates a short circuit between them. A true short between wires would be indicated by a very low ohm reading, typically around 1 ohm or less.

Ground is a term often used in conjunction with shorts and is usually a reference to a short to ground. Initial testing can be conducted by using a DVOM to carry out resistance checks of the circuit or by disconnecting the load. For example, if testing the blower motor, first disconnect the blower motor. If the short is still in place, then the wiring between the

FIGURE 10-52 Wires exhibiting insulation that has been chaffed and burned, resulting in an electrical short.

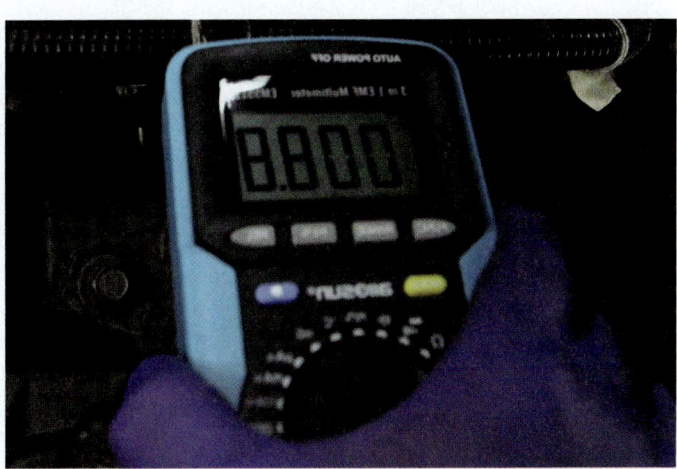

FIGURE 10-53 Circuit breaker and gauze gage short circuit tester in operation.

fuse or circuit breaker and the load must be at fault. To further narrow down the site of the short to ground, inspect the wiring harness, looking for obvious signs of damage, such as chaffing of the insulation or burning of the wire (**FIGURE 10-52**). Another test can be conducted by connecting a test lamp or buzzer in place of a fuse. Current will flow through the test lamp or buzzer and find a ground through the short. Parts of the circuit can then be disconnected along the wiring harness to narrow down the location of the short. Specialized short circuit detection tools are also available. They work by sending a signal through the wiring harness where a short is suspected. A receiving device is then moved along the wire loom and will indicate when a short is located. This type of device can be very useful in situations where it is difficult to access the wiring, such as within large wire looms or under vehicle trim. Another device that is useful for finding the location of a short circuit in a wiring harness involves a circuit breaker and a compass or gauze gauge. The circuit breaker is installed in place of the fuse; the circuit will then power up until the circuit breaker overloads and shuts off power, then the circuit breaker cools, and power is restored; this cycle continues as long as the circuit breaker is connected to power. This powering up and shutting down of the circuit causes magnetic lines of flux to be generated in the circuit up to the source of the short. To locate the short, simply move a compass along the circuit in question while the circuit breaker is cycling, and when you reach the point where the needle stops twitching with the circuit breaker's cycle, you have found the location of the short circuit (**FIGURE 10-53**).

Short to power refers to a condition where power from one circuit leaks into another circuit, causing erratic operation. A short to power situation usually causes strange electrical issues. In some cases, one or more circuits will operate when they should not. Or in the case of sensor wires, the incorrect signals caused by the short to power can cause the computer to make wrong decisions based on the faulty data. In this case, the engine, transmission, or other computer-controlled component can react strangely. Shorts to power are diagnosed first with a voltmeter to check for the unwanted voltage. Next, an ohmmeter is used to isolate the problem in the wire harness.

High resistance refers to a circuit where there is unintended resistance, which then causes the circuit to not perform properly. It can be caused by a number of faults, including corroded or loose harness connectors, incorrectly sized cable for the circuit current flow, incorrectly fitted terminals, and poorly soldered joints. The high resistance causes an unintended voltage drop in a circuit when the current flows. This drop reduces the amount of voltage that can be used by the load. The high resistance fault will also reduce the current flow in the circuit. The reduction in voltage and current to the load reduces the amount of electrical power to load (Power = Voltage × Current), affecting its performance. Unwanted high resistance can best be located by conducting a voltage drop test in the power and ground circuits.

If the high resistance is within the load, such as a relay coil, then the resistance can be checked with an ohmmeter and compared to specifications. Some devices, such as a fuel injector or ignition coil, may need further testing using an oscilloscope. In this way, the waveform can be evaluated, which can indicate issues that an ohmmeter cannot identify as easily.

Inspecting and Testing Circuit Protection Devices

N10008 Inspect and test fusible links, circuit breakers, and fuses; determine needed action.

S10008 Inspect and test circuit protection devices.

Protection devices are designed to prevent excessive current from flowing in the circuit. Protection devices like fuses and fusible links are sacrificial, meaning that if excessive current flows, they will blow or trip and have to be replaced. Circuit breakers can be reset. Once they trip, they either reset automatically or require a manual reset by pushing a button or moving a lever. Fuses, fusible links, and circuit breakers are available in various ratings, types, and sizes, and must always be replaced with the same rating and type.

In most vehicles, protection devices are situated in the power or feed side of the circuit. A blown or faulty fuse can be tested using a DVOM or test lamp. A good fuse will have virtually the same voltage on both sides. A blown fuse will typically have battery voltage on one side of the fuse and 0 volts on the other side. They can also sometimes be visually inspected. This may require the removal of the fuse from the fuse holder. The fusible element should be intact and, if measured by an ohmmeter, should have a very low resistance. The contacts on both the fuse and the fuse holder should be clean and free of corrosion and should fit snugly together.

To inspect and test circuit protection devices, follow the steps in **SKILL DRILL 10-8**.

Inspecting and Testing Switches, Solenoids, and Relays

S10009 Inspect and test switches, solenoids, and relays.

Inspection of electrical devices and wires usually starts with a visual inspection of the electrical circuit and is followed up with electrical testing. The visual inspection looks for breakage, corrosion, or deformity and includes examination of the insulation for any worn or melted spots. In the case of switches, solenoid contacts, and relay contacts, an electrical inspection is necessary. For example, all switches would require voltage drop testing to see if they operate properly without excessive resistance. Additionally, solenoid and relay contacts can wear out and produce excessive resistance, so performing a voltage drop test on them is a valid testing procedure. Some solenoids can be disassembled and visually inspected. In this case, the solenoid cap may be removed and the contacts visually inspected. Typically, if there is an excessive voltage drop across the contacts, the contacts will be pitted and burned. Measuring resistance also comes into play when a shorted relay or solenoid winding is suspected.

Manufacturers produce diagnostic flowcharts that guide the technician through a diagnostic sequence based on test results. In complex circuits, it is good practice to gather as much information as possible about the operation of the circuit and the customer concern. With that information, and the diagnostic flowchart, formulate a testing sequence for diagnosing the fault. Going through this process will help you to understand the problem and to identify possible causes, as well as a potential sequence of testing.

SKILL DRILL 10-8 Inspecting and Testing Circuit Protection Devices

1. Identify the protection device to be inspected and tested.
2. Conduct a visual inspection.
3. Set up a DVOM to read volts or use a test lamp.
4. Energize the affected circuit, if necessary.
5. Test for voltage on both sides of the circuit protection device. Determine and perform any necessary actions.

Wrap-Up

Ready for Review

- The digital multimeter (DMM), also known as a digital volt-ohm meter (DVOMs), is an electrical measuring tool frequently used to diagnose and repair electrical faults.
- A digital volt-ohm meter (DVOM) or digital multimeter (DMM) is called a digital meter because the meter gives a numerical reading on a digital display.
- An analog meter uses a needle that hovers over a series of scales, requiring the technician to determine the numerical value of the reading.
- DVOMs and test leads also should have a CAT rating listed on the front. CAT is short for "category." Each level, or CAT, is designed to work safely on higher-powered electrical systems.
- CAT ratings were not designed initially for automotive meters; most vehicles used low voltage.
- Hybrid vehicles typically require meters and test leads rated as CAT III.
- Failure to use a meter with the sufficient CAT rating could result in injury, due to the fact that the meter is not designed for use on high-voltage circuits.
- If you are working on high-voltage systems is that you need to wear a pair of certified and tested Class "0" rubber-insulated gloves, with leather protectors over top.
- Make sure that the leather protectors are at least 4" (101.60 mm) shorter than the rubber gloves, or electrical shock can result.
- It is important to always inspect the rubber high-voltage gloves for damage before each use.
- NEVER use rubber high-voltage gloves without leather protectors.
- DVOMs read very small quantities in the one-ten thousandths of a unit range up to very large quantities in the range of millions of units in the case of resistance measurements. It is not possible for DVOMs to effectively and accurately measure such ranges with only a single range or scale; they must have multiple ranges or scales.
- The DVOM screen can only display four or five digits. This means that symbols must be used to substitute for some of the digits.
- DVOMs are used to take many different electrical measurements on electrical circuits and are one of the first tools used when conducting electrical repair or diagnosis work.
- To set up a DVOM to take accurate measurements, you need to know if you will be measuring resistance, voltage, or current. You should also know the reading that you are expecting so you can be sure to set up the meter appropriately.
- The following steps describe how to set up a DMM:
  - Know what you are testing—volts, amps, or ohms.
  - Know the value you expect to be reading (specification).
  - Select leads and probes to suit the measuring task.
  - Connect the leads to the DMM.
  - Use the function switch to select the type of measurement to be undertaken (e.g., resistance, volts, or amps; DC or AC).
  - Select the correct meter range if you are using a manual range meter.
  - Connect the leads to the circuit being tested.
  - Read the meter display.
- In the min/max setting, the DVOM will record in memory the maximum and minimum reading obtained during the time the DVOM is connected to a source to take a reading.
- The hold function allows the display to be frozen.
- Common just means that the terminal is "common" to all of the functions of the meter.
- A probing technique is the way in which the DVOM probes are connected into circuits.
- The most common measurements taken with DVOMs are voltage, resistance, and current.
- To take voltage measurements, the probing lead (red) is connected to the volts/ohms, or V/Ω, terminal, and the common lead (black) is connected to the common, or COM, terminal of the DVOM.
- Select an appropriate range or auto range is selected on either AC or DC voltage, depending on the voltage to be measured.
- The probing lead is typically connected to the positive side of the circuit being tested, and the common lead to the negative side. Watch your screen. If the "+" or "−" is not what you were expecting, check the leads to verify they are connected the way you intended.
- Most DVOMs have a four-digit display, which limits the amount of information that can be displayed at any one time. For this reason, meters are capable of displaying data in multiple ranges.
- Auto ranging meters are capable of selecting the appropriate range needed to obtain the proper resolution for the electrical measurement that is being measured.
- When current loads need to be measured that exceed the DVOM's rating, an accessory called a low-amps probe can be used in conjunction with the meter to measure the current that is flowing through the circuit.
- Voltage drop is measured with a voltmeter while the circuit is operating. It is the potential difference between two points in a circuit.
- The sum of all the voltage drops in a series circuit equals the supply voltage, while the voltage drop across all parallel circuit branches is the same.

- Unwanted voltage drop becomes a problem if it becomes excessive and occurs in parts of the circuit other than the load.
- To measure voltage drop, the DVOM needs to be used on the voltage range.
- To perform this measurement process, you will need to set the function switch to "auto range volts DC" on the DVOM (some technicians prefer to use manual range, so they will not be fooled by auto range), and connect the black lead to COM and the red (probing) lead to V/Ω.
- Voltage drop can be measured across components, connectors, or cables.
- The probing lead of the DVOM is normally connected to the point in the circuit where you want to know the voltage.
- The more rust and/or corrosion that is present in an electrical circuit, the higher the resistance at these points will be.
- Ohm's law can be used in two ways to help in diagnosing electrical circuit faults. The first is by using it to perform the math to predict and verify measurements. The second method is by using the relationships it demonstrates to guide you through the diagnosis process.
- Vehicle wiring diagrams or schematics may be available as paper-based manuals, computer programs, or online resources; they are produced by manufacturers and some aftermarket publishing companies.
- Reading a wiring diagram is like reading a road map.
- Non-powered test lamps are useful in determining if electrical power is present in a part of a circuit. But you should always first test the test light on a known good power and ground before using it to test a circuit, as the bulb in the test light could be burned out, leading to a misdiagnosis of the circuit.
- Jumper leads can be used in a number of ways to assist in checking circuits. They can be created by the technician or purchased in a range of sizes, lengths, and fittings, or connectors.
- An open circuit is a break in the electrical circuit where either the power or ground circuit has been interrupted.
- Shorts, or short circuits, can occur anywhere in the circuit and can be difficult to locate, especially if the short is intermittent.
- A short is a circuit fault in which current travels along an accidental or unintended route and can be thought of as a shorter path for current to flow.
- Ground is a term often used in conjunction with shorts and is usually a reference to a short to ground. Initial testing can be conducted by using a DVOM to carry out resistance checks of the circuit or by disconnecting the load.
- Short to power refers to a condition where power from one circuit leaks into another circuit, causing erratic operation. A short to power situation usually causes strange electrical issues. In some cases, one or more circuits will operate when they should not.
- High resistance refers to a circuit where there is unintended resistance, which then causes the circuit to not perform properly.
- Protection devices are situated in the power or feed side of the circuit. A blown or faulty fuse can be tested using a DVOM or test lamp. A good fuse will have virtually the same voltage on both sides.
- A blown fuse will typically have battery voltage on one side of the fuse and 0 volts on the other side. They can also sometimes be visually inspected.
- Inspection of electrical devices and wires usually starts with a visual inspection of the electrical circuit and is followed up with electrical testing.
- The visual inspection looks for breakage, corrosion, or deformity and includes examination of the insulation for any worn or melted spots.
- Manufacturers produce diagnostic flowcharts that guide the technician through a diagnostic sequence based on test results.

Key Terms

digital multimeter (DMM) A test instrument with a digital display for measuring voltage, resistance, and current. Also called a digital volt ohm meter (DVOM).

digital volt-ohm meter (DVOM) A test instrument with a digital display for measuring voltage, resistance, and current. Also called a digital multimeter (DMM).

high resistance A term that describes a circuit or components with more resistance than designed.

hold function A setting on a DVOM to store the present reading.

min/max setting A setting on a DVOM to display the maximum and minimum readings.

open circuit A circuit that has a break that prevents current from flowing.

oscilloscope A test instrument that graphs voltage over time and displays the results on a screen.

probing technique The way in which test probes are connected to a circuit.

short circuit A condition in which the current flows along an unintended route.

short to power A condition in which current flows from one circuit into another.

Review Questions

1. A DMM is used to measure all of the following in a circuit *except*:
   a. voltage.
   b. resistance.
   c. current.
   d. capacitance.
2. All of the following statements are true *except*:
   a. Measuring source voltage is the first inspection point of any electrical system concern.
   b. The DMM on the volt setting is measuring the difference in voltage between the positive test lead and the ground.

c. Measuring available voltage and voltage drop in the circuit are performed as part of the diagnostic process.
d. Most modern DMMs can also measure frequency and temperature.

3. Which is the recommended range to be set on a DMM to get the most accurate reading?
   a. Highest range possible for the value being measured
   b. Lowest range possible for the value being measured
   c. Always the highest range on the DMM
   d. Always the lowest range on the DMM

4. Which category of DMM should be used while measuring three-phase utility connections?
   a. CAT I
   b. CAT II
   c. CAT III
   d. CAT IV

5. Which of the following is used to designate a reading of 400 megavolts?
   a. 400 mv
   b. 400 MV
   c. 40.0 Mv
   d. 400 µV

6. Which of the following precautions should be taken while measuring the resistance/current in a series circuit using a DVOM?
   a. The circuit should be disconnected.
   b. The components should be disconnected from each other.
   c. Class "0" gloves shouldn't be worn.
   d. The highest range should be selected to measure.

7. Conducting a voltage drop test in the power and ground circuits is best for locating:
   a. high resistance faults.
   b. open circuit.
   c. short circuit.
   d. parasitic draw.

8. What is the total resistance for two 2 ohm resistors connected in parallel?
   a. 1 ohm
   b. 2 ohms
   c. 3 ohms
   d. 4 ohms

9. What would be the voltage drop when a current of 0.2 A passes through a 100 Ω resistor?
   a. 200 V
   b. 20 V
   c. 2 V
   d. 0.2 V

10. When a corroded or loose connection introduces excessive resistance to the circuit:
    a. it results in less electrical power.
    b. it blows the fuse.
    c. it burns up a wire or component.
    d. it shortens the life of the load device.

ASE Technician A/Technician B Style Questions

1. Tech A says that all DVOMS are analog devices, and therefore cannot be used to effectively test modern automotive circuits. Tech B says that DVOMs serve many uses, and are required to perform many electrical diagnostics. Who is correct?
   a. Tech A
   b. Tech B
   c. Both A and B
   d. Neither A nor B

2. Tech A says that if a DVOM has a CAT III rating, it is able to safely test less voltage than a CAT I meter. Tech B says that meters and test leads should be inspected for damage before each use. Who is correct?
   a. Tech A
   b. Tech B
   c. Both A and B
   d. Neither A nor B

3. Tech A says that DVOMs can measure amps, ohms, and volts. Tech B says that if the leads are in the wrong test port of the DVOM, the internal meter fuse will be blown. Who is correct?
   a. Tech A
   b. Tech B
   c. Both A and B
   d. Neither A nor B

4. Tech A says that 100 mA is greater than 10MA. Tech B says that you should always be sure of a circuit's voltage before you test it to ensure that your meter has a high enough CAT rating for the test that you wish to perform. Who is correct?
   a. Tech A
   b. Tech B
   c. Both A and B
   d. Neither A nor B

5. Tech A says that a voltage-drop can occur if wires are improperly routed on an incline through the vehicle's chassis. Tech B says that a test light can be used to effectively perform a voltage drop test. Who is correct?
   a. Tech A
   b. Tech B
   c. Both A and B
   d. Neither A nor B

6. Tech A says if you need to test a circuit whose voltage exceeds the CAT rating of the meter, you can simply install a set of test leads with a higher CAT rating. Tech B says that the CAT rating of a DVOM and test leads is only applicable if the meter and leads are not damaged. Who is correct?
   a. Tech A
   b. Tech B
   c. Both A and B
   d. Neither A nor B

7. Tech A says that the MIN/MAX function of the meter can be used to test the meter's internal battery. Tech B says that the MIN/MAX function can be used as a low resolution flight-recording-device, which can be helpful in finding intermittent failures. Who is correct?
   a. Tech A
   b. Tech B
   c. Both A and B
   d. Neither A nor B
8. Tech A says that the HOLD function of a DVOM can be used as a mute button to isolate electrical interference from a circuit. Tech B says that the HOLD function allows you to pause circuit operation so that you can diagnose a circuit easier. Who is correct?
   a. Tech A
   b. Tech B
   c. Both A and B
   d. Neither A nor B
9. Tech A says that if the test leads' polarity is reversed in relation to the circuit being tested, a (−) will be displayed in form of the readout. Tech B says that is the meter is hooked up backward in a circuit, a fire can result. Who is correct?
   a. Tech A
   b. Tech B
   c. Both A and B
   d. Neither A nor B
10. Tech A says that the best way to test a connector is to probe the terminals from the front with the test leads of a meter. Tech B says that it is best to back probe an electrical connector. This eliminates that chance of damaging the delicate electrical contacts. Who is correct?
    a. Tech A
    b. Tech B
    c. Both A and B
    d. Neither A nor B

CHAPTER 11
Wires and Wiring Harnesses

NATEF Tasks

- **N11001** Repair data bus wiring harness.
- **N11002** Inspect, test, repair, and/or replace components, connectors, terminals, harnesses, and wiring in electrical/electronic systems (including solder repairs); determine needed action.
- **N11003** Use wiring diagrams during the diagnosis (troubleshooting) of electrical/electronic circuit problems.

Knowledge Objectives

After reading this chapter, you will be able to:

- **K11001** Explain the fundamentals of automotive wiring.
- **K11002** Use vehicle-specific wiring diagrams to understand circuit operation, wire sizes, components used, and decipher diagnostic information.
- **K11003** Describe wire maintenance and repair.

Skills Objectives

After reading this chapter, you will be able to:

- **S11001** Remove a terminal from its connector body.
- **S11002** Strip wire insulation.
- **S11003** Install a solderless terminal.
- **S11004** Apply heat-shring tubing.
- **S11005** Solder wires and connectors.

You Are the Technician

Today's vehicles contain miles of wire. With more onboard electronic devices than ever before, the electrical complexity of modern vehicles far surpasses even what was seen five years ago. As with all systems, these electrical networks can, and do, fail. In order to properly diagnose these faults, the technician must be able to properly read a wiring diagram, perform wiring harness repairs, and effectively diagnose electrical faults.

1. True or False: Modern vehicles have a bit less wiring than earlier vehicles due to more efficient circuit designs.
2. True or False: It is not required to read a wiring diagram if you have access to factory repair information.

K11001 Explain the fundamentals of automotive wiring.

S11001 Remove a terminal from its connector body.

N11001 Repair data bus wiring harness.

N11002 Inspect, test, repair, and/or replace components, connectors, terminals, harnesses, and wiring in electrical/electronic systems (including solder repairs); determine needed action.

▶ **TECHNICIAN TIP**

Under no circumstances should residential solid core wire be used for automotive applications. There are a couple of reasons for this. First, the solid copper core is more rigid than a multi-stranded copper wire, which makes it less flexible, and much more likely to break when flexed. Second, the insulation on most residential wiring is not designed to stand up to the chemicals, heat, and vibration that is encountered in the automotive environment. This can cause the insulation to fail, thereby creating an electrical short circuit.

▶ Introduction

Today's vehicles contain more wires, connections, peripheral devices, and computerized modules than was ever encountered by technicians a decade ago. These increasingly complex machines require a very specific skill set in order for them to be understood, diagnosed, and repaired properly. Yes, you read that properly. It is critical that today's technician is able to *understand* the vehicle that they are diagnosing; the technician must speak the language of the vehicle. When it comes to understanding electrical systems, a technician's success, or demise, can often be traced to their ability to read, understand, and use vehicle system wiring diagrams.

Think of the vehicle's wiring diagram as a map; in order for the technician to reach the destination (the diagnosis), they must be able to understand the map (the wiring diagram). Like all skills that we as human beings master, proficiency in reading wiring diagrams demands practice. The best advice that can be offered is to make a regular practice of finding vehicle wiring diagrams in the service information and studying them. With the information in this chapter, practice, and determination, the technician will be well on their way to becoming more accurate and efficient in their diagnoses.

▶ Wire Fundamentals

Wires and wiring harnesses are the arteries of the vehicle's electrical system, and as such they need to be kept in good condition, free of any damage or corrosion. They carry the electrical power and signals through the vehicle to control virtually all of the systems on a vehicle. As the application of technology in vehicles has increased, so has the number of wires and cables installed on these vehicles. Although wireless communications are used in some vehicle security systems, entertainment systems, and tire pressure monitoring systems, wires are still the dominant signal carriers in a vehicle. To help protect them and keep them organized, wires are bundled together in a wiring harness. A number of wiring harnesses are located throughout the vehicle.

Wires

Electrical wires are used to conduct current around the vehicle (**FIGURE 11-1**). Wire can also be referred to as cable, although cable typically refers to large-diameter wire. Automotive **wire** is commonly a multi-stranded copper core wrapped with seamless plastic insulation. Typically, the more strands that are present in the wire for a given wire size, the more flexible the wire is. Wire that contains many fine strands of copper, as opposed to a few larger strands, is referred as "hi-flex" wire. Such wire is less prone to breakage when flexed and so is typically used in places where the wire will encounter a great deal of motion. Copper is typically used as a conductor in wires as it offers low electrical resistance and remains flexible even after years of use. The insulation component of wires is designed to protect the wire and prevent leakage of the current flow so that it can get to its intended destination. **Ribbon cable** is a series of wires that are formed side by side and joined along the wire insulation; they are flat like a ribbon (**FIGURE 11-2**). Ribbon cable works well when several wires run from one component to another. The ribbon design groups them so they can be routed neatly and easily. Ribbon cable is often found inside computers and other electronic components. It is used for connections between printed circuits or between printed circuits and other components. Some wires, especially signal wires and communication wires, are shielded, which helps to prevent electromagnetic interference, also referred to as "noise."

Shielding

In certain locations within a vehicle and in environments where strong electromagnetic interference (EMI) is present, wiring harnesses are subjected to a situation where unwanted electromagnetic induction occurs. This interference is referred to as electrical

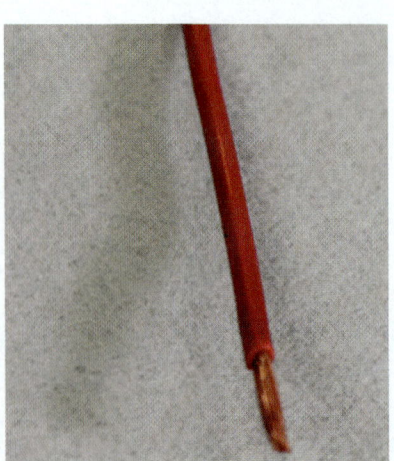

FIGURE 11-1 Standard wire (stripped).

FIGURE 11-2 Ribbon cable (stripped).

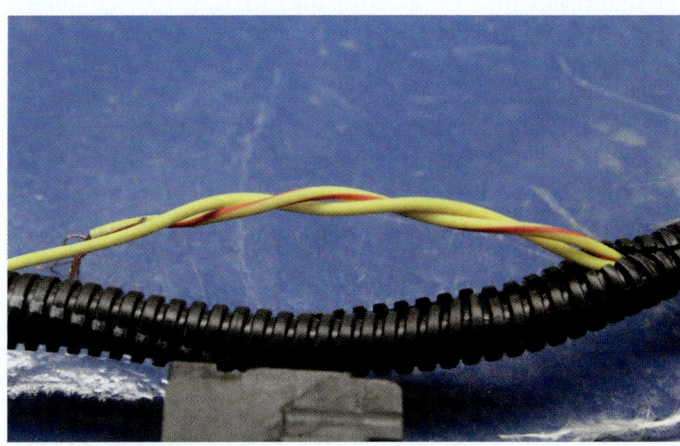

FIGURE 11-3 Twisted pair.

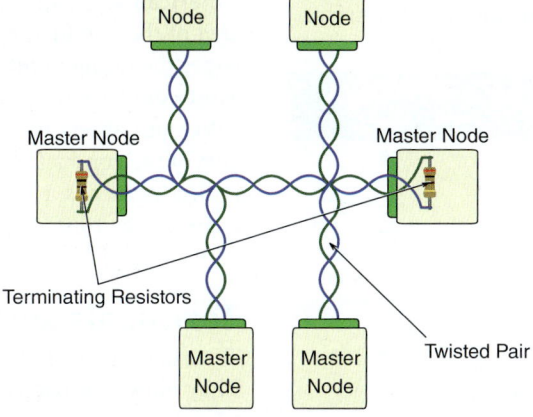

FIGURE 11-4 Wiring diagram showing a terminating resistor.

noise, or EMI noise. To prevent EMI noise, some vehicles use shielded wiring harnesses. The type of shielding used can be one of three forms: twisted pair, Mylar tape, and drains lines.

Twisted Pair

Twisted pair uses two wires delivering signals to a common component (**FIGURE 11-3**). The wires are uniformly twisted through the entire length of the harness and end at a terminating resistor (**FIGURE 11-4**). The twisted wires along with the terminating resistor have the effect of canceling any noise that occurs in the wires, reducing the loss of data in the transmitted signals. The controller area network (CAN) bus in a modern vehicle may use one or more twisted pairs to connect all the vehicle control units with one or more common data lines to share information.

Mylar Tape

Mylar tape is an electrically conductive material that is wrapped around a wiring harness inside the outer harness layer (**FIGURE 11-5**). Any noise that attempts to reach the wires inside the shield is absorbed by the Mylar, where it is conducted to ground via a ground connection. The shielding is important to prevent electrical noise from penetrating into the electrical wiring. If the harness is exposed, the Mylar will have to be re-wrapped so that noise cannot penetrate into the harness.

Drain Lines

A drain line is a non-insulated wire that is wrapped within a wiring harness (**FIGURE 11-6**). The drain wire is connected to ground at the wiring harness source end and conducts any noise to ground, negating the noise effect. If the drain wire is cut, it will be inoperative, so it is important that the wire not be cut or left disconnected.

Wire Sizes

Wire size is very important for the correct operation of electrical circuits. Selecting a wire gauge that is too small for an application will have an adverse effect on the operation of the circuit. This will cause voltage drop and poor performance, or, in extreme cases, the wire will get hot enough to melt the insulation. Selecting a wire gauge that is too large increases the cost, weight, and size of a wiring harness.

The resistance of a wire affects how much current it can carry. Even good conductors have a slight amount of resistance. The resistance of a wire is determined by its length, diameter, construction material, and temperature. The longer the wire and the smaller the diameter, the higher the resistance. The shorter the wire and the larger the diameter, the lower the resistance.

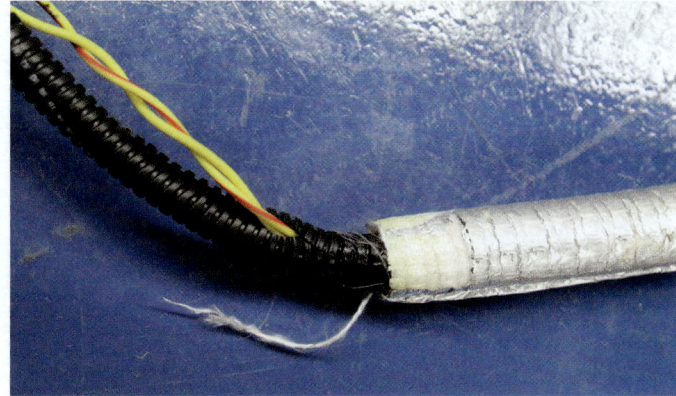

FIGURE 11-5 Mylar tape shielding on a wiring harness.

FIGURE 11-6 Wiring harness with an integral drain wire.

TABLE 11-1 Metric Wire Size Comparison and American Wire Gauge (AWG)

Metric Wire Sizes	AWG Wire Sizes
0.22	24
0.35	22
0.5	20
0.8	18
1.0	16
2.0	14
3.0	12
5.0	10
8.0	8
13.0	6
19.0	4
32.0	2

There are two scales used to measure the sizes of wires: the metric wire gauge and the **American wire gauge (AWG)** (**TABLE 11-1**). The metric system measures the cross-sectional area of the conductor in square millimeters. The AWG system uses a rating number; the larger the rating number, the smaller the wire and the lower its current-carrying capability. Most countries use the metric scale. American manufacturers are split, with some using AWG and others using the metric scale. The metric system measures the cross-sectional area of the conductor in square millimeters. A wire may be described in metric size as 5.0, indicating it has a cross-sectional area of 5.0 millimeters squared (mm²). It can also be expressed as 10/0.5, indicating there are 10 strands of wire, each with a cross-sectional area of 0.5 mm². The same system can be applied to the AWG rating.

Terminals and Connectors

Terminals are installed to the ends of wires to provide low-resistance termination to wires. They allow electricity to be conducted from the end of one wire to the end of another wire. In many cases, they allow the wires to be disconnected and reconnected. They come in many different types and sizes to suit various wire sizes and termination requirements (**FIGURE 11-7**).

Terminals can be installed as a single terminal on a wire or grouped together in a wiring harness with a connector housing, also called **wiring harness connectors** (**FIGURE 11-8**). Connector housings have male and female sides and are usually shaped so that they can be connected in only one way. They will often incorporate a locking mechanism so the plug cannot work itself loose over time. Connectors are widely used in automotive electrical systems (**FIGURE 11-9**). Electrical connectors rely on a tension fit between male and female terminals to achieve a low-resistance electrical connection. These terminals are held in place within a plastic connector body. The plastic shell ensures that terminals are mated positively, keeps them free of dirt and moisture, and protects them from vibration (**FIGURE 11-10**). There are many different types of connector bodies and terminals that are used in modern automotive electrical systems, but they all rely on a tension fit between the male and female terminals to provide a solid electrical connection.

Connector Types

The most common types of connectors are **push-on spade terminals** (**FIGURE 11-11**), **eye ring terminals** to accommodate screws (**FIGURE 11-12**), **butt connectors** (**FIGURE 11-13**), and **male and female terminals** that are designed to be separated and reconnected. Most terminals are the solderless crimp-on type, which require the use of special tools to crimp the terminal to the end of

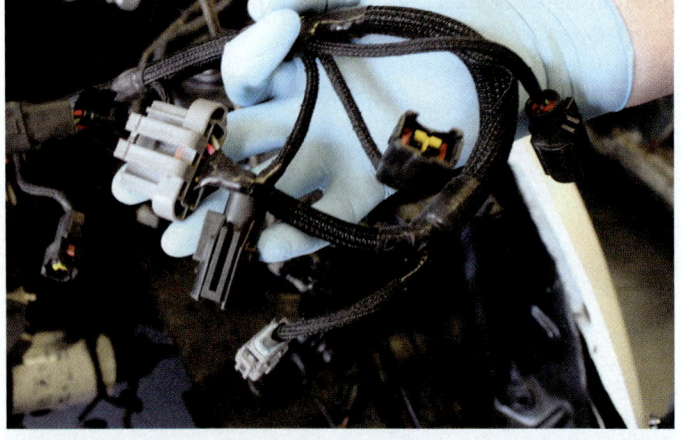

FIGURE 11-7 Terminals and connectors are installed to the ends of wires to provide low-resistance termination to wires.

FIGURE 11-8 Wiring harness connector.

FIGURE 11-9 Electrical connectors.

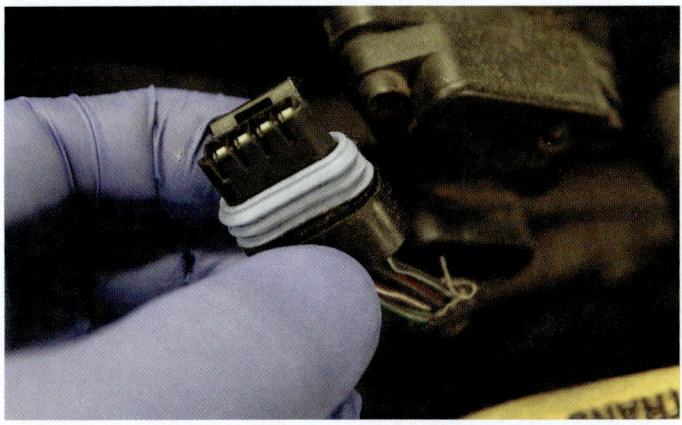

FIGURE 11-10 Weather-tight seals on an electrical connector.

FIGURE 11-11 Spade terminal.

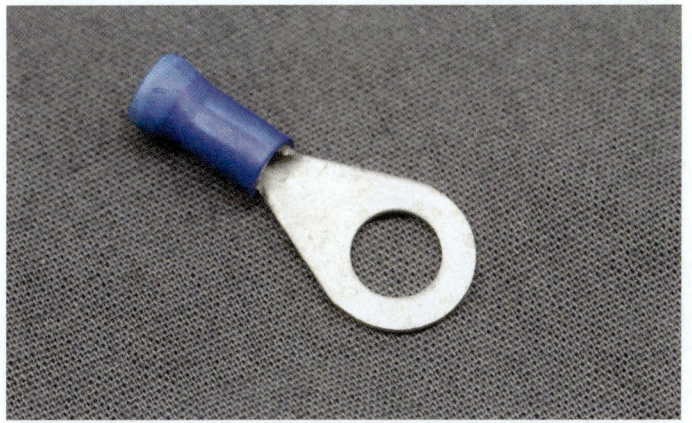

FIGURE 11-12 Eye ring terminal.

FIGURE 11-13 Butt connector.

the wire (**FIGURE 11-14**). They can be insulated or non-insulated (**FIGURE 11-15**). Some **solder-type terminals** are still used in automotive applications. These terminals require the use of electric or gas soldering irons, flux, and solder to make the connection.

Diagnosing Connector Failures

Because the environments in which connectors operate can be especially harsh, connector failure can occur over time. One means of connector failure occurs when the terminals lose their tension, causing a loose fit between the male and female terminals. Terminals can wear out when they are plugged and unplugged many times. They are subjected to the constant vibration that is present during vehicle operation, which is another degrading force. Terminals are easily damaged by improper testing techniques, such as probing the connector from the terminal end instead of properly back-probing the connection (**FIGURE 11-16**). Another cause of connector failure is a loose connection, which increases resistance and may also create a condition in which an electrical path is achieved through the connector on an intermittent basis. An increase in resistance of the terminal will cause the connector to get hot, and in some cases, melt (**FIGURE 11-17**). Melted connectors can also be caused by a component downstream from the connector that is drawing too much current. For this reason, it is a good practice to always test the current draw of components downstream from the connector when a melted connection is discovered. Another possible cause of a melted connector is that the plastic connector body is located too close to a heat source, such as an exhaust manifold. The body may melt, or it may become brittle and

FIGURE 11-14 Crimping pliers.

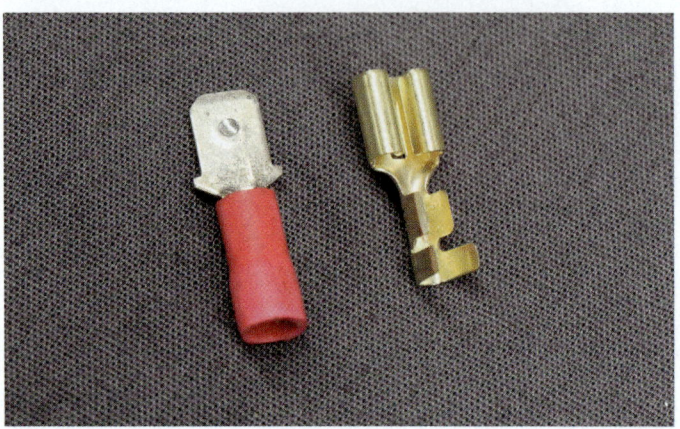

FIGURE 11-15 Insulated and non-insulated crimp on terminals.

FIGURE 11-16 Connector being probed at the terminals with meter leads, and a connector being properly back probed.

FIGURE 11-17 Connector that has melted down.

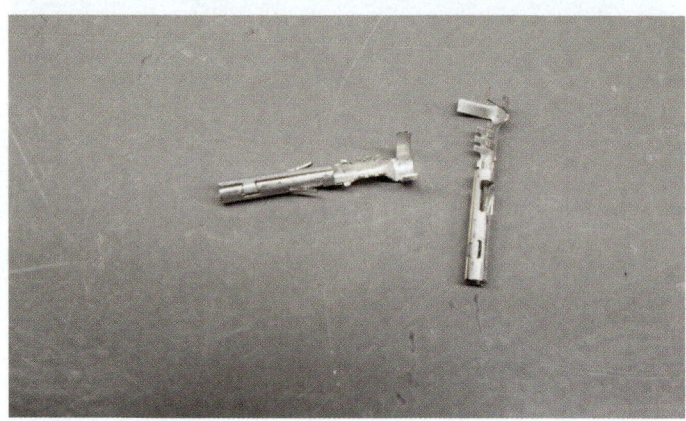

FIGURE 11-18 Electrical terminals showing locking tabs.

break. Each one of these occurrences can lead to connection problems and/or an electrical short, in the event that the terminals are allowed to contact surrounding metal.

Another common failure that is observed is when water and dirt make their way into the connector. Water intrusion will cause the terminals to oxidize, which increases their resistance. This will cause a voltage drop to be present across the connector. While it may seem easy to disconnect the connector to search for signs of corrosion, doing so may disturb any corrosion present that was impairing the electrical connection, and allow the connector to function properly for a short time. This will prevent the technician from diagnosing the root cause of the problem, due to the fact that the symptoms have temporarily subsided.

The most accurate way to test for a poor connection and/or high resistance across a connector is to back probe the wires going into a connector, and then back probe the corresponding wires coming out of the connector. This test will show the technician if electrical connection is being lost through the connector. If the desired signal is measured on one side of the connector, but not on the other, the connection is faulty, and must be diagnosed further. A voltage drop test performed across the connector while it is connected and the circuit is operating is also an effective way to identify high resistance.

In addition to a voltage drop test, an effective means of finding intermittent electrical connections is to perform the "wiggle test" on the wires and connectors of the system experiencing an intermittent problem. This test is performed by physically shaking the wires and connectors in the malfunctioning circuit while looking for signs of circuit failure. Once an area is found that causes a reaction in the circuit when it is wiggled, a loose connection has been located.

Disassembling Electrical Connectors

Terminals are held in a connector by the use of a lock tang (**FIGURE 11-18**).

To remove a terminal from its connector body, follow the steps in **SKILL DRILL 11-1**.

SKILL DRILL 11-1 Removing a Terminal from Its Connector Body

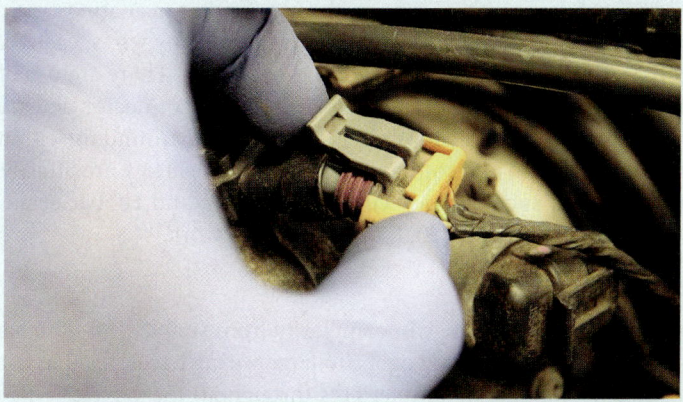

1. Separate the male and female connector halves by opening the lock.

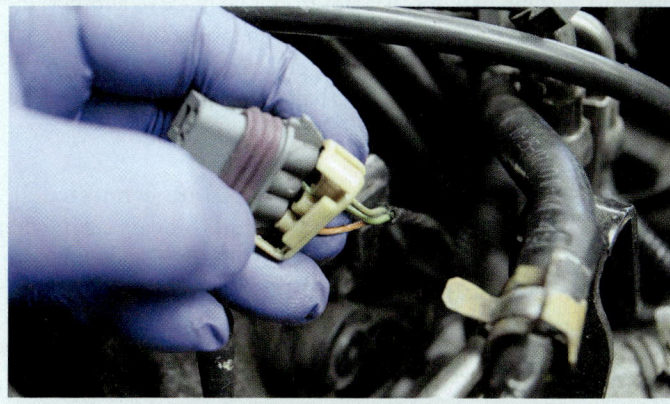

2. Inspect the connector, then release the terminal lock. It may be located on the front or back of the connector.

3. Using a connector disassembly tool, depress the lock tang on the terminal.

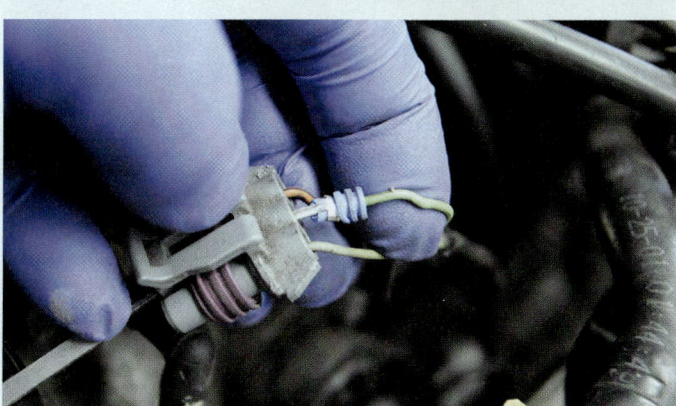

4. Carefully remove the terminal from the connector body. Great care must be taken when attempting to remove terminals from a connector body. If too much force is used, damage to the terminal or connector body may result.

Wiring Harnesses

Wiring harnesses, also known as wiring looms or cable harnesses, are used throughout the vehicle to group two or more wires together within a sheath of either insulating tape or tubing (**FIGURE 11-19**). Often, harnesses on modern vehicles contain many wires, each terminating at crimped terminals inserted into connector or harness plugs. There are usually a number of harnesses within the vehicle interconnecting with various connector plugs, as required to form the wiring system of the vehicle. Wiring harnesses run around the engine bay, through the dash and interior cabin, and to the rear of the vehicle. They are attached to the vehicle with harness fasteners such as body clips or wire ties, and rubber sealing grommets are used when the harness passes through the metal bodywork.

Wiring Harness Construction

In automotive applications, wiring harnesses range in complexity from simple, two-wire harnesses that provide electrical current to a set of fog lights, to large wiring harnesses that connect the various electronic systems, modules, and interfaces throughout the vehicle. Wiring harnesses are more than simply just a group of wires that run from one place to another; they contain splices, grounding points, protective sheathing, and more. The wires composing a wiring harness are also color coded, and each terminal, connector, device, module, splice, ground, and circuit has a unique identification number that allows the technician to locate components.

Splices and Grounds

Splices in a wiring harness indicate a point where multiple wires are connected at a central point. Splices are common points shared by multiple circuits, and as such, when multiple circuits are malfunctioning at the same time, splices should be located in the wiring harness and tested for proper electrical continuity. It is not uncommon for splice points to be the source of circuit malfunctions. Automotive manufacturers have released many Technical Service Bulletins that alert technicians of potential problems that they may encounter on various vehicles. Ground terminals in a wiring harness are points in which the wiring harness connects to ground. These connections are just as important to proper circuit operation as the power feeds of the circuit; they should not be overlooked during the diagnostic process. In fact, ground connections are usually more exposed to the elements than the power feeds in many wiring harness, and as such, are much more susceptible to loosening, corroding, and/or becoming damaged (**FIGURE 11-20**).

K11004 Use vehicle-specific wiring diagrams to understand circuit operation, wire sizes, components used, and decipher diagnostic information.

N11003 Use wiring diagrams during the diagnosis (troubleshooting) of electrical/electronic circuit problems.

▶ Wiring Diagram Fundamentals

Wiring diagrams, also known as electrical schematics or electrical diagrams, use abstract graphical symbols to represent electrical circuits and their connection or relationship to other components in the system (**FIGURE 11-21**). They are essentially a map of all of the electrical components and their connections (**FIGURE 11-22**). The wiring on modern

FIGURE 11-19 Typical wiring harness.

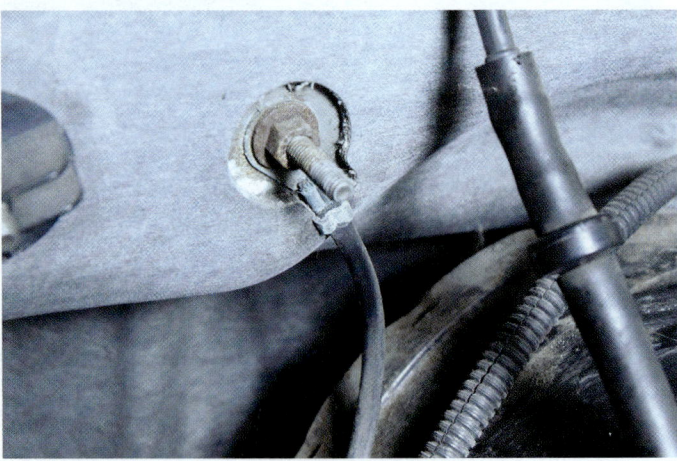

FIGURE 11-20 Ground terminal bolted to a body surface.

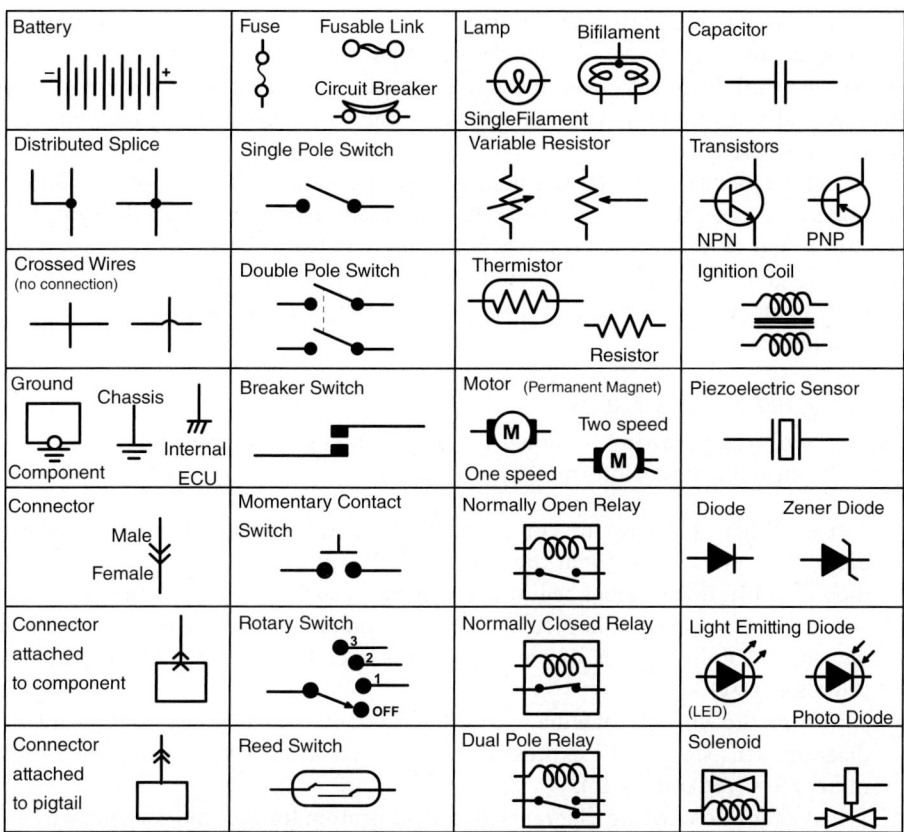

FIGURE 11-21 All of the commonly used automotive wiring harness symbols and the correlating labels.

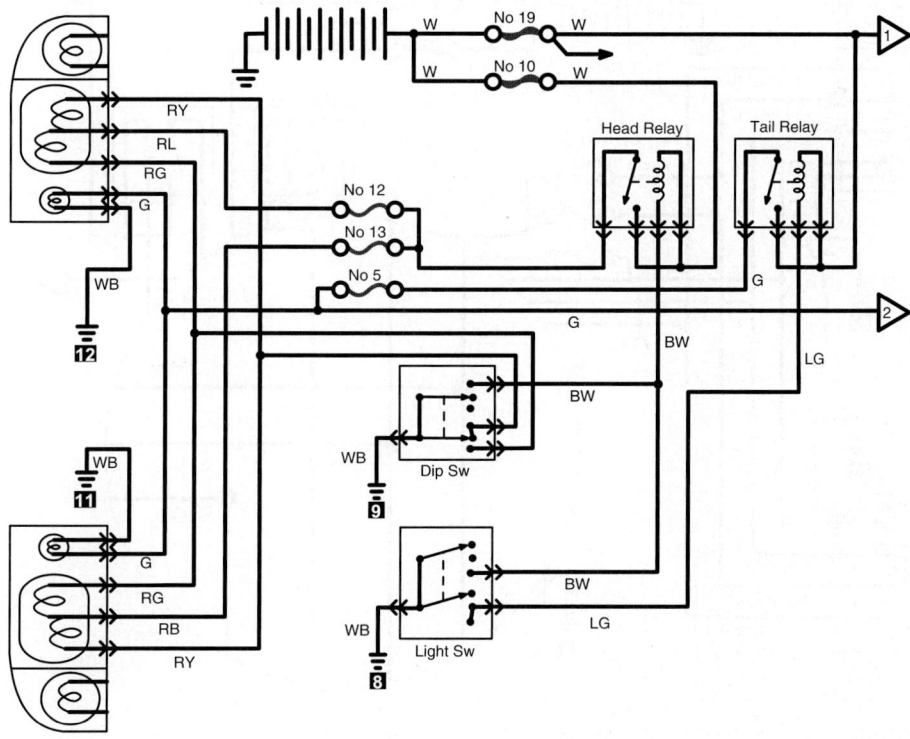

FIGURE 11-22 Electrical wiring diagram.

vehicles is very complex, with many wires and components interconnected. As such, a single wiring diagram for the whole vehicle would be very confusing and difficult to read. To make electrical diagnosis easier and more efficient, the wiring diagrams are split up by system and subsystem. This helps to reduce the complexity of each wiring diagram, making it easier to interpret.

Wiring diagrams contain large amounts of information in the form of lines and symbols. The technician will need to decode this information by interpreting the symbols and connections and relating them to the actual components on the vehicle. To assist in the understanding of wiring diagrams, manufacturers supply keys on the diagrams, which are lists of the component symbols and their names, wiring color codes, harness connectors, and pin numbers.

Wiring Harness Labels

In electrical diagrams, wiring is shown in straight lines with numbers, color codes, and letters. The meaning of this information is as follows:

- **Splices**: When two wires are connected electrically, the junction point on the wiring diagram shows a black dot at the point of their connection. Splices are identified by their identification number beginning with an "S", such as S102 (**FIGURE 11-23**). When two wires pass over one another in a wiring diagram, but they are not electrically connected, one of the wires is shown to pass over the other wire in a "U" shaped fashion (**FIGURE 11-24**).
- Wire Size: the wires in each diagram will have their size listed next to the wire on the diagram. This size can be expressed in either AWG, such as 22 gauge, or in square millimeters (**FIGURE 11-25**).
- Wire Color: Most wiring diagrams give an indication for the color of each wire in the circuit. The name of the color is usually abbreviated (**TABLE 11-2**). Whichever color code comes first is the primary color of the wire, and the codethat comes second is the color of the tracer that runs along the center of the wire (**FIGURE 11-26**).

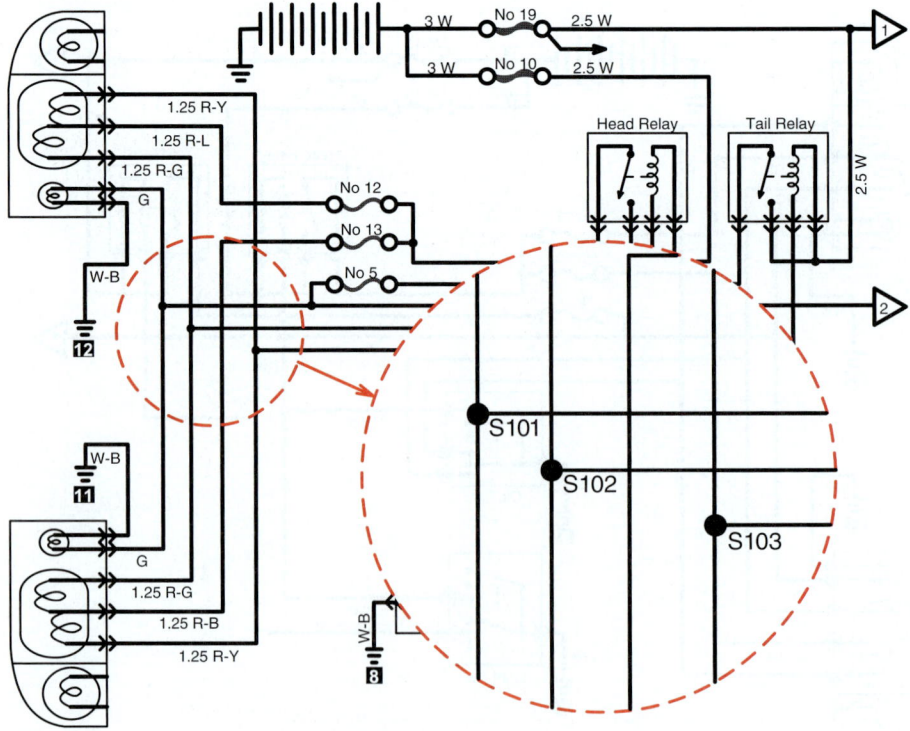

FIGURE 11-23 Wiring diagram showing a splice and the associated splice ID number.

Wiring Diagram Fundamentals 273

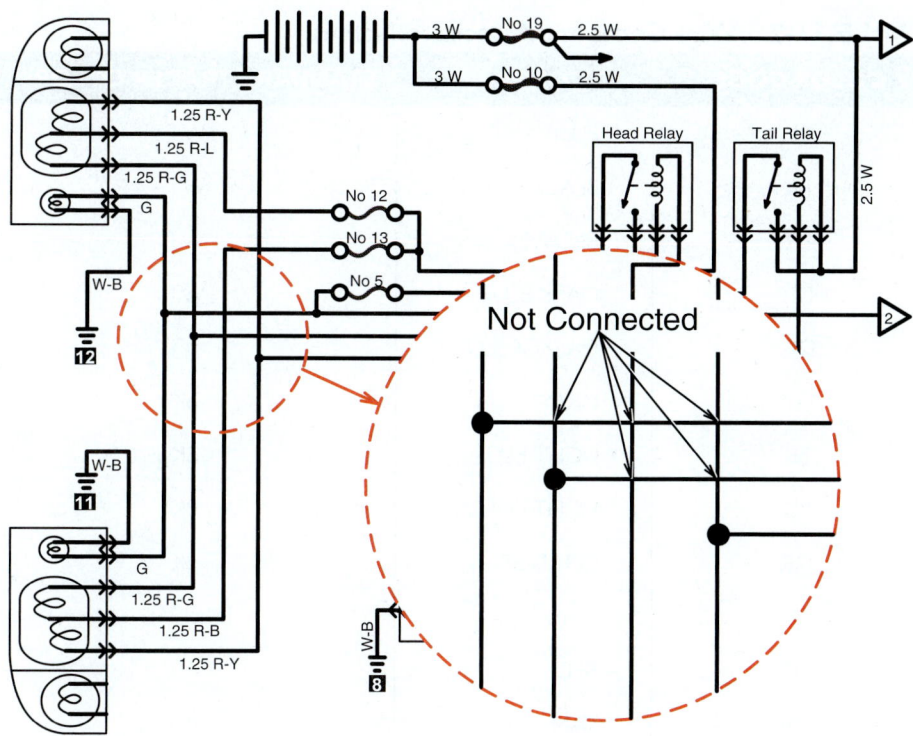

FIGURE 11-24 Wiring diagram showing two wires passing over one another that are not electrically connected.

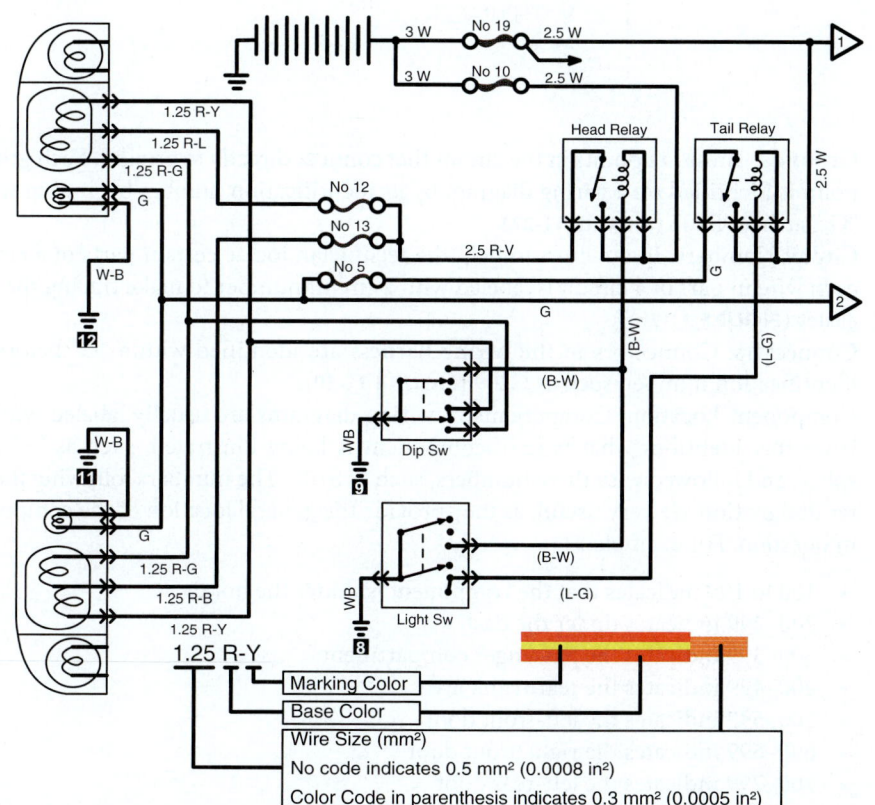

FIGURE 11-25 Wiring diagram showing the wire size.

TABLE 11-2 Wire Codes and Associated Names

COLOR CODE	COLOR	STANDARD TRACER COLOR
BL	BLUE	WT
BK	BLACK	WT
BR	BROWN	WT
DR	DARK BLUE	WT
DG	DARK GREEN	WT
GY	GRAY	BK
LB	LIGHT BLUE	BK
LG	LIGHT GREEN	BK
OR	ORANGE	BK
PK	PINK	BK or WT
RD	RED	WT
TN	TAN	WT
VT	VIOLET	WT
WT	WHITE	BK
YL	YELLOW	BK
*	WITH TRACER	

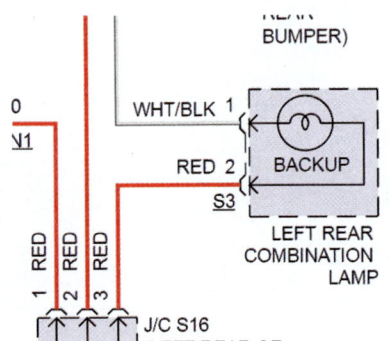

FIGURE 11-26 Wire colors represented by their associated color code.

- **Grounds**: These are points in the circuit that connect directly to ground. Each ground point is identified on a wiring diagram by its identification number beginning with a "G," such as G206 (**FIGURE 11-27**).
- **Circuit Numbers**: In an effort to help the technician locate certain parts of a circuit, each wire in part of a circuit is labeled with a circuit number to make tracing the wire easier (**FIGURE 11-28**).
- **Connectors**: Connectors in the wiring harness are identified with a "C" before the identification number, such as C209 (**FIGURE 11-29**).
- **Component Location**: Components in wiring diagrams are usually labeled with the letter that identifies what type of component is being illustrated, such as "S" for a splice, and followed with three numbers, such as S102. The numbers following the letter designation are very useful, as they provide the general location of the component in question. For example:
    - 100 to 199 indicates that the component is under the hood
    - 200–299 indicates under the dash
    - 300–399 indicates the passenger compartment
    - 400–499 indicates the rear trunk area
    - 500–599 indicates the left-front door
    - 600–699 indicates the right-front door
    - 700–799 indicates the left-rear door
    - 800–899 indicates the right-rear door
    - Even numbers are located on the right (passenger) side
    - Odd numbers are located on the left (driver's) side

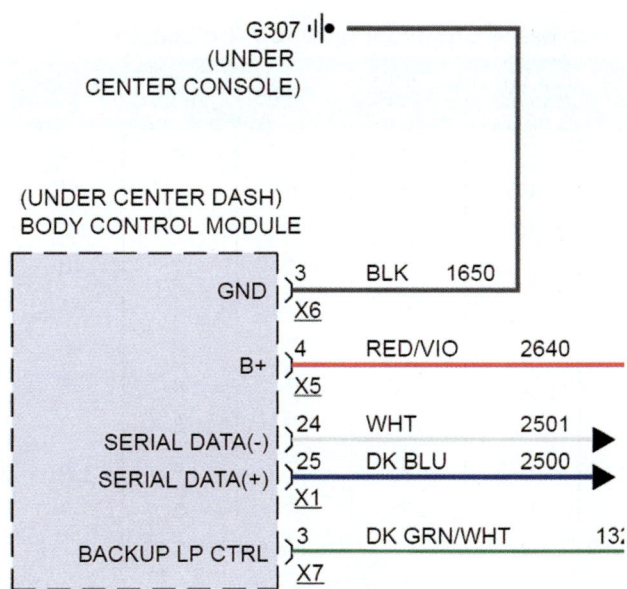

FIGURE 11-27 Wiring diagram showing a ground connection and associated ground ID number.

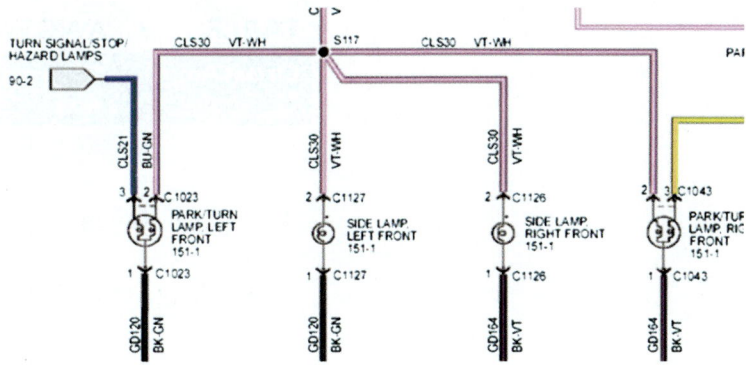

FIGURE 11-28 Wiring harness displaying a circuit number.

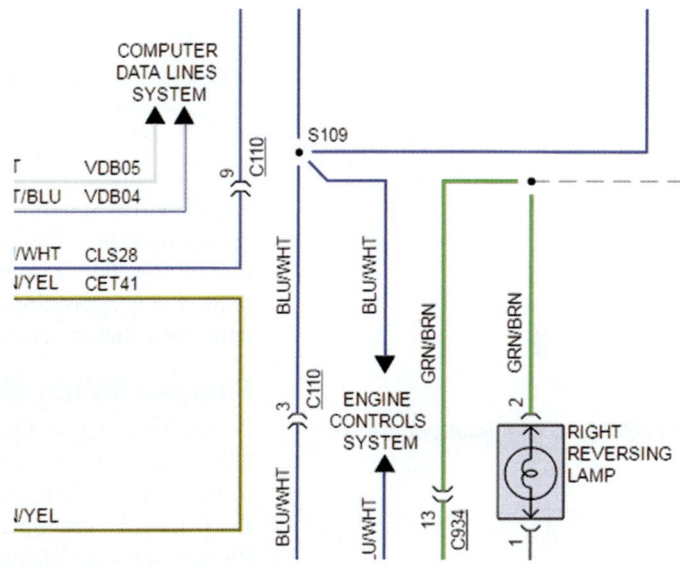

FIGURE 11-29 Wiring diagram showing a connector and associated connector ID number.

▶ Wire Maintenance and Repair

Wires are generally trouble free and long lasting. They can, however, be damaged. Generally speaking, any issues with wires are more likely to be with the terminals than with the wires themselves. Terminals can corrode, lose their tension, or push back up inside of the connector, leading to poor connections and voltage drops. When you suspect a problem with a wire, first inspect the ends. If no problems are found, look for mechanical damage to the wire or wiring harness itself. When a wire is damaged, it is usually due to one of several conditions. One possibility is that the wire has been physically broken. This could occur if a wire was not disconnected before removing a major assembly, such as the engine or transmission. Wire breakage is common in places where wires are subjected to a great deal of flex, such as passing through a door jam. Wires may become pinched between components, such as between the engine and transaxle when a clutch is replaced. The pinched wire can cause either a short circuit or an open circuit. Wires can also be misrouted so that they lay on a hot surface such as an exhaust manifold, which melts the insulation and causes the wire to short. In all of these cases, the problem may be spotted visually. A problem that may be harder to spot is wires that are melted together inside of a wiring harness, causing a short circuit. In this case the wiring harness would have to be opened up to find the source of the short. Another wiring problem that a visual inspection may not uncover is internal corrosion. This condition usually occurs when moisture seeps into the insulation of a wire. Internal corrosion will increase the resistance of the wire, so a voltage drop test of the wire in question is the most effective means of locating this type of fault.

K11003 Describe wire maintenance and repair.

Selecting the Proper Wire to Make Repairs

To select the correct wire gauge for any given application, it is best to refer to a wire chart. Manufacturers and standards organizations use wire gauge charts to define how much current each wire gauge can carry safely and efficiently. A vehicle uses a variety of wire sizes depending on the requirements of each particular circuit. The correct wire size for an application can be located on a wire size chart if you know the amperage of the circuit and the

TABLE 11-3 AWG Wire Sizes Based on Amperage and Wire Length*

Circuit Amps	Wire Length from Battery to Load						
	2 feet	5 feet	7.5 feet	10 feet	15 feet	20 feet	25 feet
2	20	20	20	18	18	18	16
5	18	18	18	18	16	14	14
8	18	16	16	14	14	12	12
10	16	16	16	14	12	12	10
12	16	16	14	14	12	12	10
15	16	16	14	12	10	10	8
18	16	14	12	12	10	8	8
20	14	14	12	10	10	8	8
25	14	12	12	10	8	8	6
30	12	12	10	10	8	6	6

*Chart is based on a maximum 0.4-volt drop per wire size, shorter distances are less than a 0.4-volt drop.

length of the wire. But be careful of the chart you use, since many of them allow up to a 10% voltage drop over the length of the wire, which is far more than is allowed in most automotive circuits. For example, if a 12-volt circuit is designed for a maximum current flow of 10 amps and is approximately 20′ (6.1 m) long, using the AWG table as a reference, you can determine that the correct wire gauge to use is 12 AWG (**TABLE 11-3**).

Proper Wire Stripping Techniques

An insulating layer of plastic covers the electrical wire used in automotive wiring harnesses. When electrical wire is joined to other wires or connected to a terminal, the insulation needs to be removed. Wire stripping tools come in various configurations, but they all perform the same task. The type of tool you use or purchase will depend on personal preference and the amount of electrical wire repairs you perform. A good pair of wire strippers removes the insulation without damaging the wire strands. Never use a knife or other type of sharp tool to cut away the insulation, as it often cuts away some of the strands of wire as well. This is known as ringing the wire; it effectively reduces the current-carrying capacity of the wire.

To strip wire insulation, follow the steps in **SKILL DRILL 11-2**.

S11002 Strip wire insulation.

SKILL DRILL 11-2 Stripping Wire Insulation

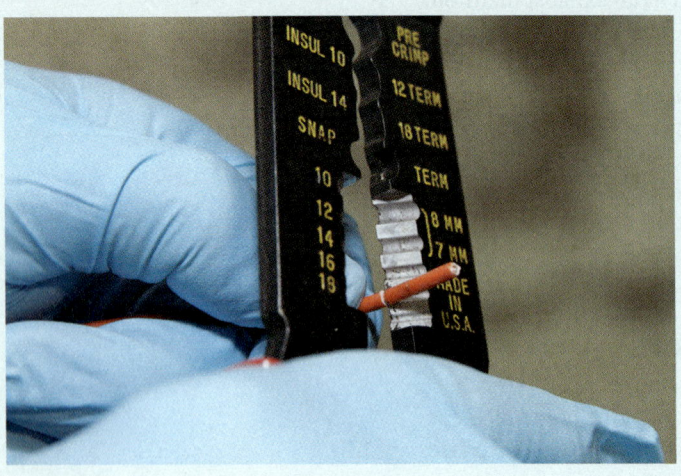

1. Choose the correct stripping tool.

Continued

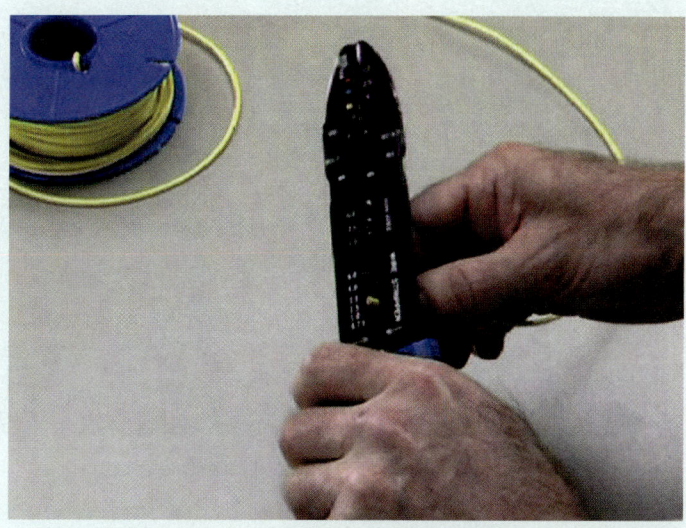

2. Select the hole that matches the diameter of the wire to be stripped. Place the wire in the hole and close the jaws firmly around it to cut the insulation.

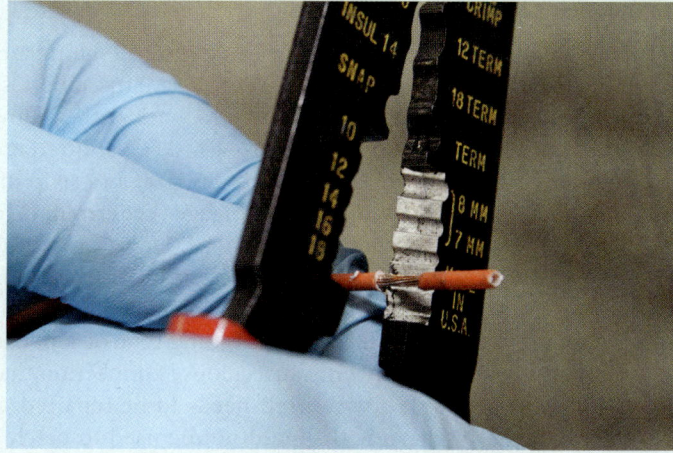

3. Remove the insulation. To keep the strands together, twist them lightly.

Improper Wire Stripping Techniques and Resultant Problems

Stripping the insulation from wire may seem like a simple and foolproof task, but if it's done improperly, wire damage and circuit failure can be the result. The most common mistake encountered when stripping wire insulation is using a stripper die that is smaller than the size of the wire's copper core. This results in strands of copper being cut from the core of the wire, effectively making the wire smaller at that point. Smaller wires are not able to carry as much current as larger wires and they have higher resistance. Therefore, cutting strands from the copper core during the insulation stripping process will result in a choke point in the circuit. The point will have higher resistance than the portions of the circuit that have not be damaged, and as such, this portion will have a lower current carrying capacity. Choke points can cause their portion of the circuit to overheat and possess the potential to spark an electrical fire.

Another commonly observed mistake is when too much insulation is stripped from the wire. This exposes more of the copper core than is required and leaves the wire open to water intrusion. Once water and oxygen are allowed to accumulate inside of a wire, the oxidation process begins to take place (**FIGURE 11-30**). The oxidation, or corrosion, that results creates a point of high resistance inside of the wire, which can make diagnosis difficult.

Using Solderless Terminals

Solderless terminals are used by the factory throughout the vehicle, primarily at connectors. They are quick to install and effective at conducting electricity across joints that are designed to be disconnected. Solderless terminals require a clean, tight connection.

S11003 Install a solderless terminal.
S11004 Apply a heat-shrink tubing.

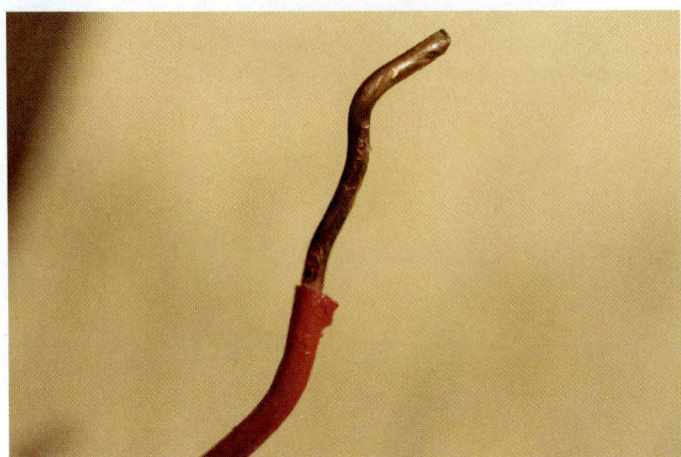

FIGURE 11-30 Corroded copper wire.

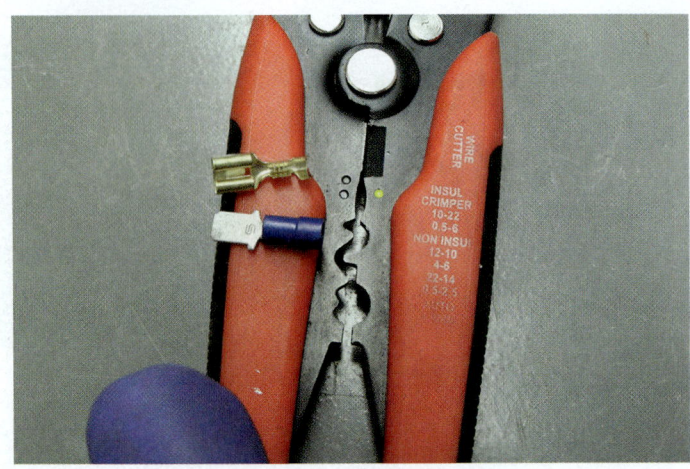

FIGURE 11-31 Insulated and non-insulated terminals, and their associated crimping tools.

It is important to make sure the wire and the connection are clean before attaching any terminals. You should use connections that match the size of the wire. Many solderless connectors are color coded for the size of wire they are designed to work with: yellow connectors are meant to be used for 12-10 AWG, blue are for 16-14 AWG, and red are for 22-18 AWG. It is very important to use the correct wire stripper size, and strip off only as much insulation as is required to allow the wire to fully engage the terminal. To keep the wires together after stripping them, give them a slight twist. Do not twist the wire too much; otherwise you risk a poor wire-to-terminal connection. Use the correct crimping tool for the connection; insulated terminals and non-insulated terminals each require a specific crimping tool (**FIGURE 11-31**). Using the wrong type of tool can compromise the connection's grip on the wire, which will cause a loose connection that will result in high resistance and voltage drop. If a wire itself needs to be repaired, it should be soldered back together; a solderless terminal should not be used to patch a broken wire.

To install a solderless terminal, follow the steps in **SKILL DRILL 11-3**.

SKILL DRILL 11-3 Installing a Solderless Terminal

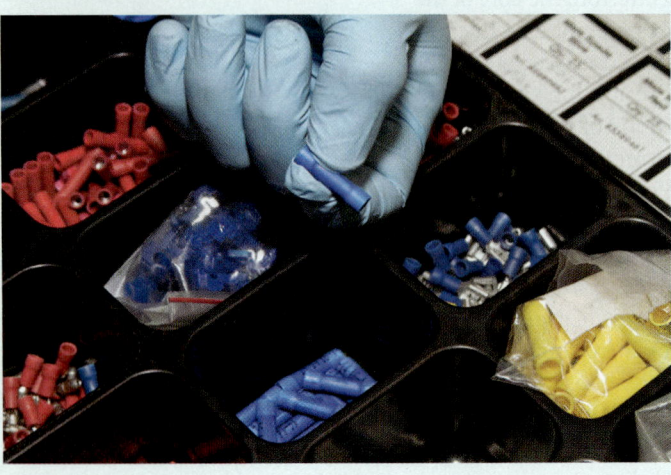

1. Make sure you have the correct size of terminal for the wire to be terminated and the terminal has the correct volt/amp rating. Remove an appropriate amount of the protective insulation from the wire.

Continued

Wire Maintenance and Repair 279

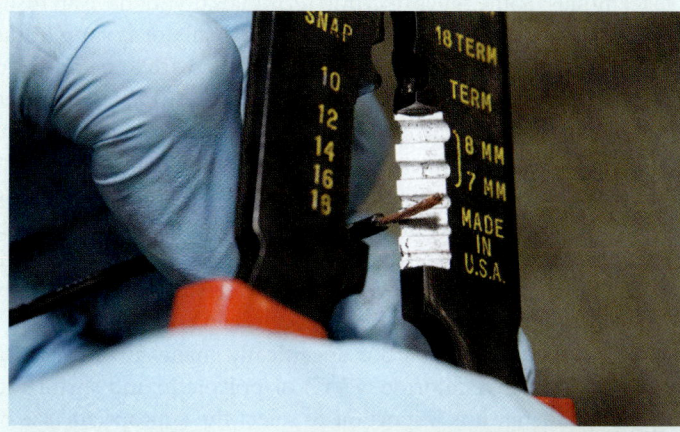

2. Lightly twist the wire strands and place the terminal onto the wire.

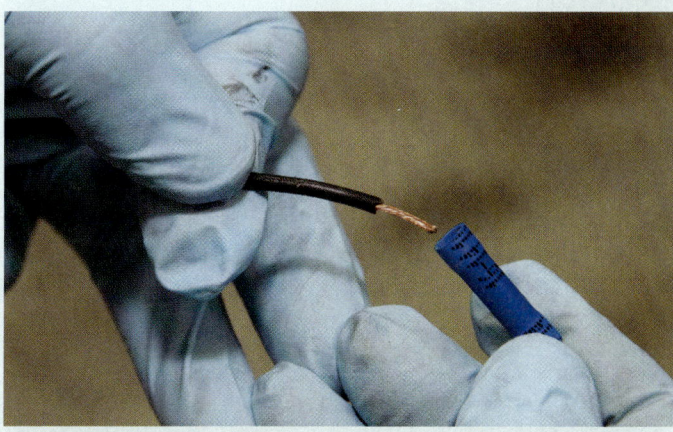

3. Use a proper crimping tool for the terminal you are crimping. Do not use pliers, as they have a tendency to cut through the connection. Select the proper anvil.

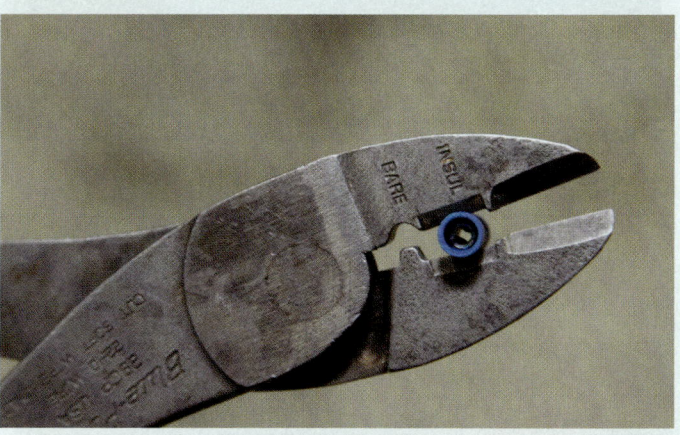

4. Crimp the core section first. Use firm pressure so that a good electrical contact will be made, but not excessive force, as this can bend the pin or terminal.

5. If crimping a non-insulated terminal, lightly crimp the insulation tabs so that they hold the insulation firmly.

A note about alternative terminals: Some types of crimp terminals do not have an insulation component fixed to them. These come in two parts, and the insulator is supplied as a separate component. In these cases, always make sure the core of the wire to be crimped extends through the core tabs in the terminal. Regardless of the type of terminal used, if the terminal is not contained in a connector body, it must be insulated and sealed to prevent current leakage and water intrusion. Connectors and wire repairs can be sealed with electrical tape, liquid rubber sealant, or heat-shrink tubing. Electrical tape is applied by taking a 2" (50.80 mm) length of it and, starting on one end of the connector, tightly wrapping it around the connector. As a note, if one piece of tape does not entirely cover the connection, apply another (**FIGURE 11-32**). Liquid rubber sealant, or liquid electrical tape, is simply brushed over the ends of the terminal and allowed to dry (**FIGURE 11-33**).

Heat-shrink tubing comes in many sizes that are specifically made for a range of wire sizes. It is usually made from polyvinyl chloride (PVC) or polyolefin and shrinks to approximately half of its original diameter. It is important to select the appropriate size of heat-shrink tubing for the job at hand.

To apply heat-shrink tubing, follow the steps in **SKILL DRILL 11-4**.

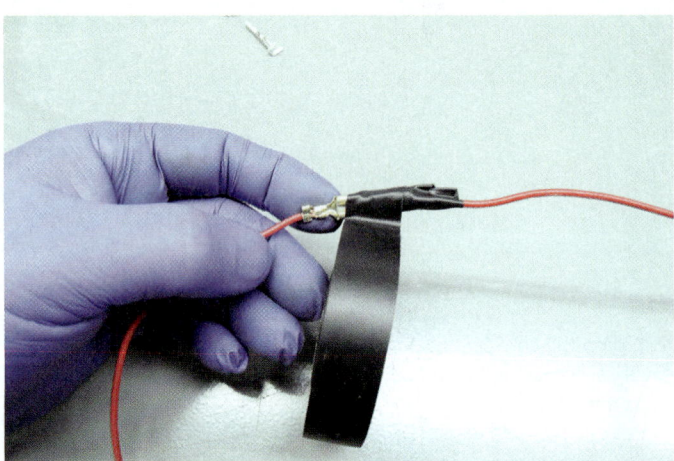

FIGURE 11-32 Electrical tape being applied to a connector.

FIGURE 11-33 Liquid electrical tape applied to a connector.

SKILL DRILL 11-4 Applying Heat-Shrink Tubing

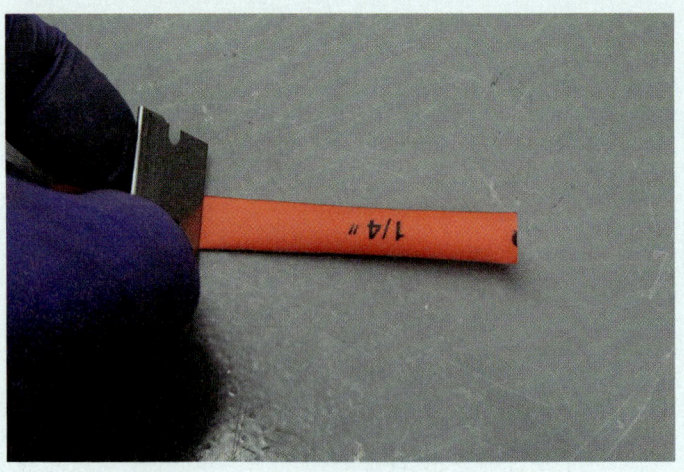

1. Cut a length of heat-shrink tubing that will extend 0.500" (12.70 mm) past each end of the connector if a butt splice is being used, or 0.500" (12.70 mm) past only the wire side when terminals are used that terminate the wire to a spade-type, pin-type, eyelet, or other form of terminal.

Continued

Wire Maintenance and Repair 281

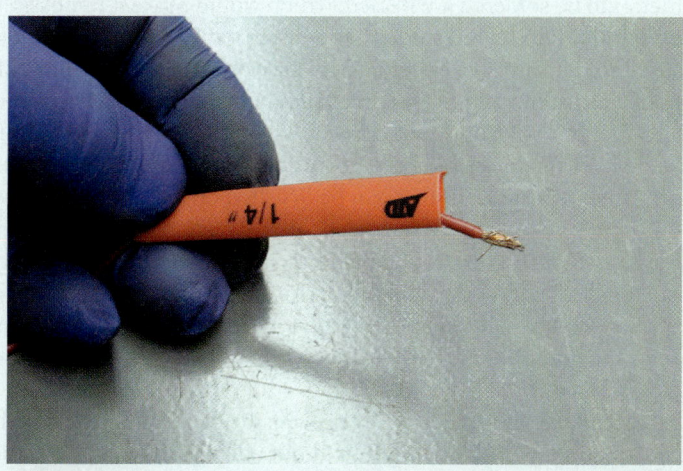

2. Place the length of heat-shrink tubing on the wire before performing the repair.

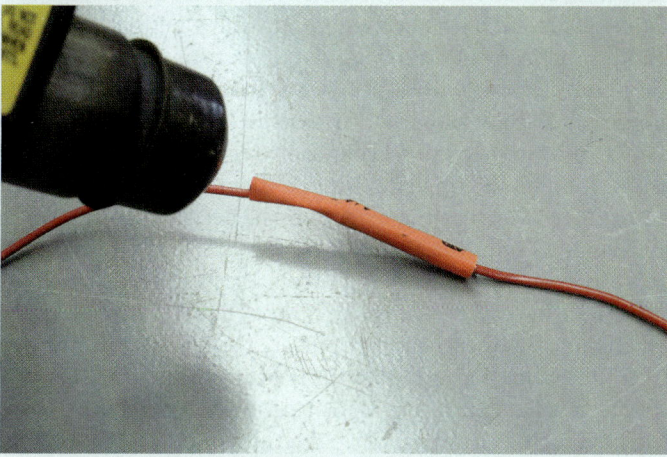

3. After the terminal has been installed, place the heat-shrink tubing over the repair or connector and use a heat-gun, or specially designed butane torch, to apply heat to the heat-shrink tubing. The tubing should tightly seal the wire and connection.

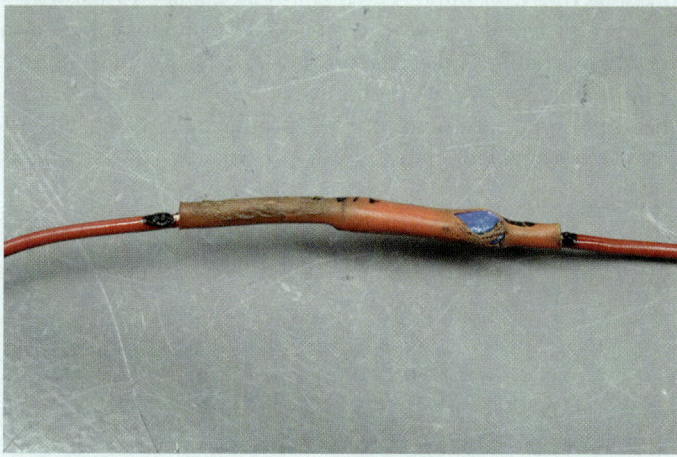

4. Never use an open flame to activate heat-shrink tubing, as it causes the tubing to become brittle and fail.

Soldering Wires, Connectors, and Terminals

Solder used in automotive electrical applications is an alloy typically made up of 60% tin and 40% lead. Solder needs to change easily from a solid state into a liquid state and then quickly return to a solid state. Solder is available as solid or flux cored. Solid solder requires an external flux to be applied in the soldering process. Flux is needed to prevent oxidation

S11005 Solder wires and connectors.

from occurring when the metals to be soldered are heated, which would make it impossible to achieve a properly soldered joint. Flux-cored solder has a bead of flux within the center of the solder. Flux in flux-cored solder can have either an acid base or a rosin base. Acid flux is designed to be used on nonelectrical metal joints such as radiators and must be removed after the soldering process so that the joint does not corrode. Rosin flux solder is used on electrical connections because it is much less likely to corrode the metals than acid flux. Acid flux and rosin flux also come in paste form that can be brushed onto the joint if using solid-core solder.

Solder is applied with a hot soldering iron. The soldering iron is heated electrically or by an external source such as a butane or oxyacetylene torch. The soldering iron tip absorbs heat that is then applied to the materials to be joined. Once they are hot enough, solder can be melted between the components. It solidifies as it cools, "gluing" the metal pieces together.

For a connection to be successful, the soldering iron needs to be clean and "tinned" (**FIGURE 11-34**). A dirty, oxidized tip on a soldering gun will not allow sufficient heat to be transferred into the wire, preventing a good solder joint from being formed (**FIGURE 11-35**). Cleaning the soldering iron may be as simple as heating the tip and wiping it on a damp cloth or sponge. If this method is ineffective in restoring the tip of the soldering iron to a bright silver appearance, additional cleaning will be necessary. To more aggressively clean the tip of the soldering iron, first ensure that the soldering iron is cold. You will need to mechanically remove the oxidation that has built up on the tip using either a file or a piece of sandpaper (**FIGURE 11-36**). Once the oxidation has been removed from the soldering iron tip, the tip must be tinned before the soldering iron can be used. The tinning process assists in transferring heat to the wire by leaving a small amount of liquid solder on the tip, which increases the surface area where the tip contacts the wires. To tin the soldering iron, the tip is heated and a small amount of solder is applied to the tip (**FIGURE 11-37**). Excess solder is removed with a cloth rag. Once the tinning process is complete, the soldering iron tip is heated and then applied to the wire so heat is transferred to the wire. The solder is then applied to the wire opposite the soldering iron (**FIGURE 11-38**). Once the wire is up to soldering temperature, it will melt the solder and pull the solder into the strands of wire, producing a strong, effective joint. If you apply too much heat to the wire, two negative results

FIGURE 11-34 Clean and tinned soldering iron tip.

FIGURE 11-35 Dirty, black, oxidized soldering iron tip.

FIGURE 11-36 Soldering iron tip that has been cleaned down to the base metal.

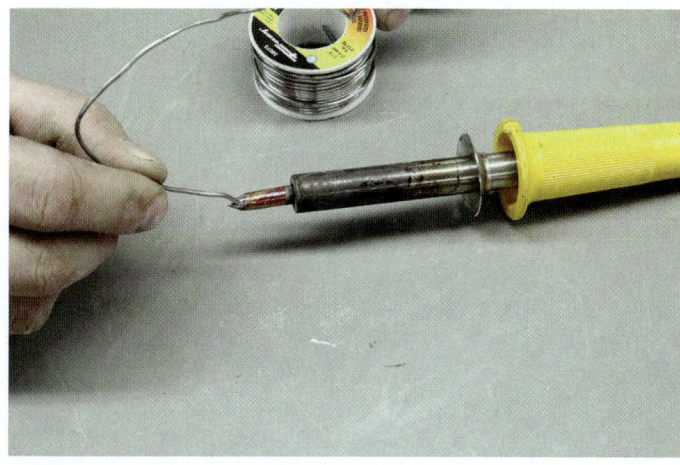

FIGURE 11-37 Soldering iron being "tinned."

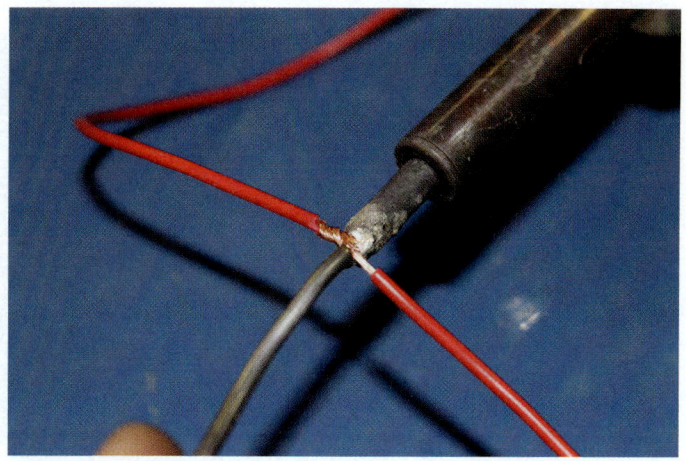

FIGURE 11-38 Soldering a wire connection.

can follow. First, the solder will be drawn too far up the strands of wire, creating a long, inflexible joint that is subject to breaking. The second problem is that the insulation may overheat and melt.

To solder wires and connectors, follow the steps in **SKILL DRILL 11-5**.

SKILL DRILL 11-5 Soldering Wires and Connectors

1. Safely position the soldering iron while it is heating up. While the soldering iron is heating, remove an appropriate amount of the protective insulation from the wires with wire strippers.

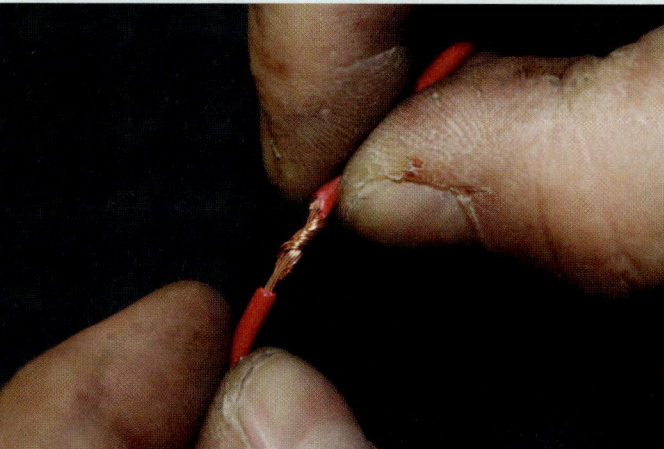

2. Twist the wires together to make a good mechanical connection between them.

Continued

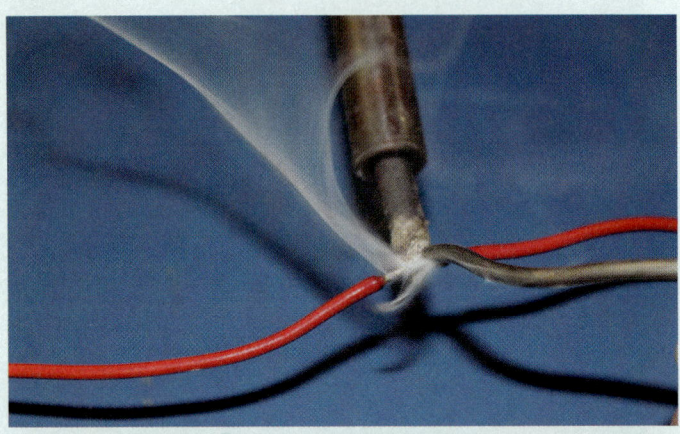

3. Tin the soldering iron tip and gently heat up the wires while placing the solder opposite of the soldering iron. Allow the solder to be drawn into the joint.

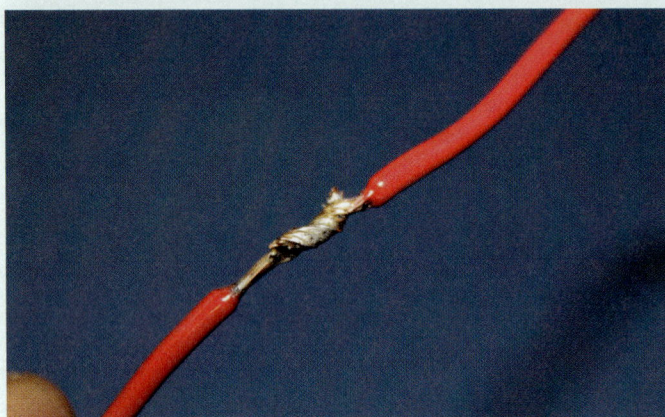

4. A good solder joint where the solder has been drawn in.

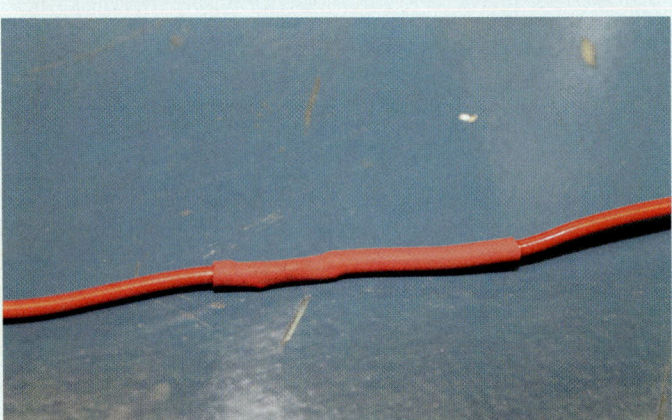

5. Once the electrical connection has been made and it has cooled enough for you to handle it, slide the insulator sleeve cover over the joint and use a heat gun to shrink the tubing around the joint.

6. To solder a wire to a terminal connector, it is best to crimp it in place as before and use the solder to "glue" the joint together. Place the heated iron onto the terminal to get it hot enough to melt the solder applied to the end of the crimped wire tabs. Some solder will be pulled between the terminal and the wire. Cover the terminal with heat-shrink tubing.

Wrap-Up

Ready for Review

- Wires and wiring harnesses are the arteries of the vehicle's electrical system, and need to be kept in good condition, free of any damage or corrosion.
- Copper is typically used as a conductor in wires as it offers low electrical resistance and remains flexible even after years of use.
- The insulation component of wires is designed to protect the wire and prevents leakage of the current flow so that it can get to its intended destination.
- Ribbon cable is a series of wires that are formed side by side and joined along the wire insulation; they are flat like a ribbon.
- To prevent EMI noise, some vehicles use shielded wiring harnesses. The type of shielding used can be one of three forms: twisted pair, Mylar tape, and drains lines.
- Twisted pair uses two wires delivering signals to a common component. The wires are uniformly twisted through the entire length of the harness and end at a terminating resistor. The twisted wires along with the terminating resistor have the effect of canceling any noise that occurs in the wires, reducing the loss of data in the transmitted signals.
- Mylar tape is an electrically conductive material that is wrapped around a wiring harness inside the outer harness layer. Any noise that attempts to reach the wires inside the shield is absorbed by the Mylar, where it is conducted to ground via a ground connection.
- A drain line is a non-insulated wire that is wrapped within a wiring harness.
- The drain wire is connected to ground at the wiring harness source end and conducts any noise to ground, negating the noise effect.
- Selecting a wire gauge that is too small for an application will have an adverse effect on the operation of the circuit. This will cause voltage drop and poor performance, or, in extreme cases, the wire will get hot enough to melt the insulation.
- The resistance of a wire is determined by its length, diameter, construction material, and temperature.
- The longer the wire and the smaller the diameter, the higher the resistance.
- The shorter the wire and the larger the diameter, the lower the resistance.
- There are two scales used to measure the sizes of wires: the metric wire gauge and the American wire gauge (AWG).
- The metric system measures the cross-sectional area of the conductor in square millimeters.
- The AWG system uses a rating number; the larger the rating number, the smaller the wire and the lower its current-carrying capability.
- Most countries use the metric scale.
- Terminals are installed to the ends of wires to provide low-resistance termination to wires and allow electricity to be conducted from the end of one wire to the end of another wire.
- Terminals can be installed as a single terminal on a wire or grouped together in a wiring harness with a connector housing, also called wiring harness connectors.
- Connector housings have male and female sides and are usually shaped so that they can be connected in only one way and often incorporate a locking mechanism so the plug cannot work itself loose over time.
- Connectors are widely used in automotive electrical systems.
- The most common types of connectors are push-on spade terminals, eye ring terminals to accommodate screws, butt connectors, and male and female terminals that are designed to be separated and reconnected.
- Most terminals are the solderless crimp-on type, which require the use of special tools to crimp the terminal.
- Terminals can wear out when they are plugged and unplugged many times. They are subjected to the constant vibration that is present during vehicle operation, which is another degrading force.
- An increase in resistance of the terminal will cause the connector to get hot, and in some cases, melt.
- Melted connectors can also be caused by a component downstream from the connector that is drawing too much current.
- It is a good practice to always test the current draw of components downstream from the connector when a melted connection is discovered.
- Another possible cause of a melted connector is that the plastic connector body is located too close to a heat source, such as an exhaust manifold.
- A voltage drop test performed across the connector while it is connected and the circuit is operating is also an effective way to identify high resistance.
- An effective means of finding intermittent electrical connections is to perform the "wiggle test" on the wires and connectors of the system experiencing an intermittent problem.
- Wiring harnesses, also known as wiring looms or cable harnesses, are used throughout the vehicle to group two or more wires together within a sheath of either insulating tape or tubing.
- Splices in a wiring harness indicate a point where multiple wires are connected at a central point. Splices are common points shared by multiple circuits, and as such, when multiple circuits are malfunctioning at the same time, splices should be located in the wiring harness and tested for proper electrical continuity.

- Ground terminals in a wiring harness are points in which the wiring harness connects to ground. These connections are just as important to proper circuit operation as the power feeds of the circuit; they should not be overlooked during the diagnostic process.
- Ground connections are usually more exposed to the elements than the power feeds in many wiring harness, and as such, are much more susceptible to loosening, corroding, and/or becoming damaged.
- Wiring diagrams, also known as electrical schematics or electrical diagrams, use abstract graphical symbols to represent electrical circuits and their connection or relationship of other components in the system.
- Wiring diagrams contain large amounts of information in the form of lines and symbols. The technician will need to decode this information by interpreting the symbols and connections and relating them to the actual components on the vehicle.
- When two wires pass over one another in a wiring diagram, but they are not electrically connected, one of the wires is shown to pass over the other wire in a "U" shaped fashion.
- The wires in each diagram will have their size listed next to the wire on the diagram. This size can be expressed in either AWG, such as 22 gauge, or in square millimeters.
- Most wiring diagrams give an indication for the color of each wire in the circuit.
- The name of the wire color is usually abbreviated.
- Each ground point is identified on a wiring diagram by its identification number beginning with a "G," such as G206.
- Each wire in part of a circuit is labeled with a circuit number to make tracing the wire easier.
- Connectors in the wiring harness are identified with a "C" before the identification number, such as C209.
- Components in wiring diagrams are usually labeled with the letter that identifies what type of component is being illustrated, such as "S" for a splice, and followed with three numbers, such as S102.
- To select the correct wire gauge for any given application, it is best to refer to a wire chart.
- Manufacturers and standards organizations use wire gauge charts to define how much current each wire gauge can carry safely and efficiently.
- Wire stripping tools come in various configurations.
- The type of tool you use or purchase will depend on personal preference and the amount of electrical wire repairs you perform.
- A good pair of wire strippers removes the insulation without damaging the wire strands.
- Never use a knife or other type of sharp tool to cut away the insulation, as it often cuts away some of the strands of wire as well.
- You should use connections that match the size of the wire.
- Many solderless connectors are color coded for the size of wire they are designed to work with: yellow connectors are meant to be used for 12-10 AWG, blue are for 16-14 AWG, and red are for 22-18 AWG.
- It is very important to use the correct wire stripper size, and strip off only as much insulation as is required to allow the wire to fully engage the terminal.
- Solder used in automotive electrical applications is an alloy typically made up of 60% tin and 40% lead.
- Solder needs to change easily from a solid state into a liquid state and then quickly return to a solid state.
- Solder is available as solid or flux cored.
- Solid solder requires an external flux to be applied in the soldering process.
- Flux is needed to prevent oxidation.
- Solder is applied with a hot soldering iron. The soldering iron is heated electrically or by an external source such as a butane or oxyacetylene torch.
- The soldering iron tip absorbs heat that is then applied to the materials to be joined. Once they are hot enough, solder can be melted between the components. It solidifies as it cools, "gluing" the metal pieces together.

Key Terms

American Wire Gauge (AWG) Standardized wire gauge used in North America. The higher the AWG number, the smaller the wire is, and the lower the current carrying capacity.

butt-connectors Crimp-type connectors that can be used to electrically join two pieces of wire.

female terminals Terminals that accept the protruding male terminals. These are easily damaged by improper probing with test lights and meter leads.

ground A point where the circuit connects to the negative side of the electrical system.

male terminals Terminals that protrude into the receiving terminal.

ribbon cable Cable with many conducting wires running parallel to one another.

solder A metal with a low melting temperature that is used to fuse metal components.

solder-type terminals Terminals that are soldered on instead of being crimped.

splice A point in the wiring harness where multiple wires are connected electrically.

terminals Metal connectors that are attached to wire ends. They are used to create electrical connections that can be disconnected and reconnected.

wire Flexible metal, usually made of copper and wrapped in insulation; used to transmit electricity within circuits.

wiring harness The network of wires, connectors, and terminals that make up an electrical circuit.

Review Questions

1. All of the following statements are true *except*:
   a. Wires are the dominant signal carriers in a vehicle.
   b. A number of wiring harnesses are located throughout the vehicle.
   c. As technology has increased, usage of wires and cables has declined.
   d. Wires and wiring harnesses are the arteries of the vehicle's electrical system.
2. Which of the following describes the characteristic of a "hi-flex wire" in automotive wiring?
   a. A wire that contains many fine strands of copper.
   b. A wire with a single thick strand and with high flexibility.
   c. A group of small wires that run next to each other.
   d. It is used in places where there is less motion.
3. All of the following statements with respect to ribbon cables are true *except*:
   a. They are a series of wires that are formed side by side.
   b. They work well when several wires run from one component to several other components.
   c. The design helps them to be routed neatly and easily.
   d. They are often found inside computers and other electronic components.
4. Which of the following type of shielding is used in wiring to prevent electromagnetic interference (EMI)?
   a. Special plastic coated wires
   b. Silicon tape
   c. Nylon tape
   d. Mylar tape
5. What is the equivalent AWG size for a 0.5 metric standard with 10 wires in a bundle?
   a. 20
   b. 2
   c. 0.2
   d. 2.0
6. A wire can melt due to all of the following reasons *except*:
   a. an increase in the resistance of the terminal.
   b. a downstream component that is drawing too much current.
   c. when the plastic connector body is located too close to a heat source.
   d. when the distance between the components it connects is long.
7. Choose the correct statement.
   a. Wire size does not affect the operation of electrical circuits.
   b. The longer the wire and smaller the diameter, the lower the resistance.
   c. The shorter the wire and larger the diameter, the higher the resistance.
   d. The resistance of a wire affects how much current it can carry.
8. All of the following can be cited as reasons for connector failures *except*:
   a. terminals being plugged and unplugged many times.
   b. tension fit between the male and female terminals.
   c. being subjected to constant vibration that is present during vehicle operation.
   d. improper testing techniques.
9. Choose the correct statement.
   a. Terminals are installed to the ends of wires to provide high-resistance termination.
   b. A drain line is an insulated wire that is wrapped within a wiring harness.
   c. Acid flux and rosin flux also come in powder form that can be brushed onto the joint if using solid-core solder.
   d. Wiring diagrams use abstract graphical symbols to represent electrical circuits and their connection or relationship to other components in the system.46
10. Which of the following is not labeled in a wiring diagram?
    a. Splice
    b. Current-carrying capacity
    c. Circuit number
    d. Connectors

ASE Technician A/Technician B Style Questions

1. Tech A says that wiring harnesses can contain miles of wire. Tech B says that wiring harnesses are complex networks of conductors and components, and require a wiring diagram to properly understand and diagnose their function. Who is correct?
   a. Tech A
   b. Tech B
   c. Both A and B
   d. Neither A nor B
2. Tech A says that when using the AWG wire scale, the larger the number, the higher the current-carrying capacity. Tech B says that when using the metric wire scale, the wire size is its cross-section in millimeters; therefore, the larger the number assigned to the wire, the higher the current carrying capacity of the wire. Who is correct?
   a. Tech A
   b. Tech B
   c. Both A and B
   d. Neither A nor B
3. Tech A says that twisted-pair wires are mainly used on luxury vehicles, and the twist in the wire allows the wiring harness to be much small, and allows the designers to design more stylish interiors. Tech B says that twisted-pair wires are used in communication networks, and the twisting of the wires provides a form of shielding from electromagnetic interference. Who is correct?
   a. Tech A
   b. Tech B
   c. Both A and B
   d. Neither A nor B
4. Tech A says that all conductors have inherent resistance in them, and therefore, the longer a length of wire or conductor is, the higher the resistance, and corresponding voltage drop, will be across the length of the wire or conductor.

Tech B says the resistance of wire is a constant, and is only affected by the gauge or size of the wire. Who is correct?
a. Tech A
b. Tech B
c. Both A and B
d. Neither A nor B

5. Tech A says that connector terminals are usually made of a special stainless steel, and therefore, are not susceptible to corrosion. Tech B says that connector terminals can wear over time due to the vibration, heat, and stress that they are exposed to, and that this damage can result in an intermittent electrical connection across the terminal. Who is correct?
a. Tech A
b. Tech B
c. Both A and B
d. Neither A nor B

6. Tech A says that it is a good practice to apply a good synthetic chassis grease to the terminals of electrical connectors; this seals the connector and prevents corrosion. Tech B says that most electrical connectors that are exposed to the elements utilize special rubber seals to prevent water and dirt from entering the connector. Who is correct?
a. Tech A
b. Tech B
c. Both A and B
d. Neither A nor B

7. Tech A says that wiring harnesses are specially engineered for their intended purposes, and as such, they should only be repaired according to the manufacturer's specifications. Tech B says that there is no need to seal repairs made to wiring harnesses inside of the vehicle, because they will not be exposed to moisture. Who is correct?
a. Tech A
b. Tech B
c. Both A and B
d. Neither A nor B

8. Tech A says that splices in wiring harnesses are the points where twisted-pair communication wires are secured together, but no electrical connection exists at these points. Tech B says that splices are points in the wiring harness where two or more conductors are connected electrically. Who is correct?
a. Tech A
b. Tech B
c. Both A and B
d. Neither A nor B

9. Tech A says that all wiring diagrams use the same symbols and wire color identification, and that wiring diagrams from nearly all manufacturers look identical to one another. Tech B says that each manufacturer has their own way of designing wiring diagrams, and as a result, each one is different. Who is correct?
a. Tech A
b. Tech B
c. Both A and B
d. Neither A nor B

10. Tech A says that wiring circuit and connector numbers are only supplied so that the technician is aware of how many electrical circuits and components are present within the wiring harness. Tech B says that each component, splice, ground, and circuit number provides locational data for the item in question, which assists the technician in locating a specific component or circuit. Who is correct?
a. Tech A
b. Tech B
c. Both A and B
d. Neither A nor B

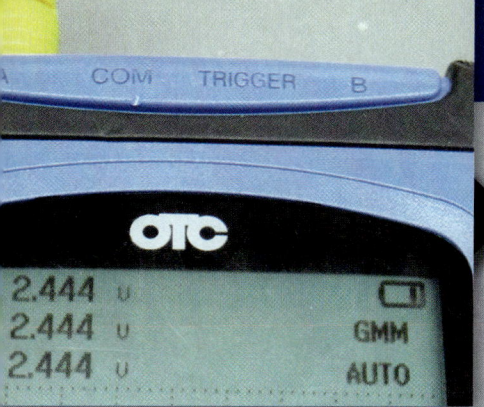

CHAPTER 12
Electrical Testing Procedures

NATEF Tasks

- **N12001** Check electrical/electronic circuit waveforms; interpret readings and determine needed repairs.

Knowledge Objectives

After reading this chapter, you will be able to:

- **K12001** Understand the basic differences between graphing multimeters and DVOMs.
- **K12002** Understand the basic differences between oscilloscopes and graphing multimeters.

Skills Objectives

After reading this chapter, you will be able to:

- **S12001** Check circuit waveforms.

You Are the Technician

Throughout your career, you will purchase many special tools, devices, and pieces of test equipment to make you more efficient as an automotive repair technician. However, one of the most important tools to master for today's technician is the digital volt-ohm meter (DVOM), also known as a multimeter. With this tool, you will be able to test electrical resistance, current flow through a circuit, voltage, and much more. When the time comes for you to purchase your own DVOM, care and concern should be given to the meter that you choose. There are many meters offered at varying price points from low to high; since you will be relying on this tool to diagnose complex circuits, it is worth your while to invest in a quality meter that will stand the test of time and produce accurate, repeatable readings. Given the complex nature of today's vehicles, it is critical that the diagnostic repair technician learns to master the use of the DVOM.

1. True or False? The DVOM is a simple, "set it and forget it" device that rarely requires maintenance.
2. True or False? All DVOMs are made to the same standards; therefore, any meter that is chosen can be relied upon to produce accurate and repeatable test results.

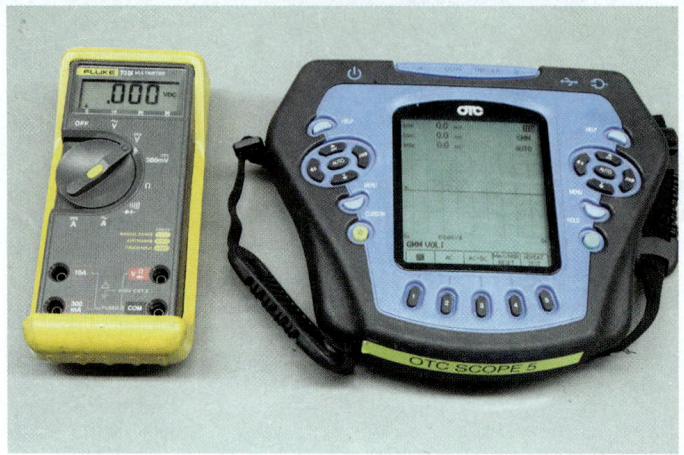

FIGURE 12-1 DVOM and graphing multimeter.

▶ Introduction

This chapter is to serve as an introduction to GMMs and oscilloscopes. As an automotive technician, it is important to have a thorough understanding of the test equipment used during the process of fault diagnosis. The knowledge gained here will provide a basic understanding of these diagnostic tools and the terminology associated with their use.

Just as digital volt-ohm meters (DVOMs) are used to diagnose electrical problems, graphing multimeters (GMMs) and oscilloscopes are used to perform more precise diagnosis of electrical circuits and signals (**FIGURE 12-1**). These tools are most often reserved for use in diagnosing problems pertaining to electronic control module communication, sensor input signals, and advanced drivability concerns. In reality, a DVOM will provide you with the proper diagnosis of an electrical problem 7 out of 10 times. There will be, however, more tenacious problems that are encountered that require a more precise measuring tool in order to identify and correct the issue. In this chapter, we will cover the basic differences between DVOMs, GMMs, and oscilloscopes.

K12001 Understand the basic differences between graphing multimeters and DVOMs.

▶ Graphing Multimeters

Graphing multimeters, or (GMMs), are used to measure electrical properties. Their basic function is not at all unlike that of a DVOM, which displays electrical values on a digital display. GMMs can measure volts, ohms, and amps. In addition, they can test diodes and perform voltage drop tests, as well as every other test you can perform on a DVOM. So if GMMs are so similar to DVOMs, what unique purposes do they serve, you ask? GMMs differ in two distinct ways from DVOMs: They have a faster sample rate, and they are able to display electrical data in a graphical manner, as well as provide a digital readout simultaneously for reference (**FIGURE 12-2**).

GMM Applications

Graphing multimeters are used to diagnose electrical problems that occur quickly and do not linger for very long. GMMs have several advantages over DVOMs. The first advantage is that they display electrical information graphically; this graphical display of electrical data allows the technician to observe small changes in electrical values that may not show up on a DVOM's digital display. The second difference is that GMMs are capable of recording electrical data in great detail, far beyond the MIN/MAX data gathered by a DVOM. GMMs are also able to play the recording back on the graphical display, allowing the technician to search for a problem that occurs quickly. This playback feature can be a useful tool when attempting to diagnose an electrical problem that occurs intermittently. The third difference is that GMMs are usually able to monitor two channels of information at one time and display information from both channels on the screen simultaneously (**FIGURE 12-3**). The fourth difference is that aside from using a high-end oscilloscope, GMMs display the next best thing to "live" data on their screens. This is made possible by GMMs' relatively high sample rate, which can be as fast as 300 microseconds. On the other hand, DVOMs sample information at a much slower rate, and the value on their screen is commonly referred to as "processed information." This means that the information being viewed on the digital display of a DVOM is an aggregate of all the data drawn from several samplings that have occurred over a period of milliseconds. This phenomenon has no effect on a DVOM's ability to perform basic electrical testing, but when chasing elusive electrical problems, the technician needs to view data in as close to real time as possible. To reiterate, GMMs provide live data, while DVOMs provide processed information with a slight delay. This display of live data is especially helpful when testing variable output signals, such as those generated by throttle position sensors (TPSs). Such devices contain at least one potentiometer, which is a variable resistor, and as such, the potentiometer's output, or electrical signal, is a variable DC voltage that changes in direct relation to the sensor's movement. Due to the variable

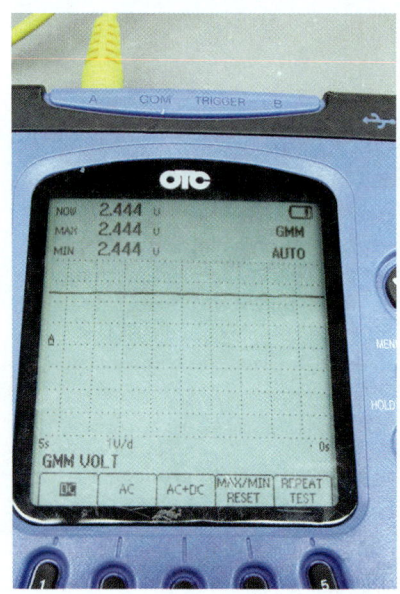

FIGURE 12-2 GMM displaying electrical data.

nature of this sensor, a GMM or oscilloscope is the only accurate way to test this sensor. The graphical display of the voltage signal provided by these testers will allow you to view the sensor's output signal as it moves through its range of motion; you should be able to view the throttle position sensor's output signal rise and fall smoothly as it is moved through its range of motion, something that cannot be accomplished on the digital display of a DVOM.

Another feature that sets most automotive-oriented graphing multimeters apart from DVOMs is the vehicle and component database contained in the tester itself, which allows the technician to select the year, make, and model of the vehicle being serviced. For instance, say that you are trying to test a throttle position sensor on a 2008 Chevrolet Malibu. You are unsure of exactly where the sensor is located, or which wires in its connector to select to test its output signal. If you are using an automotive-type GMM, such as those offered from Snap-On, Mac Tools, or MATCO, to name a few, you can enter the vehicle's year, make, and model into the tester itself. This will open the vehicle-specific testing information that is contained in the tester's memory. Contained in this information will usually be a list of tests, examples of acceptable sensor test results, examples of failed tests, descriptions of components and their locations on the vehicle, and the instructions needed to hook up the tester and conduct an array of component-specific tests. This information is exceedingly helpful, as it will tell you where the component you need to test is located and what the connector will look like, as well as provide a wiring diagram for the component's connector.

It should be noted at this point, that while the vehicle information that is contained in the tester is a useful tool to increase the speed and efficiency of the automotive technician, be advised that you should always supplement your findings by consulting the appropriate vehicle service manual to insure that you are testing the proper components and to prevent a misdiagnosis.

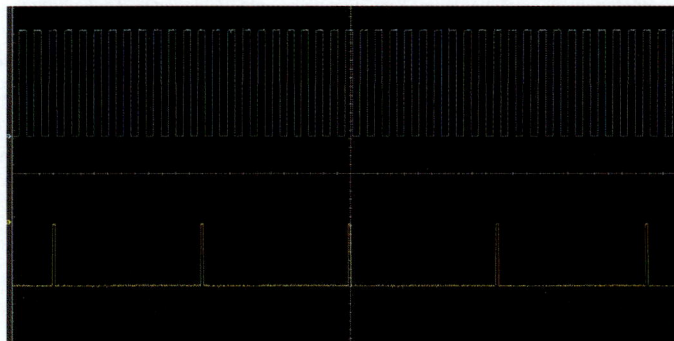

FIGURE 12-3 GMM showing two channels of electrical information on its screen.

Properly Setting Up Your GMM for Electrical Concerns

Hooking up a GMM to test for electrical concerns is very similar to hooking up a DVOM (**FIGURE 12-4**). These devices have a "COM" port, which serves the same purpose as the common port on the DVOM. In addition to the common port, there will usually be a port for each of the two channels. On most testers, all of the electrical data, such as volts, ohms, amps, frequency, and duty cycle, can be measured without changing the position of the leads on the tester. You select what unit of electrical measure that you would like to view from a graphical menu in the tester (**FIGURE 12-5**). It is important to note that some testers contain a dedicated port to test amps. Such ports are usually protected with fuses.

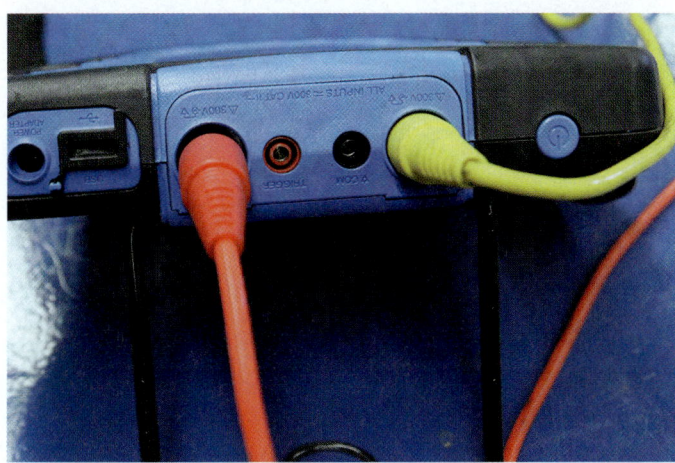

FIGURE 12-4 GMM showing leads connected to test ports.

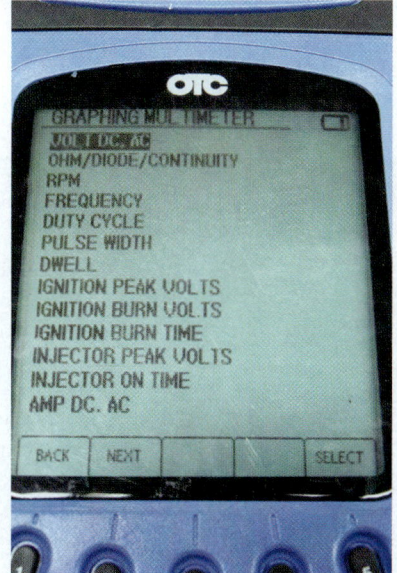

FIGURE 12-5 Graphical menu of a GMM.

Waveforms: The Good, the Bad, and the Ugly

As already stated above, a principle benefit of using a GMM instead of a DVOM is the ability to view electrical information graphically. This allows the technician to view the real-time electrical signature of the devices being tested. This signature, or waveform, is what the savvy technician can use to determine if the component being tested is good or bad.

Another benefit of a GMM or oscilloscope is that the technician can adjust the range of information being displayed on the screen and the display rate of sampled information (**FIGURE 12-6**). The range of data, such as voltage, will be displayed on the vertical axis of the screen, usually located on the right side of the screen, and the display refresh rate is usually scaled out on the bottom, horizontal axis of the screen. One benefit to adjusting the display range of the information being displayed is to make it appear larger or smaller. For instance, if testing a voltage signal that varies from 2 to 5 volts, the "bottom" scale of the GMM should be set to 0 volts, and the top range of the display should be set to 5 or 6 volts (**FIGURE 12-7**).

On a GMM, in addition to being able to adjust the display range, you can adjust the display refresh rate. This adjustment controls the number of information signatures, or "traces," that are displayed on the screen. For example, when a longer amount of time is selected for the display rate, the screen will show a wide sampling of information, but due to the slow refresh rate and the amount of information that is being displayed, it will be difficult to determine if a waveform passes or fails when the GMM is set to the high sample rates (**FIGURE 12-8**). In the opposite direction, if the display rate is set to the fastest setting, a very small portion of the information being gathered will be shown, making it difficult to ascertain any useable information in most cases (**FIGURE 12-9**). It is important to note that the "sweet spot" for viewing the information will often be found toward the middle of the refresh rate settings. It is wise to adjust the refresh rate time settings up and down until you have obtained a view of the data that you are pleased with. In **FIGURES 12-10** through **12-17**, you will find several examples of waveforms produced by common automotive sensors. Familiarize yourself with the "good" traces, as well as the "bad" traces, as this knowledge will help you spot potential problems during your diagnosis.

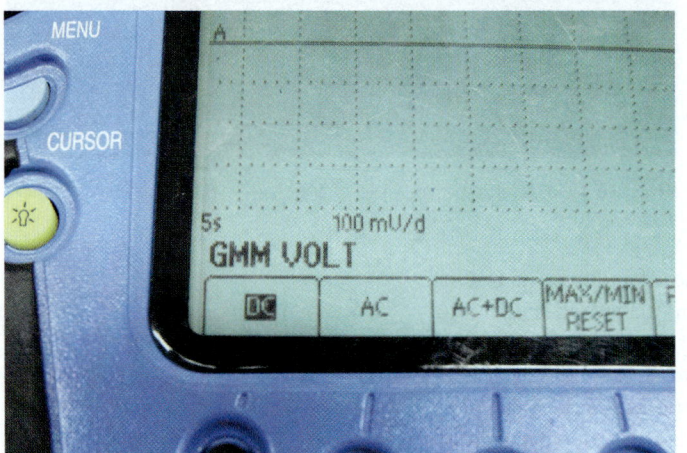

FIGURE 12-6 Data scale being shown on the screen of a graphing multimeter.

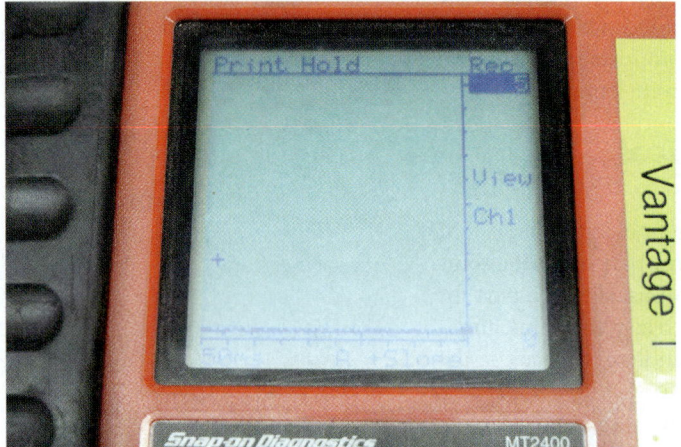

FIGURE 12-7 GMM screen showing voltage scales.

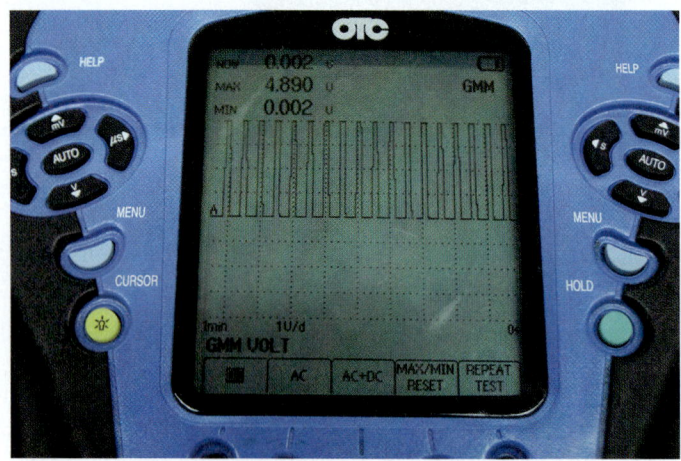

FIGURE 12-8 GMM displaying waveform data at a slow refresh rate.

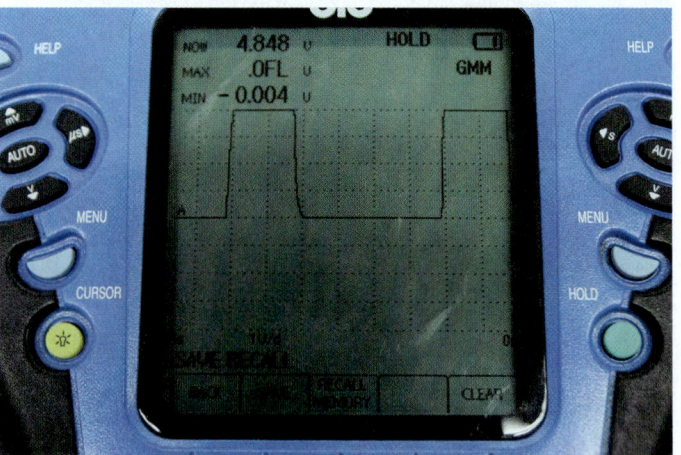

FIGURE 12-9 GMM displaying waveform data at a fast refresh rate.

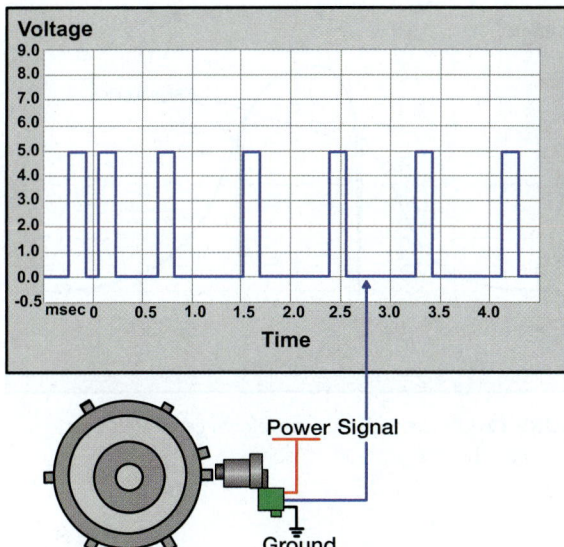

FIGURE 12-10 Screenshot of waveform generated by a known good camshaft position sensor.

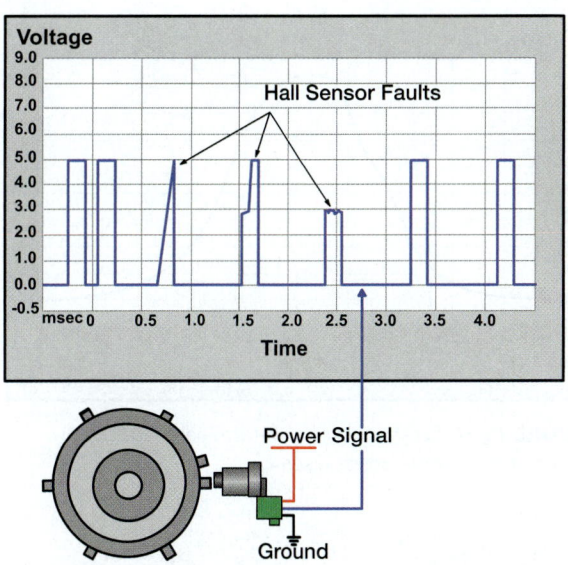

FIGURE 12-11 Screenshot of waveform generated by a known bad camshaft position sensor.

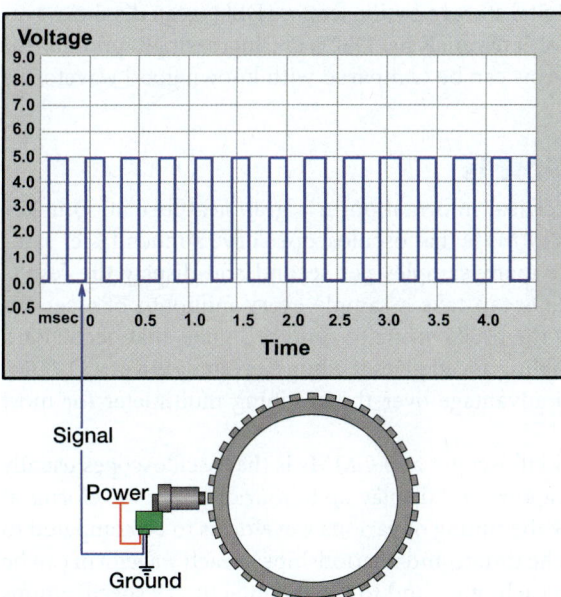

FIGURE 12-12 Screenshot of waveform generated by a known good crankshaft position sensor.

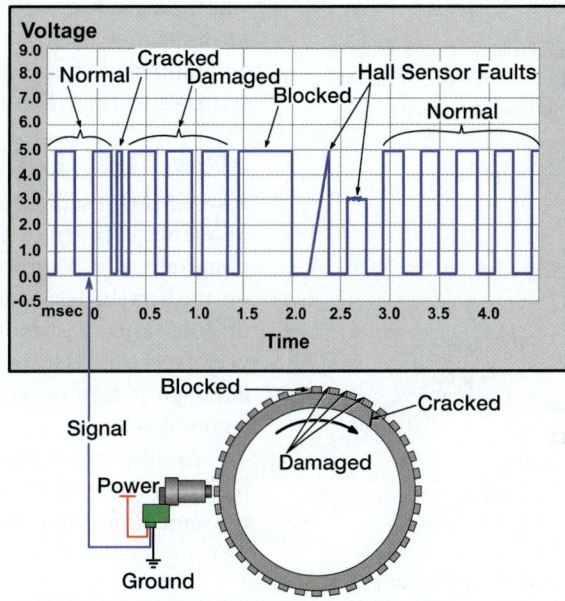

FIGURE 12-13 Screenshot of waveform generated by a known bad crankshaft position sensor.

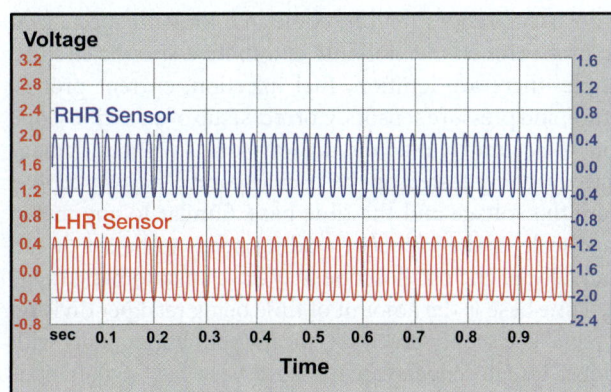

FIGURE 12-14 Screenshot of waveform generated by a known good wheel speed sensor.

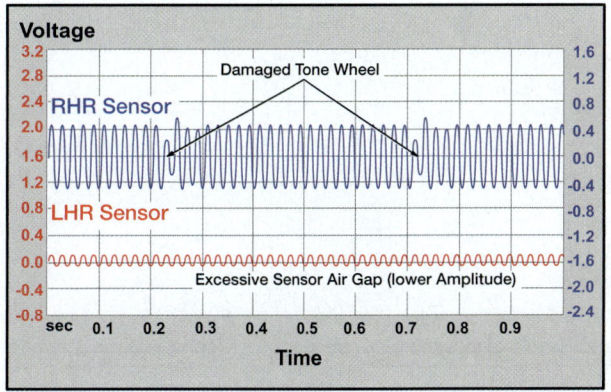

FIGURE 12-15 Screenshot of waveform generated by a known good wheel speed sensor.

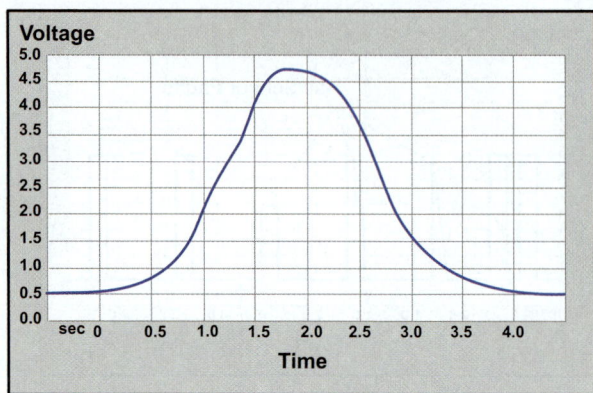

FIGURE 12-16 Screenshot of waveform generated by a known good throttle position sensor.

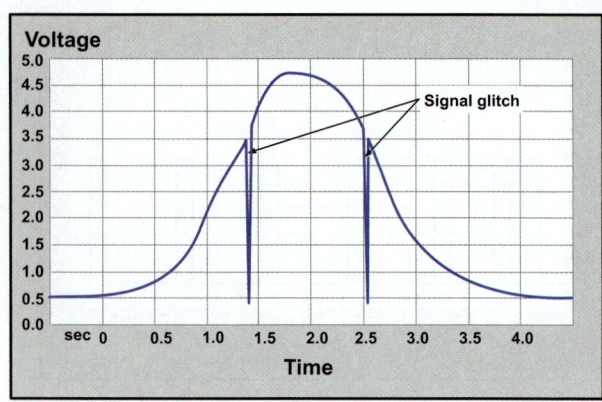

FIGURE 12-17 Screenshot of waveform generated by a known bad throttle position sensor.

▶ Oscilloscopes

K12002 Understand the basic differences between oscilloscopes and graphing multimeters.

N12001 Check electrical/electronic circuit waveforms; interpret readings and determine needed repairs.

Oscilloscopes are commonly referred to as lab scopes, or just scopes, and are very important test instruments for vehicle diagnostics. Oscilloscopes may be analog, displaying waveforms as they occur, or digital. Digital storage oscilloscopes (DSOs) can display waveforms as they occur and store them for later analysis. DSOs are increasingly popular for automotive use because stored waveforms can be compared with known good waveforms (**FIGURE 12-18**).

How They Differ from GMMs

Like GMMs, oscilloscopes display electrical information in a graphical format. One key difference between oscilloscopes and GMMs is that oscilloscopes have a much faster sampling rate, which means they can take many samples per second and display the results on the screen. For example, some DSOs can take a sample every millionth of a second or even faster. Fast sample rates allow the oscilloscope to capture signals that occur for a very short time. The oscilloscope's sampling speed gives it a huge advantage over a DVOM, although it may provide only a small advantage over the graphing multimeter for most automotive tests.

Another key difference between oscilloscopes and GMMs is that oscilloscopes usually have four channels through which to capture and display up to four different waveforms at the same time. Multiple channels allow the timing of various waveforms to be compared to each other. The timing and relationships of each waveform can be compared to each other and to the manufacturer's specifications (**FIGURE 12-19**).

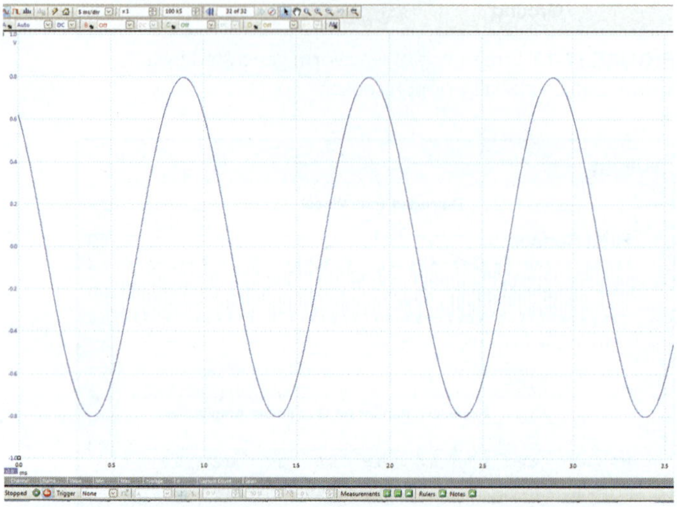

Checking Circuit Waveforms

Automotive oscilloscopes tend to come in a kit with specialized leads and probes for connection into automotive-specific circuits. For example, there are ignition, fuel injection, current probes, temperature, and pressure sensors. Correct setup of an oscilloscope is essential to accurately read circuit waveforms without damaging the circuit being tested or the oscilloscope. The two critical settings are the voltage and the time base. Similar to any meter, the voltage scale of the expected reading needs to be set. It can be set manually, or the oscilloscope may have an automatic range function. Time base is the amount of time being read per division on the screen. The scope screen is typically split horizontally with 10 divisions. Oscilloscopes can measure very fast signals; it is not unusual to have time base ranges from 100 nanoseconds per division to 200 seconds per division.

FIGURE 12-18 Digital storage oscilloscope (DSO) showing a waveform.

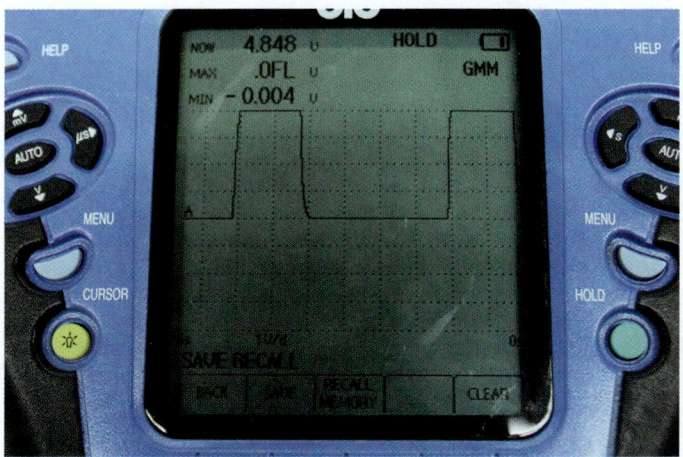
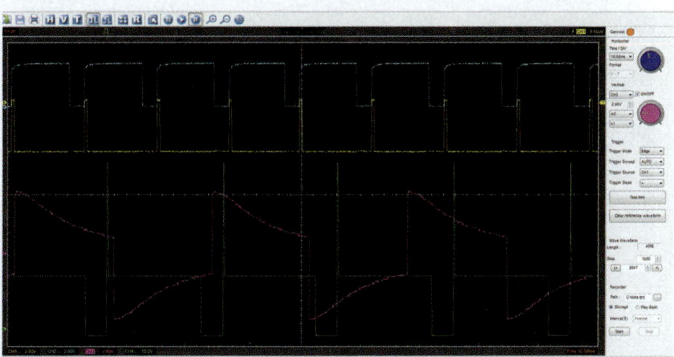

FIGURE 12-19 GMM and digital storage oscilloscope (DSO) side by side.

Oscilloscopes are capable of reading voltages as low as 50 millivolts up to maximum voltages of 50 or 100 volts. Sometimes on vehicles, higher voltages need to be measured. For example, the primary voltage of an ignition coil requires a higher voltage reading. In these circumstances, **attenuators** can be fitted to the probe leads to reduce the maximum voltage to safe levels for the oscilloscope to measure. An attenuator is a device that reduces the amount of voltage passing through it. For example, a 20:1 attenuator divides the amount of voltage passing through it by a factor of 20. Thus, 20 volts input will appear as an output of 1 volt. Always connect the oscilloscope leads according to the manufacturer's specifications. The correct connection of ground leads is important to ensure minimal interference in the waveform. In most cases, the ground lead should be connected directly to the negative battery terminal.

To check circuit waveforms, follow the steps in **SKILL DRILL 12-1**.

SKILL DRILL 12-1 Checking Circuit Waveforms

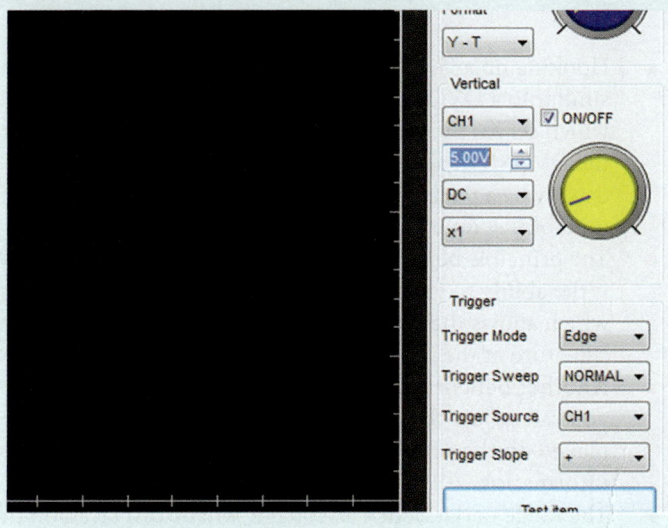

1. Determine the circuit to be tested and the likely voltages and frequency of the waveform to be measured. Set the voltage level and time base.

Continued

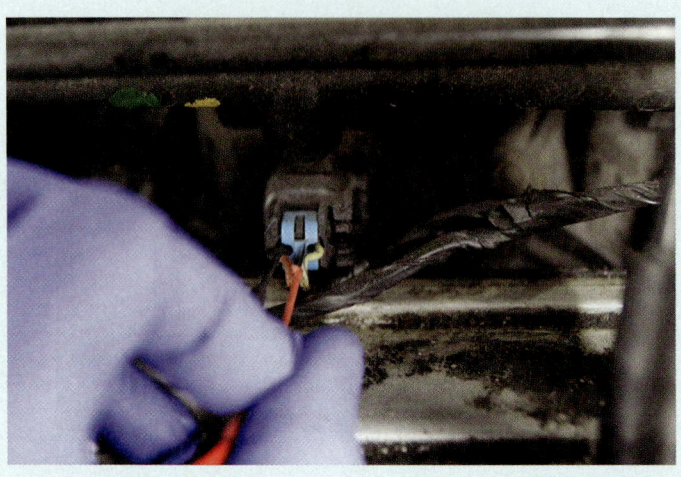

2. Connect the leads to the points in the circuit to be measured.

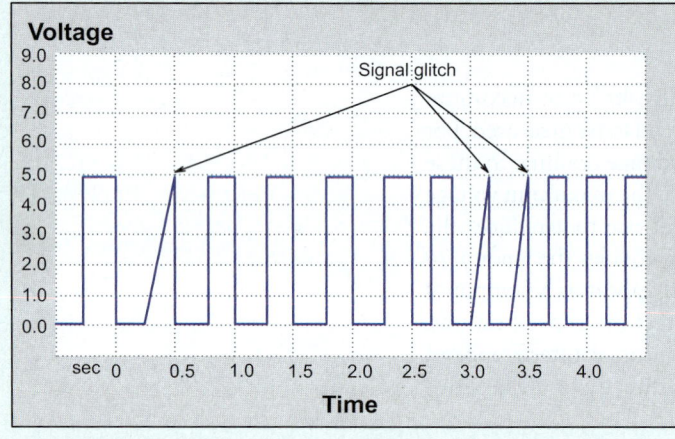

3. Capture waveforms from the circuit being tested. Analyze the waveform, comparing it to the manufacturer's specifications or known good waveforms.

Wrap-Up

Ready for Review

- Graphing multimeters (GMMs) and oscilloscopes are used to perform more precise diagnosis of electrical circuits and signals.
- A DVOM will provide you with the proper diagnosis of an electrical problem 7 out of 10 times.
- Graphing multimeters, or (GMMs), are used to measure electrical properties. Their basic function is not at all unlike that of a DVOM, which displays electrical values on a digital display.
- GMMs can measure volts, ohms, and amp plus test diodes and perform voltage drop tests, their uniqueness lies in that they have a faster sample rate, and can display electrical data in a graphical manner.
- GMMs are capable of recording electrical data in great detail, far beyond the MIN/MAX data gathered by a DVOM.
- The GMM can test an input sensor under load.
- Hooking up a GMM to test for electrical concerns is similar to a DVOM. These devices have a "COM" port, which serves the same purpose as the common port on the DVOM. In addition to the common port, there will usually be a port for each of the two channels, if two are used or just one.
- The principle benefit of using a GMM instead of a DVOM is the ability to view electrical information graphically which allows the technician to view the real-time electrical signature of the devices being tested.
- Another benefit of a GMM or oscilloscope is that you can adjust the range of information being displayed on the screen and the display rate of sampled information.
- The range of data, such as voltage, will be displayed on the vertical axis of the GMM screen, and the display refresh rate is usually scaled out on the bottom, horizontal axis of the screen.

- Familiarize yourself with the "good" traces, as well as the "bad" traces, as this knowledge will help you spot potential problems during your diagnosis.
- Oscilloscopes may be analog, displaying waveforms as they occur, or digital.
- Digital storage oscilloscopes (DSOs) can display waveforms as they occur and store them for later analysis.
- DSOs are increasingly popular for automotive use because stored waveforms can be compared with known good waveforms.
- One key difference between oscilloscopes and GMMs is that oscilloscopes have a much faster sampling rate, which means they can take many samples per second and display the results on the screen.
- Oscilloscopes usually have four channels through which to capture and display up to four different waveforms at the same time. Multiple channels allow the timing of various waveforms to be compared to each other. The timing and relationships of each waveform can be compared to each other and to the manufacturer's specifications.
- Correct setup of an oscilloscope is essential to accurately read circuit waveforms without damaging the circuit being tested or the oscilloscope. The two critical settings are the voltage and the time base.
- Oscilloscopes can measure very fast signals; it is not unusual to have time base ranges from 100 nanoseconds per division to 200 seconds per division.
- Oscilloscopes are capable of reading voltages as low as 50 millivolts up to maximum voltages of 50 or 100 volts.
- The primary voltage of an ignition coil requires a higher voltage reading. In these circumstances, attenuators can be fitted to the probe leads to reduce the maximum voltage to safe levels for the oscilloscope to measure.
- An attenuator is a device that reduces the amount of voltage passing through it. For example, a 20:1 attenuator divides the amount of voltage passing through it by a factor of 20. Thus, 20 volts input will appear as an output of 1 volt.
- Always connect the oscilloscope leads according to the manufacturer's specifications.

Key Terms

attenuator A device that weakens, or attenuates, a high-level input signal.

display refresh rate The rate at which a GMM or oscilloscope can display new electrical information. The higher the display refresh rate, the higher the resolution of the test device.

graphing multimeter (GMM) Used to display electrical data in graphical fashion. The base functionality of a GMM is similar to an oscilloscope, but GMMs typically have slower sampling rates and lower display resolution than oscilloscopes.

oscilloscope Instrument used to display the waveform of electrical signals. These have much higher resolution and sample rates than GMMs.

sampling speed How many measurements can be taken in a specified measure of time.

waveform A graphical representation of how electrical signals vary with time.

Review Questions

1. In a GMM, the "sweet spot" for viewing the information will often be toward the:
   a. middle of the refresh rate settings.
   b. higher end of the refresh rate settings.
   c. lower end of the refresh rate settings.
   d. default setting.
2. A graphing multimeter (GMM) has all of the following advantages over a traditional digital multimeter (DMM) *except*:
   a. it can present processed data after a few milliseconds.
   b. it can present electrical information on a graphical display.
   c. it is capable of performing playback on graphical display.
   d. it has a relatively high sample rate.
3. The major difference between a graphing multimeter (GMM) and a digital volt-ohm meter (DVOM) is that a GMM:
   a. contains the vehicle and component database in itself.
   b. can measure amps, volts, and ohms precisely.
   c. can diagnose harness problems.
   d. can measure the electricity output of a battery.
4. Which of the following statements is true regarding graphing multimeters?
   a. They are used only in high-end circuits.
   b. The device can sometimes give inaccurate readings.
   c. It is particularly useful in diagnosing an electrical problem that occurs intermittently.
   d. The device may fail to perform when the volume of information is large.
5. Which of the following can be accurately done using graphing multimeters (GMMs) but cannot be achieved using DMMs?
   a. Measuring resistance in the connectors
   b. Diagnosing issues in the electronic communication system in the vehicle
   c. Determining the position of a sensor using live data
   d. Checking the battery life in the vehicle
6. All the following statements with respect to oscilloscopes are true *except*:
   a. Oscilloscopes display electrical information graphically.
   b. Oscilloscopes may be analog or digital.
   c. Oscilloscopes are increasingly popular for their storage of waveforms.
   d. Oscilloscopes have a very slow sampling rate compared to GMMs.
7. Which of the following devices can store the waveforms for later reference and can compare with known good waveforms?
   a. Graphing multimeters (GMM)
   b. Digital volt-ohm meters (DVOM)
   c. Digital storage oscilloscopes
   d. Digital multimeters

8. What is the lowest voltage that an oscilloscope is capable of reading?
   a. 50 millivolts
   b. 1 volt
   c. 50 microvolts
   d. 500 millivolts
9. Which of the following describes the function of an attenuator fitted to the probe of an oscilloscope?
   a. It increases the amount of voltage passing through it.
   b. It reduces the resistance passing through the device.
   c. It controls the heat generated in the device during testing.
   d. It reduces the amount of voltage passing through it.
10. Choose the correct statement.
    a. Oscilloscopes have time base ranges from 10 nanoseconds to 100 nanoseconds.
    b. Oscilloscopes are capable of measuring max voltage of 200 volts.
    c. The scope screen is split horizontally with 10 divisions.
    d. The ground lead should be connected to the positive battery terminal.
11. The DSO uses which type of test lead?
    a. Only a positive lead, similar to a DMM lead
    b. Only a negative lead, similar to a DMM lead
    c. Color-coded DMM type leads so they are not confused with negative or positive when used
    d. A lead with a BNC-type connector

ASE Technician A/Technician B Style Questions

1. Tech A says that GMMs are the most accurate diagnostic tool for testing electrical circuits. Tech B says that Oscilloscopes have better resolution than GMMs. Who is correct?
   a. Tech A
   b. Tech B
   c. Both A and B
   d. Neither A nor B
2. Tech A says that GMMs can detect circuit glitches that would be missed with a DVOM. Tech B says that the only tool that a competent technician needs is a good DVOM. Who is correct?
   a. Tech A
   b. Tech B
   c. Both A and B
   d. Neither A nor B
3. Tech A says that a GMM provides too much information, and therefore complicates the diagnostic process unnecessarily. Tech B says that the best way to discover an intermittent component fault is to use a GMM or oscilloscope. Who is correct?
   a. Tech A
   b. Tech B
   c. Both A and B
   d. Neither A nor B
4. Tech A says that a sensor either works, or it does not. Tech B says that sometimes a sensor will initially produce a proper output signal, but as the sensor heats up, the signal will degrade. Who is correct?
   a. Tech A
   b. Tech B
   c. Both A and B
   d. Neither A nor B
5. Tech A says GMMs can have a vehicle database programmed into their memory. Tech B says that the vehicle database programmed into many GMMs contains component locations, connector pinouts, and predetermined test parameters. Who is correct?
   a. Tech A
   b. Tech B
   c. Both A and B
   d. Neither A nor B
6. Tech A says that most GMMs have four channels. Tech B says that many oscilloscopes have four channels. Who is correct?
   a. Tech A
   b. Tech B
   c. Both A and B
   d. Neither A nor B
7. Tech A says that GMMs are great tools, but they can only be used within a very narrow range of measurement. Tech B says that you can adjust the unit scales on GMMs, which allows you to be able to test a wide variety of signal strengths and ranges. Who is correct?
   a. Tech A
   b. Tech B
   c. Both A and B
   d. Neither A nor B
8. Tech A says that GMMs are great tools, but usually are only able to measure voltage. Tech B says that GMMs are capable of measuring everything that a DVOM can. Who is correct?
   a. Tech A
   b. Tech B
   c. Both A and B
   d. Neither A nor B
9. Tech A says that when looking at the screen of a GMM or oscilloscope, time is placed on the vertical axis of the screen, and the units of measure is placed on the horizontal axis of the screen. Tech B says that time is always displayed on the horizontal axis of the screen. Who is correct?
   a. Tech A
   b. Tech B
   c. Both A and B
   d. Neither A nor B
10. Tech A says that if you can master the use of GMMs and oscilloscopes, you can diagnose nearly any electrical problem. Tech B says that a scan tool is more useful than a GMM or oscilloscope in diagnosing intermittent electrical problems. Who is correct?
    a. Tech A
    b. Tech B
    c. Both A and B
    d. Neither A nor B

CHAPTER 13
Batteries

NATEF Tasks

- **N13001** Identify safety precautions for high voltage systems on electric, hybrid, hybrid-electric, and diesel vehicles.
- **N13002** Perform battery state-of-charge test; determine needed action.
- **N13003** Inspect and clean battery; fill battery cells; check battery cables, connectors, clamps, and hold-downs.
- **N13004** Perform slow/fast battery charge according to manufacturer's recommendations.
- **N13005** Jump-start vehicle using jumper cables and a booster battery or an auxiliary power supply.
- **N13006** Maintain or restore electronic memory functions.
- **N13007** Identify electrical/electronic modules, security systems, radios, and other accessories that require reinitialization or code entry after reconnecting vehicle battery.

Knowledge Objectives

After reading this chapter, you will be able to:

- **K13001** Describe battery types, ratings, sizes, and construction.
- **K13002** Describe what happens in the battery charging and discharging cycles.
- **K13003** Describe lead acid, gel cell, and AGM batteries.
- **K13004** Explain the battery testing procedure.

Skills Objectives

After reading this chapter, you will be able to:

- **S13001** Perform a battery state-of-charge test.
- **S13002** Charge a battery.
- **S13003** Jump start a vehicle.
- **S13004** Load test a battery.
- **S13005** Identify electronic modules, security systems, radios, and other accessories that require reinitialization.
- **S13006** Maintain or restore electronic memory functions.
- **S13007** Measure parasitic draw.

You Are the Technician

The battery is the heart of the vehicle's electrical system. It is responsible for starting the engine, providing the power necessary to run the many electronics systems, and providing a reserve capacity that allows the vehicle to be operated for a finite period of time in the event of a charging system failure. The battery is often overlooked over the course of vehicle operation, but when the battery develops a problem, the owner becomes very aware of the battery's existence, and its importance. The simple fact of the matter is that when the battery fails, the car is rendered inoperable. It is for this reason that regular battery testing and maintenance are critical to reliable and safe vehicle operation.

1. True or False: Battery testing and inspection should be performed at every oil change to ensure trouble-free vehicle operation.

CHAPTER 13 Batteries

K13001 Describe battery types, ratings, sizes, and construction.

▶ Introduction

The electrical systems on modern vehicles have grown more and more complex. Vehicles increasingly rely on electrical power to assist in the management and control of not only the engine and emissions system, but also almost all aspects of the vehicle, including brakes and suspension, navigation, entertainment, information retrieval, and much more. Almost every system on modern vehicles relies on electrical and electronic components and theory, electronic control modules, and networking systems to connect everything together.

▶ What Is a Battery?

As you learned in the Principles of Electrical Systems chapter, electricity is a very flexible and useful energy that can be used easily in a variety of ways. But one drawback to electricity is that it cannot be stored easily in its electrical form for later use. This means that electricity must be stored in another form of energy and reconverted to electricity when needed. This is where batteries enter the discussion.

Batteries were developed in the early 1800s, and since that time many varieties and designs have been developed. The battery is part of everyday life and is widely used in modern electrical and electronic devices. Batteries store electricity in chemical form, which is possible because electricity causes a chemical reaction within the battery. In other words, the electrical energy is transformed into chemical energy. The chemical reactions change the composition of the chemicals, which then are stored until the electrical energy is needed. When electricity is needed, the chemicals react with each other, transforming the chemical energy back into electrical energy.

A battery consists of two dissimilar metals, an insulator material separating the metals, and an **electrolyte**, which is an electrically conductive solution (**FIGURE 13-1**). The strength of the battery depends on the materials used in its construction.

The traditional automotive battery type is the lead-acid battery. It is available in many different shapes, sizes, and designs to meet the requirements for various applications. For example, the battery used for starting a vehicle's engine is different from the marine deep-cycle battery. Each requires different design characteristics for obviously different applications. Vehicle batteries are designed to provide high current draws for short periods of time, while deep-cycle batteries supply smaller, continuous loads over longer periods of time. Although the size, case configuration, and design may change, the fundamental components and their operation remain the same.

Battery Condition

Batteries do not last forever. Over time and use, the plates start to lose effectiveness as lead paste falls off the plates and collects in the bottom of the battery case. This type of damage is especially likely to develop in a battery subjected to high vibration and shock, as when subjected to off-road travel. As plate material becomes deposited in the bottom of the battery case, it eventually shorts out the plates and renders that cell useless, which causes the battery to fail under load.

A battery may also become sulfated, meaning the surfaces of the plates harden, making it more difficult for the acid to be absorbed into the plate. As a result, the battery can no longer readily accept a charge or cannot adequately discharge under load. Leaving a battery on a slow charger overnight and then finding that it fails a load test in the morning is an indication of a sulfated battery. Sulfated batteries cannot create the necessary current flow and should be replaced and recycled.

Parasitic Draw

Parasitic draw refers to the current draw that occurs once the vehicle has been turned off and the systems have shut down. All

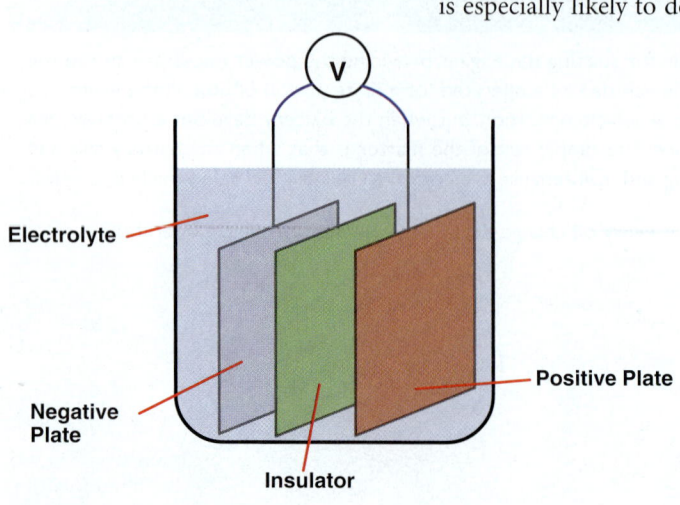

FIGURE 13-1 Components of a simple battery.

modern vehicles have a certain minimum amount of parasitic current draw that is used by the vehicle's keep alive memory (KAM) circuits. KAM systems maintain memory functions and monitor systems. For example, the vehicle's theft-deterrent system is part of this parasitic current draw. Excessive parasitic draw will discharge the battery prematurely. Typically, when the ignition is switched off, the vehicle's engine stops and the vehicle systems begin to shut down, which is called "going to sleep" or "entering sleep mode." Vehicle systems will turn off at different rates and do not necessarily immediately turn off as soon as the ignition is off. For example, some vehicles may take up to a couple of hours before the last system goes to sleep and the parasitic draw reaches its minimum value. A low-amp clamp capable of measuring milliamps or an ammeter are tools commonly used to measure parasitic draw from the battery. Manufacturers will specify how much parasitic draw is acceptable and the waiting period required after the ignition is switched off before the measurement can be accurately taken.

Jump Starting

Jump starting of a vehicle requires the use of a slave battery along with high-current –capacity leads made with clamps on each end that can connect over the battery terminals. The slave battery is connected in parallel from the slave (charged) battery to the host (discharged) battery to provide additional capacity to crank and start the vehicle. Jump starting should be performed with caution, as incorrect polarity or voltage spikes can cause damage to the sensitive electronic components installed on either of the vehicles. The safest way to deal with a discharged battery is to remove and charge it or replace it. If jump starting must be undertaken, make sure the slave battery is fully charged. Use jumper leads equipped with electrical spike protection. Spike-protected leads have a built-in device to prevent damaging electrical spikes from reaching electronic equipment. It is important to ensure that the leads are connected with the correct polarity to prevent damage to electronic components and to connect and disconnect leads in the correct sequence.

SAFETY TIP

Never allow a spark or flame around a battery, and never try to jump start a frozen, faulty, or open-circuit battery; doing so could cause the battery to explode, causing injury.

Battery Service Precautions

When servicing batteries, always ensure that you have the right personal protective equipment (e.g., safety eyewear, gloves, and shop clothing) (**FIGURE 13-2**). Never wear any conductive jewelry, such as neck chains, watches, or rings, when working on or near batteries. Such items may provide an accidental short-circuit path for high currents. If the battery is charging, or has been charged recently, ensure that the space around the battery is well ventilated. Avoid making any sparks, which may ignite the gassing hydrogen–air mixture. Never create a low-resistance connection or short across the battery terminals. Always remove the negative or ground terminal first when disconnecting battery cables, as doing so reduces the possibility of a wrench creating an accidental short to ground from the positive terminal to any grounded metal surface, which would cause a very large spark.

Low Maintenance and Maintenance Free Batteries

There are many types of batteries, with variations of cell design available for vehicles. Some batteries are specifically designed for starting, some for extended-load marine usage, some for low-maintenance or maintenance-free applications. Low-maintenance batteries require little, if any, topping off of the water in the electrolyte. The plates and venting system are designed so that they do not normally gas and release water vapor to the atmosphere. Low-maintenance batteries still have removable caps so that electrolyte can be checked and topped off if necessary (**FIGURE 13-3**).

Maintenance-free batteries are fully sealed and do not require the electrolyte to be topped off. In some cases they use

FIGURE 13-2 Battery terminal being removed with battery wrench.

FIGURE 13-3 Battery with caps removed.

a gel-type electrolyte instead of a liquid. **Absorbed glass mat (AGM)** batteries have the electrolyte absorbed within a mat of fine glass fibers. The plates in this type of fully sealed battery can be pressed flat or wound into a cylindrical cell. Because the electrolyte in absorbed glass mat batteries is a gel, which does not spill, this type of battery is especially handy for rough handling or tipping. In fact, this type of battery can even be mounted on its side and still perform well. Thus, the absorbed glass mat battery is especially suited for off-road and racing vehicles. Due to the unique construction of AGM batteries, they require a special battery charger to be used during recharging. Failure to use an AGM-rated battery charger will result in damage to the battery, and possibly the vehicle.

A sealed, or maintenance-free, battery typically has no removable cell covers, so you cannot adjust or test the fluid levels inside. However, some of these batteries do have a visual indicator (a single-cell hydrometer float) that provides information on the status of the charge and condition of one of the battery cells. Each manufacturer provides details of these visual indicators; refer to these when performing an inspection.

Battery Recycling Procedures

Batteries contain many chemicals and metals that can damage the environment. If batteries find their way into a landfill, the acid and metals within them will seep out and contaminate the soil and waterways. Correct disposal of batteries—recycling them—is not just good for the environment: the precious metals within can be reclaimed for reuse. Many municipalities require battery recycling and levy a "core charge" on every new automotive battery sold. The core charge is refunded if an old battery is brought in and exchanged for the new one. This process helps prevent batteries from being discarded in the trash or left lying around. Check local laws and regulations to ensure that batteries are disposed of correctly.

K13002 Describe what happens in the battery charging and discharging cycles.

▶ Battery Charging and Discharging Cycles

In a discharged lead-acid cell, the active material of both plates becomes lead sulfate, and the electrolyte becomes mostly water as the acid is absorbed into the plates. The result is a very weak sulfuric acid solution in the electrolyte. When being charged, the battery is connected to a direct current (DC) electrical supply with electrical pressure (voltage) higher than that of the battery's total cell voltage. The charging device acts like an electron pump, forcing electrons to move from the positive plates to the negative plates in the battery.

At the negative plates, sulfate is discharged, which changes the chemical composition of the plates back into sponge lead and also creates a stronger solution of sulfuric acid in the electrolyte. At the same time, lead peroxide is formed at the positive plates, which helps to restore the cell's electrical potential or voltage.

The charging process increases the amount of acid in the electrolyte, making the electrolyte stronger. When further charging no longer makes the electrolyte stronger, charging is complete. Connecting a lead-acid battery to a load causes chemical changes as the battery discharges. At the positive plate, sulfate from the sulfuric acid in the electrolyte joins with lead to form lead sulfate, and oxygen from the plate joins the hydrogen from the electrolyte to form water. Lead sulfate also forms at the negative plate, as sponge lead forms with sulfate from the electrolyte (**FIGURE 13-4**). Overall, the percentage of acid in the electrolyte falls, and the percentage of water rises, reducing the strength of the electrolyte. As the cell discharges, the plates develop the same composition, which reduces the potential of the cell. Recharging the battery restores the difference between its sets of plates.

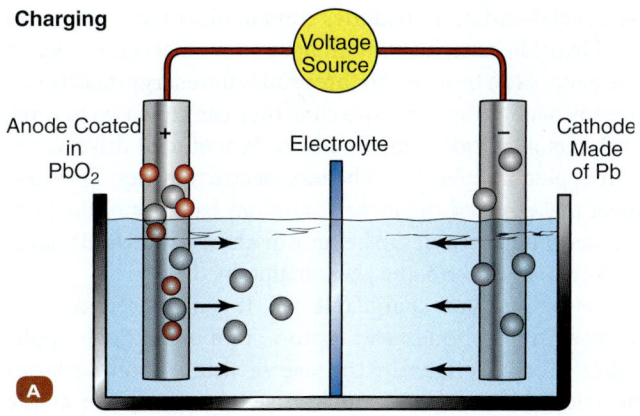 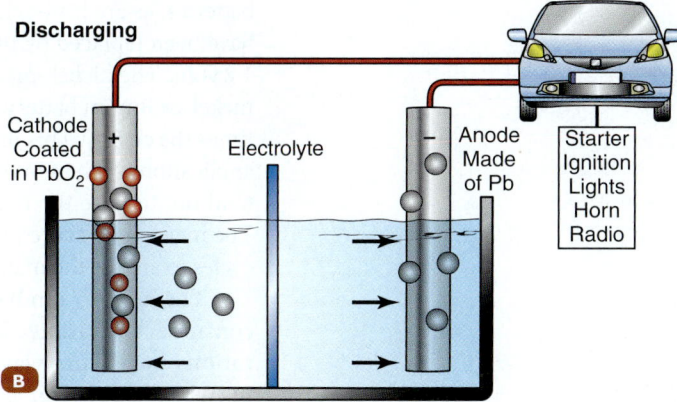

FIGURE 13-4 **A.** Charging cycle. **B.** Discharging cycle.

Battery Temperature Monitoring

Battery temperature plays an important role in the performance of a battery. Every battery has an ideal operating temperature range. In cold weather, battery performance drops dramatically due to a slowing of the chemical reaction process. At hotter temperatures, battery performance increases, but if the temperature increases too much, battery life may be compromised. Some vehicles monitor the battery temperature by placing a temperature probe in contact with the battery case (**FIGURE 13-5**). The power train control module (PCM) then controls the levels of current flow and the voltage at which the battery is charged. Generally speaking, the colder the battery temperature, the higher the rate of charging; the hotter the battery temperature, the slower the rate of charging. Smart chargers and battery management systems use the feedback from the temperature probe to manage the charge rate of a battery, ensuring the battery is at capacity and the temperature is within the optimal range, thereby promoting longer battery life.

FIGURE 13-5 Battery temperature sensor.

Battery Operating Conditions

Batteries have a range of operating conditions under which the best possible performance and lifetime can be achieved. They should be kept clean, dry, and fully charged; have minimal vibration and the correct level of electrolyte; be kept at a moderate temperature (e.g., 77°F [25°C]); and be well secured. A battery's lifetime is shortened by the following conditions:

- Being fully discharged or having deep discharge cycles
- Remaining overcharged or undercharged
- Experiencing high discharge rates for extended periods
- Experiencing excessive vibration
- Being exposed to extremes of temperature
- Having dirt or moisture around the case
- Developing corrosion

Rechargeable Cell Batteries

Batteries and cells continue to be developed to create more efficient, higher-density batteries. Consumer electronic devices such as mobile phones and laptop computers have been driving the development of new battery technologies such as nickel cadmium (Ni-Cd), nickel–metal hydride (Ni-MH), and lithium ion (Li-ion). All of these batteries are types of rechargeable cell batteries, as are lead-acid batteries. Nickel-cadmium batteries contain older technology and have been replaced by nickel–metal hydride batteries, as both have a nominal cell voltage of 1.2 volts. The nickel–metal hydride battery can have two to three times the energy density of a nickel-cadmium battery, which means that for the same size case, they can store two to three times the energy. This makes nickel–metal hydride batteries especially useful for drive motor applications such as in hybrid electric, plug-in hybrid, and battery electric vehicles. They also tend not to have the memory effect that plagued the nickel-cadmium batteries of the past. The memory effect required nickel-cadmium batteries to be fully discharged between charge cycles to ensure the maximum performance of the battery was maintained.

The lithium-ion battery is a newer type of rechargeable cell. It is now used in many consumer electronic devices such as smartphones and laptop computers. Such applications paved the way for further breakthroughs in the use of lithium-ion batteries in hybrid electric or battery electric vehicles; these vehicles use what is known as the rechargeable energy storage system (RESS). A lithium-ion battery has one of the highest energy density ratios of common batteries in production today. This high-energy density means the battery can store more energy than comparable batteries of other types. This is a real advantage for vehicle applications. Lithium-ion batteries also have a low self-discharge rate, which means they can sit on the shelf for long periods without discharging. However, they do have a shelf life, even if they are not used. Lithium-ion batteries typically last for up to five years from the date of manufacture, whether they are used or not.

The actual cell voltage depends on the final materials used to make the cell. The typical cell voltage for a lithium-ion battery is 3.6 volts. In contrast, cell voltage of a nickel-cadmium or nickel–metal hydride battery is 1.2 volts, and a typical lead-acid battery is 2.1 volts. Like all batteries, the lithium-ion cell has an anode, a cathode, and an electrolyte. When discharging, the lithium ions are removed from the anode and added into the cathode. When the cell is charging, the reverse occurs.

Lithium-ion batteries may suffer from **thermal runaway** and cell rupture if overheated or overcharged. In extreme cases, thermal runaway may result in an explosion. Extreme care should be taken when handling or charging lithium-ion batteries; always follow the manufacturer's recommendations. In hybrid or electric vehicle applications, many small, dry cell, battery-sized cells are connected in series and in parallel arrangements to form a battery pack that delivers the power requirements of the vehicle. Following safety precautions is extremely important: RESS battery packs develop voltages in excess of 200 volts.

Advantages of lithium-ion batteries include:

- High energy density: There is more power per pound.
- Low self-discharge: Self-discharge is typically less than half of that of nickel cadmium.
- Low maintenance: No periodic discharge is required.
- No memory.
- Low internal resistance: Suitable for high-current requirements.

Disadvantages of lithium-ion batteries include:

- Need for circuit protection to ensure current and voltage are within safe limits.
- Limited shelf life, even if not used.
- Sensitivity to high temperatures.
- Increased cost of manufacturing (however, these costs are being reduced as research improves the technology).
- Potential for damage if completely discharged.

Battery Maintenance

Batteries require regular maintenance, which consists of inspection, cleaning, testing, and charging when discharged. Good times to check batteries are during oil changes and in the fall, prior to the arrival of cold weather, which makes the battery work harder.

Ensure the battery's electrolyte level is correct by checking the markings on the case or by looking in each cell to ensure the plates are well covered (**FIGURE 13-6**). Make sure the exterior case is dry and free of dirt. Dirt on top of the battery can actually cause

FIGURE 13-6 Battery with caps removed showing the electrolyte level over the plates.

FIGURE 13-7 Dirty battery top.

premature self-discharge of the battery as current "leaks" across the path of dirt or grime (**FIGURE 13-7**). This mixture can become conductive and drain the battery over time. To tell if the surface of the battery needs to be cleaned, use a digital volt-ohm meter (DVOM) set to volts to measure the voltage on the surface of the top of the battery. You can do this by placing the black lead on the negative battery post and rubbing the red lead around the top of the battery, measuring the voltage present there (**FIGURE 13-8**). Any voltage over about 0.2 volts means the battery should be cleaned with a mixture of baking soda and water. Take care not to get any of that mixture down inside of the battery: it may neutralize the electrolyte, which ruins the battery.

Keeping the battery and terminals clean is one of the most cost-effective maintenance tasks for vehicles (**FIGURE 13-9**). The lead in the battery posts and terminals oxidizes over time. Unfortunately, lead oxide is an insulator. Therefore, if the lead surfaces of the post and terminal oxidize, the insulator effect can cause the vehicle to not crank over, stranding the driver. Lead oxide can be identified by its dark gray or black color. The only effective ways of removing it are with a special battery terminal scraper or a wire brush, which will restore the lead to its shiny silver color (**FIGURE 13-10**). Once the battery terminal is reinstalled and tightened, the terminals should be coated with a battery oxidation inhibitor spray to help slow down the oxidation process (**FIGURE 13-11**).

Knowing how to properly diagnose and recharge a battery is an important task that you will need to accomplish regularly. Performing a battery state-of-charge test with a hydrometer or refractometer is a good indicator of whether the battery needs to be charged

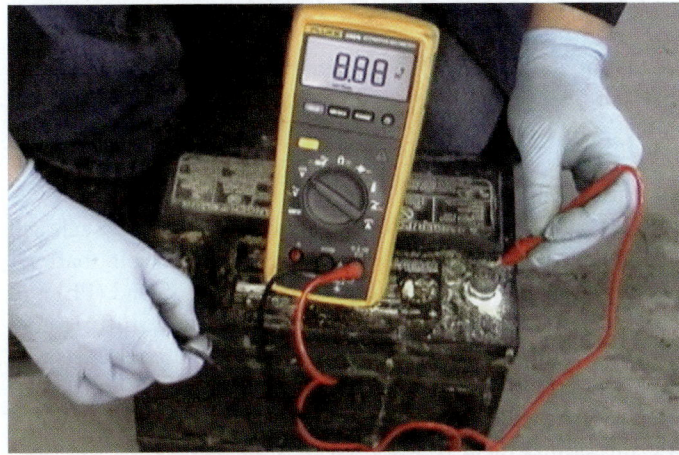

FIGURE 13-8 DVOM being used to test for voltage leaks across the battery top.

FIGURE 13-9 Battery terminals with felt corrosion inhibitors installed.

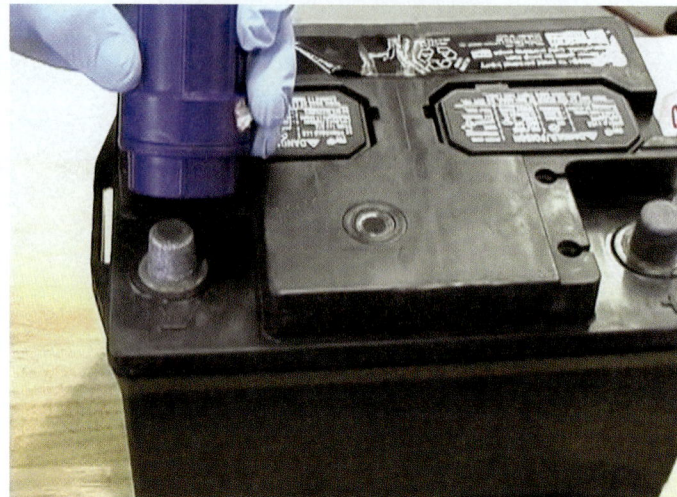

FIGURE 13-10 Battery terminals being cleaned.

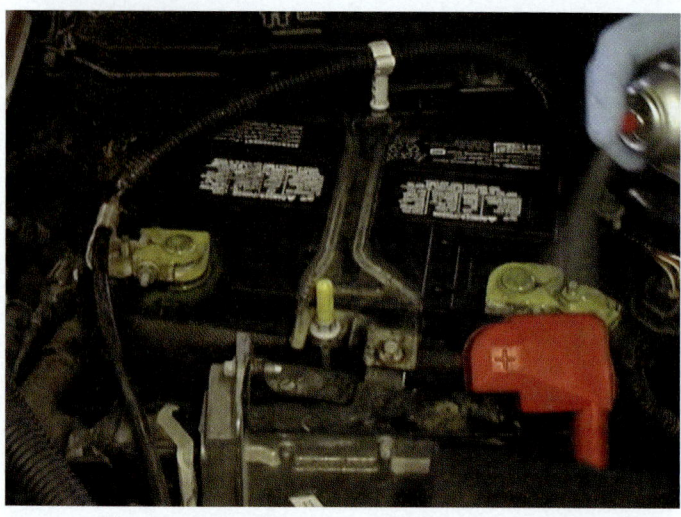

FIGURE 13-11 Battery terminals coated with corrosion inhibitor spray.

FIGURE 13-12 Refractometer being used to determine battery state of charge.

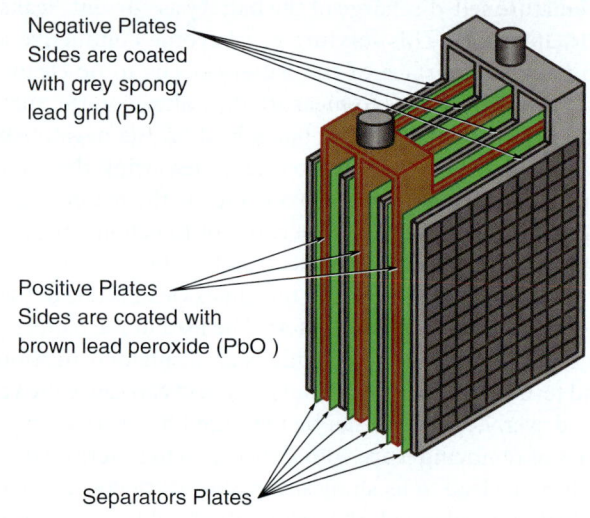

FIGURE 13-13 Typical plate arrangement in a wet cell battery.

or not (**FIGURE 13-12**). Another maintenance check of the battery is the conductance test, which indicates the capacity of the battery and how much life is left in the battery.

▶ Lead Acid, Gel Cell, and AGM Batteries

K13003 Describe lead acid, gel cell, and AGM batteries.

N13001 Identify safety precautions for high voltage systems on electric, hybrid, hybrid-electric, and diesel vehicles.

The wet cell lead-acid battery is the main storage device in automotive use. It is called a flooded cell battery because the lead plates are immersed in a water–acid electrolyte solution. An automotive battery can supply very high discharge currents while maintaining a high voltage, which is useful when cold starting. It gives a high power output for its compact size, and it is rechargeable.

The standard 12-volt car battery consists of six cells connected in series. Each cell has a nominal 2.1 volts, for a total of 12.6 volts for a fully charged "12-volt" battery. Each cell contains at least two sets of electrodes (called plates), one set of lead (Pb) and the other set of lead dioxide (PbO_2), in an electrolyte solution of diluted sulfuric acid (H_2SO_4) (**FIGURE 13-13**). In between these plates is a separator, which prevents a short circuit across the cells through physical movement, as well as dendrite growth protruding from the cells as the battery ages (**FIGURE 13-14**).

As the battery discharges, the sulfuric acid is absorbed into the lead plates and both of the plates slowly turn into lead sulfate. At the same time, the strength of the electrolyte becomes less acidic as the acid is absorbed into the plates. Recharging the battery reverses this

FIGURE 13-14 Separator between the plates of a lead acid battery.

FIGURE 13-15 Exploded lead acid battery.

process. In a conventional open wet cell battery, charging generates hydrogen and oxygen gas by separating them from the water in the electrolyte, creating a highly explosive mix. Overcharging or rapid charging causes some gas to escape from the battery; this is called **gassing**. It is important to keep all sources of spark or flame away from the battery, especially while the battery is charging. Failure to do so could result in the ignition of the oxygen/hydrogen gas, causing the battery to explode and spray sulfuric acid, lead, and battery case material in a violent fashion (**FIGURE 13-15**). While this can obviously damage a vehicle, more importantly, it poses a serious danger to the technician. Sulfuric acid contained in batteries is highly corrosive and can be very harmful to metal, painted surfaces, and skin. For this reason, it is critical to always wear protective clothing and use extra care when handling batteries. In the event of a battery-acid spill, a solution of water and baking soda will neutralize the acid and allow for safe cleanup.

The nominal 2.1 volts of each cell does not depend on the size of the cell; however, its current capacity does. The surface area of the plates in a cell determines the cell's current capacity. In a lead-acid battery, positively and negatively charged plates are assembled so that they are alternated. All of the positive plates are connected to each other in parallel, and all of the negative plates are connected to each other in parallel (**FIGURE 13-16**). The more plates that are connected in series, the greater the current capacity of the cell. Since the plates are arranged alternately, they need to be close to each other, but not touching. If they touch, an internal short circuit in the cell will occur, which is typically what has happened in a battery with a "dead" cell; such a condition would result in a battery voltage of less than 10.5 volts. Once the positive and negative plates have shorted together, the cell discharges completely. Normally, the plates are kept from touching each other by separators, which are usually made of plastic. The battery's six 2.1-volt cells are connected in series to form the battery, which gives the battery a nominal voltage of 12.6 volts. The cells are sealed from each other and filled with dilute sulfuric acid. The battery case is usually made of plastic or hard rubber. With the cells in series, one end of the battery is connected to the negative post and the other end is connected to the positive post (**FIGURE 13-17**).

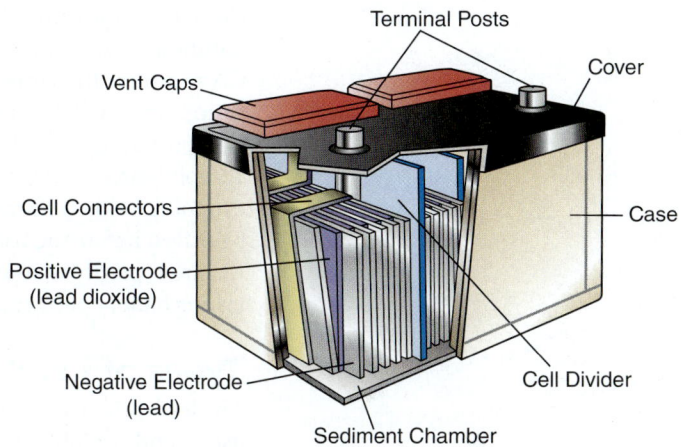

FIGURE 13-16 Typical plate arrangement in a wet cell battery.

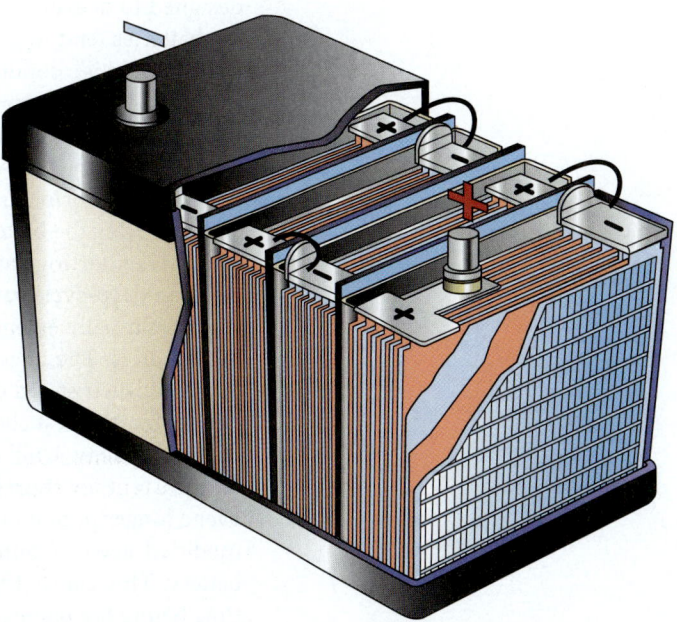

FIGURE 13-17 Interconnections between all six cells in a battery showing the most negative and positive points of the battery.

A relatively new battery technology is the hybrid lithium-ion supercapacitor. This technology has been developed to phase out lead-acid batteries, which are heavy, toxic, and pose a danger to the environment. These "hybrid" batteries combine the quick charge/discharge characteristics of capacitors and the high energy density of lithium-ion batteries. This is a promising technological development that is being implemented in Europe. Should the technology prove effective, you can expect to see these batteries in your service bay in future.

Battery Ratings

The **electrical capacity**, or the amount of charge a typical lead-acid battery can store, is determined primarily by the total surface area of the plates, as well as their thickness. The more plate surface area there is, the higher the electrical capacity of the battery. Automotive battery plates are usually manufactured in a standard size, so a common method of increasing surface area is to increase the number of plates per cell. For example, a 12-volt, 11-plate cell will have a higher capacity than a 12-volt, 9-plate cell.

There are several methods used to rate automotive battery capacity. The three most common are **cold cranking amps (CCA), cranking amps (CA),** and **reserve capacity (RC)**. CCA measures the load in amps that a battery can deliver for 30 seconds while maintaining a voltage of 1.2 volts per cell (7.2 volts for a 12-volt battery) or higher at 0°F (−18°C). CA measures the same thing, but at a higher temperature—32°F (0°C). This can cause confusion since a 500-CCA battery has about 20% more capacity than a 500-CA battery, so it is important to keep the ratings straight. RC is the time in minutes that a new, fully charged 12-volt battery at 80°F (27°C) will supply a constant load of 25 amps without its voltage dropping below 10.5 volts. This rating approximates the amount of time that a vehicle can be driven before the battery dies, if the charging system fails completely. The CCA, CA, and RC are typically marked on automotive batteries, and technicians use these ratings when testing battery performance and when selecting batteries for particular vehicle applications.

Types of Lead Acid Batteries

The lead-acid battery has been around for more than 150 years. This type of battery is very useful and flexible. There are several types of lead-acid batteries that are in wide use today. The two main types of lead-acid batteries that are encountered in the automotive industry are starting batteries and deep-cycle batteries.

Starting batteries are used in most automotive applications. This type of battery is designed to delivered high current in short bursts, usually for one minute at a time. Starting batteries tend to give the longest service life when they are not deep cycled repeatedly. Deep cycling is a condition which occurs when the batteries are taken from fully charged to fully discharged. Deep cycling a starting battery too many time will almost certainly lead to premature battery failure. Starting batteries are designed so that their plates do not extend all the way to the bottom of the battery. This prevents the lead particles that shed from the plates during normal use from shorting out a battery cell, thereby reducing battery capacity and rendering the battery unserviceable.

Unlike starting batteries, deep-cycle batteries are designed to be cycled repeatedly; the plates in deep-cycle batteries are much more robust than those found in starting batteries, and they usually extend further down into the battery case than the plates found in starting batteries. The majority of deep-cycle batteries are designed to be cycled down to 45% to 75% of their stated capacity; deep-cycle batteries can be cycled down to a lower state of charge, but the best charging/use strategy for battery longevity is to limit cycling to 45% of battery capacity. Unlike starting batteries, deep-cycle batteries are not designed to deliver high current for short bursts. Instead, they are intended to deliver a lower current output over a longer period of time. In some automotive applications, such as vehicles with highly modified stereo systems, it is desirable to install a deep-cycle battery in place of the starting battery. This allows the vehicle's accessories to be operated for a much longer period of time before the battery discharges to a point that prohibits vehicle starting. It is important to note that since deep-cycle batteries cannot deliver the same engine cranking current as a similarly sized starting battery, it may be necessary to install a deep-cycle battery that is physically larger than the starting battery being replaced in order to maintain acceptable starting performance.

No matter the type of lead-acid battery being discussed, proper mounting of the battery is crucial, both for safety reasons and for battery longevity. Batteries must be fastened down securely to prevent them from becoming airborne in the event of a collision. Batteries that aren't well secured incur needless excessive vibration, which can damage them. It is important to note that those battery hold-downs are there for a reason! A battery should never be installed in a vehicle without appropriate hold-downs in place. A loose battery can bounce up when a vehicle goes over bumps, which can cause the terminals to short out on the vehicle's hood; this can also cause excessive material to fall off from the plates, leading to premature battery failure.

Gel Cell Batteries

Gel cell batteries, also called valve-regulated lead-acid (VRLA) batteries, contain a gel-like electrolyte consisting of sulfuric acid mixed with fumed silica. This mixture transforms the once-liquid sulfuric acid into an immovable gel-like mass, which prevents electrolyte spillage and greatly reduces battery evaporation. Gel cell batteries are called "valve-regulated" because they contain a safety that is designed to vent internal gas pressure if levels raise much above 2 psi. In all other circumstances, the vented gases are contained inside of the battery (**FIGURE 13-18**).

A key benefit to the gel-like electrolyte is that the compound allows gas regeneration to take place; essentially, this is when the oxygen and hydrogen released from the electrolyte during battery charging is trapped in the gel electrolyte, instead of being vented from the battery as occurs in a typical flooded cell lead-acid battery. Gel cell batteries also use significantly less electrolyte than comparably sized lead-acid batteries, which gives them the moniker of an "acid-starved" design. Chemically, these batteries are similar to traditional flooded cell lead-acid batteries, except that the antimony in the lead plate is replaced with calcium.

This technology has been around in one form or another since the early 1930s, when it was discovered that adding silica to sulfuric acid eliminated electrolyte drainage, and made the batteries much more tolerant of vibrational shock. These early batteries were used to power portable field radios used for military communication. Gel cell batteries can be mounted in virtually any position, they can withstand severe levels of abuse, and they are must less sensitive to temperature extremes than traditional flooded cell lead-acid batteries. This attribute makes these batteries very useful for custom installations, off-road oriented vehicles, and other applications requiring shock and temperature resistance.

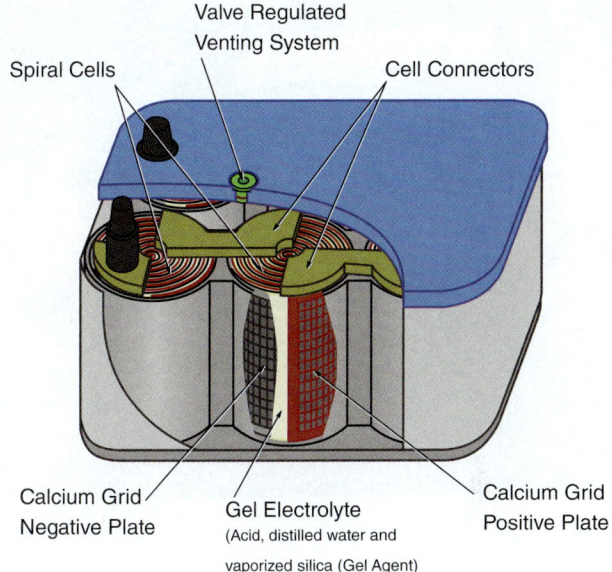

FIGURE 13-18 Gel cell battery cutaway showing gelled electrolyte.

AGM Batteries

Another emergent battery technology that is being incorporated into automotive design is the absorbent glass mat (AGM) battery, which is also known as a sealed lead-acid (SLA) battery. AGM batteries are commonly mistaken for gel cell batteries, even though the internal construction of these two battery technologies is very different. AGM batteries use an extremely absorbent micro-pore fiberglass material that traps the gases vented during the discharging and recharging processes; this differs from a gel cell battery, which uses the gel electrolyte to trap the gases being produced (**FIGURE 13-19**). For a given battery size, AGM batteries contain more energy than gel cell batteries. Because AGM batteries have a lower internal resistance than gel cell batteries, they are better suited to deliver high bursts of current.

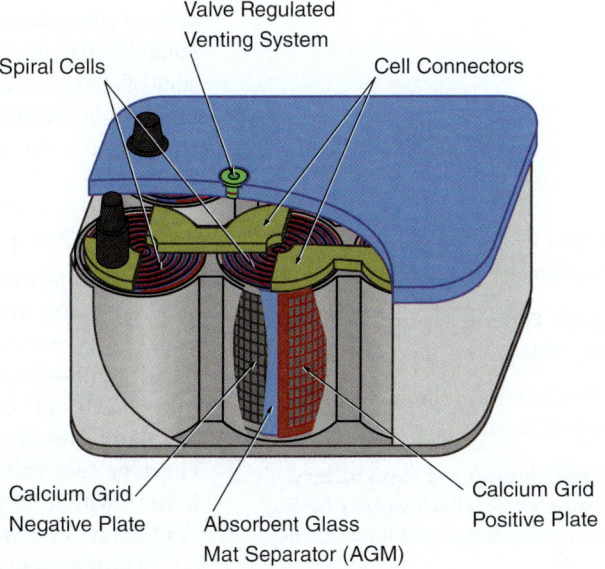

FIGURE 13-19 Cutaway of an AGM battery.

Like gel cell batteries, AGM batteries contain less electrolyte than a comparable flooded cell lead-acid battery. AGM batteries are also extremely tolerant of vibration and shock, and can be mounted in any position. AGMs are filled with just enough liquid electrolyte to completely wet the absorbent glass mat sheets. Like gel cell batteries, this allows AGM batteries to spill no electrolyte if the case should be broken for any reason. Typically, AGM batteries are cheaper to produce than gel cell batteries, and as such, this type of battery is more prevalent in automotive applications than gel cell batteries.

Special Charging Instructions Associated with Gel Cell and AGM Type Batteries

Due to the sensitive nature of AGM and gel cell batteries, special procedures and equipment are needed to ensure maximum battery service life. The first step to take is to avoid pushing too many amps into the battery during charging. This applies to any battery, be it AGM, gel cell, or flooded cell lead-acid. Overloading amperage causes the battery to overheat, which boils the electrolyte, resulting in the electrolyte filling less of the case. Once enough electrolyte has boiled off, portions of the plates can be exposed to air, which causes instant sulfation of the battery plates. Optima Batteries, in accordance with various other producers of AGM and gel cell batteries, recommends that battery charge current not exceed 10 amps.

The best practice to avoid ruining a costly battery is to seek and purchase a SMART battery charger that is capable of three-stage charging and rated for AGM battery use (**FIGURE 13-20**). The three stages of charging are as follows: The first stage, bulk charging, reintroduces 80% of the battery's energy capacity back into the battery. This is accomplished by charging the battery at exactly 14.4 volts, and delivering the maximum amount of amperage to the battery that the charger's software deems to be safe. The second stage is called the absorption stage. This stage takes the battery to a 98% state of charge by maintaining a voltage level of 14.4 volts and gradually reducing the amperage being fed into the battery as the state of charge approaches the 98% mark. The third and final charging stage is the float stage. During this stage, the charger voltage is regulated to 13.4 volts, and less than 1 amp is fed into the battery. This takes the battery's state of charge to the 100% mark. The float stage can maintain a battery at 100% state of charge for quite some time without heating the battery and causing the electrolyte to boil. While smart chargers are expensive, they are required pieces of equipment to properly charge and maintain AGM and gel cell batteries.

FIGURE 13-20 SMART battery charger hooked up to an AGM battery.

K13004 Explain the battery testing procedure.

S13001 Perform a battery state-of-charge test.

N13002 Perform battery state-of-charge test; determine needed action.

N13003 Inspect and clean battery; fill battery cells; check battery cables, connectors, clamps, and hold-downs.

▶ Battery Testing Procedure

When it comes to determining a battery's state of health, one of the tests used is the state-of-charge test. State-of-charge (SOC) testing tells us how charged or discharged a battery is, not how much capacity it has. The degree of the battery's charge is critical information when testing starting and charging system issues, as well as just about any other electrical issue. An easy and effective way to test for state of charge is to leave the headlamps on for three minutes to take any surface charge away from the battery, then check battery voltage with a voltmeter. A fully charged battery will measure 12.6 volts, a battery with a 75% SOC will measure 12.4, and a battery with a 50% or lower SOC will read below 12.2 volts.

Another way to accurately measure a battery's SOC is to measure the acid content, or specific gravity of the electrolyte; remember that the acid level drops as the battery

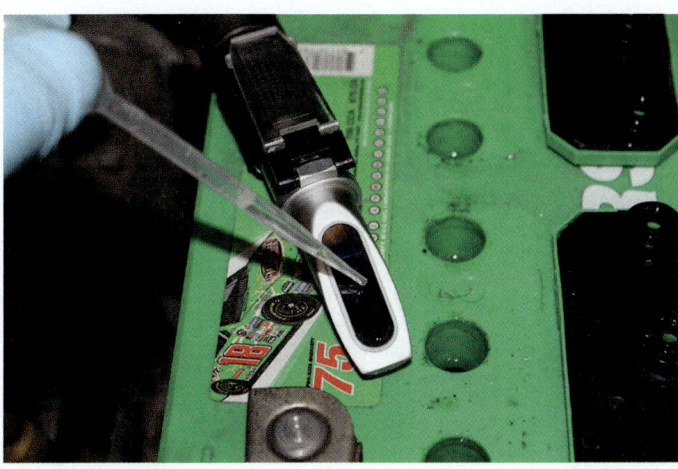

FIGURE 13-21 Specific gravity of an electrolyte being tested.

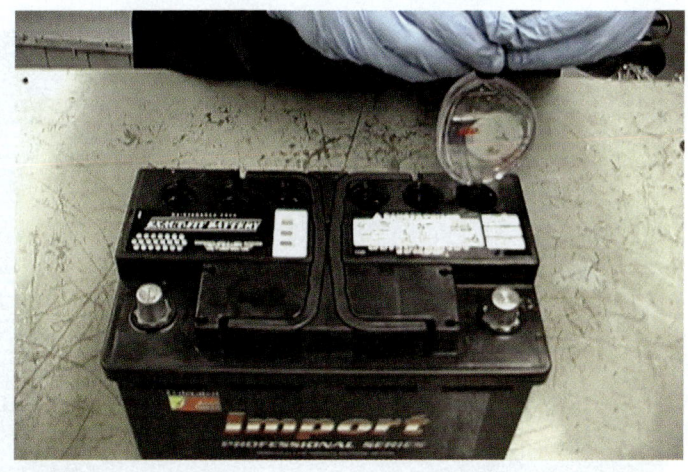

FIGURE 13-22 Hydrometer being used to test battery SOC.

becomes discharged. One of the most accurate ways of testing the electrolyte's specific gravity is to use a hydrometer or refractometer to measure the specific gravity of the electrolyte in each of the battery's cells. The higher the specific gravity, the higher the percentage of acid in the electrolyte, which corresponds to a high battery state of charge (**FIGURE 13-21**). A refractometer is able to give a precise measurement of specific gravity in one easy step. Hydrometers, on the other hand, must be corrected for electrolyte temperature, which adds a step to the process and introduces another variable (**FIGURE 13-22**). Also, since the hydrometer draws electrolyte into it to raise a float, the electrolyte level must be at least slightly above the top of the plates. If it is not, then distilled water will need to be added and the battery will need to be fully charged. It is important to note that while the refractometer and hydrometer are capable of producing very accurate SOC measurements, they are all but useless on the modern, sealed lead-acid batteries of today, which are often maintenance free. In some cases, the caps on the top of maintenance-free batteries can be pried off, which allows access to the electrolyte, however, removing the caps presents the risk of damaging the battery and spilling electrolyte.

To perform a battery state-of-charge test, follow the steps in **SKILL DRILL 13-1**.

SKILL DRILL 13-1 Performing a Battery State-of-Charge Test

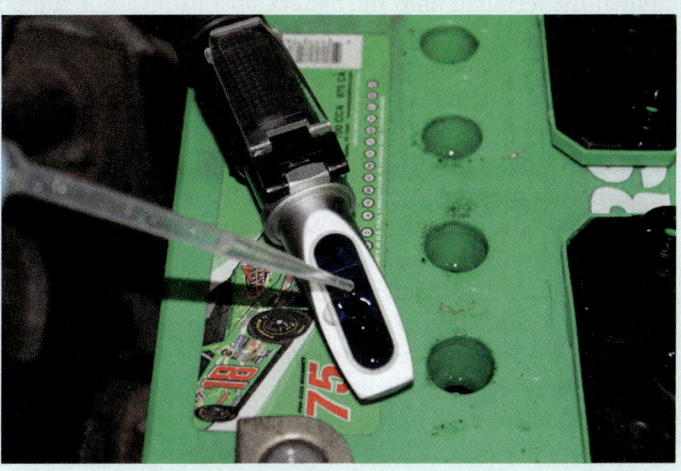

1. Check and adjust the fluid level. Charge the battery if distilled water is added. Test the specific gravity of each of the cells by using a hydrometer designed for battery testing. Draw some of the electrolyte into the tester and read the scale.

Continued

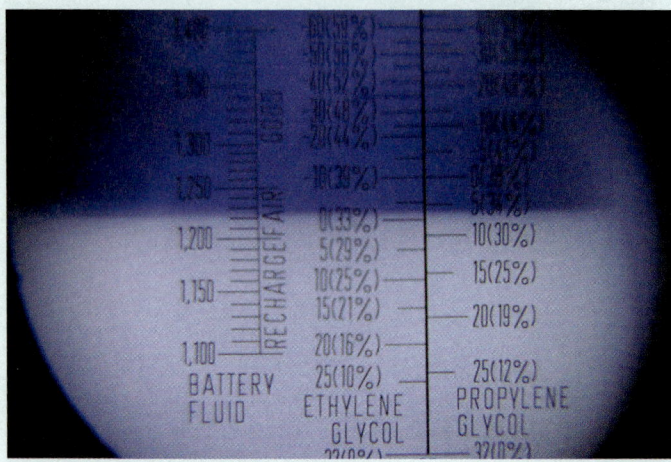

2. To test the specific gravity with a refractometer, place a drop or two of electrolyte on the specimen window and lower the cover plate. Look into the eyepiece with the refractometer under a bright light. Read the scale for battery acid. The point where the dark area meets the light area is the reading.

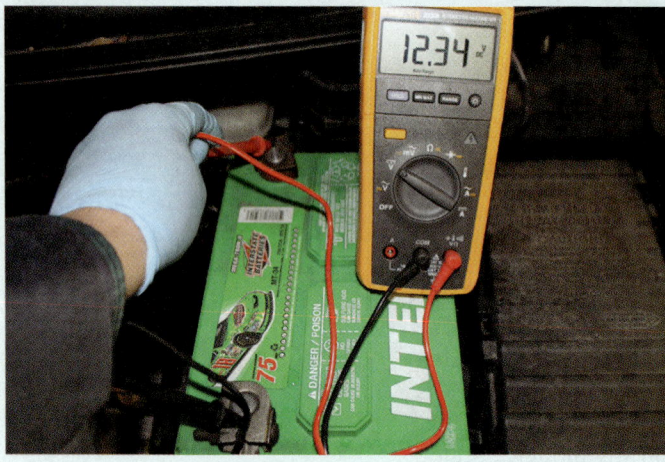

3. To conduct an open circuit voltage test, select the "volts DC" position on your DVOM and attach the probes to the battery terminals (red to positive, black to negative).

Battery Cables, Clamps, and Connections

Battery cables and terminals are designed to carry the high discharge currents that are required during cranking of the automotive engine. Battery cable terminals are usually made of lead, zinc-plated brass, or zinc alloy (**FIGURE 13-23**). There are a number of designs that allow the battery cables to be connected to the starting battery, with the most common being a cone design that provides a large surface contact area with the ability to tighten the terminal onto the battery post using a nut and bolt (**FIGURE 13-24**). The tapered cone allows easy removal of the clamp when the bolt is loosened. The larger positive and smaller negative terminals are slightly different in size to ensure correct connection of the cables when installing or servicing the battery (**FIGURE 13-25**).

Another type of battery terminal is the side terminal. It gets its name because the battery connection is on the side of the battery. The terminal is a flat circle with a center bolt that bolts the terminal tightly to the battery connection (**FIGURE 13-26**). A benefit of side-terminal batteries is that they seem to avoid some, but not all, of the oxidation and corrosion issues of a top-post battery. One shortcoming of a side-terminal battery is that it is harder to get a good connection when using jumper cables or a jumper box to jump start the vehicle. One less-common battery terminal is a flat terminal with a hole through the center (**FIGURE 13-27**). The post on the battery sticks up vertically with at least one side

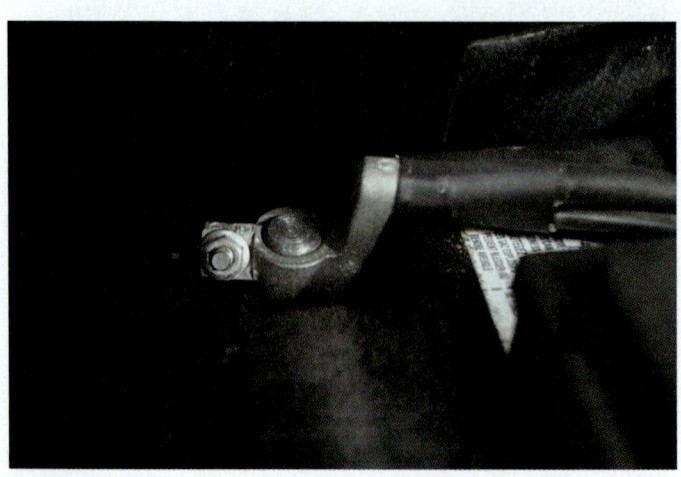

FIGURE 13-23 Types of battery cable terminals.

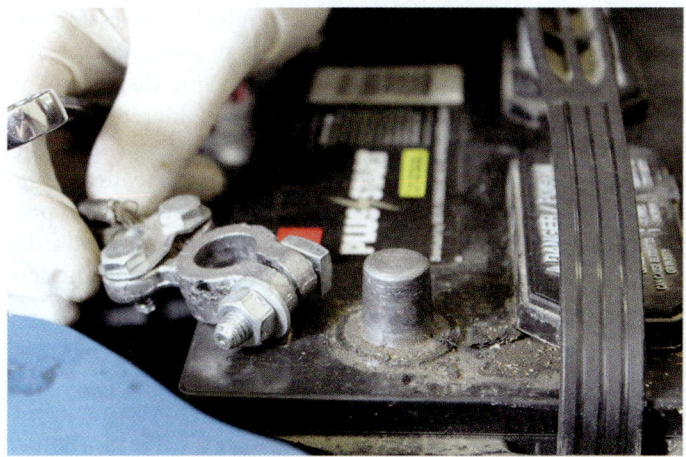

FIGURE 13-24 Cone-type battery post and associated clamps.

FIGURE 13-25 Battery terminals.

FIGURE 13-26 Side terminal battery. **A.** With battery cables attached. **B.** Without cables attached.

FIGURE 13-27 Flat post battery terminal.

FIGURE 13-28 Flat post battery terminal with battery cable installed.

flat and a matching hole through the post. The battery terminal butts up against the flat post, and a bolt inserted through the battery post and terminal holds them firmly together (**FIGURE 13-28**).

Battery cables are usually made of many fine strands of copper wire bound tightly together and insulated to make a cable with high current capacity. There are a number of cable sizes available to handle various current capacities. For example,

FIGURE 13-29 Crimped-on battery terminal.

FIGURE 13-30 Voltage drop test being performed on a battery cable.

a small-displacement spark ignition engine requires far less current to crank than, for example, a high-compression diesel (compression ignition) engine of the same displacement. Battery terminals are usually crimped or soldered onto the battery cables to ensure strong, low-resistance connections (**FIGURE 13-29**). Often, heat-shrink tubing with sealing adhesive is used over the joint to ensure it is kept clean and protected from corrosion. Battery cables should be protected from chafing or damage, and terminals should be kept clean and free of corrosion.

Testing for Defective Battery Cables

Battery cables are designed to carry great amounts of current, and typical service life is several years. However, like most other components on a vehicle, battery cables are exposed to severe operating conditions throughout their service life, and as such, they will eventually fail and need to be replaced. There are a couple types of ways in which battery cables can fail. The first failure mode is an open cable, which is often present in areas where the cable flexes repeatedly. This is most often encountered in the sections of cable that carry battery current to the engine assembly. Another way in which a battery cable can fail is when the insulation surrounding the brass conductor is chaffed or damaged. In these cases, battery positive may be allowed to short to ground where the insulation has failed, potentially causing a vehicle fire. Note that battery cables that have shorted will often be charred at the contact point. Any breaches in the cable's insulation also leave it vulnerable to the elements. This brings us another mode of failure that plagues battery cables: corrosion. Most often we think that corrosion is something that is easily seen, but in the case of battery cables, it is possible for a cable to be filled with resistance-inducing corrosion inside of the insulating jacket, but have no visible flaws when viewed from the outside. The most effective way to test for any and all types of battery cable failures is to perform a voltage drop test of each battery cable that stretches from the battery to the engine block and starter (**FIGURE 13-30**). Remember that voltage drop tests MUST be conducted with the circuit under a working load; in this case, the engine must be cranking before accurate voltage drop tests can be performed on the battery cables. Also note that the areas of the battery cable that drop excessive voltage will have high resistance, and as a result, will get much hotter during attempted engine cranking than a cable that is in good condition.

Testing for Defective Battery Clamps

Battery clamps provide both a mechanical connection and an electrical connection between the battery and the battery cables. In order for all of the current required for starting to reach the engine starter, the battery clamps must be in good condition; tightened securely with no play between battery and clamp; and free of debris, corrosion, grease, and oil. Problems associated with the battery clamps can often be identified by performing a voltage drop test between the battery terminal and the battery clamp terminal (**FIGURE 13-31**).

Battery Corrosion and Related Concerns

As in other electrical circuits, corrosion and oxidation result in electrical resistance. It is critical that the current path from the battery to the starter have as little resistance as possible; essentially, it must be free of corrosion. Battery corrosion occurs for many reasons. It can happen when the dissimilar metals contained in battery clamps and battery cables are exposed to road salt and water. This causes galvanic corrosion, which eats at the metals of the battery cables and clamps. Another source of battery corrosion is the electrolyte itself. Battery electrolyte is generally sealed within the battery case, but as the battery ages, it can seep out around the sealed vent caps and the battery terminal posts, causing the fuzzy growths that occur so often on battery posts and clamps (**FIGURE 13-32**). This corrosion creates electrical resistance, which results in heat buildup and causes excessive voltage drop at the site, which in turn lead to slow engine cranking.

Another type of harmful battery corrosion occurs when battery acid spills from the battery case and begins to eat away at the materials located below and behind the battery. Materials below the battery are damaged when the electrolyte drips down; objects behind the battery are damaged when acid is blown back from the headwind created by driving the vehicle at speed. Objects that are likely to be damaged by wayward battery electrolyte are intake piping, wiring harnesses, body sheet metal, and battery cables and clamps. In addition to causing physical damage, electrolyte seepage can cause corrosion to spread across the battery between the battery posts. This corrosion is a mix of dirt, rubber particles from the road, salt, electrolyte, and other environmental substances. It can lead to a high resistance short across the battery terminals, which will discharge the battery if left unaddressed (**FIGURE 13-33**).

Inspecting, Cleaning, Filling, and Replacing the Battery and Cables

Batteries, cables, ground connection points, and terminals should be inspected regularly for corrosion and wear. Once corrosion is found, the first order of business should be to neutralize the electrolyte that has leaked from the case. This can be accomplished by using an approved battery cleaner, or with a 50/50 solution of baking soda and water (**FIGURE 13-34**). Once the battery acid has been neutralized, flush the area thoroughly with water, then carefully dry with compressed air.

Now you may begin removing any corrosion that is present on the battery terminals, clamps, cables, and other connections. This is accomplished by the use of a file, a battery terminal brush, and/or a nylon bristle brush. A file is used to remove heavy corrosion from battery cable terminals when needed. Corrosion on

FIGURE 13-31 Voltage drop test being carried out between a battery post and battery clamp.

FIGURE 13-32 Corroded battery post and clamp.

FIGURE 13-33 Battery with corrosion across the top between the battery posts.

FIGURE 13-34 Battery being cleaned.

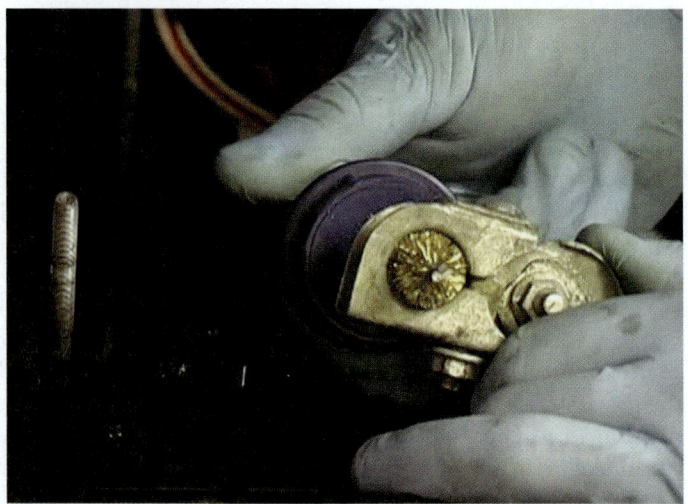

FIGURE 13-35 Battery terminal brush being used to clean battery terminal.

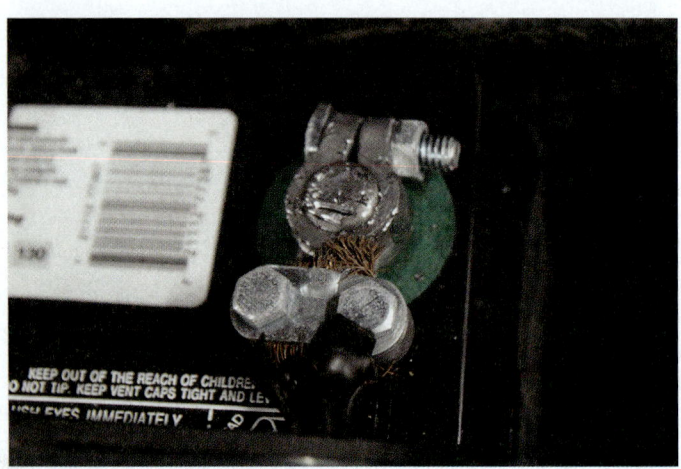

FIGURE 13-36 Generic top-post replacement battery clamp.

the battery terminals or posts is usually best removed with a battery terminal brush (**FIGURE 13-35**). This is accomplished by first removing the negative battery cable and then the positive cable. After the cables have been removed from the battery, the brush can be used to remove corrosion.

Frequently, corrosion will eat away at the bolts that hold the battery clamps tight, as well as the battery clamps themselves. This condition usually becomes evident when attempting to remove the battery cable clamps for cleaning. It may happen that the clamp bolt breaks when attempts are made to loosen the clamp, but the battery clamp body remains in serviceable condition. In these cases, it is a far better practice to replace the bolt in the existing clamp than to replace the clamp with a generic replacement. Cutting the factory clamp off and installing a replacement allows yet another place for corrosion to creep into the battery cables (**FIGURE 13-36**). In most cases, when the battery clamps are found to be faulty, the preferable solution is to replace the entire battery cable. This will ensure the lowest possible voltage drop and the highest amount of starting current at the starter.

When the battery is found to be below recommended capacity, it is best practice to replace the battery. It is important that the battery is replaced with one of the same size, type, and rating in order to ensure proper vehicle operation. Once the appropriate battery is obtained, the battery should be securely mounted in the battery tray and the battery cables should be connected: positive first, then negative.

Battery Charging

S13002 Charge a battery.

N13004 Perform slow/fast battery charge according to manufacturer's recommendations.

Vehicle batteries may become discharged and require charging, particularly if the vehicle has been sitting idle without starting for more than a couple of weeks or if the charging system is faulty. Ideally, a battery is fully charged before being tested for faults. If charging is needed, slow charging is less stressful on a battery than fast charging is; proper charging can take more than 20 hours if the state of charge is very low. Some manufacturers recommend removing the negative battery terminal while charging a battery to reduce risk to the vehicle's electronics. If the battery needs to be disconnected during charging, remember to verify if the vehicle's adaptive memory needs to be maintained. After charging the battery, it is good practice to clean the battery posts and cable terminals.

To charge a battery, follow the steps in **SKILL DRILL 13-2**.

SKILL DRILL 13-2 Charging a Battery

1. Verify whether the battery needs to be disconnected during charging and, if so, whether the adaptive memory needs to be maintained. Inspect the battery casing and ensure that the battery has not been frozen. Verify that the charger is unplugged and off. Connect the red lead to the positive terminal and the black lead to the negative terminal.

2. Turn the charger on to a slow, automatic charge.

3. Once the battery is charged, turn the charger off. Allow the battery to stand for at least 5 minutes before testing it. Using a capacitance tester, load tester, refractometer, or hydrometer, test the charged state of the battery.

Jump Starting Procedures

Before attempting to jump start a vehicle, you need to make sure the battery is not frozen, as you cannot jump start a frozen battery. Before you connect the service battery to the discharged battery, it is good practice to place a load across the discharged battery (this may be accomplished by turning on the headlamps) to absorb any sudden rise in voltage that may occur as the alternator suddenly increases its output. There are many sensitive electronic devices in most modern vehicles that are very susceptible to voltage surges. One method

S13003 Jump start a vehicle.

N13005 Jump-start vehicle using jumper cables and a booster battery or an auxiliary power supply.

of reducing the risk of damage to such devices is using jumper leads that have a built-in surge protector. Separate surge protector devices can also be used to reduce the possibility of such surges and voltage spikes.

An option that is less risky when jump starting a vehicle is the use of a jump box. A jump box does not involve a charging system, thereby lessening the risk of voltage spikes. This device contains a relatively large battery and has short cables and spring-loaded clamps attached to it that can be connected directly to the dead battery. It produces enough electrical energy to start most vehicles (**FIGURE 13-37**).

Another risk to jump starting is overheating and damaging the vehicle's alternator. With all of the electrical draws on modern vehicles, today's alternators must work harder than their counterparts from the past. Adding the job of fully recharging the battery on top of the regular electrical loads can push an alternator into the danger zone. Because of the risk of damaging the electronics and alternator, many manufacturers (and tow companies) are now refusing to jump start newer vehicles or authorize that practice. They now recommend either externally charging the battery or replacing the battery with a charged battery.

To jump start a vehicle, follow the steps in **SKILL DRILL 13-3**.

Testing Battery Conductance

When you doubt that a battery can meet adequately the demands placed upon it, it should be tested. The use of electronic battery testers has largely replaced the practice of hydrometer and other invasive types of battery testing. In fact, many manufacturers now require the use of a conductance test instead of a high-amperage load test in order for warranty coverage to be considered (**FIGURE 13-38**). Many of the testers have integrated printers; for the battery to be warranted, a printout of the test result must accompany the replaced battery.

FIGURE 13-37 Jump pack connected to a battery.

S13004 Load test a battery.

The conductance tester sends low-frequency signals into the cells to determine the battery's ability to conduct current. The greater the ability to conduct current, the higher the CCA capacity of the battery. Since batteries deteriorate over time, their CCA capacity also deteriorates. The conductance tester is able to predict a battery's CCA capacity by measuring its conductance, which provides a near-linear comparison between the two. Since conductance testing takes only a minute or two to complete, it is a good way to show customers the status of their battery and how that condition changes over time. If the conductance test shows that the battery no longer meets specifications, then customers will know that the battery is at the end of its useful life and must be replaced to avoid it failing and leaving them stranded.

Testing Battery Capacity

The load test has been used for years to test a battery's capacity and internal condition, but some manufacturers indicate that their batteries should not be load tested, and instead should be conductance tested. They claim that load testing can damage the battery. Therefore, always check the service information before performing a load test.

As the name suggests, the load test subjects the battery to a high rate of discharge, and the voltage is then measured after a set time to see how well the battery creates that current flow. In other words, if the battery can maintain a high rate of discharge for a specified time and the voltage is still relatively high, then you know the battery is in good shape. Conversely, if the voltage falls off fairly quickly, the battery is not in very good condition. Imagine two people running a mile sprint. The one who does it in 5 minutes is likely in pretty good shape. The one who takes 20 minutes is likely in poor shape. Batteries are much the same. The faster the rate at which a battery can create current flow, the higher its voltage, indicating a higher

SKILL DRILL 13-3 Jump Starting a Vehicle

1. Position the vehicle with the charged battery close to the vehicle with the discharged battery, but not touching. Connect the red or orange lead to the positive terminal of the discharged battery.

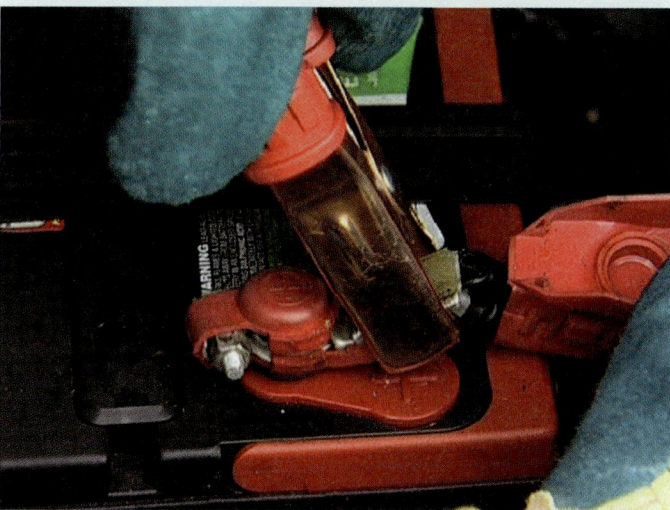

2. Connect the other end of this red or orange lead to the positive terminal of the charged battery.

3. Connect the black lead to the negative terminal of the charged battery.

Continued

4. Connect the other end of the black lead to a paint-free ground on the engine block of the vehicle with the discharged battery.

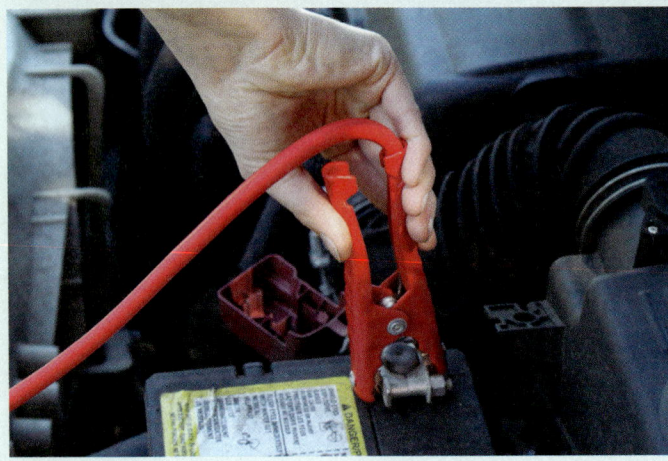

5. Start the vehicle with the discharged battery. Turn the headlights on to prevent a possible voltage spike which could damage the electronic equipment. Disconnect the jumper leads in the reverse order that you connected them. If the charging system is working correctly and the battery is in good condition, the battery will be recharged while the engine is running.

capacity. When we conduct a load test, we are testing the battery's ability to produce the necessary high starting current while it maintains enough voltage to operate the ignition and electronic control systems.

You will remember that CCAs reflect the load in amps that a battery can deliver for 30 seconds while maintaining a voltage of 7.2 volts or higher at 0°F (−18°C). Since vehicle and battery manufacturers specify the CCA rating for every vehicle and battery, that rating is used to calculate the load placed on the battery when load testing. The battery can be either in or out of the vehicle, but must be at or near a full state of charge for the test to be accurate. The electrolyte temperature should be approximately 70°F (21°C) for the most accurate results, because a cold battery cannot produce current flow as efficiently and will show a false fail result. When load testing a battery, use the following testing parameters:

- Test load = half the CCA of the battery you are testing (verify it is sized correctly for the vehicle)
- Load test time = 15 seconds
- Results: Pass = 9.6 volts or higher; Fail = less than 9.6 volts

FIGURE 13-38 Battery conductance tester installed on battery.

If the battery fails the load test, one further test is required before condemning the battery. The battery should be slow charged until it is at 100% SOC, then the load test should be repeated. If it still fails, the battery is sulfated and needs to be replaced. If it passes, test the vehicle's charging system to see if it is charging the battery properly.

To load test a battery, follow the steps in **SKILL DRILL 13-4**.

SKILL DRILL 13-4 Load Testing a Battery

1. With the tester controls off and the load control turned to the off position, connect the tester leads to the battery. Place the inductive amps clamp around either the black or the red tester cables in the correct orientation.

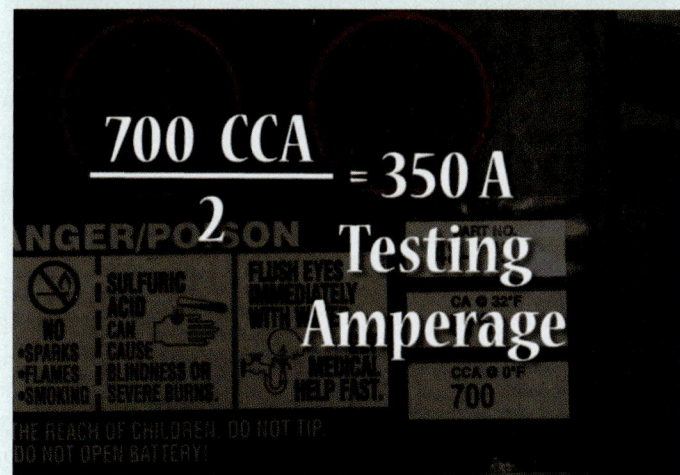

2. Verify that the temperature of the battery is within the testing parameters. Use an infrared temperature gun to determine the battery temperature by measuring the temperature of the side of the battery. If you are using an automatic load tester, enter the battery's CCA and select "test" or "start." If you are using a manual load tester, calculate the test load, which is usually half of the CCA.

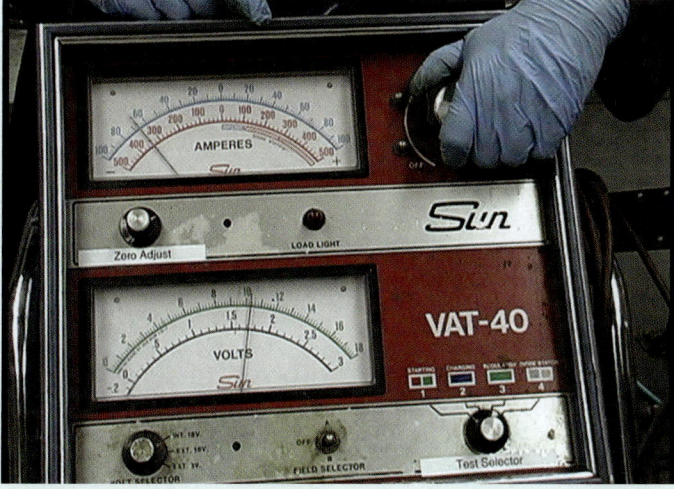

3. Maintain this load for 15 seconds while watching the voltmeter. Read the voltmeter and immediately turn the control knob off. At room temperature, the voltage should be 9.6 volts or higher at the end of the 15-second draw. If the battery is colder than room temperature, look up the compensated minimum voltage. Determine any necessary action.

Identifying Modules That Lose Their Initialization with Battery Disconnect

S13005 Identify electronic modules, security systems, radios, and other accessories that require reinitialization.

S13006, N13006 Maintain or restore electronic memory functions.

N13007 Identify electrical/electronic modules, security systems, radios, and other accessories that require reinitialization or code entry after reconnecting vehicle battery.

Many electronic modules in vehicles, such as radio and driver-specific presets, require a small amount of power to maintain their keep alive memory (KAM). When the battery is disconnected from the vehicle, the memory of specific electronic systems is usually lost. For some systems, this can be annoying. The PCM may lose its adaptive learning data, which means the vehicle will have to relearn this information during a period of driving that could take several days. For other systems, such as the security system, loss of memory may prevent the vehicle from being restarted or the radio from being used. At the very least, it may require the dealer or manufacturer to be contacted for vehicle-specific codes to reinitiate the vehicle systems. Check the manufacturer's and owner's information to determine what systems will be affected by the power loss. You will need to identify any system that will require security or initialization codes to be reentered, and ensure the procedure and equipment are available to reinitialize systems or modules.

In some cases, it may be possible to use a 9-volt memory minder or memory saver to maintain the vehicle's memory while the vehicle battery is disconnected. This should supply enough power to maintain the memory for small jobs such as changing the battery—provided the vehicle doors or trunk are not opened, causing the interior lights to illuminate and essentially drain the 9-volt battery. Many technicians advocate using an external 12-volt DC power supply connected to the data link connector with a suitable cable. If the cigarette lighter socket is always powered on, this too can be used for supplying power to the vehicle while the battery cables and connectors are being serviced. Just remember that providing power back into the circuit makes the system susceptible to short circuits by providing a ground to any of the powered wires or terminals.

To identify electronic modules, security systems, radios, and other accessories that require reinitialization or code entry following battery disconnect, follow the steps in **SKILL DRILL 13-5**.

To maintain or restore electronic memory functions, follow the steps in **SKILL DRILL 13-6**.

Parasitic Draw

S13007 Measure parasitic draw.

Parasitic draw refers to the current draw that occurs once the vehicle has been turned off and the systems have shut down. All modern vehicles have a certain minimum amount of parasitic current draw that is used by the vehicle's **keep alive memory (KAM)** circuits; a good rule of thumb is that the parasitic draw rate should be around 75mA once all of the modules have gone to sleep. KAM systems maintain memory functions and monitor

---

SKILL DRILL 13-5 Identifying Electronic Modules, Security Systems, Radios, and Other Accessories That Require Reinitialization

1. In the appropriate service information or owner's manual, find the section that lists the information on the electronic modules, security systems, radios, and other accessories that require reinitialization or code entry. For example, the immobilizer, radio, engine, and transmission each have their own section.
2. From the service information, list the systems and modules that may require initialization.

SKILL DRILL 13-6 Maintaining or Restoring Electronic Memory Functions

1. Identify which modules, if any, require reinitialization or code entry when the battery is disconnected following Skill Drill 13-5.
2. Identify the correct procedure and any needed tools, and verify that initialization codes are available.
3. If maintaining memory function, install a memory minder prior to the vehicle battery being disconnected.
4. If reinitializing the electronic systems, use the correct codes to reinitialize the modules if required.

systems. For example, the vehicle's theft-deterrent system is part of this parasitic current draw. Excessive parasitic draw will discharge the battery prematurely. Typically, when the ignition is switched off, the vehicle's engine stops and the vehicle systems begin to shut down, which is called "going to sleep" or "entering sleep mode." Vehicle systems will turn off at different rates and do not necessarily immediately turn off as soon as the ignition is off. For example, some vehicles may take up to a couple of hours before the last system goes to sleep and the parasitic draw reaches its minimum value. A low-amp clamp capable of measuring milliamps or an ammeter is the tool commonly used to measure parasitic draw from the battery. We will explore another option when we get to the parasitic draw skill drill. Manufacturers will specify how much parasitic draw is acceptable and the waiting period required after the ignition is switched off before the measurement can be accurately taken.

How to Test It

Parasitic current draw can be measured in several ways, the most common being the process of using an ammeter capable of measuring milliamps and inserting it in series between the battery post and the battery terminal. The ammeter is usually put in series with the negative battery lead (**FIGURE 13-39**). If the vehicle is equipped with systems or modules that require electronic memory to be maintained, follow the procedure listed in previous paragraphs for identifying modules that lose their initialization during battery removal and maintain or restore electronic memory functions. Note that the timers may reset during the process of disconnecting the battery terminal and connecting the ammeter in series, so you may have to wait for the timers to go back to sleep. If excessive parasitic draw is measured, disconnect fuses or systems one at a time while monitoring parasitic current draw to determine the systems causing excessive draw. Also, in most cases, opening a door or trunk will cause the timers to reset.

Disconnecting the battery can be avoided if a sensitive, low-current (i.e., milliamps) clamp is available (**FIGURE 13-40**). The low-amp **current clamp** measures the magnetic field generated by a very small current flow through a wire or cable. Placing the low-amp current clamp around the negative battery cable will allow you to measure the parasitic draw. If excessive parasitic draw is measured, disconnect fuses or systems one at a time while monitoring parasitic current draw to determine the systems causing the excessive draw.

The last way to measure a parasitic draw is a bit controversial, but it is well worth trying out since it can save a lot of time and requires no special tools other than a DVOM. It is named the Chesney parasitic load test after its creator, Sean Chesney. Instead of using an ammeter to measure the draw, an ohmmeter is used. Before doing anything, set the ohmmeter to ohms (the lowest scale), touch the meter leads together, and read the screen. The reading is the resistance of the meter leads and is called the "delta" value, which is the meter's true zero when used with those leads. Typically an ohmmeter will read about 0.1 ohms when the leads are touched together. Remember this number for later. Some meters have a delta feature that recalibrates the ohmmeter to zero when the leads are placed together and the delta button is pushed. If your meter has this delta feature, you can use it so that you will not have to remember the delta reading.

Next, with the battery terminals still connected to the battery, place the black lead on the negative post of the battery and the red lead on an unpainted surface of the alternator housing. Read the

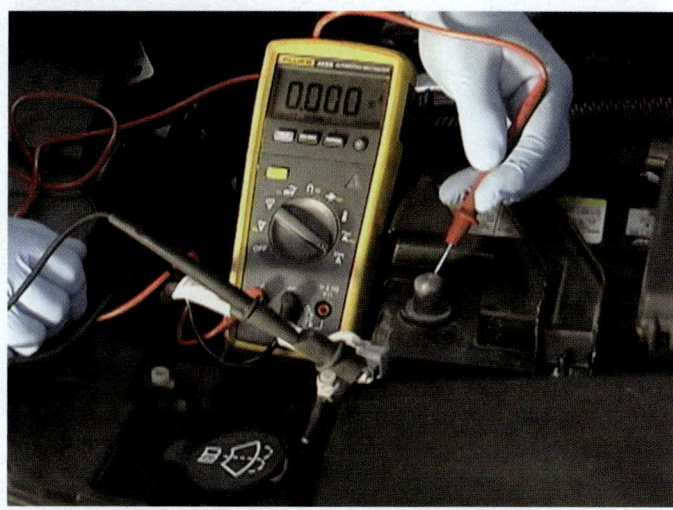

FIGURE 13-39 DVOM hooked up to measure parasitic draw.

FIGURE 13-40 Low amps probe hooked up to battery cable.

ohmmeter and subtract the delta value from the reading. This reading corresponds to the relative parasitic draw on the system.

Through testing, Chesney found that a draw of about 35 milliamps equaled an ohm reading of about 0.3 ohms delta (above the delta value) on a DVOM with 10 megohms of impedance, and about 0.6 ohms delta on a DVOM with 20 megohms of impedance. Anything above those readings indicates an excessive parasitic draw. You may be skeptical of this method, so go out and try it on any vehicle. Simulate a parasitic draw by opening the driver's door, which illuminates the dome light, and watch the ohmmeter. It went up, right? Close the door. As soon as the light went off, the ohmmeter reading went back down, right? If you use this test and find an excessive draw, you can pull fuses one at a time, watching for the ohmmeter reading to decrease. If it does not decrease after removing all of the fuses, suspect an unfused circuit such as the alternator diodes or the ignition circuit on some vehicles.

To measure parasitic draw with a standard parasitic load test, follow the steps in **SKILL DRILL 13-7**.

SKILL DRILL 13-7 Measuring Parasitic Draw

1. Research the parasitic draw specifications in the service information. Connect the low-current clamp around the negative battery cable and measure the parasitic draw. Compare the parasitic draw to specifications.

2. Disconnect the circuit fuses one at a time to determine which circuit has the excessive parasitic current draw. Determine any necessary actions.

Wrap-Up

Ready for Review

- Batteries store electricity in chemical form, which is possible because electricity causes a chemical reaction within the battery. Electrical energy is transformed into chemical energy.
- The chemical reactions in a battery change the composition of the chemicals, which then are stored until the electrical energy is needed. When electricity is needed, the chemicals react with each other, transforming the chemical energy back into electrical energy.
- A battery consists of two dissimilar metals, an insulator material separating the metals, and an electrolyte, which is an electrically conductive solution.
- The strength of the battery depends on the materials used in its construction.
- The traditional automotive battery type is the lead-acid battery.
- Batteries do not last forever. Over time and use, the plates start to lose effectiveness as lead paste falls off the plates and collects in the bottom of the battery case.
- A battery may also become sulfated, meaning the surfaces of the plates harden, making it more difficult for the acid to be absorbed into the plate. As a result, the battery can no longer readily accept a charge or cannot adequately discharge under load.
- Parasitic draw refers to the current draw that occurs once the vehicle has been turned off and the systems have shut down.
- All modern vehicles have a certain minimum amount of parasitic current draw that is used by the vehicle's keep alive memory (KAM).
- Jump starting of a vehicle requires the use of a slave battery along with high-current–capacity leads made with clamps on each end that can connect over the battery terminals.
- The slave battery is connected in parallel from the slave (charged) battery to the host (discharged) battery to provide additional capacity to crank and start the vehicle. Jump starting should be performed with caution, as incorrect polarity or voltage spikes can cause damage to the sensitive electronic components installed on either of the vehicles.
- Use jumper leads equipped with electrical spike protection.
- Spike-protected leads have a built-in device to prevent damaging electrical spikes from reaching electronic equipment.
- When servicing batteries, always ensure that you have the right personal protective equipment (e.g., safety eyewear, gloves, and shop clothing).
- Always remove the negative or ground terminal first when disconnecting battery cables, to reduce the possibility of a wrench creating an accidental short to ground from the positive terminal to any grounded metal surface.
- Low-maintenance batteries require topping off of the water in the electrolyte. The plates and venting system are designed so that they do not normally gas and release water vapor to the atmosphere.
- Low-maintenance batteries still have removable caps so that electrolyte can be checked and topped off.
- Maintenance-free batteries are fully sealed and do not require the electrolyte to be topped off. In some cases they use a gel-type electrolyte instead of a liquid.
- Absorbed glass mat (AGM) batteries have the electrolyte absorbed within a mat of fine glass fibers.
- Batteries contain many chemicals and metals that can damage the environment.
- When being charged, the battery is connected to a direct current (DC) electrical supply with electrical pressure (voltage) higher than that of the battery's total cell voltage.
- The battery charging device acts like an electron pump, forcing electrons to move from the positive plates to the negative plates in the battery.
- The charging process increases the amount of acid in the electrolyte, making the electrolyte stronger. When further charging no longer makes the electrolyte stronger, charging is complete.
- The power train control module (PCM) then controls the levels of current flow and the voltage at which the battery is charged.
- The colder the battery temperature, the higher the rate of charging; the hotter the battery temperature, the slower the rate of charging.
- Smart chargers and battery management systems use the feedback from the temperature probe to manage the charge rate of a battery.
- Batteries should be kept clean, dry, and fully charged; have minimal vibration and the correct level of electrolyte; be kept at a moderate temperature (e.g., 77°F [25°C]); and be well secured.
- The lithium-ion battery has one of the highest energy density ratios of common batteries in production today and primarily used in HEVs.
- The typical cell voltage for a lithium-ion battery is 3.6 volts.
- The cell voltage of a nickel-cadmium or nickel–metal hydride battery is 1.2 volts, and a typical lead-acid battery is 2.1 volts.
- Knowing how to properly diagnose and recharge a battery is an important task that you will need to accomplish regularly.
- Performing a battery state-of-charge test with a hydrometer or refractometer is a good indicator of whether the battery needs to be charged or not.

- Another maintenance check of the battery is the conductance test, which indicates the capacity of the battery and how much life is left in the battery.
- There are several methods used to rate automotive battery capacity. The three most common are cold cranking amps (CCA), cranking amps (CA), and reserve capacity (RC).
- CCA measures the load in amps that a battery can deliver for 30 seconds while maintaining a voltage of 1.2 volts per cell (7.2 volts for a 12-volt battery) or higher at 0°F (−18°C).
- CA measures the same thing, but at a higher temperature—32°F (0°C). This can cause confusion since a 500-CCA battery has about 20% more capacity than a 500-CA battery, so it is important to keep the ratings straight.
- RC is the time in minutes that a new, fully charged 12-volt battery at 80°F (27°C) will supply a constant load of 25 amps without its voltage dropping below 10.5 volts.
- The CCA, CA, and RC are typically marked on automotive batteries, and technicians use these ratings when testing battery performance and when selecting batteries for particular vehicle applications.
- Gel cell batteries, also called valve-regulated lead-acid (VRLA) batteries, contain a gel-like electrolyte consisting of sulfuric acid mixed with fumed silica. This mixture transforms the once-liquid sulfuric acid into an immovable gel-like mass, which prevents electrolyte spillage and greatly reduces battery evaporation.
- Gel cell batteries are called "valve-regulated" because they contain a safety that is designed to vent internal gas pressure if levels raise much above 2 psi.
- Absorbent glass mat (AGM) batteries use an extremely absorbent micro-pore fiberglass material that traps the gases vented during the discharging and recharging processes.
- The best practice to avoid ruining a costly battery is to seek and purchase a SMART battery charger that is capable of 3-stage charging and rated for AGM battery use.
- The first stage of battery charging is bulk charging, which reintroduces 80% of the battery's energy capacity back into the battery. This is accomplished by charging the battery at exactly 14.4 volts, and delivering the maximum amount of amperage to the battery that the charger's software deems to be safe.
- The second stage of battery charging is called the absorption stage. This stage takes the battery to a 98% state of charge by maintaining a voltage level of 14.4 volts and gradually reducing the amperage being fed into the battery as the state of charge approaches the 98% mark.
- The third and final battery charging stage is the float stage. During this stage, the charger voltage is regulated to 13.4 volts, and less than 1 amp is fed into the battery.
- State-of-charge (SOC) testing tells us how charged or discharged a battery is, not how much capacity it has.
- The degree of the battery's charge is critical information when testing starting and charging system issues, as well as just about any other electrical issue.
- An easy and effective way to test for state of charge is to leave the headlamps on for three minutes to take any surface charge away from the battery, then check battery voltage with a voltmeter.
- A fully charged battery will measure 12.6 volts, a battery with a 75% SOC will measure 12.4, and a battery with a 50% or lower SOC will read below 12.2 volts.
- You can test the electrolyte's specific gravity for the battery state of charge using a hydrometer or refractometer to measure the specific gravity of the electrolyte in each of the battery's cells. The higher the specific gravity, the higher the percentage of acid in the electrolyte, which corresponds to a high battery state of charge.
- The most common type of battery terminal being a cone design that provides a large surface contact area with the ability to tighten the terminal onto the battery post using a nut and bolt.
- The tapered cone allows easy removal of the clamp when the bolt is loosened. The larger positive and smaller negative terminals are slightly different in size to ensure correct connection of the cables when installing or servicing the battery.
- The side battery terminal gets its name because the battery connection is on the side of the battery. The terminal is a flat circle with a center bolt that bolts the terminal tightly to the battery connection.
- Battery cables are usually made of many fine strands of copper wire bound tightly together and insulated to make a cable with high current capacity. There are a number of cable sizes available to handle various current capacities.
- The most effective way to test for any and all types of battery cable failures is to perform a voltage drop test of each battery cable that stretches from the battery to the engine block and starter.
- The voltage drop tests MUST be conducted with the circuit under a working load; in this case, the engine must be cranking before accurate voltage drop tests can be performed on the battery cables.
- The drop should not exceed 3% of the voltage, about a maximum drop of 0.36 volts.
- It is critical that the current path from the battery to the starter have as little resistance as possible; essentially, it must be free of corrosion.
- Battery corrosion can happen when the dissimilar metals contained in battery clamps and battery cables are exposed to road salt and water. This causes galvanic corrosion, which eats at the metals of the battery cables and clamps. Another source of battery corrosion is the electrolyte itself.
- When cleaning battery terminal neutralize the electrolyte that has leaked from the case. This can be accomplished by using an approved battery cleaner, or with a 50/50 solution of baking soda and water.
- Some manufacturers recommend removing the negative battery terminal while charging a battery to reduce risk to the vehicle's electronics.

- If the battery needs to be disconnected during charging, remember to verify if the vehicle's adaptive memory needs to be maintained.
- The conductance tester sends low-frequency signals into the cells to determine the battery's ability to conduct current. The greater the ability to conduct current, the higher the CCA capacity of the battery.
- The load test has been used for years to test a battery's capacity and internal condition, but some manufacturers indicate that their batteries should not be load tested, and instead should be conductance tested.
- The load test subjects the battery to a high rate of discharge, and the voltage is then measured after a set time to see how well the battery creates that current flow.

Key Terms

absorbed glass mat (AGM) Batteries that have the electrolyte absorbed within a mat of fine glass fibers.

cold cranking amps (CCA) The load in amps that a battery can deliver for 30 seconds while maintaining a voltage of 1.2 volts per cell (7.2 volts for a 12-volt battery) or higher at 0°F (−18°C).

cranking amps (CA) The load in amps that a battery can deliver for 30 seconds while maintaining a voltage of 1.2 volts per cell (7.2 volts for a 12-volt battery) or higher at 32°F (0°C).

current clamp Measures the magnetic field generated by current flow through a wire or cable.

electrical capacity The amount of charge a typical lead-acid battery can store; determined primarily by the total surface area of the plates.

electrolyte An electrically conductive solution.

gassing When gas escapes the battery; caused by overcharging or rapid charging a battery.

keep alive memory (KAM) A certain minimum amount of parasitic current draw that is used by the vehicle's electronic systems.

parasitic draw The current draw that occurs once the vehicle has been turned off and the systems have shut down.

reserve capacity (RC) The time in minutes that a new, fully charged 12-volt battery at 80°F (27°C) will supply a constant load of 25 amps without its voltage dropping below 10.5 volts.

thermal runaway Also referred to as venting the flame; during thermal runaway, the high heat of the failing cell will propagate to neighboring cells, causing them to become thermally unstable as well. When lithium-ion batteries enter thermal runaway, extreme overheating, and in some cases, fire can be expected.

Review Questions

1. Which of these does a battery convert into electrical energy?
   a. Heat energy
   b. Mechanical energy
   c. Chemical energy
   d. Light energy

2. The type of battery that is most common in automobiles is a:
   a. lithium-ion battery.
   b. lead-acid battery.
   c. nickel-cadmium battery.
   d. zinc-bromine battery.

3. All of the following statements are true *except*:
   a. A battery consists of two similar metals.
   b. An automotive battery can supply very high discharge currents while maintaining a high voltage.
   c. The fundamental components and operation of the battery remain the same irrespective of the size and usage.
   d. Batteries store electricity in chemical form.

4. Which of the following batteries are most suited for off-road and all-terrain vehicles?
   a. Low-maintenance batteries
   b. Lead-acid batteries
   c. Lithium-ion batteries
   d. Absorbed glass mat batteries

5. Automotive battery capacity is rated based on all of the following *except*:
   a. cold cranking amps (CCA).
   b. cranking amps (CA).
   c. reserve capacity (RC).
   d. plate type of the battery.

6. The lead plates of the lead-acid batteries are immersed in:
   a. diluted sulfuric acid solution.
   b. distilled water.
   c. mineral water.
   d. brine solution.

7. All of the following are factors in shortening battery life *except*:
   a. having deep discharge cycles.
   b. excessive vibration.
   c. a moderate external temperature.
   d. developing corrosion.

8. State of charge of a no-maintenance battery can be determined using a(n):
   a. single cell hydrometer float.
   b. TPS.
   c. oscilloscope.
   d. hygrometer.

9. Which of the following is the best and safest method to clean the battery when there is a voltage of more than 0.2 volts on the top of battery?
   a. Splash water and clean
   b. Clean with a dry cloth
   c. Coat with baking soda and leave for some time
   d. Clean with baking soda and water mixture

10. Which the following tests is performed to determine the capacity and life left in the battery?
    a. Leak test
    b. Hydrometer
    c. Conductance test
    d. Vacuum test

ASE Technician A/Technician B Style Questions

1. Tech A says that automotive batteries are very robust, and will last many years with proper maintenance. Tech B says that all lead-acid batteries are made to the same standards, and as such, are interchangeable between vehicles and applications. Who is correct?
   a. Tech A
   b. Tech B
   c. Both A and B
   d. Neither A nor B

2. Tech A says that when performing battery maintenance, it is perfectly acceptable to top off the battery with tap water as long as the electrolyte level is the same in each cell. Tech B says that you should only use distilled water to top off battery electrolyte. Who is correct?
   a. Tech A
   b. Tech B
   c. Both A and B
   d. Neither A nor B

3. Tech A says that the electrolyte of lead-acid batteries is very corrosive, and can cause burns. Tech B says that when battery electrolyte leaks from the battery case, it can cause severe corrosion damage on any metal that it comes into contact with. Who is correct?
   a. Tech A
   b. Tech B
   c. Both A and B
   d. Neither A nor B

4. Tech A says that maintenance-free batteries are permanently sealed, and as a result the electrolyte level cannot be adjusted. Tech B says that the caps of maintenance-free batteries can be pried off, which allows the electrolyte to be tested and topped off. Who is correct?
   a. Tech A
   b. Tech B
   c. Both A and B
   d. Neither A nor B

5. Tech A says that AGM batteries are larger and heavier than traditional lead-acid batteries. Tech B says that when charging AGM batteries, you must use a battery charger designed for AGM batteries, or damage to the battery will occur. Who is correct?
   a. Tech A
   b. Tech B
   c. Both A and B
   d. Neither A nor B

6. Tech A says that automotive batteries can be disposed of in a normal trash receptacle, as long as the battery is fully discharged and wrapped in plastic. Tech B says that many automotive batteries contain lead, which has been known to cause birth defects. Who is correct?
   a. Tech A
   b. Tech B
   c. Both A and B
   d. Neither A nor B

7. Tech A says that extreme care must be taken when charging automotive batteries, as they sometimes vent hydrogen gas, which is highly volatile. Tech B says that you should keep all sources of heat and flame away from automotive batteries at all times. Who is correct?
   a. Tech A
   b. Tech B
   c. Both A and B
   d. Neither A nor B

8. Tech A says that all batteries can be cycled several times and suffer no damage. Tech B says that when a lead-acid battery is left in a discharged state, sulfation occurs on the plates, which reduces the battery's electrical capacity. Who is correct?
   a. Tech A
   b. Tech B
   c. Both A and B
   d. Neither A nor B

9. Tech A says that the traction motor batteries in hybrid and electric vehicles require frequent testing and maintenance. Tech B says that nickel–metal hydride batteries are alkaline batteries, which means that their electrolyte is a base instead of an acid. Who is correct?
   a. Tech A
   b. Tech B
   c. Both A and B
   d. Neither A nor B

10. Tech A says that lithium-ion batteries have a very low energy-to-weight ratio, which makes them very bulky. Tech B says that lithium-ion batteries are very sensitive to temperature, and can be damaged by high-heat conditions such as a paint curing oven. Who is correct?
    a. Tech A
    b. Tech B
    c. Both A and B
    d. Neither A nor B

CHAPTER 14

Starting Systems

NATEF Tasks

- **N14001** Identify hybrid vehicle auxiliary (12v) battery service, repair, and test procedures.
- **N14002** Perform starter current draw tests; determine needed action.
- **N14003** Perform starter circuit voltage drop tests; determine needed action.
- **N14004** Inspect and test starter relays and solenoids; determine needed action.
- **N14005** Remove and install starter in a vehicle.
- **N14006** Inspect and test switches, connectors, and wires of starter control circuits; determine needed action.
- **N14007** Differentiate between electrical and engine mechanical problems that cause a slow-crank or a no-crank condition.

Knowledge Objectives

After reading this chapter, you will be able to:

- **K14001** Identify the starting system components and explain their operation
- **K14002** Identify starter motor components and explain starter motor construction.
- **K14003** Explain starter motor and solenoid operation.
- **K14004** Explain starter drives and the ring gear.
- **K14005** Understand starting system procedures.

Skills Objectives

After reading this chapter, you will be able to:

- **S14001** Test the starter draw.
- **S14002** Test starter circuit voltage drop.
- **S14003** Inspect and test the starter control circuit.
- **S14004** Inspect and test relays and solenoids.
- **S14005** Remove and install starter in a vehicle.

You Are the Technician

The starting system is often taken for granted. When it functions properly, it is expected to crank the engine efficiently, time after time. When problems develop in the starting system, they can manifest themselves as a slow-crank condition, a no-crank condition, or excessive noise originating from the components of the starting system. It requires a knowledgeable and skilled technician to properly diagnose and repair starter systems in today's vehicles.

1. True or False: The starting systems in today's vehicles are very simple, and as a result, diagnosis can be made with minimal inspection.

329

Introduction

No system had a greater impact on the popularity of the gasoline internal combustion engine than the electric starting motor. Before the advent of this system, vehicles had to be started by hand with a crank. This task proved to be difficult and dangerous to accomplish. Hand cranking an engine was such a chore that, before gasoline powered vehicles were outfitted with electric starter motors, the motoring public preferred to own and drive the electric vehicles of the time due to their ease of operation. Vehicles have come a long way since those early automobiles, but the basic operation of the electric starting system has remained very similar in modern vehicles. In this chapter, you will learn about the various parts and pieces associated with starting systems, and the procedures that should be followed to diagnose and repair these systems.

Engine Starting (Cranking) System

K14001 Identify the starting system components and explain their operation.

The starting system provides a method of rotating (cranking) the vehicle's internal combustion engine (ICE) to begin the combustion cycle. In early vehicles, this was done by the use of a hand-crank handle. Modern vehicles use an electric starter motor that draws its electrical power from the vehicle's battery (**FIGURE 14-1**). The starter is designed to work for short periods of time and must crank the engine at sufficient speed in order for it to start. Modern starting systems are very effective provided that they, and the battery, are well maintained.

The starting/cranking system consists of the battery, high- and low-amperage wires, a solenoid, a starter motor assembly ring gear, and the ignition switch. On PCM-activated starting systems, there is also the PCM, a relay, and all of the related sensors that feed information to the PCM. A control circuit determines when and if the cranking circuit will function.

The control circuit starts sometimes with a fuse, the ignition switch (or PCM circuitry), the starter relay, a safety switch, and/or a combination relay/starter solenoid. All vehicles equipped with automatic transmissions use a neutral safety switch or a similar device, and many vehicles equipped with a manual transmission have a clutch safety switch. An on-board computer (PCM) and a security system may also determine if and when the starting system will function.

During the cranking process, two actions occur. The pinion of the starter motor engages with the flywheel ring gear, and the starter motor then rotates to turn over, or crank, the engine. The starter motor is an electric motor mounted on the engine block or transmission. It is typically powered by the 12-volt storage battery and is designed to have high turning effort (torque) at low speeds. It should be noted that some hybrid vehicles use the high-voltage battery to operate a specialized high-voltage motor that cranks over the engine. The starter cables are the heaviest in the vehicle because they carry the high current needed by the starter motor. The starter motor causes the engine flywheel and crankshaft to rotate from a resting position and keeps them turning until the engine fires and runs on its own.

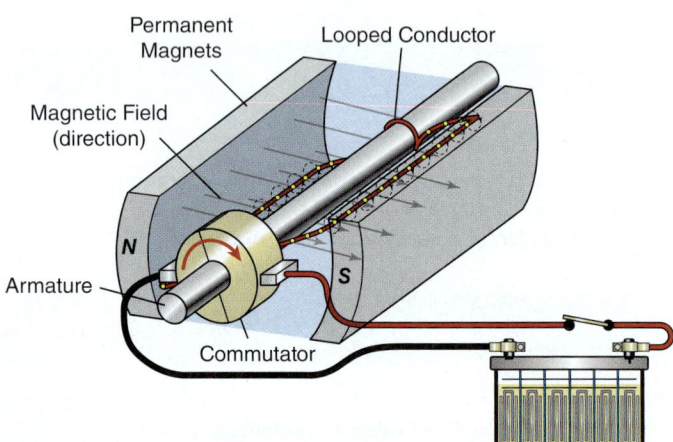

FIGURE 14-1 Simple single-loop motor and electromagnetic fields—with commutator and brushes.

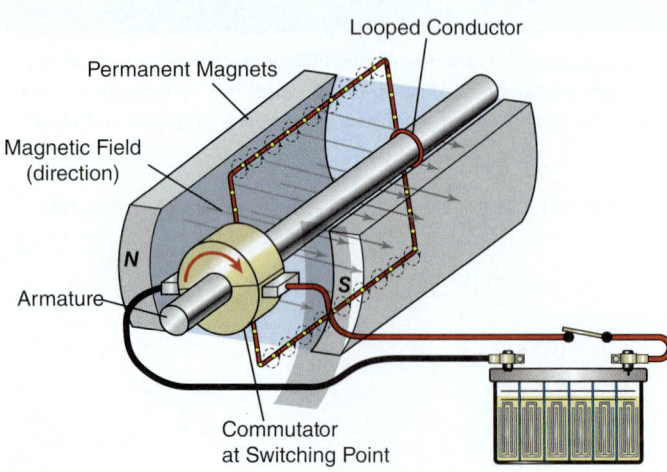

FIGURE 14-2 Simple single-loop motor and electromagnetic fields at the switching point of the commutator.

Starter Motor Principles

The starter motor converts electrical energy to mechanical energy for the purpose of cranking the engine over. There are three basic sections to the typical starter—the electric motor, the drive mechanism, and the solenoid (**FIGURE 14-2**). The starter motor is mounted on the transmission or cylinder block in a position to engage a ring gear around the outside edge of the engine flywheel, flexplate, or torque converter. Starting is usually accomplished by the operator activating a starter switch built into the ignition lock assembly, or a start button. A relatively small current flows through a neutral safety switch or clutch switch to a starter relay that controls a larger current to operate the starter solenoid, which is typically mounted on the starter motor (**FIGURE 14-3**). The solenoid plunger moves the drive pinion gear into engagement with the ring gear and also closes a set of heavy-duty contacts. This allows a very large current to flow from the battery to the starter motor, rotating the armature and drive pinion gear, causing the crankshaft to rotate. When the engine starts and is able to run on its own, the operator usually releases the key and the solenoid spring withdraws the pinion gear from the ring gear and brings the armature to a halt. On many modern vehicles, the PCM signals the relay to continue the cranking process once a crank signal is received from the operator until the vehicle starts or the starting operation times out.

Direct Drive/Gear Reduction Systems

Starter motors can be designed to drive the ring gear in one of two ways: either direct drive or gear reduction. In the direct-drive system, the starter drive is mounted directly on one end of the armature shaft. The starter drive transfers the rotating force of the armature directly to the engine flywheel (**FIGURE 14-4**). In this arrangement, the only gear reduction is the reduction between the pinion gear and the ring gear.

Gear reduction starters use an extra gear between the armature and the starter drive mechanism. They have a reduction of about 4:1, which allows the starter to spin at a higher speed while drawing less current from the storage battery (**FIGURE 14-5**). It also enables gear reduction starters to be smaller and lighter, but develop a higher torque output, than their direct drive counterparts. Two types of gearing systems are normally used in gear reduction starter motors: spur gears or planetary gears. Spur gears require the armature to be offset via a gear housing that holds the starter drive. A gear reduction system using planetary gears does not require an offset housing; the planetary gears are housed in the drive-end housing in line with the starter drive, which allows for a more compact starter motor (**FIGURE 14-6**).

FIGURE 14-3 Starter motor solenoid.

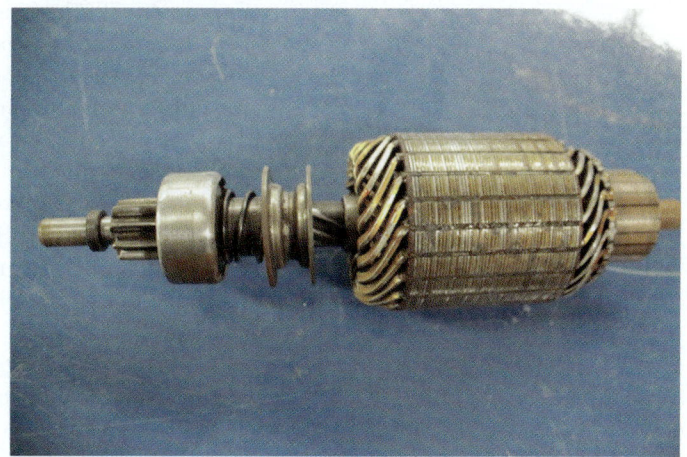
FIGURE 14-4 Armature and starter drive from a direct-drive starter.

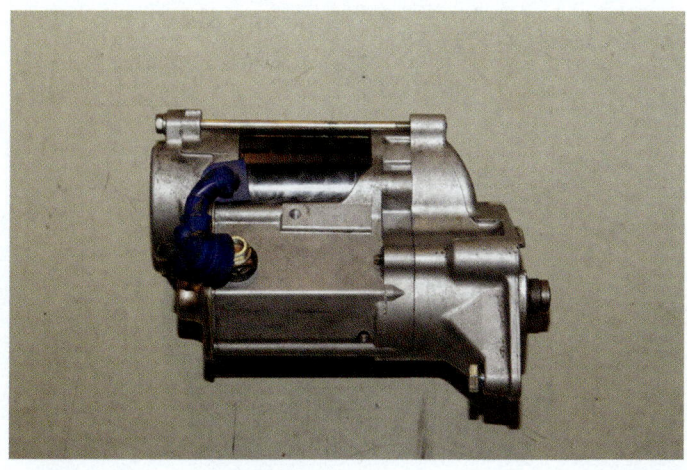
FIGURE 14-5 Gear reduction starter.

FIGURE 14-6 Spur gear and planetary gear reduction starters.

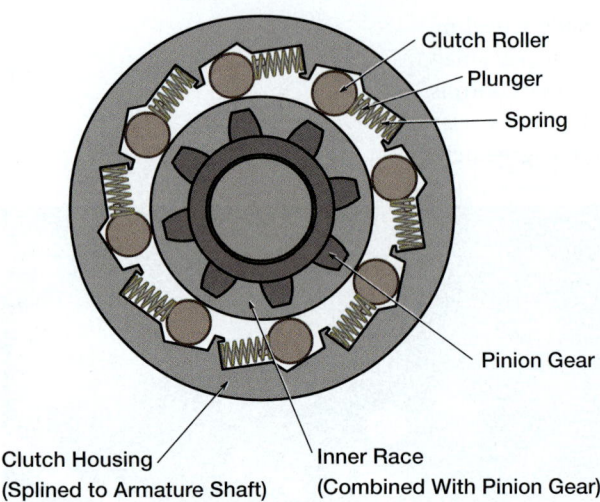

FIGURE 14-7 Starter drive one-way clutch.

Starter Motor Engagement

Engagement is provided by operation of the ignition switch in the start position or when commanded by the PCM, which activates a starter-mounted solenoid, whose plunger is connected to the end of a pinion shift lever and operating fork. Solenoid operation moves the operating fork in some systems, which causes the pinion to engage with the ring gear and also causes the plunger contacts to bridge with the main starter terminals, which allows current to flow from the storage battery. The fork is located in a guide ring on the pinion drive, which is coupled to the pinion gear via a roller-type overrunning clutch (**FIGURE 14-7**). It is designed to transmit drive in one direction only and freewheel in the opposite direction. The pinion drive gear is mounted on a slight helix that is machined onto the armature shaft to form a very coarse thread. This arrangement allows the pinion drive to rotate slightly when it is moved toward the ring gear. This feature, together with a chamfer on the leading edge of the ring gear and pinion gear teeth, is designed to assist the teeth in meshing (**FIGURE 14-8**). However, if the pinion gear teeth butt against the ring gear teeth and engagement is prevented, the guide ring continues its axial movement by sliding over the sleeve of the drive and compressing a meshing spring until the solenoid plunger contacts bridge the main terminals and the armature begins to turn (**FIGURE 14-9**).

Slight armature rotation and the force from the meshing spring push the pinion teeth into mesh with the ring gear. The meshing spring forces the pinion farther into the ring gear until the pinion contacts a stop ring on the armature shaft. This prevents further axial movement. The starter drive is locked to the shaft via the helix. The one-way clutch drives the pinion gear and transfers the armature rotation to the ring gear.

The pinion has only a small number of teeth compared to the ring gear. This creates a gear reduction that is usually around 17:1, meaning the armature will rotate 17 times for each revolution of the flywheel (**FIGURE 14-10**). The gear reduction with this ratio also multiplies the torque from the starter motor 17 times, allowing a relatively small electric motor to turn the much larger ICE. If a gear reduction starter is used, then the torque is multiplied further, giving more cranking power from the same size starter motor.

As soon as the engine starts, it may easily run at 1000 rpm or more. If still engaged, the engine would then turn the starter pinion gear about 17 times faster, or 17,000 rpm, which

FIGURE 14-8 Chamfered pinion gear teeth.

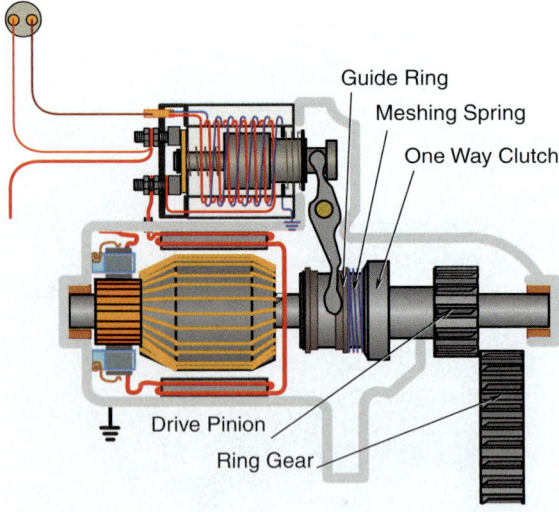

FIGURE 14-9 Improper pinion-to-ring gear alignment.

would destroy the armature due to the excessive centrifugal force created. To prevent this from occurring, the freewheeling of the overrunning clutch prevents the armature from turning too fast. Should the driver not release the ignition key from the start position in time, the starter solenoid sticks closed, or the pinion return springs fail to retract the pinion right away. The pinion remains meshed as long as the engaging lever is held in the extended position. Releasing the starter switch (or the PCM terminating the crank signal) allows the solenoid plunger return spring to disengage the pinion gear from the ring gear by returning the engaging lever, starter drive, and pinion gear to their original position.

▶ Starter Motor Construction

In today's vehicles, there are several types of starter motors that are in use, and each type of starter motor is constructed a bit differently. Keep in mind that the basic principles of operation will remain the same across each type of starter, but the parts and pieces employed to establish starter operation may vary between starter designs, be it a pole-shoe type starting motor, a permanent magnet type starting motor, a direct-drive starter, or a gear reduction unit.

FIGURE 14-10 Pinion and ring gear.

K14002 Identify starter motor components and explain starter motor construction.

Starter Magnet Types

Starter motors use two magnet types: electromagnetic and permanent magnet (**FIGURE 14-11**). Electromagnetic starter fields are formed by current flow through heavy copper windings, wound around iron pole shoes, which are then fastened to the starter case/barrel (**FIGURE 14-12**). Permanent magnet-type starters have large permanent magnets mounted in the same position as pole shoes. Since these types of magnets do not need electricity to produce a magnetic field, this type of starter motor is smaller than a starter motor that uses electromagnets. The starter case is made of iron on both types of starters, and serves to concentrate the magnetic field produced by the magnets.

Starter motors with electromagnetic field windings for light vehicle applications are typically series-wound motors, which means that the electromagnets are wired in series to the armature windings. Because the resistance of the field and armature windings is low, the current flow is high when the motor starts under load, and this generates a strong magnetic field that will produce high torque at low speeds. This high

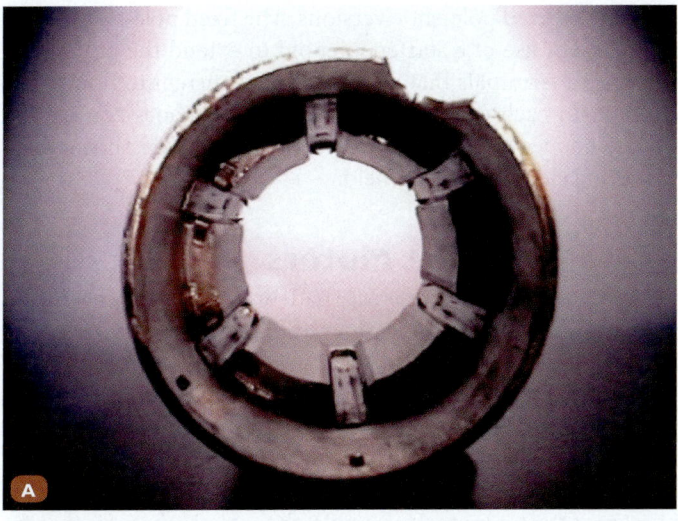

FIGURE 14-11 A. Permanent magnetic fields. **B.** Electromagnetic fields.

FIGURE 14-12 Starter pole shoes.

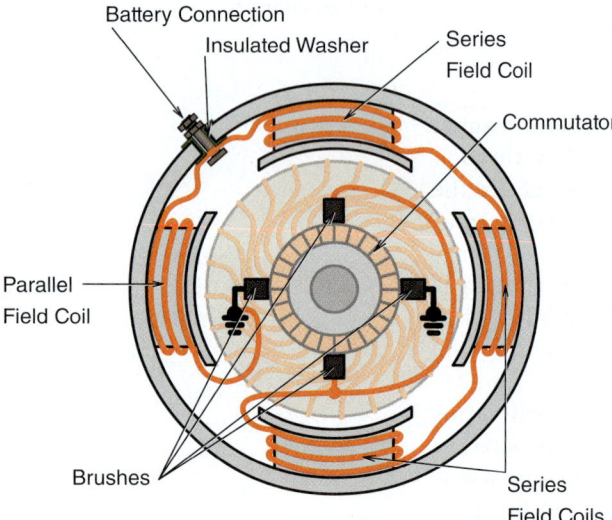

FIGURE 14-13 Electrical flow through a series-parallel-wound starter motor.

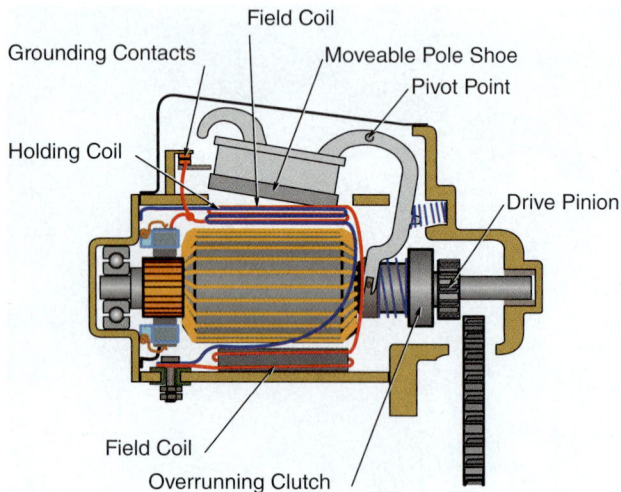

FIGURE 14-14 Internal construction of a moveable pole-shoe starter motor.

initial torque drops sharply as the motor speed increases because of the **counter-electromotive force (CEMF)** or voltage generated in the armature windings as the motor armature spins in the magnetic field. The CEMF increases with armature speed and opposes current flow. It reduces both current flow and torque output. The faster the motor turns, the less current it draws and the less torque it develops. For example, on a typical V6 engine with standard compression, the initial surge through the starter is usually over 500 amps, but as the armature starts to spin, the CEMF opposes the current flow, reducing it to around 120 amps (average) within about a second. This initial surge and subsequent drop of amperage can be observed on an oscilloscope when testing the starter motor. Some series-wound motors have parallel-wired field windings, but they are still wired in series with the armature. These starters are referred to as series-parallel-wound starter motors. By connecting the field windings in this way, more current can flow in the circuit, and an overall increase in torque is obtained (**FIGURE 14-13**).

Moveable Pole Shoe, and Fixed Pole Shoe Starter Motors

Ford is credited with the design and implementation of the moveable pole-shoe starter motor, which was a direct drive design with a moveable pole shoe (**FIGURE 14-14**). This design eliminates a starter solenoid; the pinion gear is moved into position by the moving pole shoe. A separate starter relay is used with this design to control the current delivery to the starter motor. The starter operates in the following sequence: The driver turns the key, which sends electric current to the starter relay. This closes the starter relay contacts and allows a large electric current to flow from the battery to the starter motor. At this time, one of the pole shoes, which is hinged at the front, connected to the pinion drive, and held in place by spring pressure, is pulled into position by the magnetic field created in the pole shoe. This moves the pinion gear into position to engage the ring gear and closes a set of contacts, which allow electric current to flow to the rest of the starter motor.

The main difference between moveable pole-shoe starter motors and fixed pole-shoe starter motors is that the pole shoes are mounted to the case in the fixed pole-shoe versions. The fixed pole-shoe starter motor also makes use of a starter solenoid to extend the pinion and close the pair of terminals that allow electrical current to flow to the starter motor. In each of these designs, the poles shoes consist of a heavy gauge copper conductor wound around an iron shoe, which creates a strong electromagnetic field.

Permanent Magnet Motors

Permanent magnet starter motors are widely used in today's vehicles. This design of starter motor is lighter, more compact, and requires less amperage than pole-shoe starter motors. Most permanent magnet-type starting motors are gear reduction-type starters, as they do not produce quite as much torque as the larger pole-shoe type starters.

In simplest terms, the only appreciable differences between a permanent magnet starter motor and a pole-shoe starter motor is that the field magnets do not require electrical current to produce a magnetic field, which allows for smaller starter designs. The solenoid, armature, brushes, and pinion drive function in the same manner as those found on a pole-shoe starter motor.

Starter Construction and Components

A starter motor normally consists of the following components: field coils or large permanent magnets, an armature, a commutator, brushes, a drive pinion with an overrunning clutch, and a drive pinion engagement solenoid and shift fork (**FIGURE 14-15**). The armature is the revolving component of the DC motor that is responsible for cranking over the internal combustion engine. The armature shaft is supported at each end by bushings or bearings pressed into end frames, which locate the armature centrally in the outer casing (i.e., the "barrel") of the motor and between the field coils or permanent magnets (**FIGURE 14-16**).

The commutator end frame carries copper-impregnated carbon brushes, which conduct current through the armature when it is being rotated in operation. The brushes are mounted in brush holders and are kept in contact with the commutator by tensioned spiral springs (**FIGURE 14-17**). Half of the brushes are connected directly to the end frame and ground the armature windings via the engine block to the negative battery terminal, while the other brushes are insulated from the end frame and connected to the positive battery terminal via the main starter solenoid input terminal (**FIGURE 14-18**). The connection is direct from the solenoid in the case of a permanent magnet starter; it is indirect via the electromagnetic field poles in a series-wound motor (**FIGURE 14-19**).

Commutation and Brushes

When current flows in a conductor, an electromagnetic field is generated around it. If the conductor is placed so that it cuts across a stationary magnetic field, the conductor will be forced out of the stationary field. This occurs when the lines of force of the stationary field are distorted by the electromagnetic field around the conductor and try to return to a straight-line condition. Reversing the direction of current

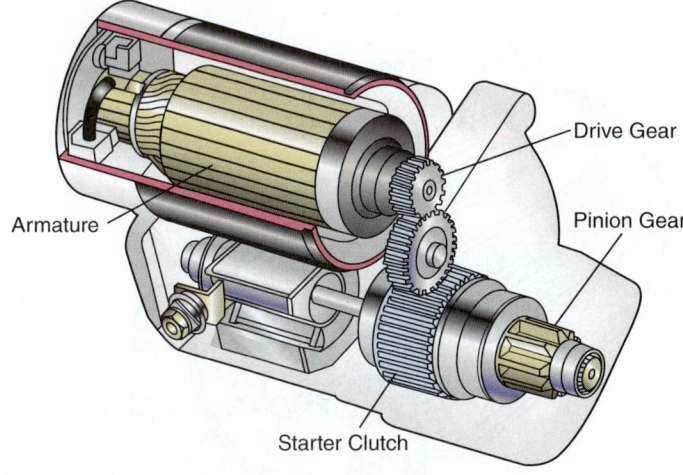

FIGURE 14-15 Exploded view of a starter motor.

FIGURE 14-16 Cutaway of an armature.

FIGURE 14-17 Starter motor brushes.

FIGURE 14-18 Positive brush connection at the starter solenoid.

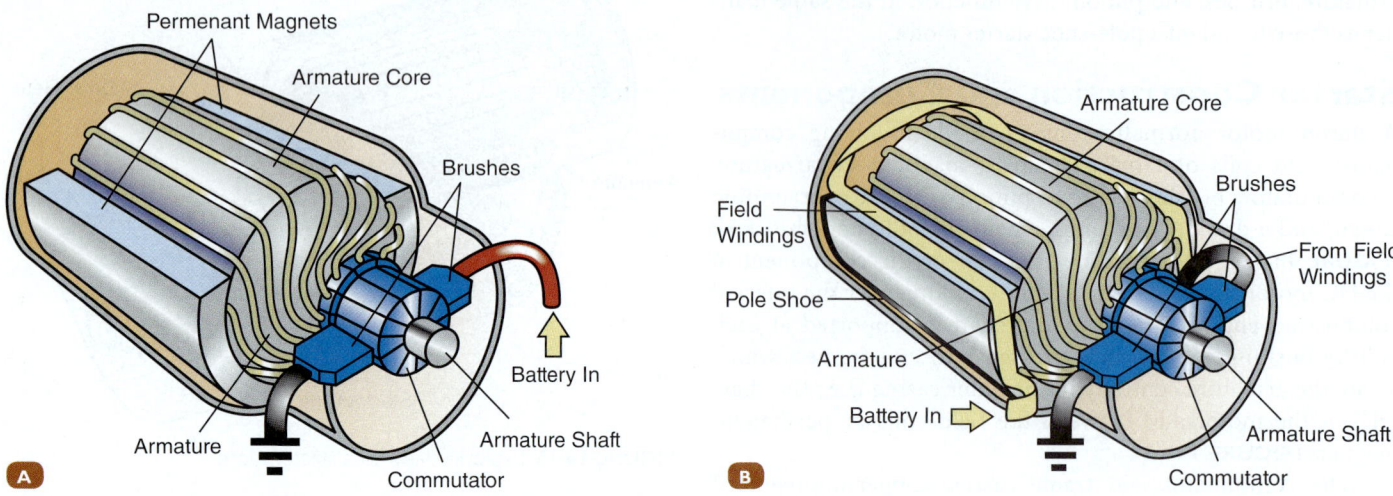

FIGURE 14-19 Electrical schematic of power flow. **A.** Permanent magnet starter. **B.** Series-wound starter.

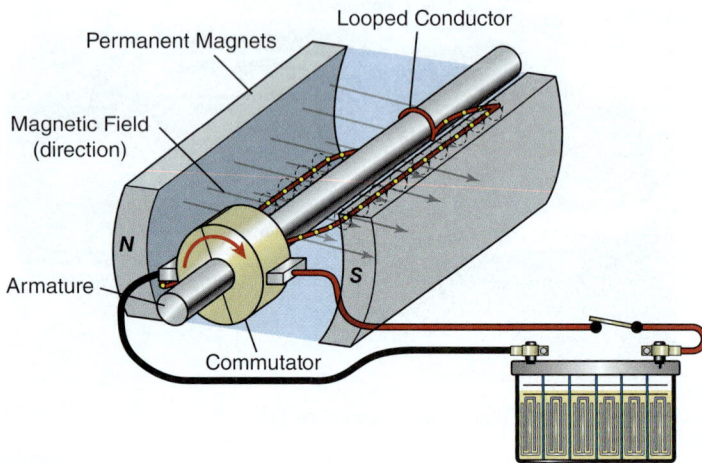

FIGURE 14-20 Simple single-loop motor and electromagnetic fields—with commutator and brushes.

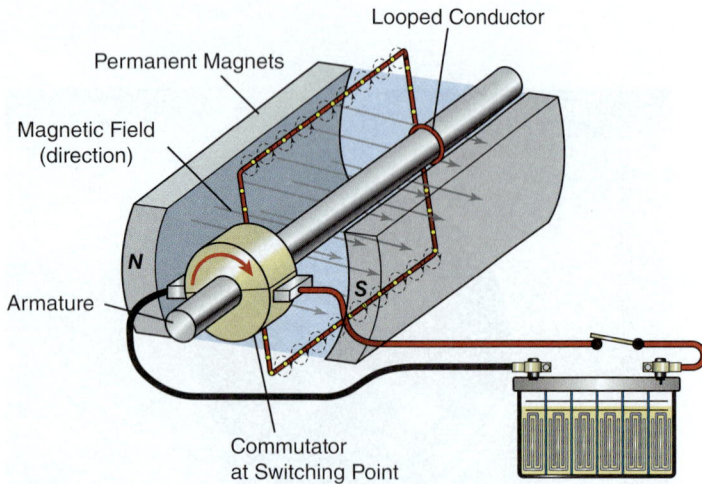

FIGURE 14-21 Simple single-loop motor and electromagnetic fields at the switching point of the commutator.

flow in the conductor will cause the conductor to move in the opposite direction. This is known as the motor effect; it is most potent when the current-carrying conductor and the stationary magnetic field are at right angles to each other.

A conductor loop that can freely rotate within the magnetic field is the most efficient motor design. In this position, when current flows through the loop, the stationary magnetic field is distorted and the lines of force try to straighten. This forces one side of the loop up and the other side of the loop down, thus turning the loop (**FIGURE 14-20**). The turning motion is called the motor effect and causes the loop to rotate until it is at 90 degrees to the magnetic field. To continue rotation, the direction of current flow in the conductor must be reversed at this static neutral point. A commutator is used to continually reverse the current flow to maintain rotation of the loop (**FIGURE 14-21**). A commutator consists of two semicircular segments that are connected to the two ends of the loop and insulated from each other (**FIGURE 14-22**). Carbon-impregnated brushes provide a sliding connection to the commutator to complete the circuit and allow current to flow through the loop.

Rotation begins with both sides of the conductor loop cutting the stationary field. When the loop passes the point where the field is no longer being cut, the momentum of rotation carries the loop and the commutator segments over so that the brushes maintain current flow in the same direction in each side of the loop relative to the stationary field. This process will maintain a consistent direction of rotation of the loop. In order to achieve a uniform motion and torque output, the number of loops must be increased. The additional loops smooth out the rotational forces. A starter motor armature has a large number of conductor loops and therefore has many segments on the commutator (**FIGURE 14-23**).

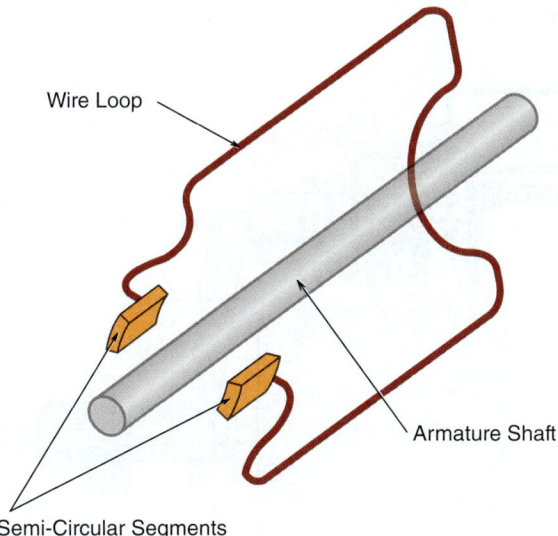

FIGURE 14-22 Commutator with two semi-circle segments.

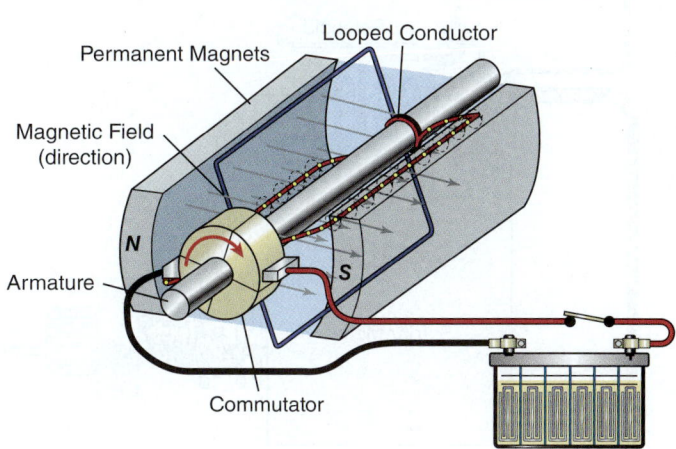

FIGURE 14-23 Simple multi-loop motor and electromagnetic fields—with commutator and brushes.

▶ Starter Motor Operation

When the starter is activated, current passes through both starter windings. The magnetic fields from both windings work together to attract the solenoid plunger toward the main starter terminals in the solenoid cap (**FIGURE 14-24**). Plunger movement also operates the shift fork lever, thus engaging the drive pinion with the flywheel ring gear. The plunger contacts a switching pin, which transfers the motion through a contact spring, closing the main solenoid terminals. This allows a large current to flow from the battery through the starter motor windings, causing armature and pinion rotation, as well as rotation of the engine crankshaft. However, closing the contacts accomplishes another very important task; it shorts to power the output wire of the pull-in winding, resulting in two sequential actions. First, the short to power means that the pull-in winding has battery voltage applied to both the input and the output of the winding, stopping the current flow through the pull-in winding and stopping that magnetic field. But the hold-in winding still has power from the control circuit, so it continues to hold the plunger in place while the starter cranks. In fact, during engine cranking, the action of the helix on the rotating armature shaft causes the pinion gear to be held firmly in mesh with the flywheel ring gear. The hold-in winding is used only to ensure that the moving contact continues to bridge the main starter terminals (**FIGURE 14-25**).

Once the engine starts, the control circuit is deactivated. This is when the second action comes into play. The current stops flowing through the control circuit to supply the input of the hold-in and pull-in windings, but the output of the pull-in winding is still activated through the bridged solenoid contacts, so current flows backwards through the pull-in winding and then forward through the hold-in winding (**FIGURE 14-26**). Since the current flows in the two windings are opposite to each other, the two magnetic fields oppose each other and tend to cancel each other out, thereby allowing the plunger return spring to retract the plunger. This disconnects the power from the pull-in and hold-in windings as well as the starter motor, causing it to stop cranking the engine. As the return spring in

K14003 Explain starter motor and solenoid operation.

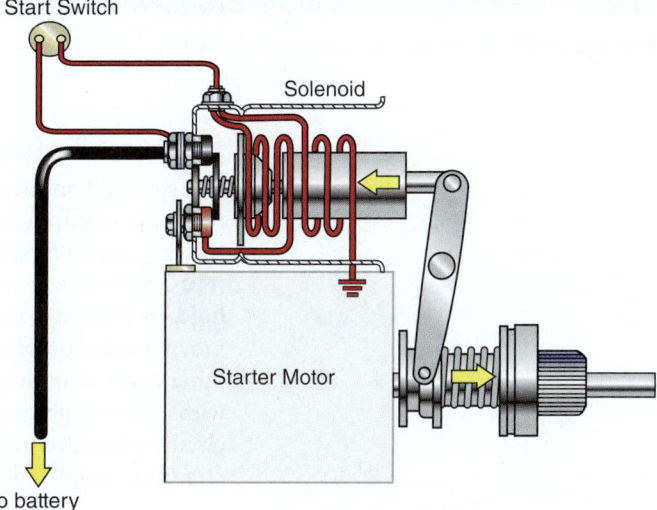

FIGURE 14-24 Both windings energized and solenoid plunger starting to move toward the cap.

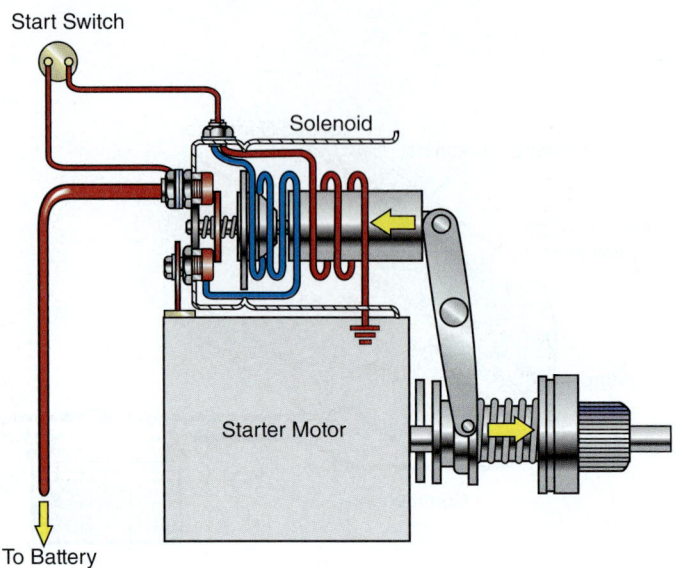

FIGURE 14-25 The solenoid plunger bridging the contacts, causing the motor to turn and shorting to power the output of the pull-in winding.

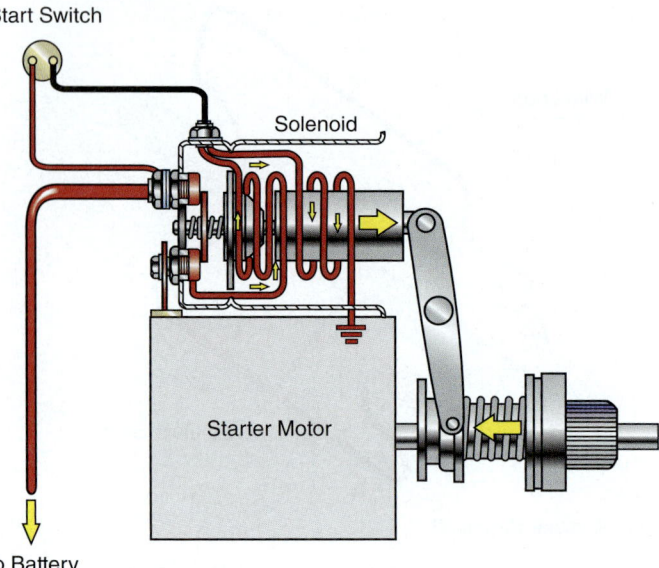

FIGURE 14-26 Current flow through the windings when the ignition key is turned from the crank position.

FIGURE 14-27 "S" terminal on a starter solenoid.

the solenoid returns the plunger, it also retracts the pinion gear to its rest position.

The starter control circuit activates the solenoid winding to draw the plunger forward. The control circuit can be activated directly by the ignition switch on some vehicles or by the PCM on other vehicles. When the control circuit is activated, it supplies battery power to the two windings in the starter solenoid. One of these is a pull-in winding, which draws a higher current and creates a stronger magnetic field than the other winding—the hold-in winding. The inputs of both windings are connected to the "S" terminal (control circuit) on the solenoid (**FIGURE 14-27**). Both windings are wound in the same direction in the solenoid housing. The output of the pull-in winding is connected to the main starter terminal leading to the field and armature windings, which provides ground to the pull-in winding until the solenoid contacts close. The output of the hold-in winding is connected to ground on the starter casing, which provides ground to the hold-in winding at all times.

Starter Solenoid Operating Principles

The solenoid on the starter motor performs two main functions: It switches the high current flow required by the starter motor on and off, and it engages the starter drive with the ring gear. The solenoid is typically a cylindrical device mounted on the starter motor (**FIGURE 14-28**). It is constructed with two electrical windings, a **pull-in winding** and a **hold-in winding**, which will be explained in a moment. One end of the solenoid has a moving soft iron plunger, which is connected to a lever that moves the starter drive. The other end has an insulated cap with electrical connections to the solenoid windings along with a set of high-current contacts and a movable copper disc that completes the cranking circuit when the solenoid plunger is drawn forward (**FIGURE 14-29**). Once the cranking circuit is completed, current flows from the battery to the starter fields and armature.

Solenoid Testing

When the starter motor fails to crank the engine, there are several parts in the system that need to be investigated to locate the cause. The first thing to check would be to test battery voltage; remember that this should be 12.6 volts on a fully charged battery. Once you have verified that the battery is fully charged, confirmed that all of the battery cable connections are tight, and double checked that the starter will not crank the engine, the first thing to test is the "S" terminal input on the starter solenoid. This terminal should receive battery voltage when attempting to start the engine. A test light is a good tool to use for this test; a small sealed beam headlamp with jumper wires attached to its terminals makes for a great test light. This device allows the technician to hook up the tester and attempt to crank over the engine. Using a brighter test light such as a small sealed beam headlamp will allow the technician to see if the light is energized from a distance away, and will serve to place a small load test on the "S" circuit to assess the circuit's electrical integrity. A DVOM set to "record" can also be used to check for voltage input to the "S" terminal upon starter operation. Using the "MIN/MAX" playback will tell the technician if battery voltage reached the "S" terminal when engine starting was commanded.

If battery voltage reaches the "S" terminal when attempting to crank over the engine, and no corresponding action takes place in the starter solenoid/motor, the fault is internal to the starter and/or solenoid. If no signal is observed at the "S" terminal, the problem is "upstream" from the starter motor. A wiring diagram should be viewed to determine how the circuit is routed and controlled; this will allow effective diagnosis of the fault.

If the problem is determined to be internal to the starter motor, it is advisable to replace the starter as a unit. Repairing a starter motor cannot only be a time-consuming process, it also exposes the repair shop or technician to liability for a failure-prone repair. In other words, there is a good chance that if you repair a starter, you might soon be working to replace it for no compensation. Quality new or rebuilt starter units usually come with a warranty that covers part price and labor, should the starter fail again. This takes the liability off of the shop or technician and places it on the part supplier.

Another failure mode associated with starter motors is when the starter "sticks," or does not stop spinning when the crank engine command stops. This can be caused by several different problems: the starter solenoid return spring fails to retract the pinion, the starter relay terminals weld closed, the solenoid terminals weld together, or something else.

Starter Control Circuit

The starter control circuit provides a means of operating the starter motor only within certain parameters, such as when the transmission is in park, the clutch is depressed, the brake pedal is applied, or the proper ignition switch is being used. These requirements help prevent accidentally starting the vehicle in gear (**FIGURE 14-30**), which could cause an accident or injury, and also help prevent the vehicle from being stolen. For many years, manufacturers have placed switches in series with the starter solenoid windings, which prevents the starter from being activated unless each of the switches is closed. If the vehicle is equipped with an automatic transmission, then a neutral safety switch is incorporated into the shifter linkage so that the switch is only closed when the transmission is in park or neutral. If the vehicle is equipped with a standard

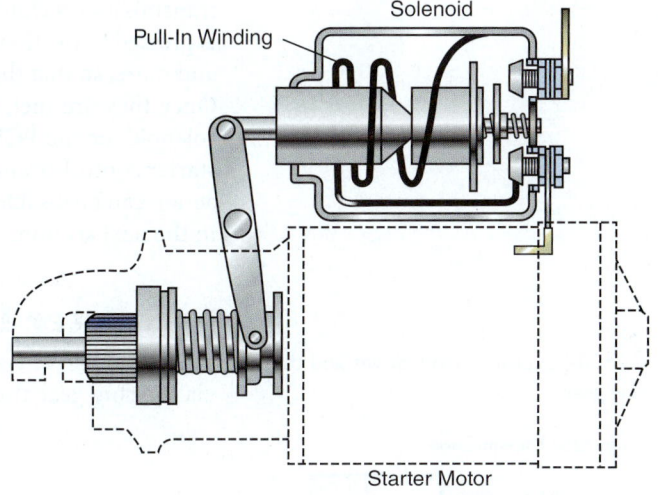

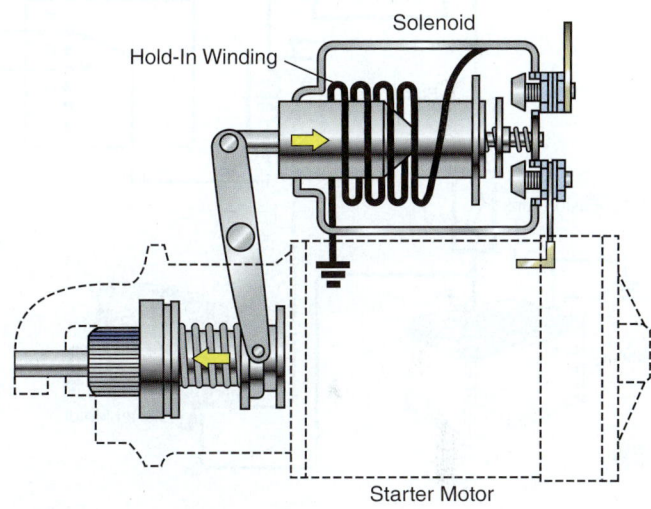

FIGURE 14-28 The solenoid uses two electrical windings: a hold-in winding and a pull-in winding.

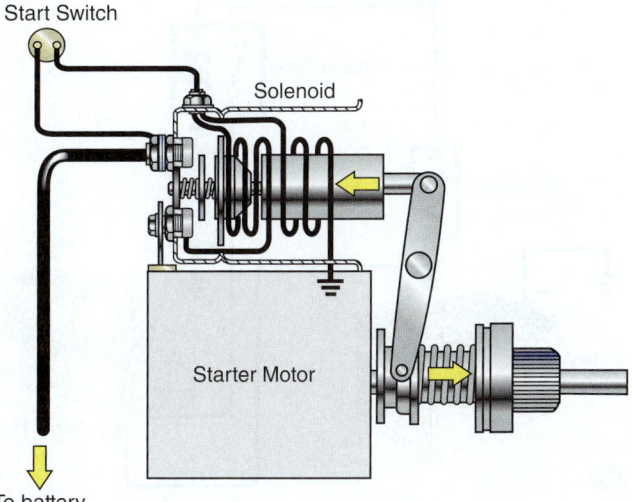

FIGURE 14-29 Solenoid starter contacts and starter drive linkage.

transmission, a clutch switch is installed in such a way that it closes when the clutch pedal is pressed to the floor. Newer vehicles use computer controls to monitor that information, and more, so that the starter will not activate until all of the required parameters are met. Once they are met, the PCM either activates a starter relay, which activates the starter solenoid, or the PCM activates the solenoid directly. With PCM-controlled starters, the starter control circuit becomes part of the vehicle theft-deterrent system, since the starter action can be disabled to prevent the vehicle from being started and stolen, as you will see in the next section.

Starter Drives and the Ring Gear

K14004 Explain starter drives and the ring gear.

The starter drive transmits the rotational drive from the starter armature to the engine via the ring gear that is mounted on the engine flywheel, flexplate, or torque converter. The starter drive is composed of a pinion gear, an internal spline that mates with the slightly curved external spline on the armature shaft, an overrunning clutch, and a return spring (**FIGURE 14-31**). The pinion gear is small in comparison to the ring gear, which means the starter turns many times faster than the engine ring gear. This also gives a large amount of mechanical advantage to the starter motor, allowing it to crank over the much larger ICE.

The overrunning clutch drives the pinion gear in one direction, while allowing it to freewheel in the opposite direction. The overrunning clutch uses roller bearings housed between an inner and an outer shell. The outer shell has tapered ramps built into it in which the roller bearings ride (**FIGURE 14-32**). Springs push the rollers toward the tapered ends of the ramps. In the forward direction, the rollers roll slightly between the tapered ramps and are pinched between the inner and outer shells, thereby locking the assembly and driving the pinion gear. In the opposite direction (once the engine starts), the rollers roll up the inclined ramps against spring pressure, unlocking the rollers and allowing the pinion gear to freewheel. The overrunning clutch prevents the starter motor from being driven by the engine once the engine starts, which would spin the armature faster than it could handle.

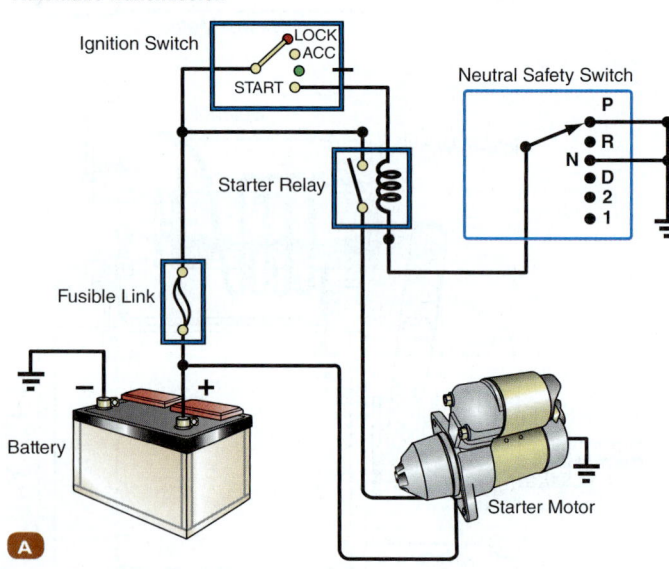

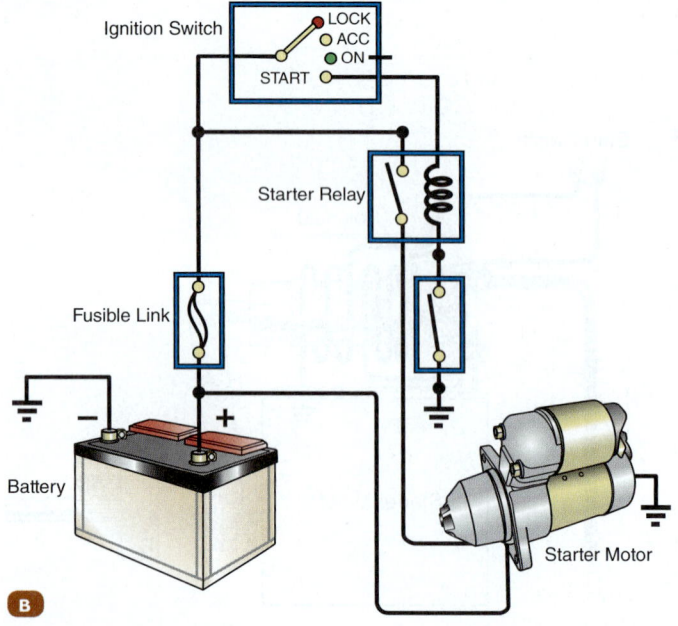

FIGURE 14-30 Basic starter control circuit. **A.** Neutral safety switch circuit. **B.** Clutch switch circuit.

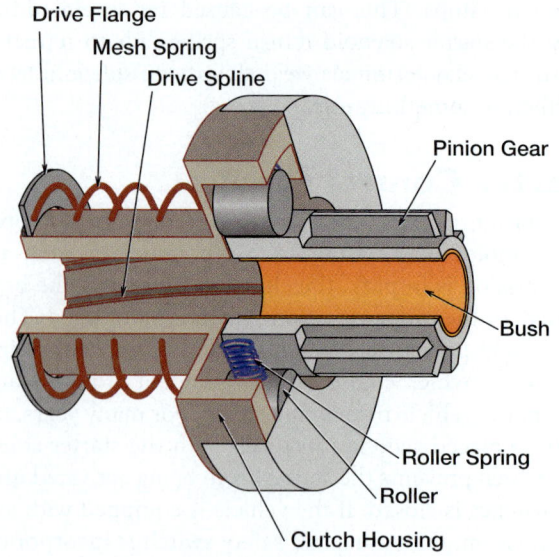

FIGURE 14-31 Exploded view of a starter drive.

Without the overrunning clutch, the vehicle engine would spin the starter motor far too quickly, causing significant damage. When the starter solenoid is activated, the starter drive teeth engage the ring gear. The solenoid plunger is connected to the starter drive by a lever and pushes the starter drive into mesh with the ring gear. It also retracts the starter drive once the solenoid has been deactivated.

Diagnosing Problems with the Starter Drive and Ring Gear

The starter drive engages the engine through the meshing of two sets of gears: the ring gear and the pinion gear. In order for the starter to properly crank the engine over, and to ensure a long service life for the starting motor, it is critical that the pinion gear and ring gear are in perfect alignment and spacing. On older vehicles, this alignment was accomplished with the use of shims that were mounted between the starter motor and the engine block (**FIGURE 14-33**). If too many shims were installed, the clearance gap between the pinion gear and ring gear would be excessive, and premature wear of the gears would result. If too little clearance were present, a loud whining noise would be emitted when the starter was cranking the engine. As a special note, when you remove any starter motor, take care to look for any shims that may be installed, and be sure that they are put back in place when the new starter motor is fitted. It should also be noted that most modern vehicles no longer use shims for starter alignment, and instead rely on dowel pins for clearance and to align the starter motor in relation to the ring gear.

In addition to alignment problems caused by improper installation of the starter motor, the ring gear and/or pinion gear can also become damaged, causing starting problems. When the pinion gear becomes stripped, a grinding noise could result when attempting to crank over the engine. If the pinion drive fails to extend, or the pinion gear is broken, you may simply hear the "whir" of the starter motor when attempting to start the engine, but the engine will not crank. Also, problems with the ring gear can be the source of a no-crank condition. For instance, if a few teeth break off of the ring gear, this creates a "dead spot" in the ring gear. Say only two teeth out of the ring gear's 100 are broken. This means that 98% of the time, the engine will stop at a point where there are enough teeth on the ring gear to start rotating the engine. Once engine rotation has started, there is usually enough momentum built up to 'jump' past the broken teeth, and continue cranking the engine. However, if the engine stops at a point where the broken ring gear teeth fall right on the pinion engagement site, the starter will not be able to engage the ring gear due to the broken teeth, and the whir of the starter motor and a failure to crank will be noted. Ring gears can also fail by cracking: the ring gear is usually press fit onto the flywheel. Should the ring gear be cracked, the press fit no longer exists, and the ring gear could spin freely of the flywheel when the starter pinion engages it.

One certainty is that there are many ways in which a starter drive can fail. It is critical that you know and understand the order of operations that must occur in order for engine cranking to take place. Once you understand these events, you can carefully observe the symptoms exhibited by the vehicle and begin to make a diagnosis. At this point, it should be noted that no diagnosis of the starter drive should be concluded until the starter is removed and a visual inspection of the pinion, pinion drive, and entire ring gear is performed.

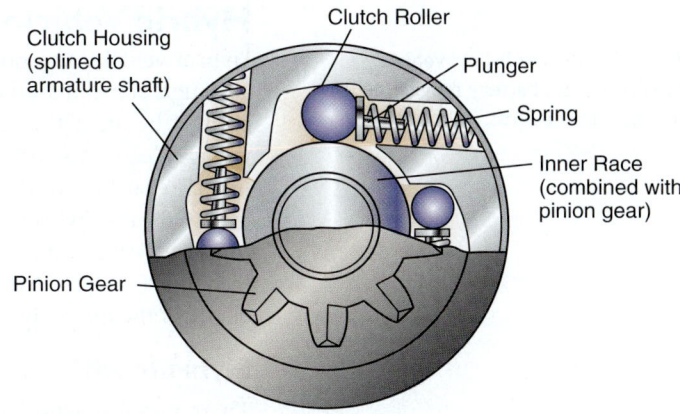

FIGURE 14-32 Starter drive one-way clutch.

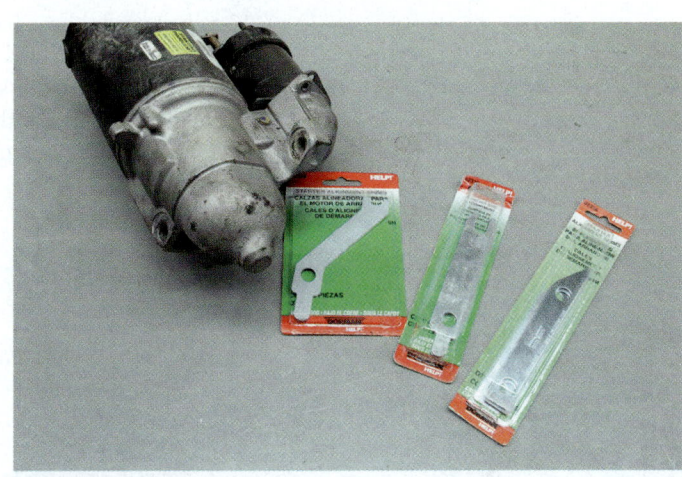

FIGURE 14-33 Starter motor and shims.

Hybrid Vehicle Starting Systems

N14001 Identify hybrid vehicle auxiliary (12v) battery service, repair, and test procedures.

Hybrid vehicles use both an ICE and electric motors to power the vehicle's drive train, although some manufacturers have released (so-called) hybrid vehicles that are driven entirely by an ICE and use only an idle-stop feature. Most hybrid vehicles use a high-voltage electric motor for engine start-up, auxiliary power, and regenerative braking functions. Regardless of the hybrid configuration, the ICE still requires a method to crank it over. On most hybrid vehicles, this task is undertaken by the hybrid's main electric drive motor, which can spin the ICE at a much faster and smoother rate, causing it to start almost instantly. On some hybrids, a more conventional ICE starter is also provided to start the engine if the main high-voltage battery bank is discharged.

Hybrids with 12-Volt Systems

There are a few hybrid vehicles that employ a traditional 12-volt starting motor as a back-up to the high-voltage starter. The most notable use of such starting motors in this capacity is on Honda vehicles; on these vehicles, the engine is cranked primarily by using a high-voltage motor and the 12-volt starter is only to be used as a back-up. In addition to the back-up 12-volt starters, there have been a number of vehicles brought to market with a "start-stop" 12-volt starting system. These vehicles are not technically hybrids, but they do allow the ICE to shut down at a stop to conserve fuel. Most of these systems employ a high-output 12-volt starting motor that either engages the ring gear or drives the ICE from the accessory belt. These systems crank the engine automatically the instant the driver's foot leaves the brake pedal.

Hybrids with High-Voltage Starters

Most hybrid vehicles use a high-speed starter motor, which is operated by the high-voltage hybrid battery pack, to crank over the engine. A big advantage of high-voltage starter motors—and one of the main reasons that they are used—is that they are able to crank the engine at a much higher speed than is possible with a standard 12-volt starter motor; engine cranking speeds on most hybrids can exceed 1500 RPM in some instances, compared to the 250 RPM cranking speed of most 12-volt starter motors. This provides a much smoother transition for starting the ICE during periods of idle-stop. In these types of vehicles, the high-voltage starter can crank the engine through the accessory belt (**FIGURE 14-34**), by a motor mounted between the engine and transmission (**FIGURE 14-35**), or by the motor generators in the hybrid transmission (**FIGURE 14-36**). As noted above, some hybrid vehicles have a 12-volt starter motor that serves as a back-up, but generally, this unit is not used, except

FIGURE 14-34 Belt-alternator starter motor (BAS).

FIGURE 14-35 Honda hybrid drive motor.

FIGURE 14-36 Hybrid transmission.

for instances where the high-voltage starting system registers a fault. These hybrid starting systems contain high voltages that can cause serious injury or death if handled improperly. Under no circumstances should service on these vehicles be attempted by unqualified personnel.

▶ Starting System Procedures

The starting system is critical to proper vehicle operation. Most drivers don't even consider the system until a component of it fails and they are left stranded. It is critical for a technician to understand how the individual parts and pieces of the system function, so that a proper diagnosis of a malfunction can be reached.

Starter Draw Testing

Testing starter motor current draw is a good indicator of overall starter motor performance. Manufacturers will specify the current draw for starter motors, and any tests must be performed with a fully charged battery of the correct capacity for the vehicle. Starter motors can be tested in two ways: on the vehicle or off the vehicle. The on-vehicle test is usually called a starter draw test, while the off-vehicle test is called a no-load test. Manufacturers will provide specifications for one or both of the tests.

Ideally, the starter motor current draw should be tested under load, which is often easier to accomplish while the starter motor is mounted in the vehicle. Remember that starter current draw is at its highest when the starter pinion gear first engages with the engine flywheel. As the starter motor and engine cranking speed increase, the current draw decreases and quickly stabilizes once the engine reaches full cranking speed. There is a variety of starter test equipment used to test starter draw. Each device will operate slightly differently, but all should have an inductive high-current ammeter to measure the cranking current flow and a voltmeter to measure the cranking voltage. The inductive ammeter is better to use than a standard ammeter because it does not require any battery cables to be removed; it is quickly clamped around the main starter cable (**FIGURE 14-37**). When conducting the test, the engine must be disabled, so it will crank but not start. The current flow and voltage will be measured during cranking and compared to specifications.

To test the starter draw, follow the steps in **SKILL DRILL 14-1**.

K14005 Understand starter system procedures.

S14001 Test the starter draw.

N14002 Perform starter current draw tests; determine needed action.

FIGURE 14-37 Inductive amp clamp attached to a battery cable for starter draw testing.

SKILL DRILL 14-1 Testing the Starter Draw

1. Research the specifications for the starter draw test. Prepare the starter tester by setting it up to measure starter current. This may require setting it to "starter," "starter testing," or something similar.

Continued

2. Connect the red lead to the positive terminal of the battery and the black lead to the negative terminal of the battery.

3. Connect the amps clamp around either the positive or the negative battery lead in the correct orientation. Make sure all of the appropriate wires are inside the clamp and the clamp is completely closed.

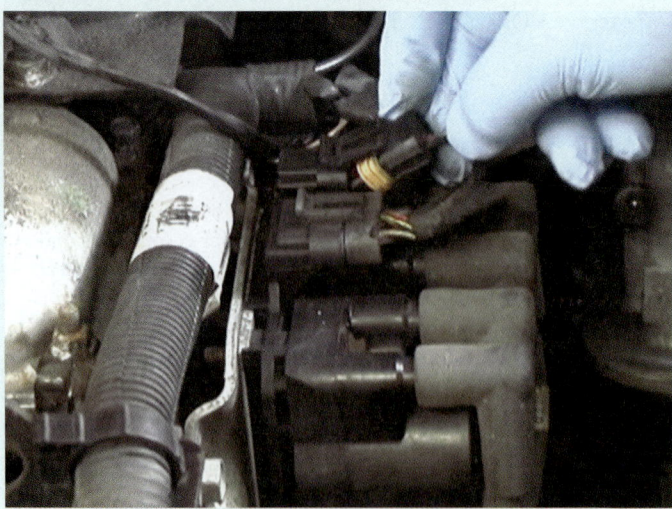

4. Disable the engine from starting by one of the following methods:
   a. Clear flood mode: This mode is programmed by manufacturers on many electronic fuel injection vehicles. It can be activated by holding the throttle down to the floor before turning the key. If the engine starts, lift your foot off the throttle and try another method.
   b. Pull the fuel pump relay and run the engine until it dies.
   c. Unplug the fuel injectors or ignition coils.

Continued

5. With the engine disabled, crank the engine and read the amps and volts as soon as the amps stabilize. Be sure not to crank the engine for more than 10 seconds at a time. Doing so can cause the starter motor to overheat, which will damage it.

6. Compare the readings to specifications and determine any necessary actions.

Starter Circuit Voltage Drop Testing

The electrical circuit of the starter motor consists of a high-current circuit and a control circuit. The high-current circuit consists of the battery, main battery cables to the starter motor solenoid, solenoid contacts, and heavy ground cables back to the battery from the engine and chassis (**FIGURE 14-38**). The control circuit activates the solenoid at the "S" terminal. It can be either PCM-controlled or non-PCM controlled. We will look

S14002 Test starter circuit voltage drop.

N14003 Perform starter circuit voltage drop tests; determine needed action.

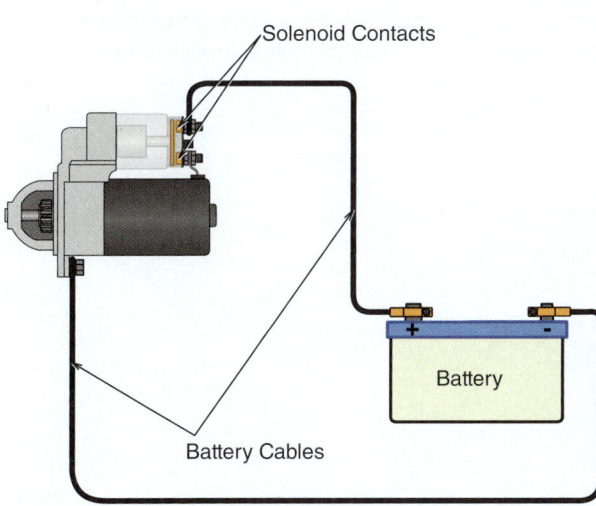

FIGURE 14-38 Components contained in the high-current side of the starting system.

FIGURE 14-39 DVOM being used to test voltage drop of a starter system.

at testing the control circuit voltage drop in the next section. Voltage drop can occur across both the high-current and control circuits; however, the high-current circuit is more susceptible to voltage drop due to the much larger amount of current flowing in the circuit. One thing to remember: When testing the voltage drop on the high-current side, for the measurement to be meaningful, the starter must be activated by the solenoid. The starter does not necessarily have to turn, but the solenoid must at least click when the ignition switch is engaged. If the solenoid does not click, then you will need to test the control side of the starter circuit.

A voltmeter or DVOM is used to measure voltage drop across all parts of the circuit. A voltmeter with a minimum/maximum range setting is very useful when measuring voltage drop because it will record and hold the maximum voltage drop that occurs for a particular operation cycle.

Voltage drop is tested while the circuit is under load. The DVOM is connected in parallel across the component or part of the circuit that is to be tested for voltage drop (**FIGURE 14-39**). Usually the most efficient method is to test large sections of the circuit first and, if required, narrow down the test to individual components to identify precisely where excessive voltage drop is located. For example, you can connect the red probe to the ground side of the starter motor and the black probe to the negative battery terminal, operate the starter with the engine disabled, and record the voltage drop across the ground circuit while the starter motor is cranked. Manufacturers will specify the maximum allowable voltage drop, but a rule of thumb is no more than 0.5 volts (500 millivolts) for a 12-volt circuit.

The same test should also be performed on the positive side of the circuit. On most vehicles, the starter cable is connected to the input of the solenoid, while the output of the solenoid is connected to the input of the starter motor. Since the heavy contacts are located between the input and the output terminals of the solenoid, where possible, it is best to measure the voltage drop from the positive battery post to the starter motor input (not the solenoid input, or "S" terminal). That way you are measuring any voltage drop across the solenoid contacts, which are a high-probability failure point. Check the service information for safe methods of disabling the engine so it will not start.

To test starter circuit voltage drop, follow the steps in **SKILL DRILL 14-2**.

SKILL DRILL 14-2 Testing Starter Circuit Voltage Drop

1. Set the DVOM to volts. Connect the black lead to the positive battery post and the red lead to the input of the starter (not the input of the solenoid, unless that is the only accessible terminal).

Continued

2. Crank the engine and read the maximum voltage drop for the positive side of the circuit.

3. Connect the black lead to the negative battery post and the red lead to the starter housing. Crank the engine and read the voltage drop.

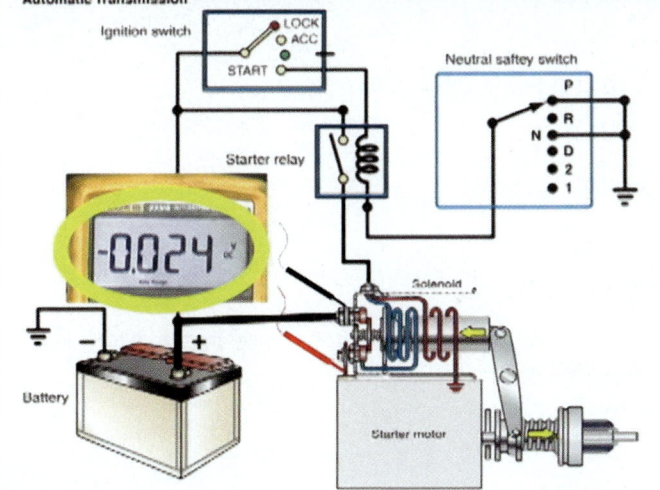

4. If the voltage drop is more than 0.5 volts on either side of the circuit, use the voltmeter and wiring diagram to isolate the voltage drop. Determine any necessary actions.

Inspecting and Testing the Starter Control Circuit

The starter control circuit activates the starter solenoid, which activates the starter motor. If there is a problem in the starter control circuit, the vehicle will likely not crank over at all, or the non-cranking condition may occur intermittently. The control circuit is made up of the battery, fusible link, ignition switch, neutral safety switch (automatic vehicles), clutch switch (manual vehicles) starter relay, and solenoid windings (**FIGURE 14-40**). If the starter is controlled by the PCM, then you must be aware of all of the circuits, such as the immobilizer circuit and the PCM itself.

S14003 Inspect and test the starter control circuit.

S14004 Inspect and test relays and solenoids.

348 CHAPTER 14 Starting Systems

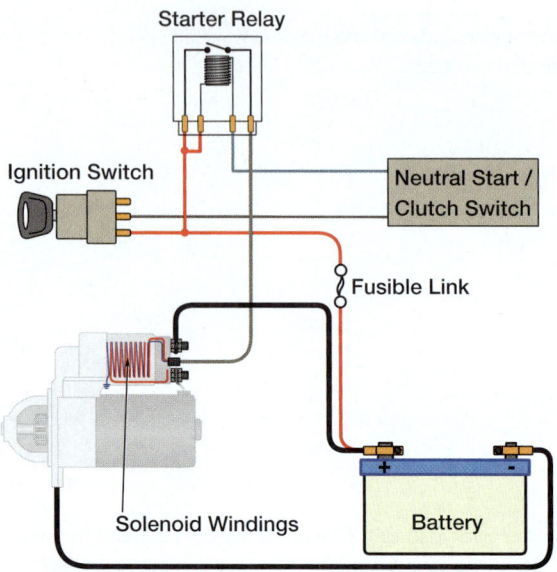

FIGURE 14-40 Wiring schematic of the control side of the starter.

Before performing any tests, you should know and confirm the customer's concern. The manufacturer's wiring diagrams should be consulted to determine the circuit operation and to identify all components in the starter control circuit.

Once an understanding of the customer concern and circuit operation is obtained, it is time to test the system. Start by placing the DVOM's red lead on the starter input terminal and the black lead on the starter housing. Measure the voltage with the key in the crank position. At that point, assuming a fault in the control circuit is present, it's likely that the voltage at the control circuit will be less than 10.5 volts. If it is, you will need to perform voltage drop tests on the power side of the control circuit to determine which side(s) of the circuit the voltage drop is located on. If the voltage drop is less than 0.5 volts, then the voltage drop on the starter ground circuit should be measured with the voltmeter. If the voltage drop is excessive, perform individual voltage drops on the ground leg. If both the control circuit power and the ground circuit voltage are within specifications, the resistance of the solenoid pull-in and hold-in windings will need to be measured. If out of specifications, the solenoid or starter motor and solenoid will need to be replaced.

To inspect and test the starter control circuit, follow the steps in **SKILL DRILL 14-3**.

SKILL DRILL 14-3 Inspecting and Testing the Starter Control Circuit

1. Use a DVOM to measure voltage between the solenoid control circuit terminal on the solenoid and the housing of the starter while the engine is cranking.

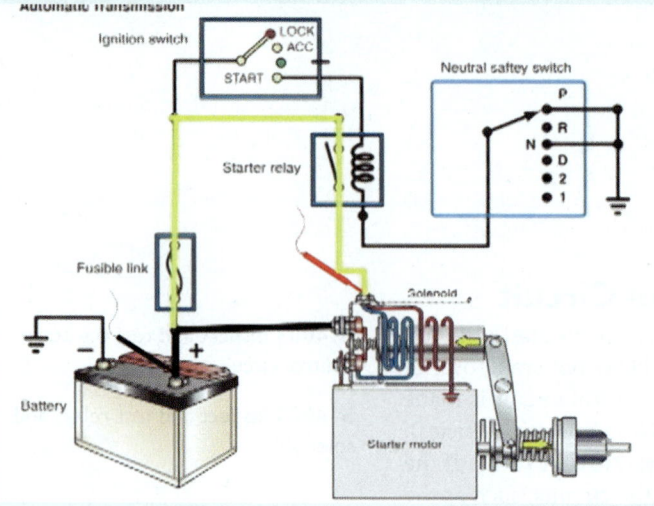

2. If the voltage is less than 10.5 volts, measure the voltage drop on the power side of the circuit.

Continued

3. If the voltage drop is less than 0.5 volts, measure the voltage drop on the ground side of the circuit.

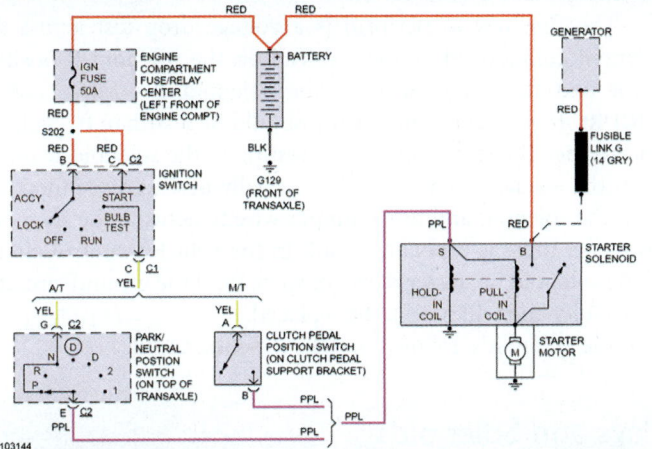

4. If the voltage drop is higher than 0.5 volts on either side of the circuit, use the wiring diagram to guide you in isolating the voltage drop on that side of the circuit.

5. If the voltage drops are within specifications on both sides of the circuit, continue to the next test.

Inspecting and Testing Relays and Solenoids

The starting system typically contains solenoids and relays that activate the control circuit. The solenoid is mounted on the starter motor, while the starter circuit relay is usually in or near the main fuse box with other similar devices. Before performing any tests, ensure that the vehicle battery is charged and in good condition. The manufacturer's wiring diagrams should be checked to determine the circuit operation, identification, and location of all components in the starter circuit.

Relays must be tested in two or three ways, depending on the relay. The simplest test is to measure the resistance of the control side/winding of the relay. If it is out of specifications, the relay will need to be replaced. Another equally simple and effective test: replace the relay with a known good part for testing purposes, and then try to crank over the engine. If the problem is corrected, replace the relay. You can also test the relay contacts for an excessive voltage drop. The best way to do this is by using an adapter that fits between the relay and the relay socket. This will allow the normal circuit current flow to flow through the contacts so that a voltage drop measurement can be taken. Any excessive voltage drop across the relay contacts will require the replacement of the relay. This last test is used only on relays with a suppression diode in parallel with the relay winding. Connect a reasonably fresh

N14004 Inspect and test starter relays and solenoids; determine needed action.

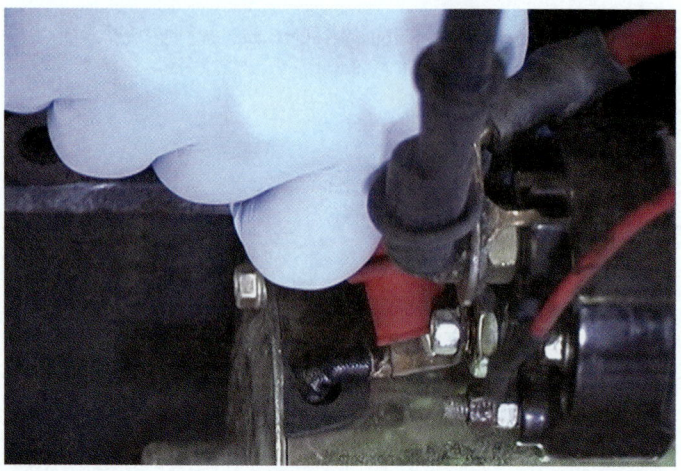

FIGURE 14-41 Testing the voltage drop between the solenoid input terminal and the solenoid output terminal.

9-volt battery across the relay winding terminals in one direction, and then switch polarity by turning the battery around. If the diode is good, the relay should click in one direction and not in the other. If it clicks in both directions, the diode is open. If it does not click in either direction, the relay winding is open or the diode is shorted.

Solenoids can be difficult to test on the vehicle due to poor access, and tests will usually be limited to voltage and voltage drop tests on the main contacts. For other tests, such as pull-in and hold-in winding tests, the starter motor will usually need to be removed. Care should be taken when testing relays and solenoids to ensure that cables are not shorted to ground and the engine is not accidentally cranked over.

The first test to perform is a voltage drop test across the solenoid contacts. Place the red lead on the solenoid B-positive input and the black lead on the solenoid B-positive output (**FIGURE 14-41**). The voltage drop should be less than 0.5 volts. If not, replace the starter assembly. Testing of the solenoid winding requires partial disassembly of the solenoid. Therefore, it is usually best to disconnect the control circuit connector from the solenoid and use a jumper wire to activate the solenoid. If the solenoid and starter operate, there is probably a fault in the vehicle's control circuit that needs further testing. If the solenoid or starter does not work (and the ground circuit is good), then the starter is likely faulty and will need to be replaced.

To inspect and test relays and solenoids, follow the steps in **SKILL DRILL 14-4**.

SKILL DRILL 14-4 Inspecting and Testing Relays and Solenoids

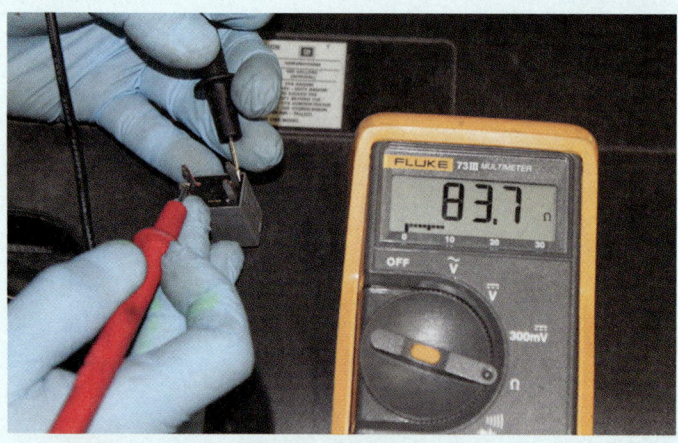

1. To test a relay, measure the resistance of the relay winding and compare to specifications. If out of specifications, replace the relay.

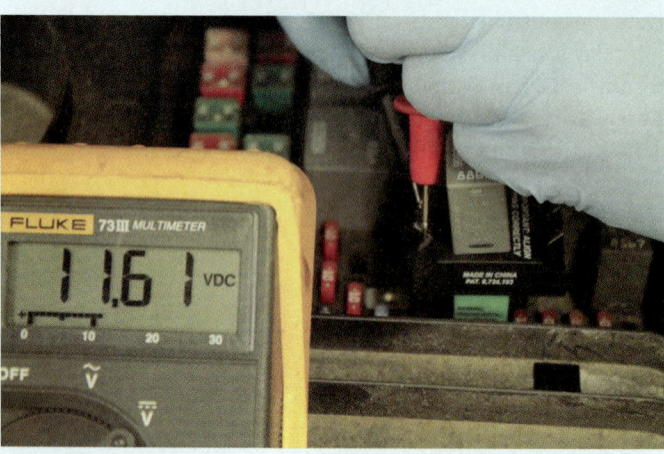

2. Use a relay adapter to mount the relay on top of the relay socket so you can check the control circuit wiring and perform voltage drop tests on the contacts. Activate the relay while measuring the voltage across the relay winding. If it is near battery voltage, the control circuit wiring is not the source of the problem.

Continued

3. Measure the voltage across the contacts with the relay NOT activated. This should read near battery voltage if both sides of the switched circuit are functional. If not, perform voltage drop tests on each side of the switch circuit. Activate the relay while measuring the voltage drop across the contacts. If it is more than 0.5 volts, the relay will need to be replaced.

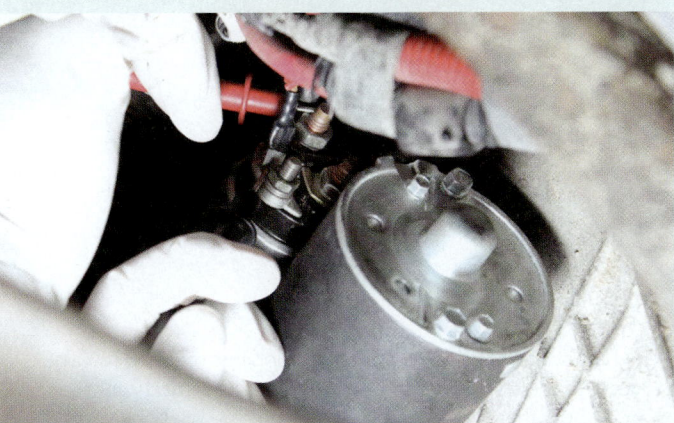

4. To test a starter solenoid, measure the voltage drop across the solenoid contact terminals with the key in the crank position. If it is more than 0.5 volts, replace the solenoid or starter assembly.

5. If the solenoid does NOT click with the key in the crank position, remove the electrical connection for the control circuit at the solenoid.

6. Use a jumper wire to apply battery voltage to the control circuit terminal on the solenoid and see if the solenoid clicks. If it does, then there is likely a fault in the control circuit wiring. If the solenoid still does NOT click (and the ground circuit is good), then the solenoid windings or starter brushes are likely worn (sometimes tapping on the starter while the key is turned to the crank position will free up the brushes enough that the pull-in winding can operate). Determine any necessary actions.

CHAPTER 14 Starting Systems

S14005 Remove and install starter in a vehicle.

N14005 Remove and install starter in a vehicle.

N14006 Inspect and test switches, connectors, and wires of starter control circuits; determine needed action.

Removing and Reinstalling a Starter

Starter motors are usually located close to the flywheel end of the engine. They can be in difficult-to-reach locations, and some engine components or covers may need to be removed to gain access. In most cases, the starter can be accessed more easily from underneath the vehicle. Always disconnect the negative battery lead before attempting to remove or install a starter motor, and be sure to use a memory minder if specified by the manufacturer.

To remove and install a starter in a vehicle, follow the steps in **SKILL DRILL 14-5**.

SKILL DRILL 14-5 Removing and Installing a Starter

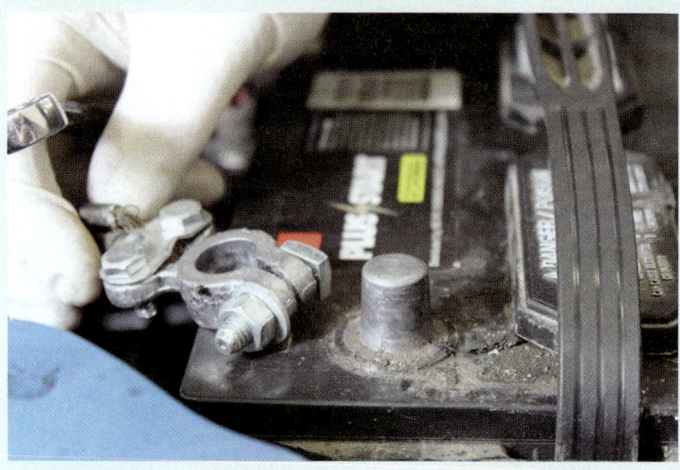

1. Disconnect the negative terminal of the battery after determining whether a memory minder is required.

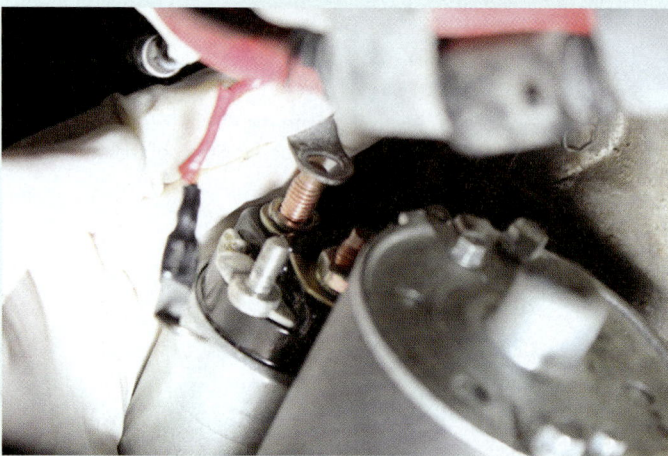

2. Remove any engine covers or components required to gain access to the starter. Remove starter motor electrical connections, noting how the wires were routed. In some vehicles, the wires cannot be accessed until the starter is removed.

3. Loosen the starter motor mounting bolts and remove them while holding the starter so it does not fall. The starter is quite heavy and can pinch fingers or break toes if it slips or falls.

Continued

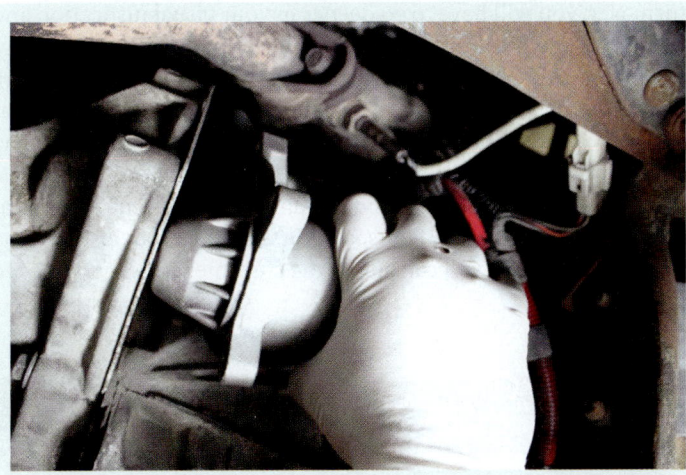

4. Remove the starter motor, being careful to catch any shims that might be between the starter and the block. Reinstall the starter motor by reversing these steps, then verify its proper operation.

Differentiating Between Electrical and Mechanical Problems

Failure to crank over properly, whether a slow-crank or a no-crank condition, can be caused by electrical or mechanical problems. For example, slow cranking could result from an electrical fault such as high resistance in the solenoid contacts. This problem could be resolved by replacing the starter with a new or remanufactured unit. But the slow-crank condition could also be caused by a mechanical engine fault, such as a spun main bearing, that is causing a lot of drag on the crankshaft, preventing the starter from cranking it over at normal speed. In this case, the entire engine will need to be rebuilt. As you can imagine, telling customers that they need a new starter motor when in fact they need a new engine (costing 10 to 20 times as much money) will not make them very happy with you. It is important to be able to differentiate between the two types of faults so that a wrong diagnosis can be avoided and the problem is fixed appropriately the first time.

Typical electrical problems that can cause starting system problems include loose, dirty, or corroded terminals and connectors, a discharged or faulty battery, a faulty starter motor, or a faulty control circuit. Mechanical problems that may cause starting system problems include seized pistons or bearings, hydrostatic lock from liquid in the cylinder(s) (e.g., a leaky fuel pressure regulator or water ingestion during off-road operation), incorrect ignition or valve timing, a seized alternator or other belt driven device, and so on. Gathering as much customer and vehicle information as possible will assist in narrowing down the options of what is causing the fault.

The order in which various tests are conducted is determined by the likelihood of a particular fault occurring given the facts you gather. For example, if the vehicle is in a parking lot, there is a better chance that the fault is an electrical fault in the starting system than if the vehicle is out in the woods or in the middle of a huge puddle of mud and water, which would tend to move the odds toward a hydro-locked or seized engine. You will also need to consider the ease of conducting a particular test or performing a visual inspection for determining the fault. For example, it is easier to perform a starter draw test than it is to pull the oil pan and inspect the main bearings. Always check for the most common and easiest faults first.

If possible, measure starter motor current draw; excessive draw may indicate a starter motor electrical fault or a mechanical engine fault. A slow crank accompanied with low draw typically points to high resistance in the main starter circuit or the starter itself. A slow crank accompanied by a high draw could be due to a fault in the starter or to engine mechanical fault. If a mechanical fault is suspected, check the oil and coolant for signs of contamination. If the coolant and oil are mixing, suspect a head gasket or cracked head/block issue. If the oil and coolant are not contaminated, turn the engine over by hand to see if it is tight compared to a similar known good engine. If it is harder to turn than it should be, remove the accessory drive belt, spin each of the accessories, and try to turn the engine

N14007 Differentiate between electrical and engine mechanical problems that cause a slow-crank or a no-crank condition.

over again. If still hard to turn over, you will have to go deeper in your visual inspection and start disassembling components based on the information you have gathered along the way. For example, if the crankshaft cannot be turned a complete revolution, remove the spark plugs and see if liquid is ejected out of one or more spark plug holes. If so, the engine was hydro-locked and you will need to determine the cause. If no liquids are ejected, then you will need to disassemble the engine further until you determine the cause of the mechanical resistance. The important thing to remember is that slow-crank and no-crank conditions can be caused by both electrical and mechanical faults, so don't jump to conclusions. You need to identify the root cause of the fault so you can advise the customer on what is needed to repair the vehicle.

Wrap-Up

Ready for Review

- The starting system provides a method of rotating (cranking) the vehicle's internal combustion engine (ICE) to begin the combustion cycle.
- The starting/cranking system consists of the battery, high- and low-amperage wires, a solenoid, a starter motor assembly ring gear, and the ignition switch.
- On PCM-activated starting systems, there is also the PCM, a relay, and all of the related sensors that feed information to the PCM. A control circuit determines when and if the cranking circuit will function.
- The control circuit starts sometimes with a fuse, the ignition switch (or PCM circuitry), the starter relay, a safety switch, and/or a combination relay/starter solenoid.
- All vehicles equipped with automatic transmissions use a neutral safety switch or a similar device, and many vehicles equipped with a manual transmission have a clutch safety switch.
- An on-board computer (PCM) and a security system called an immobilizer may also determine if and when the starting system will function.
- During the cranking process, two actions occur. The pinion of the starter motor engages with the flywheel ring gear, and the starter motor then rotates to turn over, or crank, the engine.
- The starter motor converts electrical energy to mechanical energy for the purpose of cranking the engine.
- There are three basic sections to the typical starter—the electric motor, the drive mechanism, and the solenoid.
- A relatively small current flows through a neutral safety switch or clutch switch to a starter relay that controls a larger current to operate the starter solenoid, which is typically mounted on the starter motor.
- The solenoid plunger moves the drive pinion gear into engagement with the ring gear and also closes a set of heavy-duty contacts.
- Starter motors can be designed to drive the ring gear in one of two ways: either direct drive or gear reduction.
- In the direct-drive system, the starter drive is mounted directly on one end of the armature shaft. The starter drive transfers the rotating force of the armature directly to the engine flywheel.
- Gear reduction starters use an extra gear between the armature and the starter drive mechanism. They have a reduction of about 4:1, which allows the starter to spin at a higher speed while drawing less current from the storage battery.
- Starter engagement is provided by operation of the ignition switch in the start position or when commanded by the PCM, which activates a starter-mounted solenoid, whose plunger is connected to the end of a pinion shift lever and operating fork.
- Starter motors use two magnet types: electromagnetic and permanent magnet.
- Electromagnetic starter fields are formed by current flow through heavy copper windings, wound around iron pole shoes, which are then fastened to the starter case/barrel.
- Permanent magnet-type starters have large permanent magnets mounted in the same position as pole shoes.
- Permanent magnets do not need electricity to produce a magnetic field, this type of starter motor is smaller than a starter motor that uses electromagnets.
- Starter motors with electromagnetic field windings are typically series-wound motors, which means that the electromagnets are wired in series to the armature windings.
- This high initial torque drops sharply as the motor speed increases because of the counter-electromotive force (CEMF) or voltage generated in the armature windings as the motor armature spins in the magnetic field.
- Ford is credited with the design and implementation of the moveable pole-shoe starter motor, which was a direct drive design with a moveable pole shoe that does not use a starter mounted solenoid.
- When current flows in a conductor, an electromagnetic field is generated around it.
- A conductor loop that can freely rotate within the magnetic field. When current flows through the loop, the stationary magnetic field is distorted and the lines of force

- try to straighten. This forces one side of the loop up and the other side of the loop down, thus turning the loop (starter armature). The turning motion is called the motor effect and causes the loop to rotate until it is at 90 degrees to the magnetic field. To continue rotation, the direction of current flow in the conductor must be reversed at this static neutral point.
- The solenoid on the starter motor performs two main functions: It switches the high current flow required by the starter motor on and off, and it engages the starter drive with the ring gear.
- The solenoid is typically a cylindrical device mounted on the starter motor and constructed with two electrical windings, a pull-in winding and a hold-in winding, which will be explained in a moment.
- When the starter motor fails to crank the engine, test battery voltage; remember that this should be 12.6 volts on a fully charged battery. Once you have verified that the battery is fully charged, confirmed that all of the battery cable connections are tight, and double checked that the starter will not crank the engine, the first thing to test is the "S" terminal input on the starter solenoid. This terminal should receive battery voltage when attempting to start the engine.
- The starter drive transmits the rotational drive from the starter armature to the engine via the ring gear that is mounted on the engine flywheel, flexplate, or torque converter.
- The overrunning clutch drives the pinion gear in one direction, while allowing it to freewheel in the opposite direction.
- It is critical that the pinion gear and ring gear are in perfect alignment and spacing. On older vehicles, this alignment was accomplished with the use of shims that were mounted between the starter motor and the engine block.
- If too many shims were installed, the clearance gap between the pinion gear and ring gear would be excessive, and premature wear of the gears would result.
- If too little clearance were present, a loud whining noise would be emitted when the starter was cranking the engine.
- When you remove any starter motor, take care to look for any shims that may be installed, and be sure that they are put back in place when the new starter motor is fitted.
- Newer vehicles no longer use shims for starter alignment, and instead rely on dowel pins for clearance and to align the starter motor in relation to the ring gear.
- Most hybrid vehicles use a high-voltage electric motor for engine start-up, auxiliary power, and regenerative braking functions.
- There are a few hybrid vehicles that employ a traditional 12-volt starting motor as a back-up to the high-voltage starter.
- Most hybrid vehicles use a high-speed starter motor, which is operated by the high-voltage hybrid battery pack, to crank over the engine.
- Starter motor current draw should be tested under load.
- Starter current draw is at its highest when the starter pinion gear first engages with the engine flywheel.
- The inductive ammeter is better to use than a standard ammeter because it does not require any battery cables to be removed; it is quickly clamped around the main starter cable.
- When checking starter draw, the engine must be disabled, so it will crank but not start.
- When checking starter draw, current flow and voltage will be measured during cranking and compared to specifications.
- The electrical circuit of the starter motor consists of a high-current circuit and a control circuit.
- The starter high-current circuit consists of the battery, main battery cables to the starter motor solenoid, solenoid contacts, and heavy ground cables back to the battery from the engine and chassis.
- The control circuit activates the solenoid at the "S" terminal. It can be either PCM-controlled or non-PCM controlled.
- Voltage drop is tested while the circuit is under load. The DVOM is connected in parallel across the component or part of the circuit that is to be tested for voltage drop.
- The starter control circuit activates the starter solenoid, which activates the starter motor.
- If there is a problem in the starter control circuit, the vehicle will likely not crank over at all, or the non-cranking condition may occur intermittently.
- The control circuit is made up of the battery, fusible link, ignition switch, neutral safety switch (automatic vehicles), clutch switch (manual vehicles) starter relay, and solenoid winding.
- The simplest test is to measure the resistance of the control side/winding of the starter relay. If it is out of specifications, the relay will need to be replaced.
- Another equally simple and effective test: replace the relay with a known good part for testing purposes, and then try to crank over the engine.
- Failure to crank over properly, whether a slow-crank or a no-crank condition, can be caused by electrical or mechanical problems.

Key Terms

counter-electromotive force (CEMF) Voltage generated in the armature windings as the motor armature spins in the magnetic field.

hold-in winding The winding that is responsible for holding the solenoid in the "ON" position; typically draws less current than the pull-in winding.

pull-in winding The magnetic coil in a solenoid that is responsible for creating the initial movement in the solenoid when it is powered on.

Review Questions

1. What is the main purpose of the starter system in a vehicle?
   a. To crank the vehicle's internal combustion engine
   b. To start the vehicle's electrical systems
   c. To start the HVAC system
   d. To start the conversion of chemical energy to mechanical energy
2. All of the following are part of a starter system *except*:
   a. battery.
   b. torque converter.
   c. solenoid.
   d. ignition switch.
3. Choose the correct statement with respect to gear reduction starters.
   a. They allow the starter to spin at a lower speed while drawing less current from the storage battery.
   b. They use one lesser gear between the armature and the starter drive mechanism.
   c. A gear reduction system using planetary gears requires an offset housing.
   d. They are smaller and lighter, but develop a higher torque output than their direct drive counterparts.
4. The revolving component of the DC motor that is responsible for cranking over the internal combustion engine is the:
   a. large permanent magnets.
   b. armature.
   c. commutator.
   d. brushes.
5. Which of the following components switches the high current flow required by the starter motor on and off, and engages the starter drive with the ring gear?
   a. Solenoid
   b. Armature
   c. Commutator
   d. Drive pinion
6. What would be the best way to address a problem that is determined to be internal to the starter motor?
   a. Replace the solenoid.
   b. Replace the brushes.
   c. Replace the starter unit.
   d. Replace the permanent magnets.
7. What would result if only a couple of teeth break off of the ring gear?
   a. The engine will never crank.
   b. A loud whining noise will be emitted when the starter cranks the engine.
   c. The engine will not crank if the engine stops at a point where the broken ring gear teeth fall right on the pinion engagement site.
   d. A grinding noise could result when attempting to crank over the engine.
8. Which of the following is an advantage of using high-voltage starter motors?
   a. They are safer to use.
   b. They are able to crank the engine at a much higher speed.
   c. They do not fail as easily as 12-volt starting motors.
   d. They are most beneficial in vehicles where an ICE solely powers the vehicle's drive train.
9. Which of these is used to test overall starter motor performance?
   a. Relay and solenoid testing
   b. Starter control circuit testing
   c. Starter circuit voltage drop testing
   d. Starter draw testing
10. Slow-crank and no-crank conditions can be caused:
    a. only by electrical faults.
    b. only by mechanical faults.
    c. by both electrical and mechanical faults.
    d. neither by electrical nor mechanical faults.

ASE Technician A/Technician B Style Questions

1. Tech A says that the current drawn from the starter motor is the highest after the starter has operated for a few seconds. Tech B says that starter current draw is the highest when the engine initially begins to crank, as the starter motor must overcome the inertia of the rotating engine parts. Who is correct?
   a. Tech A
   b. Tech B
   c. Both A and B
   d. Neither A nor B
2. Tech A says voltage-drop in the starter circuit should be less than 4 volts. Tech B says that total circuit voltage drop should be less than 250 mV. Who is correct?
   a. Tech A
   b. Tech B
   c. Both A and B
   d. Neither A nor B
3. Tech A says that a no-start condition could be caused by a faulty ignition switch. Tech B says that a no-start condition could be caused by a dead battery. Who is correct?
   a. Tech A
   b. Tech B
   c. Both A and B
   d. Neither A nor B
4. Tech A says that the starter solenoid moves the pinion gear into mesh with the ring gear by acting on a mechanical shift linkage. Tech B says that the starter solenoid is responsible for operating the overrunning clutch in the starter motor. Who is correct?
   a. Tech A
   b. Tech B
   c. Both A and B
   d. Neither A nor B
5. Tech A says that the starter solenoid is energized through the "S" terminal. Tech B says that the starter solenoid closes electrical contacts that allow battery voltage to flow to the starter motor. Who is correct?
   a. Tech A
   b. Tech B

c. Both A and B
   d. Neither A nor B
6. Tech A says that all starter motors are permanent magnet motors. Tech B says that it is a safe practice to strike the starter motor with a hammer in the event of a no-start condition. Who is correct?
   a. Tech A
   b. Tech B
   c. Both A and B
   d. Neither A nor B
7. Tech A says that gear-reduction starters are heavier and draw more current than traditional pole-shoe starters. Tech B says that gear-reduction type starters are smaller and draw less current than direct-drive starter motors. Who is correct?
   a. Tech A
   b. Tech B
   c. Both A and B
   d. Neither A nor B
8. Tech A says that a worn starter pinion gear may cause noisy starter operation. Tech B says that broken teeth on the ring gear could allow the engine to stop in a position where the missing teeth are above the pinion gear and thereby prevent the starter pinion from engaging and cranking the engine. Who is correct?
   a. Tech A
   b. Tech B
   c. Both A and B
   d. Neither A nor B
9. Tech A says that some automotive starter motors are liquid cooled. Tech B says that a starter motor will overheat if it is operated for more than 30 seconds at a time. Who is correct?
   a. Tech A
   b. Tech B
   c. Both A and B
   d. Neither A nor B
10. Tech A says that replacing the brushes on a starter motor are routine maintenance, much like performing an oil change. Tech B says that on many modern starters, when the unit fails, the starter is replaced, rather than being repaired. Who is correct?
    a. Tech A
    b. Tech B
    c. Both A and B
    d. Neither A nor B

CHAPTER 15
Charging Systems

NATEF Tasks

- **N15001** Perform charging circuit voltage drop tests; determine needed action.
- **N15002** Perform charging system output test; determine needed action.
- **N15003** Inspect, adjust, and/or replace generator (alternator) drive belts; check pulleys and tensioners for wear; check pulley and belt alignment.
- **N15004** Diagnose (troubleshoot) charging system for causes of undercharge, no-charge, or overcharge conditions.
- **N15005** Remove, inspect, and/or replace generator (alternator).

Knowledge Objectives

After reading this chapter, you will be able to:

- **K15001** Explain charging system theory.
- **K15002** Explain voltage regulation in an alternator or AC generator.
- **K15003** Explain the operation of the HEV (hybrid electric vehicle) charging system.
- **K15004** Describe charging system procedures.

Skills Objectives

After reading this chapter, you will be able to:

- **S15001** Perform charging circuit voltage drop tests.
- **S15002** Perform charging system output test.
- **S15003** Inspect drive belt condition, drive belt tension/adjustment, and pulley condition/alignment.
- **S15004** Replace an alternator.

You Are the Technician

Today's vehicles typically have more electrical circuits and optional equipment than comparable vehicles of a decade ago. While it is true that these available options afford the average driver levels of convenience and luxury that were not possible to achieve outside of the high-end luxury brands, these conveniences do not come without a side effect. As electrical systems become more and more complicated, they draw more current. Thus, they rely heavily on the charging system to keep the power flowing.

On the other side of this coin, the operational principles of the alternator dictate that the more amperage that an alternator is producing, the more force (horsepower) it takes to turn. This robs power from the engine and increases fuel consumption. In the race to make the most luxurious and fuel-efficient cars possible, engineers must design intricate charging system control circuits that allow the system to meet the necessary electrical demands that are placed upon it, while at the same time avoiding the creation of excess energy and thereby wasting fuel. These complex systems require high levels of skill and training to properly diagnose and repair.

1. True or False: When an alternator is producing its maximum current output, it is easiest to turn.
2. True or False: Charging systems are fairly simple systems, and as such, they require only very basic tools and knowledge to troubleshoot and repair.

▶ Introduction

Today's vehicles have more electrical accessories and convenience features than ever before. These devices put an increased load on the vehicles' electrical systems. Newer vehicles require charging systems with higher current capacities than their predecessors needed. A negative side effect of an alternator that is capable of supplying increased current loads is that it requires more mechanical energy from the engine to drive it when delivering these increased current levels, which can reduce fuel economy. Given that automakers are attempting to make vehicles as fuel efficient as possible while still providing increasing amounts of electronic accessories within them, today's charging systems must be carefully developed in order to meet both output and economy requirements.

▶ Charging System Theory

K15001 Explain charging system theory.

Alternators supply the electrical energy required for modern vehicles. Alternators produce electricity by relative movement of a magnetic field through a winding of conductors, which induces an electrical potential or voltage within the conductors. In the alternator, the magnetic field is created by the **rotor**, which rotates within the stationary **stator** windings to generate electricity there.

The charging system provides electrical energy for all of the electrical components on the vehicle. The main parts of the charging system include the battery, the alternator, the **voltage regulator** (which may be integrated into the alternator), a charge warning light, and wiring that completes the circuits.

The battery stores an electrical charge in chemical form, acts as an electrical dampening device for variations in voltage or voltage spikes, and provides the electrical energy for cranking the engine. Once the engine is running, the alternator—which is connected to the engine and driven by a drive belt—converts some of the mechanical energy of the engine into electrical energy to supply electric current to all of the electrical components of the vehicle. In addition to providing the electrical current needed to run the vehicle's accessories, the alternator also charges the battery to replace the energy used to start the engine.

▶ Alternator Principles

K15002 Explain voltage regulation in an alternator or AC generator.

The alternator converts mechanical energy into electrical energy by applying the principle of electromagnetic induction. In a simplified version of an alternator, a bar magnet rotates in an iron yoke, which concentrates the magnetic field. A coil of wire is wound around the stem of the yoke. As the magnet turns, voltage is induced in the coil, producing a current flow. When the north pole is up and the south pole is down, voltage is induced in the coil, producing current flow in one direction. As the magnet rotates, and the positions of the poles reverse, the polarity of the voltage reverses as well and, as a result, so does the direction of current flow (**FIGURE 15-1**). Current that changes direction in this way is called alternating current, or AC. In this example, the change in direction occurs once for every complete revolution of the magnet. Vehicle computers and devices do not run on alternating current, but on direct current (DC). In order for the electrical energy that is created by the alternator to be used to charge the battery and power a vehicle's systems, the AC voltage must be converted to DC voltage before leaving the alternator. This operation is carried out by a series of six diodes called the rectifier assembly.

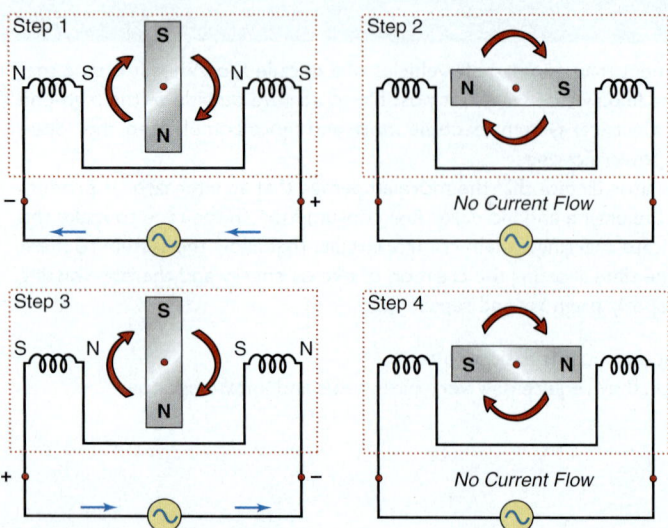

FIGURE 15-1 Electromagnetic induction.

Alternating Current

The quantity of voltage potential induced by an AC generator depends on four factors. The first is the strength of the magnetic field. Increasing the strength of the magnetic field increases the

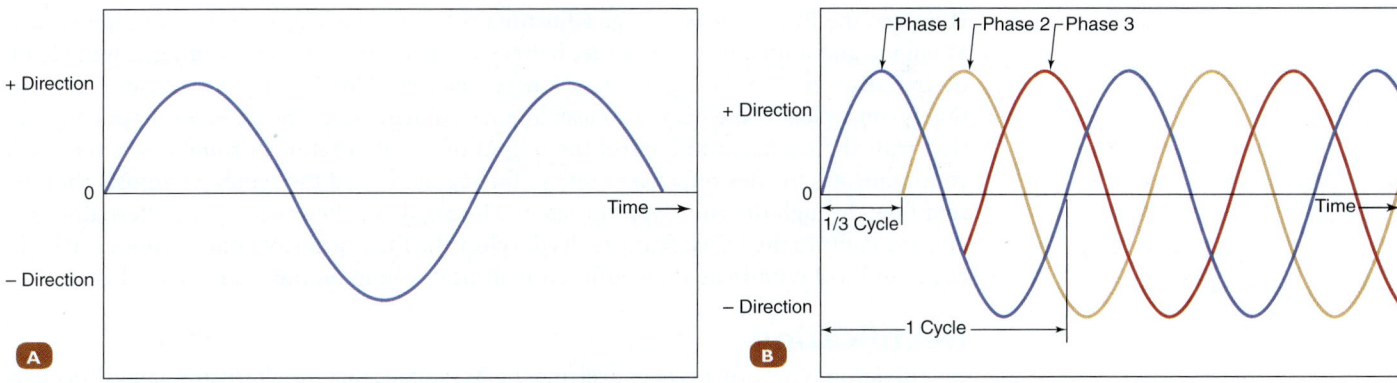

FIGURE 15-2 A. Single-phase AC signal. **B.** Three-phase AC signal.

voltage output. The second factor is the speed at which the magnet rotates. The third factor is the relative distance between the magnet and conductors. The last factor is the number of turns of wire on the stationary coil. In a typical automobile alternator, three, or possibly four, separate coils of wire, or phase windings, are common. The windings are arranged so that when the magnet is rotated, it generates a three-phase (or four-phase) output. The phases are equally spaced in time, and this results in a phase shift of, in the case of a three-phase AC generator, one phase every 120 degrees (**FIGURE 15-2**).

Alternator

The alternator's output is determined by the amount of current that flows through the rotor field coils. The greater the current flow through the rotor field coils, the stronger the magnetic field generated. The stronger the magnetic field, the stronger the alternator output. The voltage regulator monitors battery voltage and adjusts the current flow through the rotor in direct proportion to battery voltage. When the engine is running and voltage output is low, the regulator allows more current to flow through the rotor field winding. This increased current flow strengthens the magnetic field, which raises the induced voltage in the stator windings, causing alternator output to increase. If an increase is not achieved, this will cause a voltage differential to exist across the charging system warning light, thereby forcing the warning lamp to be illuminated, alerting the driver to a problem in the charging system (**FIGURE 15-3**). As the output voltage increases to the maximum regulated voltage, the voltage regulator reduces the current flow through the rotor, reducing alternator output. Regulator switching takes place in milliseconds, so the voltage to the battery is fairly constant. This process occurs quickly enough to maintain consistent system voltage even as loads are switched on or off.

The voltage regulator in modern vehicles is a solid-state electronic device that controls rotor field current with a pulse-width–modulated signal, for smoother regulation and quicker control. It may be installed inside the alternator or, for older vehicles, mounted on the fender well as a separate component. The regulator's electronic circuit senses the battery voltage and switches the rotor circuit on and off rapidly in order to maintain a constant voltage output up to the alternator's maximum output current.

Computer-Controlled Voltage Regulation

More recent developments have seen the alternator regulator function built into the powertrain control module (PCM) of the vehicle. The control principle is the same: Output voltage is controlled by switching the rotor's field circuit on and off.

FIGURE 15-3 Charging system warning light.

However, the PCM makes voltage adjustments based on a larger number of inputs, such as engine and ambient temperature, battery temperature, engine cranking, engine load, desired regeneration during deceleration, and electrical load. In most cases, an alternator that is controlled by the PCM also uses an internal regulator. The PCM works in conjunction with the regulator to control the output of the alternator. In some cases, the PCM communicates the desired charge rate to the regulator, and the regulator adjusts the current flow through the rotor appropriately. The regulator then reports the alternator load and any faults to the PCM. If any faults develop, the PCM will store one or more pertinent codes and will illuminate the malfunction indicator light on the instrument cluster.

Rectification

Rectification is the process of converting the AC voltage that the alternator naturally generates into the DC voltage that is required by the battery and nearly all of the automobile systems. To change AC voltage to DC voltage, automotive alternators use a rectifier assembly consisting of diodes in a specific configuration. Remember that a diode allows current to flow in one direction but blocks the flow of current in the other direction. A so-called three-phase "bridge" rectifier has a minimum of six diodes (three positive and three negative) to rectify the AC output of the stator windings to DC output from the rectifier (**FIGURE 15-4**). A diode bridge gets its name from two diodes in series bridged with a wire (**FIGURE 15-5**). If one end of a stator winding is connected between the diodes on the bridge, then the current flows through the ground diode into the winding and out of the winding and through the positive diode. In a three-phase alternator, one side of each winding is connected to one of the bridges. Since the other end of each winding is connected to at least one other winding, a complete path from the negative diode through a stator winding(s) and the positive diode can be completed as needed.

As the rotor rotates, each time the magnetic field changes from north to south and south to north, the polarity of each phase winding reverses and, as a result, the current changes direction. No matter what direction the current is flowing in the stator windings, the diodes in the rectifier only allow current to flow into the rectifier and out of the rectifier in one direction (DC). So the diodes guide the current flow, regardless of direction, through the stator windings and use the current flow (even if it is reversed) to push electrons out of the rectifier's positive terminal. For example, as the rotor turns, it induces a voltage in a stator winding, which generates current flow in one direction. In this position, and with this polarity, the current path is as follows: output of winding A, positive diode A, alternator terminal B-positive, positive battery terminal, battery ground (B-negative), alternator ground, negative diode B, output of winding B, neutral or star point (**FIGURE 15-6**).

When the rotor's magnet rotates further, windings B and C come under the influence of the magnetic field. The current path then is as follows: winding C, at start point, winding B,

FIGURE 15-4 Rectifier bridge.

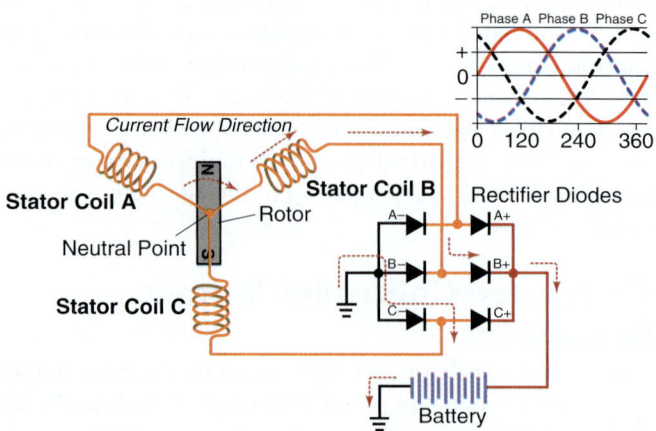

FIGURE 15-5 Current flow through a single phase in the reverse direction.

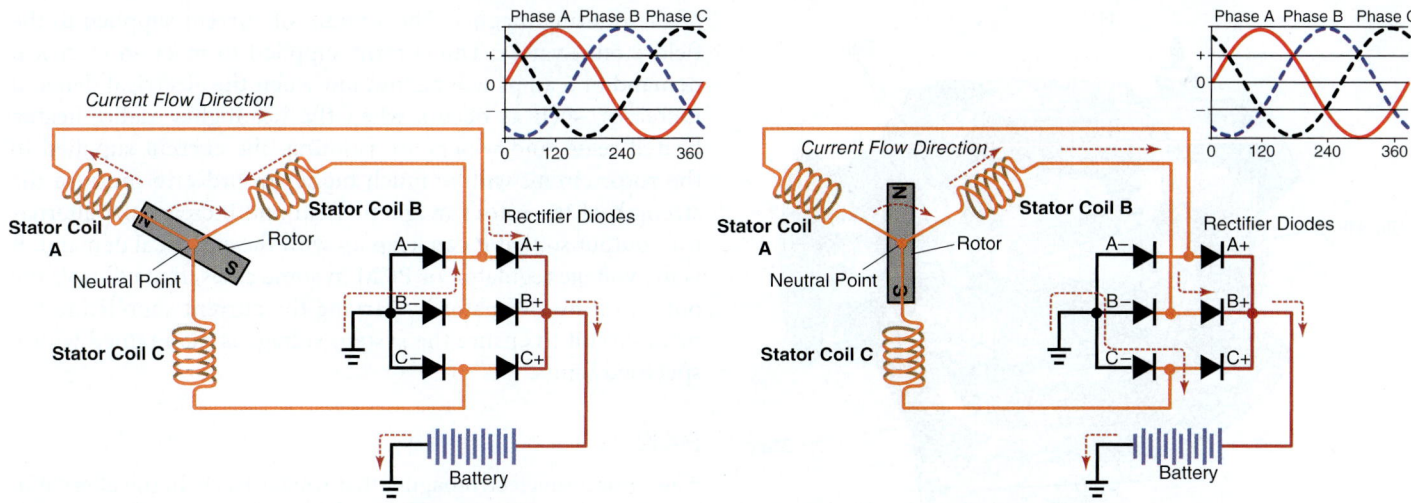

FIGURE 15-6 Current flow through a single phase in the forward direction.

FIGURE 15-7 Current flow through a single phase in the reverse direction.

positive diode B, alternator terminal B-positive, positive battery terminal, battery ground, alternator ground, negative diode C, output of winding C (**FIGURE 15-7**). As the rotor moves through its various positions, individual phase currents change in magnitude and polarity, but the output current to the battery and the electrical circuits remains in one direction only. This is because the individual phase windings are set 120 degrees apart (three-phase system).

Thus, two things are happening. First, the three phases are split so that as the voltage in one phase is falling, the voltage in the next phase is still rising. As the second phase is falling, the third phase is rising. And as the third phase is falling, the first phase is rising again (**FIGURE 15-8**).

Second, by redirecting the current so that it is flowing in the positive direction, the diodes effectively flip all of the current flow activity, from below the neutral point of the graph to the positive side, which irons out the power flow even further (**FIGURE 15-9**).

Alternator Components

The alternator consists of a stationary winding assembly called the stator; a rotating electromagnet called the rotor with a slip ring, a brush assembly, a rectifier assembly, two end frames, and a cooling fan; and a drive pulley (**FIGURE 15-10**). A voltage regulator, or in the case of modern vehicles, the PCM, monitors battery voltage and varies current flow through the rotor field circuit, thus controlling the strength of the magnetic field

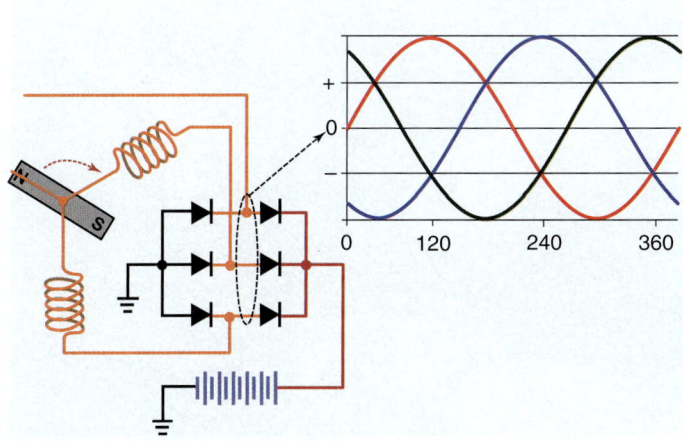

FIGURE 15-8 Three phases—not rectified.

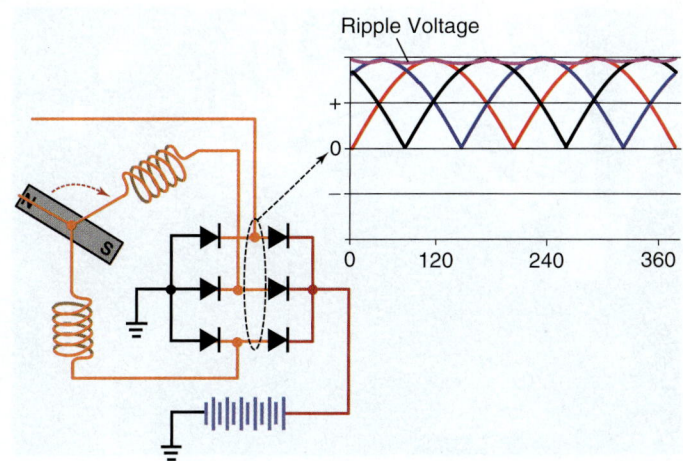

FIGURE 15-9 Three phases—rectified.

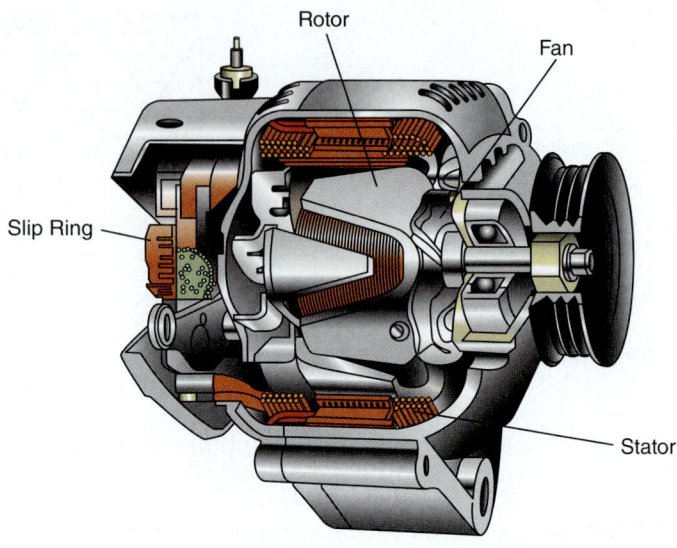

FIGURE 15-10 The alternator.

of the rotating magnet. The amount of current supplied to the field circuit varies. The current supplied to meet an electrical demand of 5 amps will be low, but when the electrical demand increases, such as occurs when the headlights, radio, heater, heated seats, and wipers are running, the current supplied to the rotor circuit will be much higher in order to increase the strength of the rotor's magnetic field and increase the alternator's output so that it can keep up with the electrical demand. It is the voltage regulator, or PCM in some cases, that controls the output of the alternator by varying the current supplied to the rotor circuit to ensure the system voltage is maintained within specified limits.

Rotor

The rotor is an electromagnet that rotates freely in the alternator. It is driven by the engine's crankshaft with the aid of a drive belt (**FIGURE 15-11**). The rotor is supported on each end by sealed ball bearings, and consists of a coil of insulated wire wound around an iron core that is pressed onto a steel shaft. A fan is either pressed or slipped onto the rotor shaft to assist in cooling the rotor and stator windings and the rectifier assembly.

An iron pole piece is fastened on each side of the coil assembly so that the projections or claws (which are bent) interlace. The ends of the rotor coil winding are connected to insulated slip rings mounted on the shaft. The spring-loaded brushes maintain contact with the slip rings at all times so that current can flow into and out of the rotor winding.

Rotor Circuit

Current can be provided to the rotor in several ways. On older vehicles, rotors were supplied power by means of three extra diodes, called a diode trio, which were connected to the bridge rectifier circuit (**FIGURE 15-12**). These extra diodes are known as field diodes or exciter diodes. This type of alternator is said to be self-exciting. However, this self-excitation can only occur when the alternator is producing an output.

On certain vehicles, current flows from the positive battery terminal through the ignition switch when it is turned to the run position. On other vehicles, this initial current is supplied to the rotor by the PCM. In either case, the circuit is completed through the slip rings and rotor field winding, and through the voltage regulator, to ground on the vehicle frame or body.

When a current is passed through the slip rings and the coil winding, it establishes strong north and south poles at the ends of the iron core and the shaft. The projections then

FIGURE 15-11 The alternator and its drive belt.

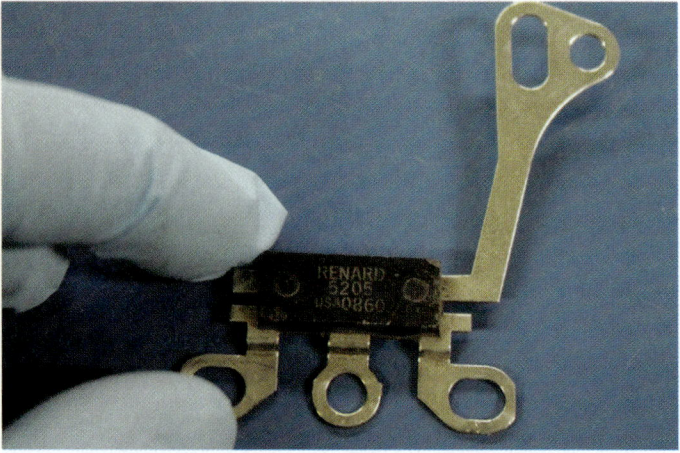

FIGURE 15-12 A diode trio supplies power to the rotor in some alternators.

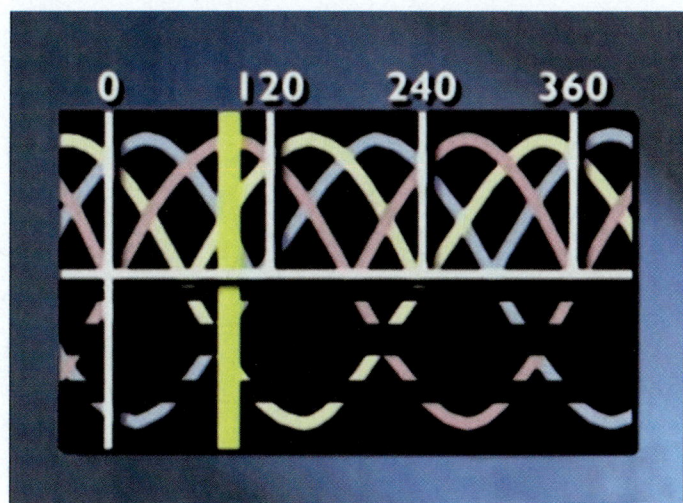

FIGURE 15-13 AC sine wave.

take on the same polarity as the end of the shaft on which they are mounted. This forms pairs of north and south poles that are alternately spaced around the rotor circumference. The rotor usually has 8 to 12 poles, which are tapered to reduce noise and create a smooth AC sine wave output as the rotor rotates (**FIGURE 15-13**).

Once the rotor is fully energized, the magnetic field requires a fair amount of mechanical energy to rotate the rotor inside of the stator. In fact, a high-current alternator can use more than 5 horsepower from the crankshaft to operate. Thus, any sizable electrical load introduced to the system will have an effect on fuel economy.

When the rotor is driven (turned by the pulley) from the engine crankshaft, the rotating magnetic field induces an AC voltage in the stator phase windings, which is then fed to the bridge rectifier to be converted to DC voltage. Stator output is also fed to the exciter diodes so that DC current can flow to the field circuit, restricted only by the resistance of the field winding (and the regulator circuit). This strengthens the magnetic field and the output voltage rises quickly.

The voltage regulator takes control of the field circuit to limit field current in order to maintain a preset, regulated output voltage at the B-positive terminal voltage of approximately 14 volts. The alternator is now charging and since the output voltage at the B-positive alternator terminal is greater than that of the battery, current flows to the battery to begin the recharging process.

Stator

The stator consists of a cylindrical, laminated iron core which carries the three- (or four-) phase windings in slots on the inside (**FIGURE 15-14**). The windings are insulated from each other and also from the iron core. They form a large number of conductor loops, which are each subjected to the rotating magnetic fields. The stator is mounted between two end housings; it holds the stator windings stationary so that the rotating magnetic fields can permeate the windings, thereby generating an electric current.

The windings can be oriented in a wye or delta configuration. These can be identified as follows: The wye configuration will have one end of each phase winding connected to a central point where the ends are connected together. This is known as the star or neutral point. The other end of each winding is connected in the bridge rectifier circuit between a positive and a negative diode. Each winding is then always part of a complete circuit, and when current is flowing, it is always flowing through two series windings at the same time. In the delta method, the windings are connected in the shape of a triangle. Connections are then taken

FIGURE 15-14 The stator consists of a cylindrical, laminated iron core which carries the three-phase windings in slots on the inside.

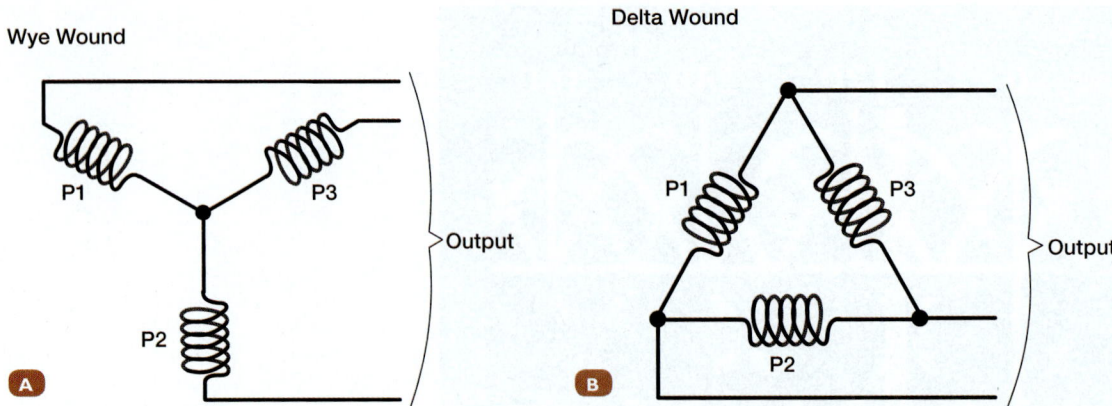

FIGURE 15-15 Phase windings. **A.** Wye. **B.** Delta.

FIGURE 15-16 Air-cooled alternator end frames.

from each point of the triangle, straight to the bridge circuit. In this arrangement, only one winding sits between two legs of the rectifier (**FIGURE 15-15**).

Alternator End Frames and Bearings

The alternator housings are typically constructed from aluminum, with vents within the frames to provide for a large amount of airflow to assist in dissipating heat (**FIGURE 15-16**). However, some alternators have coolant passages cast into the housings and are thereby liquid cooled (**FIGURE 15-17**). These types of alternators are typically found on European luxury vehicles, and do not have vents for cooling purposes. The housings accept the bearing assemblies, which support the rotor at the drive and slip ring ends.

Slip Rings and Brush Assembly

The relationship between the slip rings and brushes is what make an electrical connection to the rotating rotor assembly possible (**FIGURE 15-18**). The slip rings are normally copper bands, molded onto an insulating material, then pressed onto the steel shaft of the rotor. Each end of the rotor winding is connected to one of the copper bands, so that as the rotor rotates, the brushes maintain a constant connection with each end of the winding; the brushes carry the current from the stationary end frame to the moving rotor. Brushes are made of a combination

FIGURE 15-17 Liquid-cooled alternator.

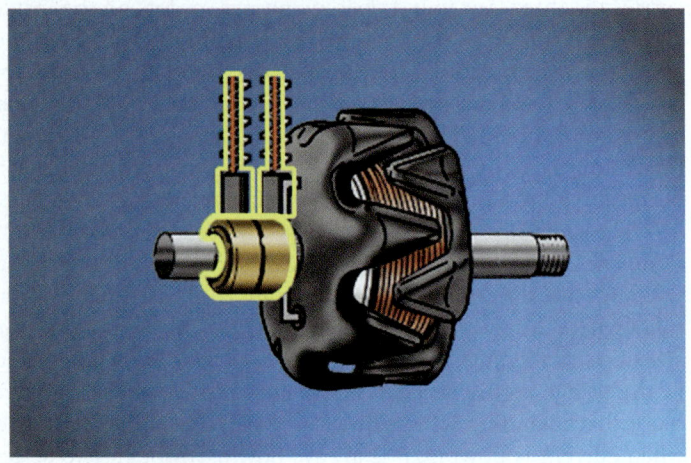

FIGURE 15-18 Slip rings and brushes aid in making an electrical connection through a rotating assembly.

of copper and carbon and are carried in brush holders mounted in the end frame of the alternator. They are spring loaded to maintain contact with the slip rings, and they wear out over time (**FIGURE 15-19**).

Rectifier Assembly

The diodes used for rectification (converting AC to DC) are mounted on heat sinks to assist in dissipating the heat generated in the diodes during the rectification process. Each diode drops approximately 0.7 volts during the conversion process, and thus tends to run hot. The diodes must be properly cooled to avoid premature failure. Three or more diodes are mounted on each heat sink. One heat sink has the positive diodes while the other contains the negative diodes (**FIGURE 15-20**). The positive diode heat sink is insulated from the frame and is connected through the output terminal to the positive battery terminal (**FIGURE 15-21**). The negative diode heat sink is connected to the frame, which allows the return circuit, via the negative battery terminal, to be completed.

FIGURE 15-19 Brushes in their holder.

Alternator Cooling Fan and Pulley

A pulley that is driven by a belt is mounted at the drive end of the alternator. Air-cooled alternators employ a powerful, centrifugal cooling fan, which is used to circulate air through the case and maintain safe operating temperatures (**FIGURE 15-22**). It is mounted on the rotor shaft and may be an integral part of the drive pulley or part of the rotor. It is essential to maintain a constant stream of air over the diodes and stator, otherwise, these components would overheat and fail. Due to the short length of the rotor, the cooling air needs to be spiraled in. This creates a long cooling path for the air and helps maintain component temperatures within the manufacturer's specifications. To achieve this spiraling effect, the cooling fins on the plate have different openings—small and large. To get the maximum cooling effect, the alternator fan must be driven in the correct direction. Refer to the manufacturer's specifications to determine the correct direction if removing and replacing the alternator cooling fan.

Over-Running Clutch Drive

On some alternators, an over-running clutch is incorporated into the alternator drive pulley. This device allows the alternator to freewheel when the engine suddenly decelerates, such as during gear changes and engine shutdown (**FIGURE 15-23**). With a direct drive pulley, when engine speed decreases rapidly, the alternator's inertia fights to maintain its rotational speed, often causing slippage, internal component stress, and noise. Over-running clutches are installed as a means of improving operating smoothness, decreasing belt slippage and noise, and increasing component lifespan.

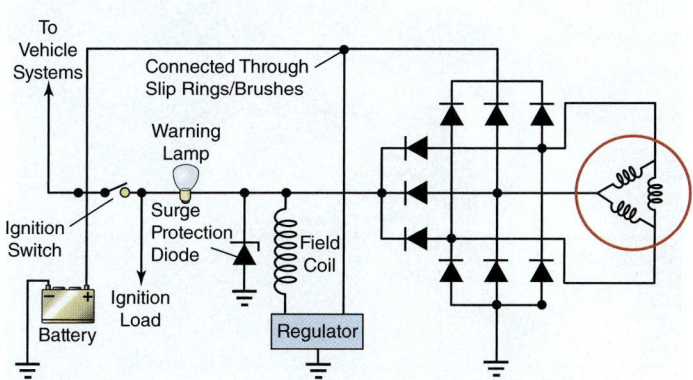

FIGURE 15-20 Positive and negative diode schematic, including diode trio.

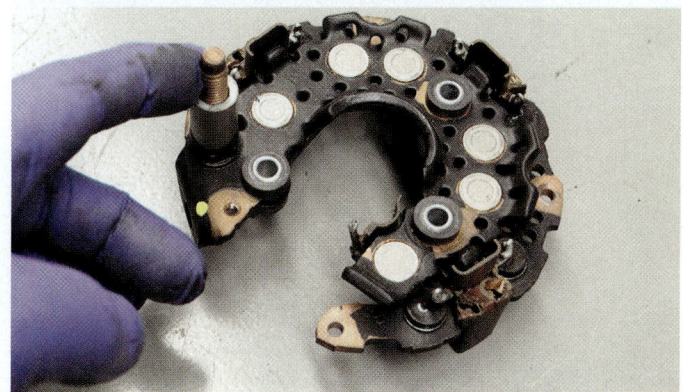

FIGURE 15-21 Positive rectifier assembly.

FIGURE 15-22 Alternator fan and pulley.

FIGURE 15-23 Typical overrunning alternator decoupler (OAD).

Component Failure Diagnosis

In an alternator, there are several parts and pieces that must work together in order for the device to function properly. Like any mechanical device, alternators can and do fail, and it is the job of the technician to be able to accurately diagnose this device so that parts are not replaced unnecessarily. It should be noted that the accepted practice today is to replace the entire alternator with a new or rebuilt unit when an internal failure is discovered. A noise concern during engine operation can be caused by faulty bearings in one of the alternator's end plates. This noise can resemble a high-pitched whine, or a low-pitched growl, and will vary with engine speed if it is an alternator problem, or a problem with any other accessory driven at crankshaft speed. When diagnosing a noise concern, the first step is to verify that the noise is present. From this point, steps should be taken to home in on the source of the noise problem; this is where a mechanics' stethoscope comes in handy (**FIGURE 15-24**). This device allows noises from a very small area to be amplified, making it easier for the technician to locate the source of the concern. Remember that alternator bearings are in the end plates, so the stethoscope will be placed on the pulley side of the alternator, as well as the back side of the alternator. Special care should be taken to stay clear of the battery output terminal and the rotating belt (**FIGURE 15-25**). At this point, it is a good idea to remove the alternator drive belt and spin the alternator by hand to feel for any catching or roughness. Before you finalize your diagnosis of a faulty alternator, be sure to spin each belt-driven accessory and idler/tensioner pulley to ensure that the noise is not being "telegraphed" from another component through the metal of the engine and accessory mounts. If the noise is

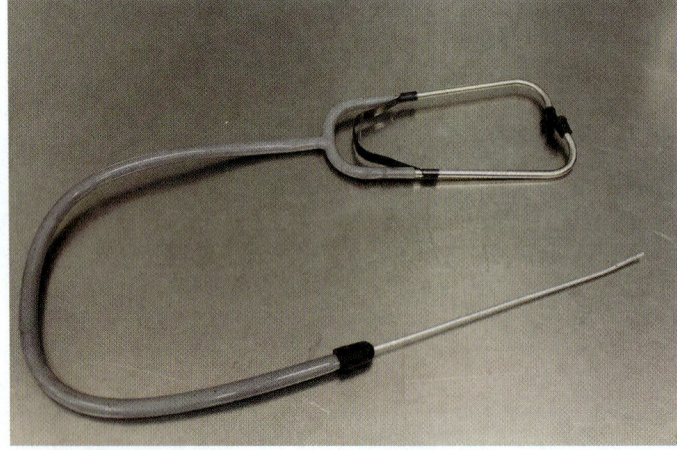

FIGURE 15-24 Mechanic's stethoscope.

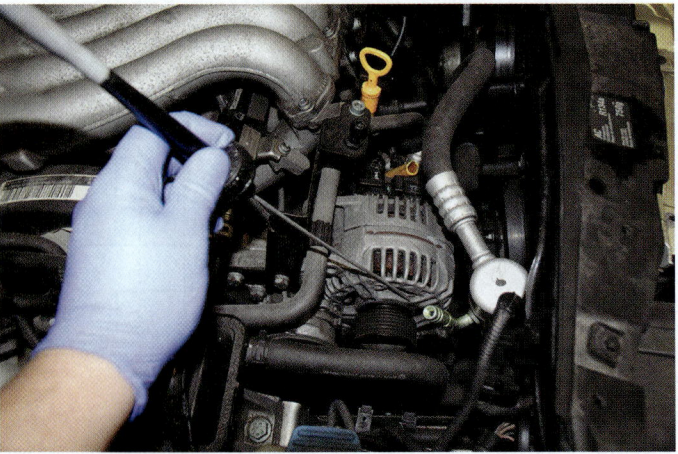

FIGURE 15-25 Stethoscope being used to test alternator end plates.

discovered to be the loudest at the alternator, and or roughness or looseness is felt when turning the alternator by hand, the problem is internal to the alternator, and the alternator should be replaced. Component failure leading to a charging problem will be discussed in the next few paragraphs.

▶ Hybrid Vehicle Charging Systems

Hybrid vehicle technology has been a driving force of automobile evolution for quite some time. The main goal of this initiative to offer more and more hybrid vehicles is this technology allows vehicles to consume much less fuel than their traditional, ICE-powered counterparts, and hybrids produce far fewer exhaust emissions compared to traditional non-hybrid vehicles. In its simplest form, a hybrid is a vehicle that uses one or more high-powered electric motors—which are powered by a high-voltage battery pack—to start the engine, provide power assist to aid acceleration, and permit the ICE to shut off when the vehicle comes to a stop (called "idle stop"), and then start up instantaneously when the driver lifts their foot off of the brake pedal. During the period of time in which the ICE is in idle stop, the vehicle and its electrical systems are still turned on and operating, but since the ICE is not running, a traditional alternator is rendered useless. To accomplish the duties of recharging the lead-acid storage battery and providing the current required to run all of the vehicle's electrical accessories, hybrid vehicles employ a device called the DC-DC converter. This device will be discussed in the next paragraph.

K15003 Explain the operation of the HEV (hybrid electric vehicle) charging system.

DC-DC Converter Operation

DC-DC converters are essentially transformers that convert high DC voltage from the hybrid battery pack to the 13.5 to 14 volts DC that is required to charge the lead-acid storage battery and power the vehicles electric accessories (**FIGURE 15-26**). In most hybrid vehicles on the market today, the DC-DC converter completely replaces the alternator as a means of supplying the DC voltage needed to power the vehicle's systems. DC-DC converters are nothing new. They have been in use inside of PCMs since the early eighties. In this role, the DC-DC converter takes the 14-volt DC supplied by the alternator and converts it to the 5-volt DC reference voltage that is used by most of the sensors on the engine.

The DC-DC converter will either be an air-cooled unit, such as those found on most Honda hybrid vehicles, or a liquid-cooled unit, such as those used on Ford, Toyota, and many other hybrid vehicles. By the nature of their design, DC-DC converters are not nearly as robust as a traditional alternator. They are typically capable of producing only 45 to 50 amps at full output. For this reason, you should never use a hybrid vehicle to jump start a vehicle with a dead battery.

SAFETY TIP

It is important to note that hybrid vehicles in general, and DC-DC converters in particular, use very high levels of DC voltage. The information provided on these systems is for reference purposes only. These systems can contain more than 300 volts DC, which can prove fatal to the technician if handled improperly (**FIGURE 15-27**). Under no circumstances should a technician attempt to diagnose or service a hybrid vehicle without specialized hybrid vehicle safety and repair training.

FIGURE 15-26 DC-DC converter in a hybrid vehicle.

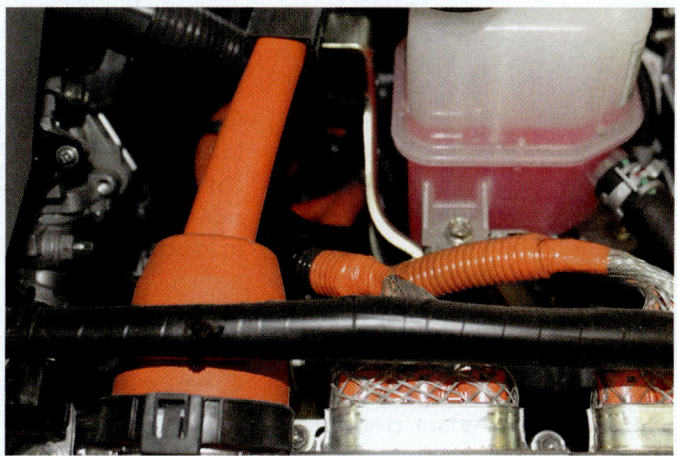

FIGURE 15-27 Orange high-voltage cables in a hybrid vehicle.

K15004 Describe charging system procedures.

▶ Charging System Procedures

Charging systems on today's vehicles are more complicated than ever. Accurately and efficiently diagnosing them requires an in-depth knowledge of their components and operation. The technician should develop a thorough, consistent diagnostic strategy that surveys all of the systems and components of the vehicle's electrical system. Once a technician develops a diagnostic strategy based on an in-depth understanding of the charging system, diagnostic effectiveness will increase at a rapid pace.

Testing for Circuit Voltage Drop

S15001, N15001 Perform charging circuit voltage drop tests.

An excessive voltage drop in the charging system output and ground circuit tends to cause one of two problems: (1) The battery is not allowed to fully charge, because although the alternator is creating the specified voltage, the voltage drop is reducing the amount of voltage supplied to the battery, or (2) the battery is fully charged, but the alternator is working at a higher voltage to accomplish this, potentially overheating the battery and causing premature failure. The determining factor of which of these issues occurs depends on where the voltage is sensed. If it is sensed at the alternator, then the battery will generally be undercharged when a voltage drop in the charging system is present. If the voltage is sensed at the battery, then the alternator will work at higher voltage levels to overcome the voltage drops present. Knowing how the system you are servicing operates will help you diagnose voltage drop issues in the output and ground circuits of the charging system.

The external alternator output circuit components consist of the wires and fuse or fusible link between the positive battery post and the alternator output terminal, and the ground wire back to the battery from the engine and chassis. Voltage drop may occur anywhere in the output current circuit and ground circuit, but is especially common at the terminals and connectors due to the charging system's high current flowing through them. Even a small amount of resistance can cause significant voltage drop when conducting these higher levels of current flow.

As with testing for voltage drop at the starter circuit, a voltmeter or DVOM is used to measure voltage drop across all parts of the circuit. A voltmeter with a minimum/maximum range setting is very useful when measuring voltage drop, as it will record and hold the maximum voltage drop that occurs for a particular operation cycle. Voltage drop tests are only valid when the circuit is under load, because voltage drops can only occur when current is flowing. The greater the flow, the greater the voltage drop, if resistance is present. Therefore, when testing for voltage drop, always perform the test when the circuit is being operated.

To measure for voltage drop, the DVOM is connected in parallel across the component, cable, or connection that is to be tested. Usually it is most efficient to first measure the voltage drop on both the entire positive side and the entire negative side of the circuit, and if required, narrow down the test to individual components to identify precisely where excessive voltage drop is located (individual voltage drops in a simple circuit add up). For example, you can connect the black probe to the output side of the alternator and the red probe to the positive post of the battery. Then, operate the charging system under a heavy load by turning on as many electrical items as possible or by using a load tester to load down the battery. Record the voltage drop across the output side of the circuit while the alternator is fully charging. Then perform the same test on the ground side of the circuit from the negative battery terminal to the frame of the alternator. Manufacturers will typically specify the maximum allowable voltage drop, but a rule of thumb is no more than 0.5 volts (500 millivolts) for each side of a 12-volt circuit.

To perform a charging circuit voltage drop test, follow the steps in **SKILL DRILL 15-1**.

Charging System Output Test

S15002, N15002 Perform charging system output test.

Vehicle charging systems are voltage regulated, which means that the alternator will try to maintain a set voltage across the electrical systems. As electrical load current increases in the vehicle systems, voltage starts to drop. The voltage regulator senses this voltage drop and

SKILL DRILL 15-1 Performing a Charging Circuit Voltage Drop Test

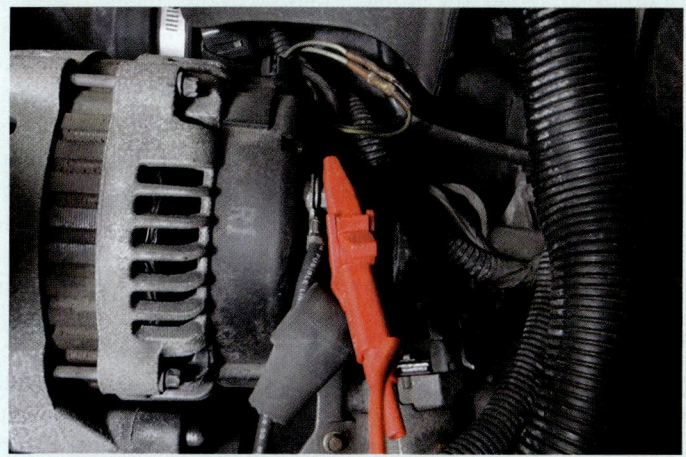

1. Set the DVOM to measure voltage, and select min/max if available. Connect the red probe of the DVOM to the output terminal of the alternator and the black probe to the positive post of the battery.

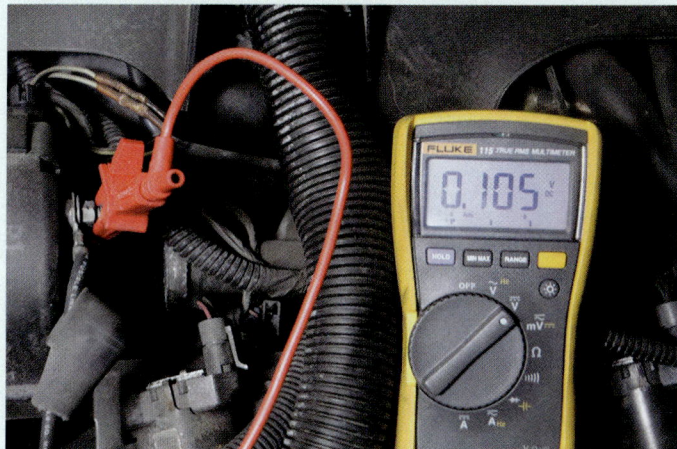

2. Start the engine and turn on as many electrical loads as possible or use an external load bank to load the battery. Read the maximum voltage drop for the output circuit.

3. Move the leads to measure the voltage drop on the ground circuit by placing the black probe on the alternator case and the red probe on the negative terminal of the battery. Read the maximum voltage drop for the ground circuit. If the measurements are excessive, check each part of the circuit for excessive voltage drops by slowly bringing the probes closer together on each section of the circuit. Determine any necessary actions.

increases the current output of the alternator, which in turn increases system voltage to try to maintain the correct voltage in the system. The testing of an alternator output initially involves the testing of the system's regulated voltage using a voltmeter. Regulated voltage is the voltage at which the regulator is allowing the alternator to create only a small charge due to the battery being relatively charged, as evidenced by the greatly reduced current output.

Regulated voltage should be between the manufacturer's specified minimum and maximum regulated voltage. If it is incorrect, verify that there are no voltage drops on the alternator and regulator. If no voltage drops are found, the problem may be in the voltage regulator, or on PCM-regulated alternators, a fault code may be recorded in the PCM that explains why charging is not being allowed. It is important to refer to the manufacturer's service information in order to determine if all of the required inputs are present at the alternator. If all inputs are present, no trouble codes are recorded on PCM-controlled alternators, and voltage drop testing of the system has been carried out, it can be deduced that the regulator is faulty. On vehicles with external voltage regulators, you can simply replace the voltage regulator, but on vehicles with an internal voltage regulator, the alternator should be replaced as a unit.

Once the regulated voltage is confirmed, it is time to check the charging system output; this is done by using an external electrical load such as a carbon pile to reduce the battery voltage, thereby tricking the regulator into full-fielding the alternator, making it produce maximum amperage output. This output is read using an inductive ammeter and should then be compared to the manufacturer's rated output specifications. An alternator that puts out within 10% of its rated output is functional. Less output than that indicates a faulty regulator, faulty alternator, or excessive voltage drop(s) on the alternator output or ground circuits.

To perform a charging system output test, follow the steps in **SKILL DRILL 15-2**.

SKILL DRILL 15-2 Performing a Charging System Output Test

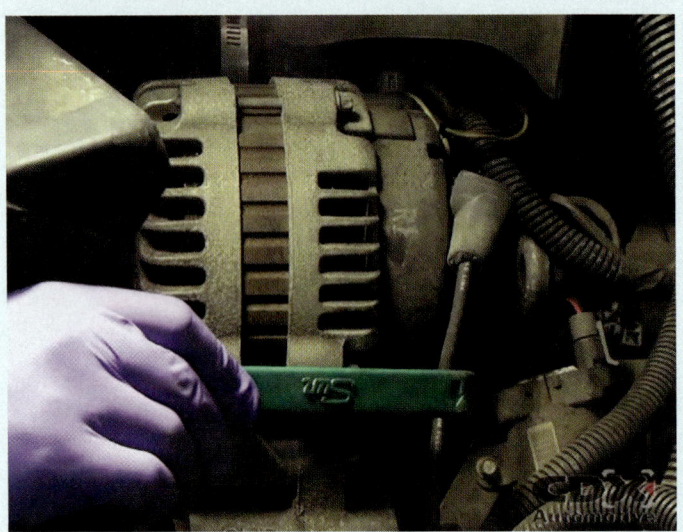

1. Connect the red lead on the charging system tester to the battery's positive post, the black lead to the negative post, and the amps clamp around the alternator output wire.

2. Start the engine, turn off all accessories, and measure the regulated voltage at around 1500 rpm. The regulated voltage is the highest voltage the system achieves once the battery is relatively charged, as evidenced by the ammeter reading less than about 15 to 20 amps when the amps clamp is around the alternator output cable. Typical regulated voltage specifications are wider than they used to be due to the ability of the PCM to adjust the output voltage for a wide range of conditions. Thus, typical readings could be about 13.2 to 15.5 volts on newer vehicles and 13.5 to 14.5 volts on older vehicles. Always check the specifications.

Continued

3. Operate the engine at about 1500 rpm and either manually or automatically load down the battery just enough to obtain the maximum amperage output without pulling battery voltage below 12.0 volts. This reading should be compared against the alternator's rated output. Normally, readings more than 10% out of specifications indicate a problem.

Diode Test

As mentioned previously, rectifier diodes are responsible for converting the AC voltage that is normally produced by the alternator to the DC voltage that is used to charge the lead-acid storage battery and power the vehicle's electrical systems. Only a small amount of AC "ripple" voltage can be tolerated by the vehicle's systems.

To test for diode problems, follow these steps:

- Set your meter to measure AC volts.
- Start the engine and increase its speed to 2000 rpm.
- Connect your DVOM across the positive and negative battery terminals.
- Turn on all of the vehicle's electrical devices to load the system, then record the AC voltage reading being picked up by the DVOM.
    - A reading of 400 mV (0.4 volts) AC indicates that the diodes are good.
    - An AC reading of 500 mV (0.5 volts) AC or greater indicate the rectifier diodes have failed, and the alternator needs to be replaced.

Inspecting Drive Belt Condition, Drive Belt Tension/Adjustment, and Pulley Condition/Alignment

Vehicles will either use a traditional V-belt (**FIGURE 15-28**), or a serpentine poly-V belt (**FIGURE 15-29**) to transmit rotational force from the crankshaft to the accessories and the alternator. V-belts have a face angle of 34 degrees, while the pulleys that they ride in have

N15003 Inspect, adjust, and/or replace generator (alternator) drive belts; check pulleys and tensioners for wear; check pulley and belt alignment.

S15003 Inspect drive belt condition, drive belt tension/adjustment, and pulley condition/alignment.

FIGURE 15-28 V-belt.

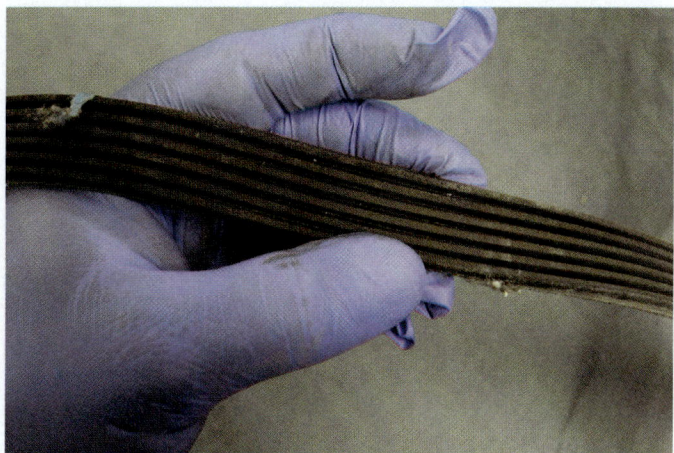

FIGURE 15-29 Serpentine poly-V belt.

FIGURE 15-30 Cracked V-belt and cracked serpentine poly-V belt.

a face angle of 36 degrees, which creates a wedging action that allows the belt to transmit rotational force. Serpentine poly-V belts use the same variations in engagement angles employed with the traditional V-belt, but the fact that they have multiple ribs means that they are capable of transmitting greater amounts of rotational force than are traditional V-belts.

During the course of vehicle diagnosis, it is critical that the technician performs a thorough visual inspection of the drive belt, taking special care to look for frayed edges, missing chunks of belt, greasy residue on the belt, and/or hard rubber and "shiny" areas, which indicates that the rubber has become glazed and will not be able to transmit the rotational force required to properly drive the alternator. Check for cracks on the driving side of the belt (the side with the "V" or ribs); any cracks on a V-belt, or more than three cracks on any one rib in a three-inch span on a serpentine poly-V belt, indicate that the belt should be replaced (**FIGURE 15-30**). It should be noted that newer serpentine belts are constructed of EPDM rubber, and will not crack like the earlier serpentine belts constructed of neoprene. These EPDM belts, however, still wear at the points where the "V" pattern in the belt engages the pulleys; they require replacement every four to seven years like all of the other belts mentioned above.

The pulleys on which the drive belt rides play an equally important role in the transmission of rotational torque from the crankshaft to the driven accessories. It is the pulleys that are responsible for keeping the belt captured so that it does not fly off when engine speed increases. As a rule of thumb, all of the pulleys associated with a single belt should be in perfect alignment with one another. When performing your diagnosis, it is important to pay special attention to the orientation of the pulley. Doing so will allow the technician to spot a pulley that may be crooked or bent, causing belt retention, slipping, and/or noise problems. All of the pulleys should also be inspected for damage and turned by hand with the belt removed to assess the condition of their bearings. Any pulleys found to have an issue should be replaced.

Proper belt tension is critical in order for the drive belt to properly engage the pulleys and transmit rotational torque. Belt tension is achieved either by the use of a tensioner, or, on modern vehicles, a "stretchy" belt is used. These stretchy belts have elastic properties that allow them to be self-tensioning, eliminating the need for an additional belt tensioner. There are several ways to measure belt tension. It can be measured with a belt tension gauge (**FIGURE 15-31**) by using the manufacturer-supplied marks on the tension (**FIGURE 15-32**) (refer to vehicle service information). Belt tension may be determined

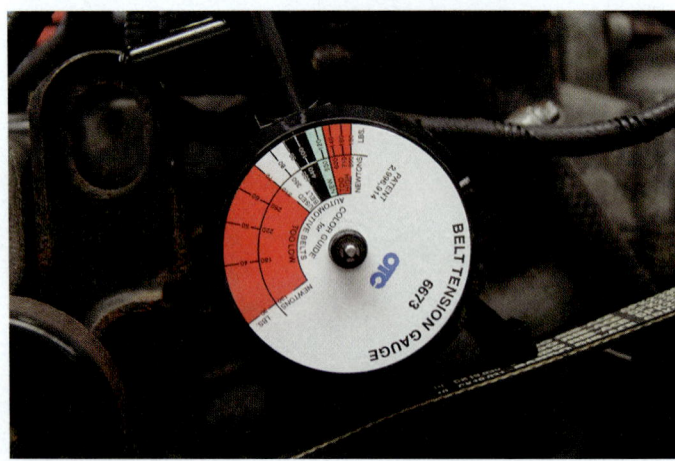

FIGURE 15-31 Belt tension gauge.

FIGURE 15-32 Belt tension marks on spring-loaded tensioner.

by using a beam-type torque wrench to measure the force required to deflect the spring-loaded belt tensioner (refer to service information for specifications). Belt tension can also be found by measuring the deflection of the belt between the two pulleys that are the farthest apart. The flex, or deflection, should be no more than 0.500" (12.70 mm) on a properly tensioned belt.

Diagnosing Undercharge and Overcharge Conditions

Undercharging and overcharging are both bad for a vehicle. Undercharging leads to compromised function of electrical systems and a less-than-fully-charged battery, a condition which leads to sulfation and, consequently, early battery death. Overcharging can shorten the lifespans of bulbs and other electrical devices, while at the same time overcharging the battery, which can increase gassing and loss of water from the electrolyte. Maintaining a proper charge is critical to long life and proper operation of the electrical system.

The voltage regulator maintains a constant voltage output from the alternator up to the maximum output current. Always check the manufacturer's specifications for alternator regulated voltages; more and more manufacturers are using PCM-controlled charging systems, which means the voltages can have a wider range than non-PCM-controlled systems.

As a rule of thumb, for a typical non-PCM-controlled 12-volt charging system, the regulated alternator output voltage should be between 13.5 and 14.5 volts. If the voltage measured is outside of the manufacturer's specifications, further testing will need to be performed to determine the cause.

For this task, we will assume that the charging system is charging (not a no-charge condition) and that it is either not charging fully or is overcharging. If the system is overcharging, it can almost always be tracked to either a faulty voltage regulator or a faulty regulator ground, both of which can cause the alternator to overcharge.

An undercharge condition has many more potential causes, such as a loose drive belt, voltage drop in the charging system wiring, faulty regulator, worn brushes in the alternator, high resistance in the rotor, open or shorted diodes, open or shorted stator windings, or even a shorted cell on a battery. Always check for the easiest and most common faults first. For example, check that the drive belts are not slipping, particularly under load. Also verify that the battery does not have a shorted cell and that its capacity is adequate.

Replacing an Alternator

Alternators need to be replaced whenever they are electrically or mechanically faulty. Electrical faults include no-charge, undercharge, or overcharge conditions. Mechanical faults include worn and noisy bearings, or other internal or external mechanical damage. When replacing an alternator, there are hazards. Battery voltage is always present at the output terminal of the alternator. This means that if the battery terminal is not removed from the battery and a wrench is placed on the output terminal of the alternator, a very large spark could occur if the wrench touches something metal. Always disconnect at least one battery terminal prior to removing the alternator. Also, check the manufacturer's information regarding maintaining the adaptive memory, if necessary.

The alternator must not be operated with the battery disconnected or with the terminals at the back of the alternator disconnected. Doing so will damage the alternator. Even though most vehicle electrical systems are described as "12-volt," they typically operate at between 13.5 to 14.5 volts. If the system is not operating in this range, the drive belt may have become loose, or excessive voltage drops in the charging or voltage-sensing circuit may be the cause. If all of these components check out, the alternator, external voltage regulator, internal regulator, or even the built-in voltage regulating circuit in the ECM/BCM may be faulty. Make sure you have identified the cause of the fault before removing the alternator.

To replace an alternator, follow the steps in **SKILL DRILL 15-3**.

> ▶ **TECHNICIAN TIP**
>
> Since a large amount of mechanical energy is required to rotate the alternator when it is energized, a hard or loose drive belt could slip, failing to deliver enough torque to spin the alternator. This will cause a no-charge or low-charge condition, and may not even make a squeaking belt noise! For this reason, when investigating alternator charging issues, always be sure to thoroughly inspect the alternator drive belt, ensuring that it is in acceptable shape and tensioned properly before continuing your diagnosis of the charging system.

N15004 Diagnose (troubleshoot) charging system for causes of undercharge, no-charge, or overcharge conditions.

S15004 Replace an alternator.

N15005 Remove, inspect, and/or replace generator (alternator).

SKILL DRILL 15-3 Replacing an Alternator

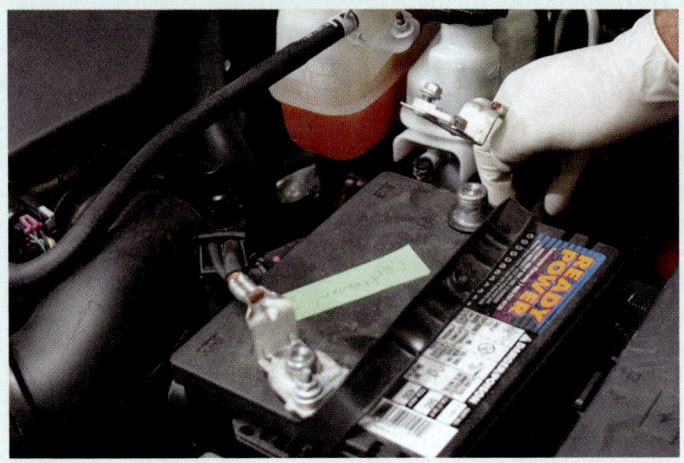

1. Install fender covers. Verify any memory issues and remove the negative terminal of the battery.

2. Loosen the drive belt and remove it from the alternator pulley. Check the condition of the belt to see if it is still serviceable.

3. Locate the electrical connections at the rear of the alternator and note their positions. Loosen any securing fasteners or covers, and remove terminals one at a time.

Continued

Charging System Procedures 377

4. Loosen the securing fasteners that hold the alternator to its mounting bracket(s), making sure the alternator is supported. Remove the alternator.

5. Reinstall the alternator. Situate the alternator in the mounting bracket(s) and, while still supporting the alternator, loosely start the securing fasteners that hold the alternator to its mounting bracket(s).

6. Reinstall the electrical wires to their correct terminals, referring to the manufacturer's information. Check the security of any fastening devices.

Continued

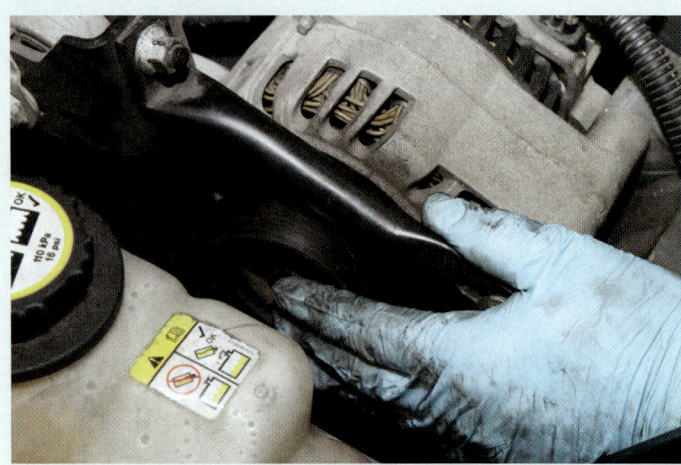

7. Install the drive belt over the alternator drive pulley and, using the correct tools, adjust the belt to the correct tension.

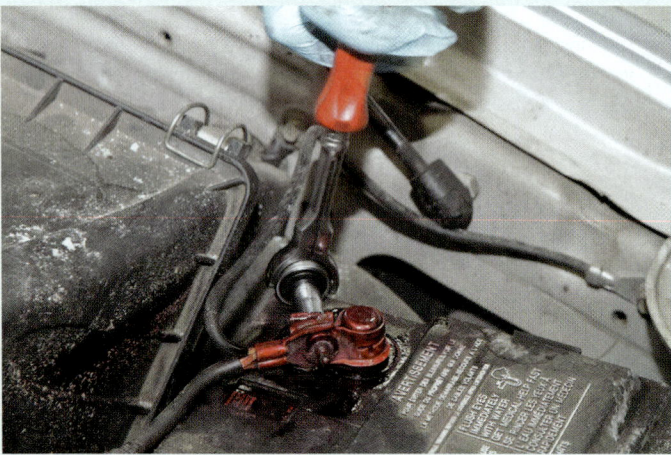

8. Reattach to the negative post of the battery. Make sure the fastener is tight and replace any battery terminal covers.

9. Turn the ignition to the on position and make sure the charge light on the instrument panel illuminates. Start the engine and see if the charge light goes off. Measure the regulated voltage and maximum alternator amperage output. Remove the fender covers and return any tools used to their correct place.

Wrap-Up

Ready for Review

- Alternators produce electricity by relative movement of a magnetic field through a winding of conductors, which induces an electrical potential or voltage within the conductors.
- In the alternator, the magnetic field is created by the rotor, which rotates within the stationary stator windings to generate electricity.
- The main parts of the charging system include the battery, the alternator, the voltage regulator (which may be integrated into the alternator, or controlled by the vehicle computer), a charge warning light, and wiring that completes the circuits.
- The battery stores an electrical charge in chemical form, acts as an electrical dampening device for variations in voltage or voltage spikes, and provides the electrical energy for cranking the engine.
- The alternator—which is connected to the engine and driven by a drive belt—converts some of the mechanical energy of the engine into electrical energy.
- In addition to providing the electrical current needed to run the vehicle's accessories, the alternator also charges the battery.
- The alternator converts mechanical energy into electrical energy by applying the principle of electromagnetic induction.
- Alternators work when a magnetic field in the alternator rotor crosses the stator, which is the conductor, a voltage is produced in the stator.
- The quantity of voltage potential induced by an AC generator depends on four factors:
  - Strength of the magnetic field. Increasing the strength of the magnetic field increases the voltage output.
  - Speed at which the magnet rotates.
  - Distance between the magnet and conductors.
  - Number of turns of wire on the stationary coil. In a typical automobile alternator, three, or possibly four, separate coils of wire, or phase windings, are common.
- The alternator's output is determined by the amount of current that flows through the rotor field coils.
- The greater the current flow through the alternator's rotor field coils, the stronger the magnetic field generated. The stronger the magnetic field, the stronger the alternator output.
- The voltage regulator monitors battery voltage and adjusts the current flow through the rotor in direct proportion to battery voltage.
- As the output voltage increases to the maximum regulated voltage, the voltage regulator reduces the current flow through the rotor, reducing alternator output.
- The voltage regulator in modern vehicles is a solid-state electronic device that controls rotor field current with a pulse-width–modulated signal, for smoother regulation and quicker control.
- The regulator's electronic circuit senses the battery voltage and switches the rotor circuit on and off rapidly in order to maintain a constant voltage output up to the alternator's maximum output current.
- More recent developments have seen the alternator regulator function built into the powertrain control module (PCM) of the vehicle.
- The PCM works in conjunction with the regulator to control the output of the alternator.
- Rectification is the process of converting the AC voltage that the alternator generates into the DC voltage.
- To change AC voltage to DC voltage, automotive alternators use a rectifier assembly consisting of diodes in a specific configuration.
- Diode allows current to flow in one direction but blocks the flow of current in the other direction.
- A three-phase "bridge" rectifier has a minimum of six diodes (three positive and three negative) to rectify the AC output of the stator windings to DC output from the rectifier.
- A diode bridge gets its name from two diodes in series bridged with a wire.
- As the rotor rotates, each time the magnetic field changes from north to south and south to north, the polarity of each phase winding reverses and, as a result, the current changes direction.
- No matter what direction the current is flowing in the stator windings, the diodes in the rectifier only allow current to flow into the rectifier and out of the rectifier in one direction (DC).
- The alternator consists of a stationary winding assembly called the stator; a rotating electromagnet called the rotor with a slip ring, a brush assembly, a rectifier assembly, two end frames, and a cooling fan; and a drive pulley.
- The rotor is an electromagnet that rotates freely in the alternator. It is driven by the engine's crankshaft with the aid of a drive belt and provides the magnetic field.
- The stator consists of a cylindrical, laminated iron core which carries the three- (or four-) phase windings in slots on the inside and is the conductor.
- The windings can be oriented in a wye or delta configuration.
- The relationship between the slip rings and brushes is what make an electrical connection to the rotating rotor assembly possible.

- The alternator slip rings are normally copper bands, molded onto an insulating material, then pressed onto the steel shaft of the rotor.
- The brushes are spring loaded to maintain contact with the slip rings, and they wear out over time.
- A pulley that is driven by a belt is mounted at the drive end of the alternator.
- Air-cooled alternators employ a powerful, centrifugal cooling fan, which is used to circulate air through the case and maintain safe operating temperatures.
- On some alternators, an over-running clutch is incorporated into the alternator drive pulley, which allows the alternator to freewheel when the engine suddenly decelerates, such as during gear changes and engine shutdown.
- With a direct drive pulley, when engine speed decreases rapidly, the alternator's inertia fights to maintain its rotational speed, often causing slippage, internal component stress, and noise.
- Over-running clutches are installed as a means of improving operating smoothness, decreasing belt slippage and noise, and increasing component lifespan.
- It is the accepted practice today is to replace the entire alternator with a new or rebuilt unit when an internal failure is discovered.
- On Hybrids. DC-DC converters are essentially transformers that convert high DC voltage from the hybrid battery pack to the 13.5 to 14 volts DC that is required to charge the lead-acid storage battery and power the vehicles electric accessories.
- In most hybrid vehicles, the DC-DC converter completely replaces the alternator as a means of supplying the DC voltage needed to power the vehicle's systems.
- An excessive voltage drop in the charging system output and ground circuit tends to cause one of two problems:
  - Battery is not allowed to fully charge because although the alternator is creating the specified voltage, the voltage drop is reducing the amount of voltage supplied to the battery.
  - Battery is fully charged, but the alternator is working at a higher voltage to accomplish this, potentially overheating the battery and causing premature failure.
- Voltage drop may occur anywhere in the output current circuit and ground circuit, but is especially common at the terminals and connectors due to the charging system's high current flowing through them.
- To measure for voltage drop, the DVOM is connected in parallel across the component, cable, or connection that is to be tested.
- It is most efficient to first measure the voltage drop on both the entire positive side and the entire negative side of the circuit.
- Narrow down the voltage drop test to individual components to identify precisely where excessive voltage drop is located (individual voltage drops in a simple circuit add up).
- The testing of an alternator output initially involves the testing of the system's regulated voltage using a voltmeter.
- Regulated voltage is the voltage at which the regulator is allowing the alternator to create only a small charge due to the battery being relatively charged, as evidenced by the greatly reduced current.
- Alternator output is measured with an output test using a DVOM with an amp clamp or an AVR (alternator voltage regulator) tester.
- The regulated voltage is the highest voltage the system achieves once the battery is relatively charged, as evidenced by the ammeter reading less than about 15 to 20 amps when the amps clamp is around the alternator output cable.
- Typical regulated voltage specifications could be about 13.2 to 15.5 volts on newer vehicles and 13.5 to 14.5 volts on older vehicles. Always check the specifications.
- To test for diode problems (called AC ripple), follow these steps:
  - Set your meter to measure AC volts.
  - Start the engine and increase its speed to 2000 rpm.
  - Connect your DVOM across the positive and negative battery terminals.
  - Turn on all of the vehicle's electrical devices to load the system, then record the AC voltage reading being picked up by the DVOM.
  - A reading of 400 mV (0.4 volts) AC indicates that the diodes are good.
  - An AC reading of 500 mV (0.5 volts) AC or greater indicates the rectifier diodes have failed, and the alternator needs to be replaced.
- It is critical that a thorough visual inspection of the drive belt, taking special care to look for frayed edges, missing chunks of belt, greasy residue on the belt, and/or hard rubber and "shiny" areas, which indicates that the rubber has become glazed.
- Proper belt tension is critical in order for the drive belt to properly engage the pulleys and transmit rotational torque.
- Belt tension is achieved either by the use of a tensioner, or, on modern vehicles, a "stretchy" belt is used.
- Stretchy belts have elastic properties that allow them to be self-tensioning, eliminating the need for an additional belt tensioner.
- There are several ways to measure belt tension. It can be measured with a belt tension gauge, using the manufacturer-supplied marks on the tension.
- Belt tension may be determined by using a beam-type torque wrench to measure the force required to deflect the spring-loaded belt tensioner.
- Belt tension can also be found by measuring the deflection of the belt between the two pulleys that are the farthest apart. The flex, or deflection, should be no more than 0.500" (12.70 mm) on a properly tensioned belt.
- Undercharging leads to compromised function of electrical systems and a less-than-fully-charged battery,

- a condition which leads to sulfation and, consequently, early battery death.
▸ Overcharging can shorten the lifespans of bulbs and other electrical devices, while at the same time overcharging the battery, which can increase gassing and loss of water from the electrolyte.
▸ Maintaining a proper charge is critical to long life and proper operation of the electrical system.

Key Terms

over running clutch drive Allows the alternator to freewheel when the engine suddenly decelerates, such as during gear changes and engine shutdown.

rectification The process of converting the AC voltage that the alternator naturally generates into the DC voltage that is required by the battery and nearly all of the automobile systems.

rotor Electromagnet that rotates freely in the alternator; responsible for creating the magnetic field necessary to generate electricity.

stator A cylindrical, laminated iron core which carries the three- (or four-) phase windings.

voltage regulator Component responsible for controlling the current supplied to the rotor. As the output voltage increases to the maximum regulated voltage, the voltage regulator reduces the current flow through the rotor, reducing alternator output.

Review Questions

1. Which of the following components converts some of the mechanical energy of the engine into electrical energy to supply electric current to all of the electrical components of the vehicle?
   a. Battery
   b. Voltage regulator
   c. Transformer
   d. Alternator

2. Before leaving the alternator, AC voltage is converted to DC voltage by the:
   a. rectifier assembly.
   b. rotor.
   c. stator.
   d. brush assembly.

3. The quantity of voltage potential induced by an AC generator depends on all of the following *except*:
   a. the strength of the magnetic field.
   b. the speed at which the magnet rotates.
   c. the number of turns of wire on the stationary coil.
   d. the number of conductors.

4. How is the output voltage controlled in an alternator?
   a. By varying the current to the stator
   b. By switching the rotor's field circuit on and off
   c. By varying the position of the brushes
   d. By using the alternator drive pulley

5. Due to the short length of the rotor, the cooling air needs to be spiraled in. To achieve this spiraling effect:
   a. a centrifugal cooling fan is used.
   b. cooling fins have small and large openings.
   c. the position of heat sinks in the rectifier assembly is manipulated.
   d. a drive pulley is used.

6. Over-running clutches serve all of the following purposes *except*:
   a. improving operating smoothness.
   b. decreasing belt slippage.
   c. increasing component lifespan.
   d. maintaining safe operating temperatures.

7. Which of the following makes it easier for the technician to locate the source of noise?
   a. Oscilloscope
   b. Scan tool
   c. Stethoscope
   d. DVOM

8. In hybrid vehicles, which of these provides the current required to run all of the vehicle's electrical accessories at idle stop?
   a. AC-DC converter
   b. Computer-controlled voltage regulator
   c. Rectifier assembly
   d. DC-DC converter

9. When a faulty internal voltage regulator is diagnosed, replace the:
   a. alternator as a unit.
   b. voltage regulator.
   c. rectifier assembly.
   d. stator and rotor.

10. Which of the following is the most likely cause of overcharging?
    a. A shorted cell on a battery
    b. Worn brushes in the alternator
    c. Open or shorted diodes
    d. A faulty voltage regulator

ASE Technician A/Technician B Style Questions

1. Tech A says that alternators and generators are the same device, and the terminology can be used interchangeably. Tech B says that generators natively produce DC voltage, and alternators natively produce three-phase AC voltage, which must be rectified to DC for use in automotive circuits. Who is correct?
   a. Tech A
   b. Tech B
   c. Both A and B
   d. Neither A nor B

2. Tech A says that the alternator's only job is to keep the battery charged. Tech B says that the alternator restores the

energy to the battery that is used while cranking the engine. Who is correct?
a. Tech A
b. Tech B
c. Both A and B
d. Neither A nor B

3. Tech A says that some alternators in modern vehicles use permanent magnets instead of a rotor coil to save rotational weight and increase fuel efficiency. Tech B says that the amount of current that an alternator delivers is partly controlled by modulating the strength of the magnetic field in the rotor. Who is correct?
a. Tech A
b. Tech B
c. Both A and B
d. Neither A nor B

4. Tech A says that the stator windings are phased 120 degrees apart. Tech B says that the phases are placed 120 degrees apart to maximize the efficiency of the alternator. Who is correct?
a. Tech A
b. Tech B
c. Both A and B
d. Neither A nor B

5. Tech A says that the voltage regulation circuit is responsible for controlling the strength of the magnetic field in the rotor. Tech B says that some alternators employ a variable speed drive clutch to control the speed of alternator rotation. Who is correct?
a. Tech A
b. Tech B
c. Both A and B
d. Neither A nor B

6. Tech A says that the charging system warning light is responsible for alerting the driver to an overcharge condition. Tech B says the charging system warning light alerts the driver that the alternator is not producing sufficient output. Who is correct?
a. Tech A
b. Tech B
c. Both A and B
d. Neither A nor B

7. Tech A says that in many modern vehicles, the PCM is responsible for modulating the output of the alternator, which allows the charging system output to be tailored to operating conditions, thereby maximizing efficiency. Tech B says that problems with the charging system can activate a trouble code in the PCM. Who is correct?
a. Tech A
b. Tech B
c. Both A and B
d. Neither A nor B

8. Tech A says that the rectifier bridge assembly uses six diodes to convert the three-phase AC voltage created in the alternator into the DC voltage used in automotive electrical systems. Tech B says that rectifier diode failure can allow AC voltage to seep into the electrical system, which can cause many strange issues within the vehicle's computer systems. Who is correct?
a. Tech A
b. Tech B
c. Both A and B
d. Neither A nor B

9. Tech A says that a worn alternator drive belt can cause the alternator output to diminish due to slippage when the alternator is called upon to increase its current output. Tech B says that voltage drops in the charging system can cause the system to malfunction. Who is correct?
a. Tech A
b. Tech B
c. Both A and B
d. Neither A nor B

10. Tech A says that the stator is the part of the alternator that is driven by the belt. Tech B says that the rotor is the part of the alternator that produces the rotating magnetic field required to generate electricity. Who is correct?
a. Tech A
b. Tech B
c. Both A and B
d. Neither A nor B

CHAPTER 16
Lighting System Fundamentals

NATEF Tasks

- **N16001** Identify system voltage and safety precautions associated with high-intensity discharge headlights.
- **N16002** Inspect interior and exterior lamps and sockets including headlights and auxiliary lights (fog lights/driving lights); replace as needed.
- **N16003** Aim headlights.
- **N16004** Diagnose (troubleshoot) the causes of brighter-than-normal, intermittent, dim, or no light operation; determine needed action.

Knowledge Objectives

After reading this chapter, you will be able to:

- **K16001** Describe different types of lighting found on a vehicle and the function of each type.
- **K16002** Explain the operation and benefits of various lighting systems.
- **K16003** Describe lighting systems procedures and peripheral systems.

Skills Objectives

After reading this chapter, you will be able to:

- **S16001** Check warning device operation.
- **S16002** Check and change a headlight bulb.
- **S16003** Aim headlights.

You Are the Technician

Modern lighting systems are more effective than ever before. New lighting technologies have made it possible for manufacturers to design brighter headlamps, and therefore, increase the safety of the vehicle. These systems are becoming increasingly complex. Special care must be taken when servicing these systems, as some of them contain high voltages that can cause serious injury.

1. True or False: LED lights are not used in modern headlight systems.
2. True or False: Some lighting systems produce voltages that are high enough to cause serious injury.

K16001 Describe different types of lighting found on a vehicle and the function of each type.

Introduction

Well-designed vehicle lighting systems enhance vehicle safety by increasing the driver's visibility while operating the vehicle and by clearly signaling the driver's intent to those around the vehicle. Lighting systems are also used inside the vehicle to indicate messages to the driver and provide convenience to occupants. Manufacturers continue to improve and develop their lighting systems as new technologies become available. Modern lighting system features include the increased use of light-emitting diode (LED) and xenon lighting, electronic body control units to manage lighting, and high-intensity discharge (HID) lamps, which are much brighter than the traditional sealed-beam or halogen units. Many vehicles also have systems that warn you if lamps are not working properly.

Types of Lamps

Modern vehicles use many different kinds and sizes of lamps, also known in some places as light bulbs or light globes. There are several lamp types available, including standard incandescent lamps, halogen lamps, vacuum tube fluorescent (VTF) lighting, HID xenon gas systems, LEDs, and more. Conventional incandescent lamps are being replaced in many applications by these more-efficient types of lights.

Incandescent Lighting

Incandescent lamps consist of one or more filaments that heat up to approximately 5000°F (2760°C) and glow white hot (**FIGURE 16-1**). The filament material does not burn because most of the oxygen in the bulb has been replaced by inert gases that stop combustion from occurring. The power in watts consumed is often marked on the lamp; wattage (work performed) is found by multiplying the voltage used by the lamp by the current flowing through it. The higher the wattage, the more light that can be created when compared to a similar type of bulb. Incandescent bulbs are inefficient, converting only about 10% of the electricity to visible light, while the remaining 90% of energy is wasted to heat.

Halogen Lighting

Halogen lamps are another type of incandescent lamp, but they are filled with a halogen gas such as bromine or iodine (**FIGURE 16-2**). These lamps have a much longer life and are generally brighter and produce more light per unit of power consumed. However, they become very hot during operation. They are manufactured from highly heat-resistant materials, and the bulbs must be handled carefully because they are sensitive and can be damaged even by finger oil residue left by fingerprints. Touching the glass portion of a halogen lamp with bare skin will cause the lamp to fail prematurely.

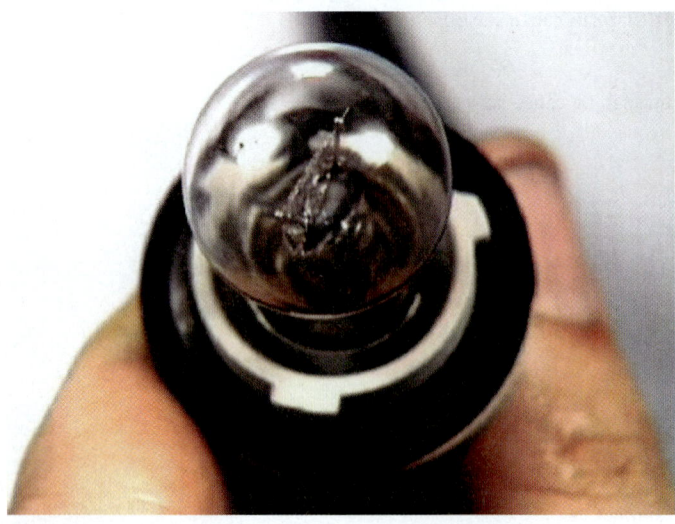

FIGURE 16-1 Incandescent bulb with single filament.

FIGURE 16-2 Halogen bulb.

Vacuum Tube Fluorescent

Vacuum tube fluorescent (VTF), also called vacuum fluorescent display (VFD), is used for instrumentation displays on vehicle instrument panel clusters. This type of lighting emits a very bright light with high contrast and can display in various colors. Usually VTF displays are of bar graphs, seven-segment numerals, multi-segment alphanumeric characters, or a dot-matrix pattern. VTF displays include different kinds of alphanumeric characters and symbols to alert drivers of various conditions.

HID Systems

High-intensity discharge (HID) headlamps produce light with an electric arc rather than a glowing filament (**FIGURE 16-3**). The high intensity of the arc comes from metallic salts that are vaporized within an arc chamber. HIDs produce more light for a given level of power consumption than ordinary tungsten or halogen bulbs. Automotive HID lamps are commonly called "xenon headlamps," though they are actually metal halide lamps that contain xenon gas. The light from HID headlamps exhibits a distinct bluish tint as compared with the yellow-white color of tungsten-filament headlamps. HID headlamp bulbs do not run on low-voltage direct current; they require a **ballast** with an internal or an external igniter that is either integrated into the bulb or included as a separate unit or part of the ballast. The ballast increases the voltage substantially and controls the current to the bulb.

HID headlamps produce between 2800 and 3500 lumens from between 35 and 38 watts, while halogen filament headlamp bulbs produce between 700 and 2100 lumens from between 40 and 72 watts at 12.8 volts. The advantage of using HID headlight systems is that they offer substantially greater luminance than halogen bulbs (about 3000 lumens versus 1400 lumens for comparable halogen bulbs) (**FIGURE 16-4**). If the higher-output HID light source is used in a well-engineered headlamp optic, the driver gets more usable light. Studies indicate that drivers react faster and more accurately to roadway obstacles when using good HID headlamps than when using halogen headlamps; therefore, good HID headlamps contribute to driving safety.

An argument against HID headlamps is that they can impact negatively the vision of oncoming traffic due to their high intensity and the "flashing" effect caused by the rapid transition between low and high illumination in the field of illumination. This potential distraction increases the risk of a head-on collision between a vehicle using HID headlamps and a blinded oncoming driver. Scientific studies of headlamp glare from HID systems has shown that for any given intensity level, the light from HID headlamps is 40% more glaring than the light from tungsten halogen headlamps.

Some countries mandate that HID headlamps may only be installed on vehicles (except motorcycles) with lens-cleaning systems (to reduce glare) and automatic self-leveling

N16001 Identify system voltage and safety precautions associated with high-intensity discharge headlights.

FIGURE 16-3 HID headlamp assembly.

FIGURE 16-4 Comparison between halogen and HID lighting.

systems (which prevent dazzling oncoming traffic). These systems are usually absent on vehicles not originally equipped with HID lamps, so if a halogen headlamp is retrofitted with an HID bulb, light distribution with illegal levels of glare will be produced. Another disadvantage of HID headlamps is that they are significantly more expensive to produce, install, purchase, and repair. However, some of this cost is offset by the longer lifespan of the HID burner relative to halogen bulbs.

HID System Safety Precautions

HID lamps produce a very bright white light. Manufacturers use various designs of HID lamps. Regardless of the design, they generally require a high-voltage spark of up to 25,000 volts to start and a high operating voltage (e.g., 40–85 volts AC) to maintain light. To generate the voltages required to operate, a transformer and electronic circuitry called a ballast are used to supply electrical components.

Safety precautions should be taken when working on HID systems. There is a risk of electrocution, burns, or shock from the high voltages generated by the HID system. If diagnosing the HID system, be very careful when working on the system when it is live. You should wear safety glasses, high-voltage safety gloves, and safety boots, and you should ensure that the vehicle, engine compartment, and ground under the vehicle are dry. Do not touch the ballast while it is operating; it will often be generating a lot of heat. Persons with active electronic implants, such as heart pacemakers, should not work on HID headlamps. If changing out the bulb, make sure the headlights are turned off. The manufacturer may specify that the battery be disconnected when replacing the bulb. If so, follow the manufacturer's instructions.

There is also a risk of injury caused by exposure to ultraviolet light produced by the HID lamp if the lamp is operated outside of its housing. Once ignited, the pressure inside an HID bulb can build up to a very high level (around 220 pounds per square inch, or 1517 kilopascals) due to the high operating temperature (about 1500°F, or 816°C). This pressure creates a potential explosion hazard, so do not attempt to power an HID bulb outside of the headlamp assembly to test it or operate it near flammable gases or liquids. Also, the bulb must be maintained in a horizontal position when it is on; otherwise it may overheat and fail. HID headlamps use various heavy metals in their construction; therefore, it is important to always dispose of the bulbs in an environmentally responsible way. Avoid breaking bulbs, as there is also a risk of poisoning caused by inhalation or skin contact of heavy metal vapors and toxic salts.

LED Lighting

Light-emitting diodes (LEDs) have been used for some time in various automotive applications, such as warning indicators and alphanumeric displays. More recent developments in LED technology have seen the production of a wider range of colors and LEDs that are brighter than previous types. It is now possible to get LEDs that emit bright red, green, blue, yellow, and clear or white light. This has made it possible to use LEDs for many new applications, such as more general lighting applications. For example, LEDs are now often used for brake lights, turn signals, and interior lighting on vehicles (**FIGURE 16-5**).

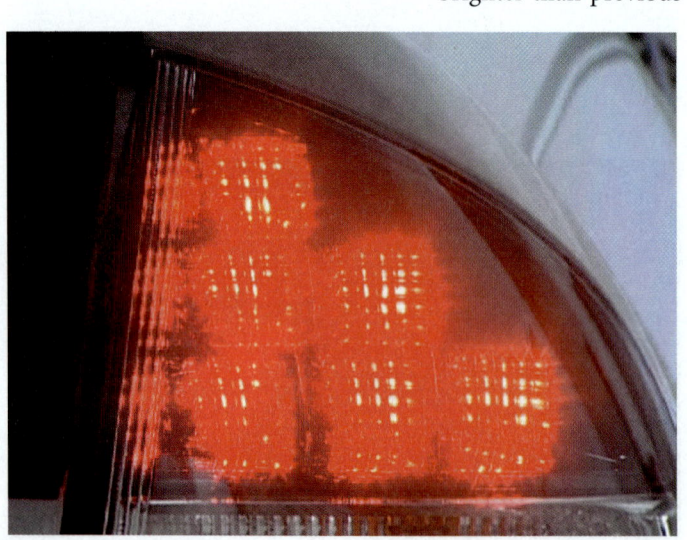

FIGURE 16-5 LEDs.

One of the advantages of LEDs is that they turn on instantly. This is particularly useful in brake lights, as they can reduce the braking light response time by two-tenths of a second. This translates to an extra 16' (4.9 m) of stopping distance for vehicles traveling at highway speeds. LEDs also have better visibility in inclement weather, operate at cooler temperatures, consume less energy, are much smaller, and can last up to 100 times longer, reducing the cost of servicing. LEDs can also be specifically designed as a direct-fit retrofit, which allows an LED replacement bulb to be installed in place of a traditional incandescent light bulb (**FIGURE 16-6**). It should be noted that since LEDs use significantly less power than their incandescent counterparts, installing LED replacement lamps in place of an incandescent light bulb may cause turn signals to blink rapidly, or cause vehicle information systems to report that a bulb is burned out.

For automotive applications, a number of LEDs are grouped together to provide the amount of light required for the application. Additionally, LED lenses are specifically designed with light-focusing prisms and lenses to focus the light generated by the LEDs. A typical LED has a voltage drop of 1.2 to 3.5 volts across it, depending upon the color, when it is forward biased and emitting light.

When used in automotive lighting, many LEDs are required to give off a specified amount of light. To do so, they are usually connected in groups called series strings. A number of series strings are then connected in parallel until enough LEDs are connected to give off the required amount of light. LEDs work best when the voltage to them and the current flow through them remains constant at a preset level. There are two main ways to achieve this; the first is via a resistor. The second and more preferred way is through the use of a voltage regulation circuit.

FIGURE 16-6 LED replacement bulb next to a traditional incandescent bulb.

Lamp/Light Bulb Information

All lamps or light bulbs have letters and numbers stamped on them that typically indicate their part number and often the operating voltage and power consumed. For instance, in a bulb marked "12V/21W," the filament will consume 21 watts of power when 12 volts is applied across the filament. While the wattage is not necessarily an indication of light output, it can be generally assumed that the higher the wattage, the greater the light output.

Lamps and light bulbs come in a variety of configurations to fit the various applications within a vehicle. One designation is how many filaments the bulb has (**FIGURE 16-7**). Single-filament bulbs are common for use as courtesy lights, dash lights, and warning lights. Dual-filament bulbs have two filaments of different wattage; one filament emits a small amount of light, and the second filament emits more light. These bulbs work well as a combination taillight and brake light. Headlights can also be dual-filament bulbs. In some cases, the low-beam filament has lower wattage than the high-beam filament, but not always. In a headlight, the filaments are positioned to give a different profile of light. Low beams emit light closer to the vehicle and angled slightly toward the side of the road, while high beams tend to focus farther down the road and straight ahead.

Another feature that differs among lights is the type of base on the lamp—in other words, what type of socket it is retained in. Bayonet-style bulbs have been around for a long time. They get their name from the two retaining pins on the side of the base (**FIGURE 16-8**).

FIGURE 16-7 Dual-filament bulb.

FIGURE 16-8 Bayonet-style bulbs.

K16002 Explain the operation and benefits of various lighting systems.

N16002 Inspect interior and exterior lamps and sockets including headlights and auxiliary lights (fog lights/driving lights); replace as needed.

The pins follow slots in the side socket and at the bottom, and the slots turn sideways into a small pocket. The pins are retained in the pocket by the spring-loaded base in the bottom of the socket, which pushes the bulb upward. This design resists vibration very well. Removal of the bulb requires carefully pushing in on the bulb and rotating the bulb slightly counter-clockwise and pulling it out. One or two electrical contacts are built into the bottom of the bulb's base. If the bulb is a dual-filament bulb with two contacts on the base, then the pins will be of unequal height so that the contacts will register properly in the socket.

Many newer bulbs use a wedge base either made from the glass bulb itself or with a built-in plastic base (**FIGURE 16-9**). The bulbs are pushed straight into the socket, and tension from the socket retains the bulb. The electrical contact on the glass-based bulbs is made by wires extending from the base of the bulb and bent over opposite sides of the wedge. Dome lights use festoon light bulbs, which have a base on each end of a cylindrical light bulb (**FIGURE 16-10**). Each end of the filament is connected to one of the bases. Generally, the bases fit in spring steel contacts, which hold the light bulb securely in place.

▶ Types and Styles of Lighting Systems

There are many different styles and types of lights. Each style and type is designed to perform specific roles. For example, warning lamps, turn indicators, stop lights, taillights, courtesy lamps, and headlamps all perform different roles. Lamp locations, color, and brightness are governed by regulations to ensure consistency and safety in the application of lighting on vehicles. Lighting regulations should always be consulted before modifying or adding to any of the vehicle lighting systems. Well-maintained lighting systems improve road safety for all drivers.

Park/Tail/Marker/License Lights

Park, tail, and marker lights are all low-intensity or low-wattage bulbs used to mark the outline or width of the vehicle. Park and tail lamps tend to be installed close to the corners of the vehicle. Park lamps are placed to the front of the vehicle, or in some cases they are incorporated in the headlight assembly and are white or yellow in color. Tail lamps are red and usually installed in a cluster assembly with the stops lamps at the rear of the vehicle.

License plate lamps produce a white light and are designed to illuminate the lettering on the license plate at night without the white light itself being seen from the rear (**FIGURE 16-11**). The bulbs are connected in parallel to each other so that the failure of one filament will not cause a total circuit failure. License plate illumination lamps are usually connected in parallel to the taillights and operate whenever the taillights are on.

FIGURE 16-9 Wedge-style bulbs.

FIGURE 16-10 Festoon-style bulbs.

FIGURE 16-11 License plate lights.

FIGURE 16-12 Park lights.

Park lights are located at the front of the vehicle and are used at night when the vehicle is parked on the side of the road and are also on anytime the headlights are on (**FIGURE 16-12**). They use low-wattage bulbs and may have a lens or diffuser that makes the emitted light widespread. In some cases, park lights are incorporated in the headlight assembly. Park lights operate when the light switch is moved to the park light position. For safety reasons, park lights and taillights continue to operate when the light switch is moved to the headlight position.

On computer-controlled lighting systems, the park lights and taillights are controlled by a body control module (BCM). The lights are supplied with power and ground through a networked light controller. The controller is connected to a network/bus system by a twisted pair of communication wires. The park light switch is hardwired to a module. When the park light switch is activated, the module sends a park light request out on the network where the appropriate controllers pick up the message and supply power or ground to the appropriate light bulb filaments to illuminate the park lights.

Marker lights are used to mark the sides of some vehicles (**FIGURE 16-13**). They can be located on the front and rear fenders. On newer vehicles, they are sometimes placed on side-view mirrors, or between the front and the rear doors on large SUVs or pick-ups (**FIGURE 16-14**). Red marker lamps face toward the rear, and yellow lamps face toward the front of the vehicle. These lights are designed to work when the park lights or headlights are selected and sometimes operate as turn signals so a driver can warn others of his or her intent to switch lanes or turn a corner.

Government regulations control the positioning of lights, including the height of the lamps' beam and their brightness. The park, tail, marker, and license plate lamps operate when the headlight switch is in both the park and the headlight-on positions. The bulbs are

FIGURE 16-13 Marker lights.

FIGURE 16-14 Marker lights on an SUV.

connected in parallel to each other so that the failure of one filament will not cause a total circuit failure. Tail and park lamps may use separate fuses, so if one circuit fails the other will continue to operate.

Brakes Lights and CHMSL

Brake lights, which may also be called stop lights, are red lights mounted to the rear of the vehicle (**FIGURE 16-15**). They are usually incorporated in the taillight cluster. As required by law, a higher additional third brake light is mounted on top of the trunk lid or on the rear window on all recent vehicles. This light is called a center high mount stop light (CHMSL), or "chimsul" (**FIGURE 16-16**). The brake lights are activated whenever the driver operates the foot brake to slow or to stop the vehicle or when a control module automatically applies the brakes. The lighting circuit consists of the battery, fusible links and fuses, a brake light switch, brake light bulbs, wiring to connect the components, and the ground circuit to return current from the light to the battery (**FIGURE 16-17**). It may also include a BCM to command the lights on when the proper inputs are present.

On older vehicle models, when the operator of the vehicle depresses the brake pedal, a switch mounted on the pedal support closes. This allows the electrical current to flow from the battery through the fuse, through the switch, to the brake lamp, and return to the battery by the ground circuit. When the driver releases the pedal, it returns to the rest position and opens the brake switch. The flow of electrical current stops and the brake lamps are extinguished. Today, computer-controlled brake lights are activated by the BCM when the computer sees an input from the brake pedal switch.

Back-Up Lights

The back-up lights, also called reverse lights, are white lights mounted at the rear of a vehicle (**FIGURE 16-18**). They provide the driver with vision behind the vehicle at night and also alert other drivers to the fact that the vehicle is in reverse. When the ignition is on, and the vehicle is placed in reverse gear, the current flows from the battery, through the ignition switch, and through the closed reverse lamp or transmission position switch on the transmission. Electrical current then flows out of the closed switch to the back-up lamps and returns to the battery by the vehicle chassis ground circuit. Modern vehicles use network/bus systems and the BCM to command the back-up lights to come on. Back-up lights are the only white lights on the rear of the vehicle. Since they operate only in reverse, other drivers can tell that the driver is backing up. This is why it is illegal to have broken tail/brake/turn

FIGURE 16-15 Brake lights.

FIGURE 16-16 Center high mount stop light.

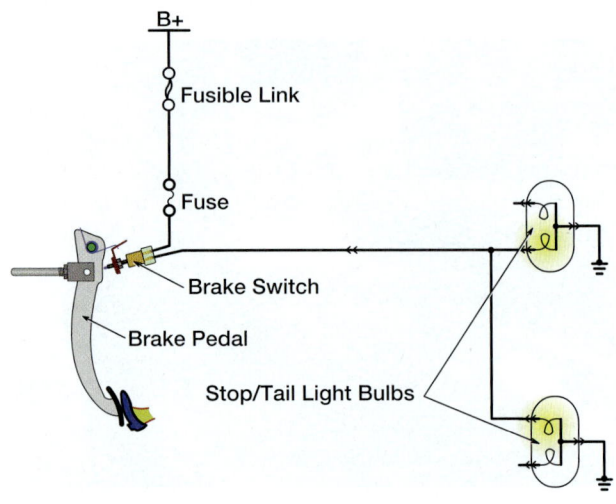

FIGURE 16-17 Wiring diagram of brake lamp circuit.

FIGURE 16-18 Back-up lights.

FIGURE 16-19 Turn signal indicators.

signal lenses in the rear of the vehicle: white light would be visible, potentially confusing other drivers.

Turn Signal and Cornering Lights

Turn signal indicators are located on the extreme corners of the vehicle. They are usually amber in the front and can be either red or amber in the rear (**FIGURE 16-19**). A column-mounted switch, operated by the driver, commands a pulsing current to the indicator lights on one side of the vehicle or the other. These pulsing lights warn other road users of the driver's intended change of direction.

Once activated, turn signal indicators continue until the switch is canceled either by the operator or by a canceling mechanism in the switch. The canceling mechanism operates to return the switch to its central or "off" position after a turn has been completed and the steering wheel is returned to the straight-ahead position. If the indicator switch is turned to indicate a right turn, current from the battery typically flows through the fusible link to the ignition switch, where it is directed through a fuse to the flasher unit. The flasher unit uses a timing circuit to pulse the current flowing out of the flasher unit 60 to 120 times per minute. This pulsing current is directed through the indicator switch to the right indicator lights at the front and rear of the vehicle, causing the lamps to flash on and off. An indicator light on the instrument cluster also blinks in sync with the turn signals. The operation of the flasher unit also produces a clicking sound to audibly inform the driver that the indicators are in operation.

When the turn signal switch is returned to the off position, no current flows through the flasher unit, so the timer circuit is switched off. When the turn signal switch is turned in the opposite direction, it directs the pulsing current to the left lights at the front and rear of the vehicle as well as the left indicator light on the instrument cluster. Older vehicles used a thermo-mechanical flasher unit that relied on heat from the current flow to cause the flasher unit to work. It is very important to use bulbs of the proper wattage on all types of flasher units, as the speed of the flash may be incorrect if incorrect bulbs are used.

Turn indicators can also be controlled by computer. The BCM commands the appropriate turn indicators to come on as it sees an input from the turn signal switch, and flashes it at the proper rate. The computer turns off the turn signals when a steering angle sensor signals the steering wheel is being centered. It can also cancel the turn indicators if the vehicle is driven for a programmed amount of time or distance without the steering wheel being turned. Often a chime will alert the driver if the turn indicator has been left on for too long. On many computer-controlled turn signals, the turn signals will not work at all when the computer senses the wrong amperage flow.

Hazard Warning Lights

All modern vehicles are equipped with hazard warning lights. This circuit is similar to the turn indicator lights except that it simultaneously causes a pulsing in all exterior indicator lights and both indicator lights on the instrument panel. Hazard lights can

FIGURE 16-20 Warning lights.

S16001 Check warning device operation.

warn other road users that a hazardous condition exists or that the vehicle is standing or parked in a dangerous position. The hazards also use a flasher unit that can be a separate unit or the same as that used for the turn signals. The BCM may control the hazard lights when it sees an input from the hazard switch or an input from the restraint's control module indicating a vehicle crash.

Warning Lights

Warning devices are installed on the vehicle to provide the driver with information about critical vehicle operating and safety systems, including the battery, supplemental restraint system (SRS), anti-lock brake system (ABS), parking brake, oil pressure, and engine temperature (**FIGURE 16-20**). In some cases, the systems use sensors to relay information to the warning device, such as low oil pressure. In this case, the oil pressure may need to be measured with an external oil pressure gauge. If the oil pressure is operational, then the oil pressure switch may be faulty and will need to be replaced. In other circuits, such as the ABS system, the PCM may send a signal through each wheel speed sensor to measure its continuity. In this case, you will need to measure the resistance of the sensor or the continuity of the wiring harness.

The warning devices may be controlled by an electronic control unit (ECU). SRS systems are very dangerous; for example, they control the deployment of airbags by using pyrotechnical charges. Check the manufacturer's procedures and properly disable the SRS system before working on it. Check vehicle faults codes and PIDs for possible faults. A multimeter or oscilloscope may be used to check signal voltages from sensors used on the vehicle warning devices.

To check warning device operation, follow the steps in **SKILL DRILL 16-1**.

Daytime Running Lights

Daytime running lights (DRLs) are an additional safety feature designed to improve a vehicle's visibility to other drivers in all weather conditions. They are existing lights that turn on when the vehicle is running and turn off when the engine stops. The lamps on the front are the headlights, which are typically operated at about a 60% power level, providing light without excessively decreasing bulb life or using full electrical power. The lamps used on the rear in DRL systems are the taillights. DRLs are mandatory on modern vehicles licensed in Canada and some other countries. In the United States, DRLs are permitted but not required. Their use is somewhat controversial. First, if they are too bright, they can cause daytime glare. Second, they tend to mask the visibility of turn signals, making it harder for other drivers to determine a vehicle's intent. Overall, they have not been proved to reduce accidents or increase safety. They also require energy to operate, which reduces fuel economy and increases carbon dioxide emissions.

Headlights, Driving Lights, and Fog Lamps

Headlights are built into the front of a vehicle to illuminate the road ahead of the vehicle when driving at night or in other conditions of reduced visibility. In headlights, most

SKILL DRILL 16-1 Checking Warning Device Operation

1. Identify the circuit to be tested. Research the correct test procedure from the manufacturer's information.
2. Check fault codes and circuit PIDs using a scan tool. Inspect the sensors, wires, and connectors.
3. Test the affected circuit and any sensors with a multimeter or oscilloscope to determine the correct output.
4. If sensors or devices test OK, check the individual circuits for continuity.
5. Identify the circuit to be tested. Research the correct test procedure from the manufacturer's information.

vehicles require two beams to provide for high beam and low beam operation. The beams are created by separate filaments; these may be in the same bulb or in separate bulbs. These filaments must be positioned correctly in relation to the highly polished reflector. This is called focusing and is carried out when designing the light assembly and lens for the vehicle. The high-beam filament is positioned at the focal point of the reflector to project the maximum amount of light forward and parallel to the reflector axis (**FIGURE 16-21**). This light is then shaped by the lens, which is made up of many small glass or plastic prisms fused together. These prisms bend the light horizontally and vertically to achieve the desired light pattern for road illumination.

The low-beam filament is often placed above and slightly to one side of the high-beam filament. Mounting the low-beam filament in this position produces a beam of light that is projected downward and toward the curb side (**FIGURE 16-22**). With this arrangement, the high-beam filament produces the most concentrated light output, while the low-beam filament gives a downward and dispersed beam that is less likely to blind oncoming drivers.

Timer and Delay Circuits

Some vehicles have a timer circuit that is integrated into the headlamp circuit. The purpose of the timer circuit is to allow a vehicle's headlamps to remain on for a period of time after the vehicle has been turned off. This period of time is usually adjustable by the use of a dial or from within a setting menu of the vehicle's driver information center. The timer function in these systems can either be carried out by a stand-alone headlamp timer or integrated into the BCM.

Automatic Lighting Systems

Many luxury vehicles have automatic lighting systems. These systems can turn the headlamps on and off without input from the driver. Some headlamp circuits may also have an "auto dimming" function, which switches from high beam operation to low beam operation when an oncoming vehicle is detected. A commonality between all automatic lighting systems is that they use a light sensor to determine whether the headlamps are needed, and/or to detect the headlamps of an oncoming vehicle. The light sensor feeds into the BCM, or lighting control module, which uses this information to switch the lights on or off via a relay. In the event that the light sensor would fail, these systems default to turning the headlamp on, and will set a trouble code in the PCM.

Driving Lights

Driving lights are used to supplement vehicle headlight systems. The driving lights are installed on the front of the vehicle and provide higher intensity illumination over longer

FIGURE 16-21 High- and low-beam filament position in relation to the reflector.

FIGURE 16-22 Low-beam filament position in relation to the reflector.

FIGURE 16-23 Vehicle with factory driving lights.

distances than standard headlight systems (**FIGURE 16-23**). Vehicle design rules and regulations specify the limitations in relation to the positioning and lens configuration of driving lights. It is essential that the local regulations are adhered to when mounting or adjusting driving lights.

There are many types of driving lights available. They come in different sizes, shapes, lens patterns, and bulb wattages. In some instances, a single driving light can be installed to suit particular applications, but lights are normally installed in pairs.

Most driving lights use quartz halogen bulbs in the 55- to 120-watt range. The quality of the reflector is extremely important in driving lights to get optimum performance. Driving lights are wired so that they operate only when the high beam is operating. This safety feature ensures that driving lights turn off when the headlights are switched from high to low beam, thus ensuring that oncoming traffic is not blinded by excessive light. While many performance vehicles come equipped with driving lights, they can be added to almost any vehicle. If they are, a relay and circuit breaker should always be used for circuit protection reasons. This arrangement should also be used whenever additional bulbs or electrical loads are installed. The purpose of installing a relay to control the light bulbs is to prevent too much current from passing through the light switch. In the setup, the switch simply turns on the relay, which handles all of the current flow to light bulbs (**FIGURE 16-24**).

Fog Lights

Fog lights are used with other vehicle lighting in poor weather such as thick fog, driving rain, or blowing snow. Because fog is made up of water droplets suspended in the air, it

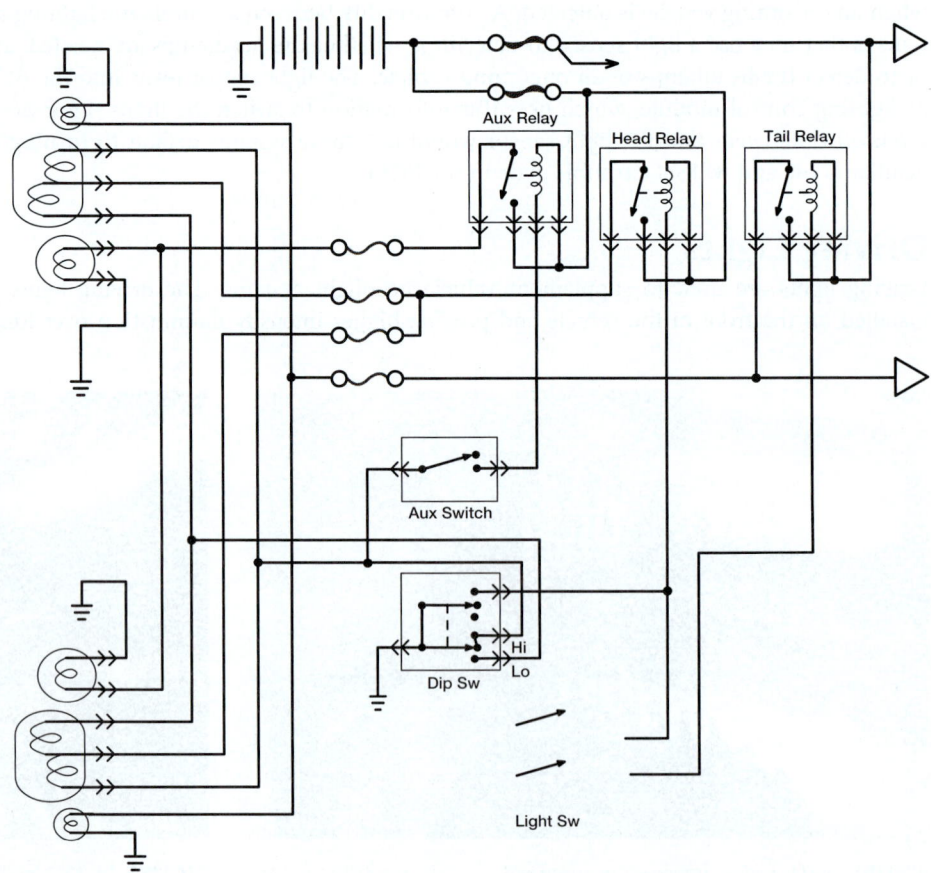

FIGURE 16-24 Auxiliary relay circuit used to control high wattage lighting systems.

can reflect headlights back into the driver's eyes at night. In such conditions, fog lights can help drivers see farther ahead and illuminate the road's edges at reasonable speeds. They are used with park lights and low beam headlights but not with high beams.

Most older fog lights have yellow-colored reflectors, although more recently, white fog lights have become more widely used because yellow lenses reduce fog light brilliance by about 30%. Fog lights typically use quartz halogen bulbs and are available in different shapes and sizes. Fog lights are usually mounted lower than headlights and tend to be aimed straight forward and low (**FIGURE 16-25**). Fog light lenses have a sharp cutoff pattern so that most of the light projected remains below the driver's eye level.

Fog lights are typically wired with a relay and circuit breaker. The method of connection of fog lights will depend on local regulations. They may be wired to work only with park lights and to turn off when headlights are used or to work when high beams are used. In modern vehicles, the BCM usually controls the function of the fog lights via a relay if they are installed as original equipment.

FIGURE 16-25 Vehicle with factory fog lights.

Types of Headlights

Headlight systems have traditionally used reflector-type lighting systems. There are several types of headlights: reflector-type lamps, which can be either sealed-beam or semi-sealed beam, and projector-type lamps. In a reflector headlight, the light from the bulb is reflected forward by a specially shaped reflector. An alternative is a projector-type headlight system. This type of headlight often has a smaller front lens; however, it produces a high-intensity forward beam. It uses a lens system to project the light forward, rather than the traditional reflector system (**FIGURE 16-26**). A projector-style light can use a standard incandescent bulb, or more commonly, an HID light.

A sealed-beam headlight has a highly polished aluminized glass reflector that is fused to the optically designed lens. It is a completely sealed unit that has the filaments accurately positioned in relation to the reflector. Most older vehicles used two dual-filament 7" (178-mm) round lamps; for a time, four single-filament 5.250" (133-mm) round lamps were used. Other older vehicles used two large dual-filament 7.875" (200-mm) rectangular lamps, or four smaller rectangular single-filament lamps. Regardless of size, when a filament fails in a sealed-beam light, the whole sealed unit must be replaced.

A semi-sealed beam headlight uses a replaceable bulb with a pre-focus collar. The collar locates the bulb in the headlight and also controls the correct positioning of the filaments to the reflector and lens (**FIGURE 16-27**). Some replaceable headlight bulbs have a partial shield below the low-beam filament. This shield prevents light from the filament from striking the lower part of the reflector, which would be reflected higher than the midpoint of the lamp. The shield provides the primary shape of the low beam. The final shaping of the beam is carried out by small cylindrical prisms in the headlight lens. This provides a low beam that is asymmetrical. The asymmetrical lens pattern causes light to be thrown upward at a 15 degree angle on the curb side to better illuminate objects, persons, or animals close to the road.

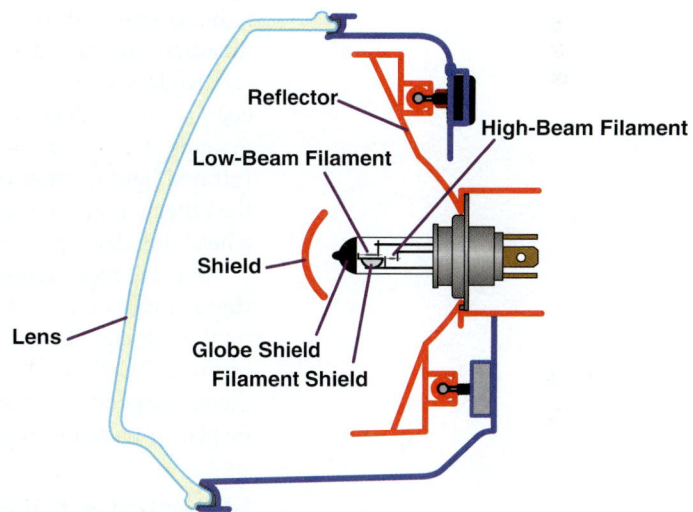

FIGURE 16-26 Projector bulb assembly.

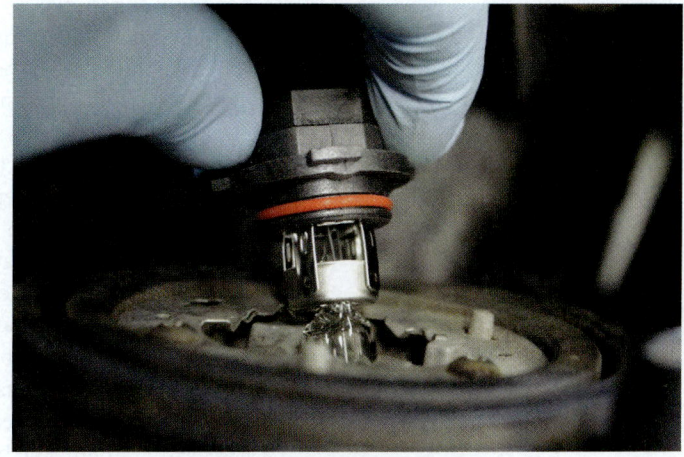

FIGURE 16-27 Replaceable halogen bulb.

FIGURE 16-28 HID headlight assembly.

HID lights use light from an electric arc rather than heating up a filament until it glows (**FIGURE 16-28**). High voltage is applied to tungsten electrodes. Xenon gas inside the bulb is then ionized and creates an electrical path between the electrodes, which lowers the resistance of the gap. As the temperature rises, metallic salts are vaporized and provide a stable arc, emitting much light. To initially jump the gap, a ballast and igniter raise the vehicle's low-voltage direct current to as much as 25,000 volts alternating current. Once the light reaches full operation, voltage is maintained between approximately 40 and 85 volts AC, depending on the system. HID lights give off a brighter and bluer light than halogen bulbs while using less electrical energy, so they help fuel efficiency slightly. They last two to four times as long as halogen bulbs but are quite a bit more expensive. Also, they take up more room in the engine compartment.

Some vehicles use LED lights as headlights. While LEDs are not currently as bright as HID lights, they are at least as bright as halogen lights. LED lights are known for low power consumption, but for LEDs to create enough light to function as headlights, they must consume almost as much energy as HID lights. Still, LED technology continues to improve. One drawback of current LED headlights is that high temperatures degrade or damage them. Thus, heat sinks and cooling measures are needed, which add complexity, cost, and space.

Night vision is another relatively new technology that enhances a driver's visual perception in the dark or in poor weather conditions. There are two types of night vision systems: active and passive. Active systems use an infrared light generator that projects infrared light in front of and to the side of the roadway ahead. A special camera picks up the reflected infrared radiation. The image is then displayed either on the windshield using a heads-up display or on an LCD screen on the dash or navigation system.

Passive night vision systems use a heat-sensing camera (thermal imaging) to pick up thermal radiation emitted by objects. This system does not have an infrared light source on the vehicle. The captured thermal image is then displayed either on a heads-up display or on an LCD screen. One benefit of the thermal system is that it can be programmed to recognize pedestrians and animals and can then either place an outline around them on the display or flash a warning symbol.

Headlight Circuit Construction

Generally speaking, the headlight circuit consists of the battery, the fusible link or maxi-fuse, headlight fuses, the headlight switch, the headlight relay(s), the beam selector switch, headlights, the high beam indicator light, wiring of a suitable size to carry the electrical current through the circuit, and the ground circuit. Modern vehicles use the BCM to manage headlight functions.

On older, non-computer-controlled systems, when the headlights are switched on, current is supplied from the battery and proceeds through the fusible link or maxi-fuse, headlight fuse(s), headlight switch, and dimmer switch, reaching either the low or the high beams. Some vehicles use relays to control the current through the low- or high-beam circuits. In this configuration, the beam selector switch either powers or grounds the relay windings in each of the relays, depending on the position of the beam selector switch. Activating the beam selector switch in the low-beam position creates a magnetic field inside the low-beam relay that closes the relay contacts, which allows electrical current to flow to the low beams. Activating the beam selector switch in the high-beam position creates a magnetic field inside the high-beam relay that closes the relay contacts, turning on the high beams.

The beam selector switch is a single-pole, double-throw switch, meaning it has one movable pole but makes contact in two positions. In the high-beam position, the current is switched to the high-beam circuit. In the low-beam position, the current is switched to the low-beam circuit. The beam selector switch can be set to switch a common power input to two outputs, or it can switch a common ground input to two outputs.

As is common with high-technology on-board systems, headlight and dimmer functions are managed with a network/bus system in modern vehicles. On many vehicles, the headlight switch and dimmer switch are just inputs to a control module that sends a headlight "on" or "off" request over the network. The appropriate light controller processes the request and actually commands the lights on or off, sometimes with the help of relays—especially for high beams.

▶ Lighting Systems Procedures and Peripheral Systems

The lighting systems on modern vehicles contain many types of lights and circuits. It is important to understand how these systems operate in order to be able to effectively diagnose them. As a general rule of thumb (which is applicable to all electrical circuits), it is important to study the wiring diagram of the lighting circuit in question before attempting to dig in and make a quick repair. This will allow the technician to have a thorough understanding of the lighting circuit and its operation. As always, follow the safety precautions and directions provided in the vehicle service information when attempting to service any system on the vehicle.

K16003 Describe lighting systems procedures and peripheral systems.

Checking and Changing a Headlight Bulb

There are many types of headlight bulbs available. Always make sure you replace a bulb with one of exactly the same type. Sealed-beam units require that the whole light be replaced when one filament has failed. If the reflector in the sealed-beam light shows signs of degradation, it also indicates that you must change the unit. If both lights operate but are not bright when switched on, start the engine to see if this solves the problem. If the lights operate at normal brightness with the engine running, the battery may be in a poor state of charge. When replacing a halogen bulb, avoid touching the glass portion of the bulb with your fingers, which can leave a residue from your fingers on the outer surface. This residue can cause the bulb to crack or shatter and burn out after a short time of operation. If you inadvertently touch the bulb, clean it with alcohol and a lint-free cloth. Do not use gasoline or paraffin to clean the bulb.

S16002 Check and change a headlight bulb.

To check and change a headlight bulb, follow the steps in **SKILL DRILL 16-2**.

Aiming Headlights

Although the principle of aiming headlights is the same in the majority of cases, the legal rules can differ from region to region. Be sure to check the requirements for your location. If you are unsure of what these are, ask your supervisor. Some manufacturers may suggest that the headlights be aimed one way for high beams and another way for low beams. In

S16003, N16003 Aim headlights.

SKILL DRILL 16-2 Checking and Changing a Headlight Bulb

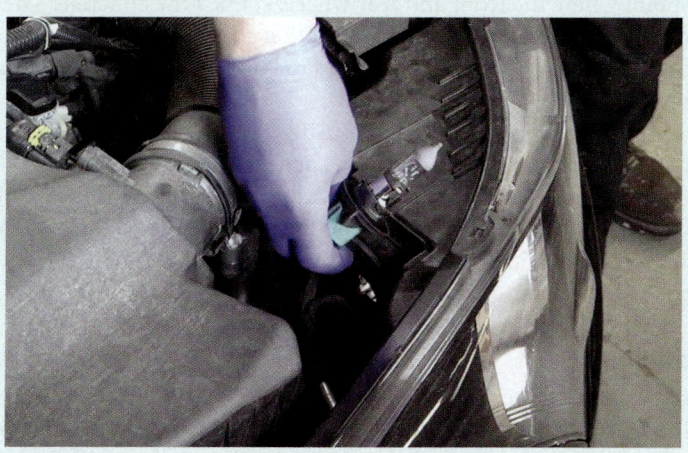

1. Test the vehicle's headlights. Obtain the replacement lamp for the vehicle. Unplug the electrical connector at the back of the lamp unit.

Continued

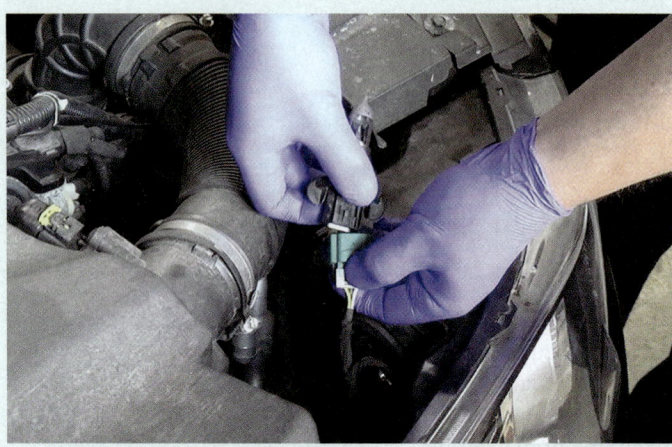

2. Remove the old bulb and replace it with the new one. Handle the new bulb only by its base or, if supplied, by the card cover.

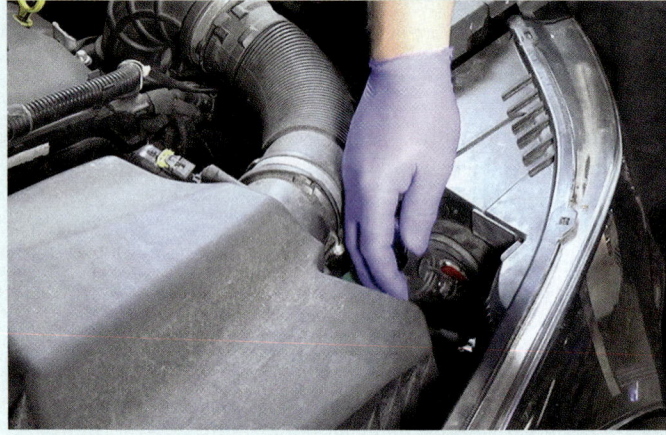

3. Replace the unit and the retaining ring or bulb assembly, then plug in the connector. Switch on the lights again to confirm that they are both operating correctly.

cases of separate high and low beams, both beams may need to be aimed independently. The manufacturer may also suggest that a load be placed in the vehicle, to simulate the ride height of the vehicle when it is traveling down the road. Headlights are typically aligned both vertically and horizontally so that as much of the road is illuminated as possible, without blinding oncoming traffic. Some vehicles have headlights that adjust automatically; do not try to adjust those unless you are diagnosing a fault in the system.

When aiming or aligning headlights, there are a couple of possible methods to utilize. The first method, one that is no longer in common use, involves an on-vehicle headlight aligner. This method was used extensively on sealed-beam headlights. The aligners are calibrated to the floor at the point where the front and rear wheels contact the concrete. Then the aligners are installed on the headlights with a built-in suction cup, which holds them in place. The aligners use a set of mirrors that can be calibrated to local regulations so that the headlights can be aligned properly.

With the introduction of aerodynamic headlights using removable halogen bulbs, off-vehicle headlight aligners became more popular. These aligners sit in front of the vehicle after being calibrated for the floor slope and orientation with the vehicle. With the headlights on, adjustments can be made by following the instructions on the aligner. Refer to the manufacturer's information for specific information regarding headlight aiming.

To aim headlights, follow the steps in **SKILL DRILL 16-3**.

Headlight Brightness

N16004 Diagnose (troubleshoot) the causes of brighter-than-normal, intermittent, dim, or no light operation; determine needed action.

Headlight brightness is critical to safe driving. Too dim, and the road ahead will not be adequately illuminated. Too bright, and the life of the headlight will be shortened, making it prone to early burnout. In most cases, the headlights are more likely to be too dim than too bright. Bulbs that are too bright indicate that the charging system voltage is too high

SKILL DRILL 16-3 Aiming Headlights

1. Make sure the tires are inflated properly, the wheels point straight ahead, and there is no extra weight in the vehicle. Position the vehicle correctly in relation to the headlamp aligner unit following the equipment manufacturer's instructions. Calibrate the aligner to any floor slope and for the vehicle being tested.

2. On the types of aligners that require the headlights to be on during alignment, turn the headlights on to low-beam setting. The center of the illuminating beams should be in the lower right quadrants of the chart or wall markings or as specified by the manufacturer.

3. The high beam should be centered, falling on the intersections of the horizontal and vertical marks or as specified by the manufacturer. If necessary, locate the adjustment screws on the headlight and turn them so the lights point to the correct places or bubbles on the levels are centered, depending on the type of aligner equipment you are using.

and needs to be measured with a multimeter. Alternatively, someone may have replaced the headlights with an aftermarket set, which may be illegal. Checking bulb numbers will verify this condition.

If the headlights are too dim, there can be several causes. The most common one is high resistance in the light circuit. This can be checked by measuring the voltage drop on both the power side of the bulb and the ground side. The voltage drop should be less than 0.5 volts on each side. If an excessive voltage drop is indicated, use a wiring diagram to research the circuit. Once you know how it is wired, use the multimeter to isolate the voltage drop by measuring each segment of the circuit. When the high resistance is located, perform the needed repair, whether that is a wire repair or a connector, switch, or relay replacement. If the voltage drop is less than 0.5 volts on each side, suspect the light bulb is wearing out. One way to verify this is with a light intensity meter. This meter can measure the amount of light energy the lamp is producing, so you can compare that to specifications. If the light is being supplied with the specified voltage, but it is not creating enough light, the bulb will need to be replaced.

▶ Wrap-Up

Ready for Review

- ▶ Modern vehicles use many different kinds and sizes of lamps, also known in some places as light bulbs or light globes.
- ▶ There are several lamp types available, including standard incandescent lamps, halogen lamps, vacuum tube fluorescent (VTF) lighting, HID xenon gas systems, and LEDs.
- ▶ Incandescent lamps consist of one or more filaments that heat up to approximately 5000°F (2760°C) and glow white hot.
- ▶ Halogen lamps are another type of incandescent lamp, filled with a halogen gas such as bromine or iodine.
- ▶ Halogen lamps have a much longer life and are brighter and produce more light per unit of power consumed. However, they become very hot during operation.
- ▶ Vacuum tube fluorescent (VTF) is used for instrumentation displays on vehicle instrument panel clusters. This type of lighting emits a very bright light with high contrast and can display in various colors.
- ▶ High-intensity discharge (HID) headlamps produce light with an electric arc rather than a glowing filament.
- ▶ The high intensity of the arc comes from metallic salts that are vaporized within an arc chamber.
- ▶ HIDs produce more light for a given level of power consumption than ordinary tungsten or halogen bulbs.
- ▶ Automotive HID lamps are commonly called "xenon headlamps," though they are actually metal halide lamps that contain xenon gas.
- ▶ The light from HID headlamps exhibits a distinct bluish tint as compared with the yellow-white color of tungsten-filament headlamps.
- ▶ HID headlamp bulbs do not run on low-voltage direct current; they require a ballast with an internal or an external igniter that is either integrated into the bulb or included as a separate unit or part of the ballast. The ballast increases the voltage substantially and controls the current to the bulb.
- ▶ HID headlamps produce between 2800 and 3500 lumens from between 35 and 38 watts, while halogen filament headlamp bulbs produce between 700 and 2100 lumens from between 40 and 72 watts at 12.8 volts.
- ▶ The advantage of using HID headlight systems is that they offer substantially greater luminance than halogen bulbs.
- ▶ Safety precautions should be taken when working on HID systems. There is a risk of electrocution, burns, or shock from the high voltages generated by the HID system.
- ▶ When diagnosing the HID system, be very careful when working on the system when it is live. You should wear safety glasses, high-voltage safety gloves, and safety boots, and you should ensure that the vehicle, engine compartment, and ground under the vehicle are dry.
- ▶ Do not touch the ballast while it is operating; it will often be generating a lot of heat.
- ▶ Persons with active electronic implants, such as heart pacemakers, should not work on HID headlamps. If changing out the bulb, make sure the headlights are turned off.
- ▶ There is also a risk of injury caused by exposure to ultraviolet light produced by the HID lamp if the lamp is operated outside of its housing.
- ▶ Once ignited, the pressure inside an HID bulb can build up to a very high level (around 220 pounds per square inch, or 1517 kilopascals) due to the high operating temperature (about 1500°F, or 816°C).
- ▶ Light-emitting diodes (LEDs) have been used for some time in various automotive applications, such as warning indicators and alphanumeric displays.
- ▶ One of the advantages of LEDs is that they turn on instantly.
- ▶ For automotive applications, a number of LEDs are grouped together to provide the amount of light required for the application.
- ▶ A typical LED has a voltage drop of 1.2 to 3.5 volts across it, depending upon the color, when it is forward biased and emitting light.

- All lamps or light bulbs have letters and numbers stamped on them that typically indicate their part number and often the operating voltage and power consumed.
- In a bulb marked "12V/21W," the filament will consume 21 watts of power when 12 volts is applied across the filament.
- Single-filament bulbs are common for use as courtesy lights, dash lights, and warning lights.
- Dual-filament bulbs have two filaments of different wattage; one filament emits a small amount of light, and the second filament emits more light.
- Park, tail, and marker lights are all low-intensity or low-wattage bulbs used to mark the outline or width of the vehicle.
- Park and tail lamps tend to be installed close to the corners of the vehicle.
- On computer-controlled lighting systems, the park lights and taillights are controlled by a body control module (BCM).
- Government regulations control the positioning of lights, including the height of the lamps' beam and their brightness.
- As required by law, a higher additional third brake light is mounted on top of the trunk lid or on the rear window on all recent vehicles. This light is called a center high mount stop light (CHMSL), or "chimsul."
- Modern vehicles use network/bus systems and the BCM to command the back-up lights to come on.
- Turn signal indicators are located on the extreme corners of the vehicle. They are usually amber in the front and can be either red or amber in the rear.
- Once activated, turn signal indicators continue until the switch is cancelled either by the operator or by a cancelling mechanism in the switch.
- If the indicator switch is turned to indicate a turn, current from the battery typically flows through the fusible link to the ignition switch, where it is directed through a fuse to the flasher unit.
- The flasher unit uses a timing circuit to pulse the current flowing out of the flasher unit 60 to 120 times per minute.
- The operation of the flasher unit also produces a clicking sound to audibly inform the driver that the indicators are in operation.
- Turn indicators can also be controlled by computer. The BCM commands the appropriate turn indicators to come on as it sees an input from the turn signal switch, and flashes it at the proper rate.
- The hazard warning light circuit is similar to the turn indicator lights except that it simultaneously causes a pulsing in all exterior indicator lights and both indicator lights on the instrument panel.
- Warning devices are installed on the vehicle to provide the driver with information about critical vehicle operating and safety systems, including the battery, supplemental restraint system (SRS), anti-lock brake system (ABS), parking brake, oil pressure, and engine temperature.
- Daytime running lights (DRLs) are existing lights that turn on when the vehicle is running and turn off when the engine stops. The lamps on the front are the headlights, which are typically operated at about a 60% power level.
- DRLs are mandatory on modern vehicles licensed in Canada and some other countries.
- In the United States, DRLs are permitted but not required.
- Headlights used as fog lamps are built into the front of a vehicle to illuminate the road ahead of the vehicle when driving at night or in other conditions of reduced visibility.
- Some vehicles have a timer circuit that is integrated into the headlamp circuit. The purpose of the timer circuit is to allow a vehicle's headlamps to remain on for a period of time after the vehicle has been turned off.
- A sealed-beam headlight has a highly polished aluminized glass reflector that is fused to the optically designed lens. It is a completely sealed unit that has the filaments accurately positioned in relation to the reflector.
- On older, non-computer-controlled systems, when the headlights are switched on, current is supplied from the battery and proceeds through the fusible link or maxi-fuse, headlight fuse(s), headlight switch, and dimmer switch, reaching either the low or the high beams.
- Some vehicles use relays to control the current through the low or high beam circuits. In this configuration, the beam selector switch either powers or grounds the relay windings in each of the relays, depending on the position of the beam selector switch.
- The beam selector switch is a single-pole, double-throw switch, meaning it has one movable pole but makes contact in two positions.
- In the high-beam position, the current is switched to the high-beam circuit. In the low-beam position, the current is switched to the low-beam circuit.
- The beam selector switch can be set to switch a common power input to two outputs, or it can switch a common ground input to two outputs.
- On high-technology on-board systems, headlight and dimmer functions are managed with a network/bus system in modern vehicles.
- On many vehicles, the headlight switch and dimmer switch are just inputs to a control module that sends a headlight "on" or "off" request over the network.
- There are many types of headlight bulbs available. Always make sure you replace a bulb with one of exactly the same type.
- When replacing a halogen bulb, avoid touching the glass portion of the bulb with your fingers, which can leave a residue from your fingers on the outer surface. This residue can cause the bulb to crack or shatter and burn out after a short time of operation.
- If you inadvertently touch the halogen bulb, clean it with alcohol and a lint-free cloth.
- Do not use gasoline or paraffin to clean the halogen bulb.

- Although the principle of aiming headlights is the same in the majority of cases, the legal rules can differ from region to region. Be sure to check the requirements for your location.
- When aiming or aligning headlights, there are a couple of possible methods to utilize.
  - The first method, involves an on-vehicle headlight aligner. This method was used extensively on sealed-beam headlights. The aligners are calibrated to the floor at the point where the front and rear wheels contact the concrete. The aligners use a set of mirrors that can be calibrated to local regulations so that the headlights can be aligned properly.
  - With the introduction of aerodynamic headlights using removable halogen bulbs, off-vehicle headlight aligners are used. These aligners sit in front of the vehicle after being calibrated for the floor slope and orientation with the vehicle. With the headlights on, adjustments can be made by following the instructions on the aligner.
- If the headlights are too dim, there can be several causes. The most common one is high resistance in the light circuit. This can be checked by measuring the voltage drop on both the power side of the bulb and the ground side. The voltage drop should be less than 0.5 volts on each side.

Key Terms

ballast Increases the voltage substantially and controls the current to the bulb.

halogen lamps A type of incandescent lamp that is filled with a halogen gas such as bromine or iodine.

high-intensity discharge (HID) Headlamps that produce light with an electric arc rather than a glowing filament.

incandescent lighting One or more filaments that heat up to approximately 5000°F (2760°C) and glow white hot.

light-emitting diodes (LEDs) Diodes that produce light when current flows across the P/N junction.

vacuum tube fluorescent (VTF) Lighting used for instrumentation displays on vehicle instrument panel clusters. This type of lighting emits a very bright light with high contrast and can display in various colors.

Review Questions

1. Which of the following alert(s) the other drivers of a change in direction?
   a. Red or amber turn signals
   b. Beam selector
   c. Brake lights
   d. Emergency flasher light
2. Which light bulbs are very efficient and sensitive, and can be damaged even by finger oil residue?
   a. Incandescent lamps
   b. Halogen lamps
   c. Xenon lamps
   d. VTF lamps
3. Which of the following lamps produce more lumens with a bluish tinge for the given wattage when compared with all other lamps?
   a. Incandescent lamps
   b. Halogen lamps
   c. High-intensity discharge lamps
   d. VTF lamps
4. Which lights are parallel to the taillights and operate whenever the taillights are switched on?
   a. Park lights
   b. Taillights
   c. Marker lights
   d. License lights
5. All of the following statements referring to driving lights are true *except*:
   a. Driving lights use quartz halogen bulbs in the 55- to 120-watt range.
   b. Quality of reflector is extremely important in driving lights to get the optimum performance.
   c. Driving lights are wired so that they operate only when the high beam is operating.
   d. Driving lights turn off when the headlights are switched from low to high beam.
6. Which of the following statements describes the purpose of back-up lights?
   a. They indicate to other drivers that the vehicle is stopping.
   b. They provide the driver with vision behind the vehicle.
   c. They alert other drivers that the vehicle is about to turn.
   d. They provide vision to other drivers who are approaching.
7. Choose the correct statement with respect to turn signal lights.
   a. They serve a decorative purpose.
   b. They warn other drivers that the vehicle is about to stop.
   c. They warn the drivers in front to slow down.
   d. They warn other road users of the driver's intended change of direction.
8. Which of the following lights use flashing control and also warn other road users if the vehicle is parked in a dangerous position on the side of the road?
   a. High-intensity flashing beam lights
   b. Turn signal lights
   c. Hazard warning lights
   d. Parking lights
9. All of the following statements with respect to the function of headlights are true *except*:
   a. They illuminate the road ahead.
   b. They help drivers at times of reduced visibility.
   c. They provide two beams, high and low, to serve different purposes.
   d. They use the same filaments in different bulbs to create different beams.

10. All of the following statements with respect to electronically controlled lighting systems are true *except*:
    a. The bus combines all the individual systems wherever possible into a multiplexed serial communications network.
    b. The common wiring harness increases the weight of the electronic system used.
    c. Troubleshooting a networked system requires more training and more tools than a simple digital volt-ohm meter.
    d. The advantage of such a system is less wire and fewer connections.

ASE Technician A/Technician B Style Questions

1. Tech A says that some circuits within the lighting systems of modern automobiles contain high voltage, and must be treated with caution to avoid injury. Tech B says that HID bulbs contain a filament, and get very hot during operation. Who is correct?
    a. Tech A
    b. Tech B
    c. Both A and B
    d. Neither A nor B
2. Tech A says that a dim headlamp could be caused by a voltage drop at the body ground connection for the light in question. Tech B says that a multimeter is needed to perform a voltage drop test. Who is correct?
    a. Tech A
    b. Tech B
    c. Both A and B
    d. Neither A nor B
3. Tech A says that the lighting system should be inspected at every oil change to ensure safe vehicle operation. Tech B says that care must be taken when installing halogen headlamp bulbs to ensure that the glass envelope of the bulb does not come into contact with your fingers. Who is correct?
    a. Tech A
    b. Tech B
    c. Both A and B
    d. Neither A nor B
4. Tech A says that there is no specification for aiming headlights and the best practice is to adjust the lights so they provide maximum illumination distance. Tech B says that headlamps should only be aimed to manufacturer's specifications; this is to avoid aiming the headlamps too high, which can blind oncoming drivers. Who is correct?
    a. Tech A
    b. Tech B
    c. Both A and B
    d. Neither A nor B
5. Tech A says that brighter than normal brake lamp operation could be the result of a short to voltage in the brake lamp wiring harness. Tech B says that when inspecting brake lamps, you must remember to check the center high mount stop light (CHMSL) for proper operation. Who is correct?
    a. Tech A
    b. Tech B
    c. Both A and B
    d. Neither A nor B
6. Tech A says that it is perfectly safe to install higher-wattage headlamp bulbs to increase nighttime visibility. Tech B says that installing higher-wattage headlamp bulbs places excess strain on the headlamp circuit, and can cause wires, switches, and components to be damaged due to the increase in current flow. Who is correct?
    a. Tech A
    b. Tech B
    c. Both A and B
    d. Neither A nor B
7. Tech A says that HID headlamps can safely be retrofitted into a natively halogen-bulb headlamp assembly. Tech B says that installing HID lights into a housing designed for halogen bulbs can cause improper light distribution and blind oncoming drivers at night. Who is correct?
    a. Tech A
    b. Tech B
    c. Both A and B
    d. Neither A nor B
8. Tech A says that LED lighting uses far less current than traditional incandescent lamps. Tech B says that LED lights can produce a wide array of light temperatures (colors). Who is correct?
    a. Tech A
    b. Tech B
    c. Both A and B
    d. Neither A nor B
9. Tech A says that LEDs produce much less heat that halogen headlamps. Tech B says that LEDs are able to turn off and on much faster than incandescent bulbs, making them ideal for use in brake lights. Who is correct?
    a. Tech A
    b. Tech B
    c. Both A and B
    d. Neither A nor B
10. Tech A says that a burned-out bulb can be replaced with any bulb that properly fits into the socket as long as the bulb colors are the same. Tech B says that a damaged or melted bulb socket can produce intermittent lamp operation. Who is correct?
    a. Tech A
    b. Tech B
    c. Both A and B
    d. Neither A nor B

APPENDIX A

2017 NATEF AUTOMOBILE ACCREDITATION TASK LIST CORRELATION GUIDE

Task List	MAST	AST	MLR	Chapter
ENGINE REPAIR				
General: Engine Diagnosis; Removal and Reinstallation (R & R)				
Complete work order to include customer information, vehicle identifying information, customer concern, related service history, cause, and correction.	P-1	P-1		
Research vehicle service information including fluid type, internal engine operation, vehicle service history, service precautions, and technical service bulletins.	P-1	P-1	P-1	
Verify operation of the instrument panel engine warning indicators.	P-1	P-1	P-1	
Inspect engine assembly for fuel, oil, coolant, and other leaks; determine needed action.	P-1	P-1	P-1	
Install engine covers using gaskets, seals, and sealers as required.	P-1	P-1	P-1	
Verify engine mechanical timing.	P-1	P-1	P-2	
Perform common fastener and thread repair, to include: remove broken bolt, restore internal and external threads, and repair internal threads with thread insert.	P-1	P-1	P-1	4
Inspect, remove and/or replace engine mounts.	P-2	P-2		
Identify service precautions related to service of the internal combustion engine of a hybrid vehicle.	P-2	P-2	P-2	
Remove and reinstall engine on a newer vehicle equipped with OBD; reconnect all attaching components and restore the vehicle to running condition.	P-3			
Cylinder Head and Valve Train Diagnosis and Repair				
Remove cylinder head; inspect gasket condition; install cylinder head and gasket; tighten according to manufacturer's specification and procedure.	P-1	P-1		
Clean and visually inspect a cylinder head for cracks; check gasket surface areas for warpage and surface finish; check passage condition.	P-1	P-1		
Inspect pushrods, rocker arms, rocker arm pivots and shafts for wear, bending, cracks, looseness, and blocked oil passages (orifices); determine needed action.	P-2	P-2		
Adjust valves (mechanical or hydraulic lifters).	P-1	P-1	P-3	
Inspect and replace camshaft and drive belt/chain; includes checking drive gear wear and backlash, end play, sprocket and chain wear, overhead cam drive sprocket(s), drive belt(s), belt tension, tensioners, camshaft reluctor ring/tone-wheel, and valve timing components; verify correct camshaft timing.	P-1	P-1		
Establish camshaft position sensor indexing.	P-1	P-1		
Inspect valve springs for squareness and free height comparison; determine needed action.	P-3			
Replace valve stem seals on an assembled engine; inspect valve spring retainers, locks/keepers, and valve lock/keeper grooves; determine needed action.	P-3			
Inspect valve guides for wear; check valve stem-to-guide clearance; determine needed action.	P-3			
Inspect valves and valve seats; determine needed action.	P-3			
Check valve spring assembled height and valve stem height; determine needed action.	P-3			
Inspect valve lifters; determine needed action.	P-2			
Inspect and/or measure camshaft for runout, journal wear and lobe wear.	P-3			
Inspect camshaft bearing surface for wear, damage, out-of-round, and alignment; determine needed action.	P-3			
Identify components of the cylinder head and valve train.			P-1	
Engine Block Assembly Diagnosis and Repair				
Remove, inspect, and/or replace crankshaft vibration damper (harmonic balancer).	P-1	P-2		
Disassemble engine block; clean and prepare components for inspection and reassembly.	P-1			

(continued)

Task List	MAST	AST	MLR	Chapter
Inspect engine block for visible cracks, passage condition, core and gallery plug condition, and surface warpage; determine needed action.	P-2			
Inspect and measure cylinder walls/sleeves for damage, wear, and ridges; determine needed action.	P-2			
Deglaze and clean cylinder walls.	P-2			
Inspect and measure camshaft bearings for wear, damage, out-of-round, and alignment; determine needed action.	P-3			
Inspect crankshaft for straightness, journal damage, keyway damage, thrust flange and sealing surface condition, and visual surface cracks; check oil passage condition; measure end play and journal wear; check crankshaft position sensor reluctor ring (where applicable); determine needed action.	P-1			
Inspect main and connecting rod bearings for damage and wear; determine needed action.	P-2			
Identify piston and bearing wear patterns that indicate connecting rod alignment and main bearing bore problems; determine needed action.	P-3			
Inspect and measure piston skirts and ring lands; determine needed action.	P-2			
Determine piston-to-bore clearance.	P-2			
Inspect, measure, and install piston rings.	P-2			
Inspect auxiliary shaft(s) (balance, intermediate, idler, counterbalance and/or silencer); inspect shaft(s) and support bearings for damage and wear; determine needed action; reinstall and time.	P-2			
Assemble engine block.	P-1			
Lubrication and Cooling Systems Diagnosis and Repair				
Perform cooling system pressure and dye tests to identify leaks; check coolant condition and level; inspect and test radiator, pressure cap, coolant recovery tank, heater core, and galley plugs; determine needed action.	P-1	P-1	P-1	
Identify causes of engine overheating.	P-1	P-1		
Inspect, replace, and/or adjust drive belts, tensioners, and pulleys; check pulley and belt alignment.	P-1	P-1	P-1	
Inspect and/or test coolant; drain and recover coolant; flush and refill cooling system; use proper fluid type per manufacturer specification; bleed air as required.	P-1	P-1	P-1	
Inspect, remove, and replace water pump.	P-2	P-2		
Remove and replace radiator.	P-2	P-2		
Remove, inspect, and replace thermostat and gasket/seal.	P-1	P-1	P-1	
Inspect and test fan(s), fan clutch (electrical or mechanical), fan shroud, and air dams; determine needed action.	P-1	P-1		
Perform oil pressure tests; determine needed action.	P-1	P-1		
Perform engine oil and filter change; use proper fluid type per manufacturer specification.	P-1	P-1	P-1	
Inspect auxiliary coolers; determine needed action.	P-3	P-3		
Inspect, test, and replace oil temperature and pressure switches and sensors.	P-2	P-2		
Inspect oil pump gears or rotors, housing, pressure relief devices, and pump drive; perform needed action.	P-2			
Identify components of the lubrication and cooling systems.			P-1	
AUTOMATIC TRANSMISSION AND TRANSAXLE **General: Transmission and Transaxle Diagnosis**				
Identify and interpret transmission/transaxle concerns, differentiate between engine performance and transmission/transaxle concerns; determine needed action.	P-1	P-1		
Research vehicle service information including fluid type, vehicle service history, service precautions, and technical service bulletins.	P-1	P-1	P-1	
Diagnose fluid loss and condition concerns; determine needed action.	P-1	P-1		
Check fluid level in a transmission or a transaxle equipped with a dip-stick.	P-1	P-1	P-1	
Check fluid level in a transmission or a transaxle not equipped with a dip-stick.	P-1	P-1	P-1	

Task List	MAST	AST	MLR	Chapter
Perform pressure tests (including transmissions/transaxles equipped with electronic pressure control); determine needed action.	P-1			
Diagnose noise and vibration concerns; determine needed action.	P-2			
Perform stall test; determine needed action.	P-2	P-2		
Perform lock-up converter system tests; determine needed action.	P-3	P-3		
Diagnose transmission/transaxle gear reduction/multiplication concerns using driving, driven, and held member (power flow) principles.	P-1	P-1		
Diagnose electronic transmission/transaxle control systems using appropriate test equipment and service information.	P-1			
Diagnose pressure concerns in a transmission using hydraulic principles (Pascal's Law).	P-2	P-2		
Demonstrate knowledge of pressure test including transmissions/transaxles equipped with electronic pressure control.		P-3		
Diagnose electronic transmission/transaxle control systems using appropriate test equipment and service information.		P-2		
Check transmission fluid condition; check for leaks.			P-2	
Identify drive train components and configuration.			P-1	
In-Vehicle Transmission/Transaxle Maintenance and Repair				
Inspect, adjust, and/or replace external manual valve shift linkage, transmission range sensor/switch, and/or park/neutral position switch.	P-1	P-1	P-2	
Inspect for leakage; replace external seals, gaskets, and bushings.	P-2	P-2	P-1	
Inspect, test, adjust, repair, and/or replace electrical/electronic components and circuits including computers, solenoids, sensors, relays, terminals, connectors, switches, and harnesses; demonstrate understanding of the relearn procedure.	P-1	P-1		
Drain and replace fluid and filter(s); use proper fluid type per manufacturer specification.	P-1	P-1	P-1	
Inspect, replace and align powertrain mounts.	P-2	P-2	P-2	
Off-Vehicle Transmission and Transaxle Repair				
Remove and reinstall transmission/transaxle and torque converter; inspect engine core plugs, rear crankshaft seal, dowel pins, dowel pin holes, and mounting surfaces.	P-2	P-2		
Inspect, leak test, flush, and/or replace transmission/transaxle oil cooler, lines, and fittings.	P-1	P-1		
Inspect converter flex (drive) plate, converter attaching bolts, converter pilot, converter pump drive surfaces, converter end play, and crankshaft pilot bore.	P-2	P-2		
Describe the operational characteristics of a continuously variable transmission (CVT).	P-3	P-3	P-3	
Describe the operational characteristics of a hybrid vehicle drive train.	P-3	P-3	P-3	
Disassemble, clean, and inspect transmission/transaxle.	P-1			
Inspect, measure, clean, and replace valve body (includes surfaces, bores, springs, valves, switches, solenoids, sleeves, retainers, brackets, check valves/balls, screens, spacers, and gaskets).	P-2			
Inspect servo and accumulator bores, pistons, seals, pins, springs, and retainers; determine needed action.	P-2			
Assemble transmission/transaxle.	P-1			
Inspect, measure, and reseal oil pump assembly and components.	P-2			
Measure transmission/transaxle end play and/or preload; determine needed action.	P-1			
Inspect, measure, and/or replace thrust washers and bearings.	P-2			
Inspect oil delivery circuits, including seal rings, ring grooves, and sealing surface areas, feed pipes, orifices, and check valves/balls.	P-2			
Inspect bushings; determine needed action.	P-2			
Inspect and measure planetary gear assembly components; determine needed action.	P-2			
Inspect case bores, passages, bushings, vents, and mating surfaces; determine needed action.	P-2			

(continued)

Task List	MAST	AST	MLR	Chapter
Diagnose and inspect transaxle drive, link chains, sprockets, gears, bearings, and bushings; perform needed action.	P-2			
Inspect measure, repair, adjust or replace transaxle final drive components.	P-2			
Inspect clutch drum, piston, check-balls, springs, retainers, seals, friction plates, pressure plates, and bands; determine needed action.	P-2			
Measure clutch pack clearance; determine needed action.	P-1			
Air test operation of clutch and servo assemblies.	P-1			
Inspect one-way clutches, races, rollers, sprags, springs, cages, retainers; determine needed action.	P-2			
MANUAL DRIVE TRAIN AND AXLES				
General: Drive Train Diagnosis				
Identify and interpret drive train concerns; determine needed action.	P-1	P-1		
Research vehicle service information including fluid type, vehicle service history, service precautions, and technical service bulletins.	P-1	P-1	P-1	
Check fluid condition; check for leaks; determine needed action.	P-1	P-1	P-2	
Drain and refill manual transmission/transaxle and final drive unit; use proper fluid type per manufacturer specification.	P-1	P-1	P-1	
Identify manual drive train and axle components and configuration.			P-1	
Clutch Diagnosis and Repair				
Diagnose clutch noise, binding, slippage, pulsation, and chatter; determine needed action.	P-1	P-1		
Inspect clutch pedal linkage, cables, automatic adjuster mechanisms, brackets, bushings, pivots, and springs; perform needed action.	P-1	P-1		
Inspect and/or replace clutch pressure plate assembly, clutch disc, release (throw-out) bearing, linkage, and pilot bearing/bushing (as applicable).	P-1	P-1		
Bleed clutch hydraulic system.	P-1	P-1		
Check and adjust clutch master cylinder fluid level; check for leaks; use proper fluid type per manufacturer specification.	P-1	P-1	P-1	
Inspect flywheel and ring gear for wear, cracks, and discoloration; determine needed action.	P-1	P-1		
Measure flywheel runout and crankshaft end play; determine needed action.	P-2	P-2		
Describe the operation and service of a system that uses a dual mass flywheel.	P-3	P-3		
Check for hydraulic system leaks.			P-1	
Transmission/Transaxle Diagnosis and Repair				
Inspect, adjust, lubricate, and/or replace shift linkages, brackets, bushings, cables, pivots, and levers.	P-2	P-2		
Describe the operational characteristics of an electronically-controlled manual transmission/transaxle.	P-2	P-2	P-2	
Diagnose noise concerns through the application of transmission/transaxle powerflow principles.	P-2			
Diagnose hard shifting and jumping out of gear concerns; determine needed action.	P-2			
Diagnose transaxle final drive assembly noise and vibration concerns; determine needed action.	P-3			
Disassemble, inspect, clean, and reassemble internal transmission/transaxle components.	P-2			
Drive Shaft and Half Shaft, Universal and Constant-Velocity (CV) Joint Diagnosis and Repair (Front, Rear, All-wheel, and Four-wheel drive)				
Diagnose constant-velocity (CV) joint noise and vibration concerns; determine needed action.	P-1	P-1		
Diagnose universal joint noise and vibration concerns; perform needed action.	P-2	P-2		
Inspect, remove, and/or replace bearings, hubs, and seals.	P-1	P-1	P-2	
Inspect, service, and/or replace shafts, yokes, boots, and universal/CV joints.	P-1	P-1	P-2	
Check shaft balance and phasing; measure shaft runout; measure and adjust driveline angles.	P-2	P-2		
Inspect locking hubs.			P-3	
Check for leaks at drive assembly and transfer case seals; check vents; check fluid level; use proper fluid type per manufacturer specification.			P-2	

Task List	MAST	AST	MLR	Chapter
Drive Axle Diagnosis and Repair				
Ring and Pinion Gears and Differential Case Assembly				
Clean and inspect differential case; check for leaks; inspect housing vent.	P-1	P-1	P-1	
Check and adjust differential case fluid level; use proper fluid type per manufacturer specification.	P-1	P-1	P-1	
Drain and refill differential case; use proper fluid type per manufacturer specification.	P-1	P-1		
Drain and refill differential housing.			P-1	
Diagnose noise and vibration concerns; determine needed action.	P-2			
Inspect and replace companion flange and/or pinion seal; measure companion flange runout.	P-2	P-2		
Inspect ring gear and measure runout; determine needed action.	P-3			
Remove, inspect, and/or reinstall drive pinion and ring gear, spacers, sleeves, and bearings.	P-3			
Measure and adjust drive pinion depth.	P-3			
Measure and adjust drive pinion bearing preload.	P-3			
Measure and adjust side bearing preload and ring and pinion gear total backlash and backlash variation on a differential carrier assembly (threaded cup or shim types).	P-3			
Check ring and pinion tooth contact patterns; perform needed action.	P-3			
Disassemble, inspect, measure, adjust, and/or replace differential pinion gears (spiders), shaft, side gears, side bearings, thrust washers, and case.	P-3			
Reassemble and reinstall differential case assembly; measure runout; determine needed action.	P-3			
Limited Slip Differential				
Diagnose noise, slippage, and chatter concerns; determine needed action.	P-3			
Measure rotating torque; determine needed action.	P-3			
Drive Axles				
Inspect and replace drive axle wheel studs.	P-1	P-1	P-1	
Remove and replace drive axle shafts.	P-1	P-1		
Inspect and replace drive axle shaft seals, bearings, and retainers.	P-2	P-2		
Measure drive axle flange runout and shaft end play; determine needed action.	P-2	P-2		
Diagnose drive axle shafts, bearings, and seals for noise, vibration, and fluid leakage concerns; determine needed action.	P-2			
Four-wheel Drive/All-wheel Drive Component Diagnosis and Repair				
Inspect, adjust, and repair shifting controls (mechanical, electrical, and vacuum), bushings, mounts, levers, and brackets.	P-3	P-3		
Inspect locking hubs; determine needed action.	P-3	P-3		
Check for leaks at drive assembly and transfer case seals; check vents; check fluid level; use proper fluid type per manufacturer specification.	P-3	P-3		
Identify concerns related to variations in tire circumference and/or final drive ratios.	P-2	P-2		
Diagnose noise, vibration, and unusual steering concerns; determine needed action.	P-3			
Diagnose, test, adjust, and/or replace electrical/electronic components of four-wheel drive/all-wheel drive systems.	P-2			
Disassemble, service, and reassemble transfer case and components.	P-2			
SUSPENSION AND STEERING				
General: Suspension and Steering Systems				
Research vehicle service information including fluid type, vehicle service history, service precautions, and technical service bulletins.	P-1	P-1	P-1	
Identify and interpret suspension and steering system concerns; determine needed action.	P-1	P-2	P-1	

(continued)

Task List	MAST	AST	MLR	Chapter
Steering Systems Diagnosis and Repair				
Disable and enable supplemental restraint system (SRS); verify indicator lamp operation.	P-1	P-1	P-1	
Remove and replace steering wheel; center/time supplemental restraint system (SRS) coil (clock spring).	P-1	P-1		
Diagnose steering column noises, looseness, and binding concerns (including tilt/telescoping mechanisms); determine needed action.	P-2	P-2		
Diagnose power steering gear (non-rack and pinion) binding, uneven turning effort, looseness, hard steering, and noise concerns; determine needed action.	P-2	P-2		
Diagnose power steering gear (rack and pinion) binding, uneven turning effort, looseness, hard steering, and noise concerns; determine needed action.	P-2	P-2		
Inspect steering shaft universal-joint(s), flexible coupling(s), collapsible column, lock cylinder mechanism, and steering wheel; determine needed action.	P-2	P-2		
Remove and replace rack and pinion steering gear; inspect mounting bushings and brackets.	P-2	P-1		
Inspect rack and pinion steering gear inner tie rod ends (sockets) and bellows boots; replace as needed.	P-1	P-1	P-1	
Inspect power steering fluid level and condition.	P-1	P-1	P-1	
Flush, fill, and bleed power steering system; use proper fluid type per manufacturer specification.	P-2	P-2	P-2	
Inspect for power steering fluid leakage; determine needed action.	P-1	P-1	P-1	
Remove, inspect, replace, and/or adjust power steering pump drive belt.	P-1	P-1	P-1	
Remove and reinstall power steering pump.	P-2	P-2		
Remove and reinstall press fit power steering pump pulley; check pulley and belt alignment.	P-2	P-2		
Inspect, remove and/or replace power steering hoses and fittings.	P-2	P-2	P-2	
Inspect, remove and/or replace pitman arm, relay (centerlink/intermediate) rod, idler arm, mountings, and steering linkage damper.	P-2	P-2	P-1	
Inspect, replace, and/or adjust tie rod ends (sockets), tie rod sleeves, and clamps.	P-1	P-1	P-1	
Inspect, test and diagnose electrically-assisted power steering systems (including using a scan tool); determine needed action.	P-2			
Identify hybrid vehicle power steering system electrical circuits and safety precautions.	P-2	P-2	P-2	
Test power steering system pressure; determine needed action.	P-2			
Inspect electric power steering assist system.		P-3	P-2	
Suspension Systems Diagnosis and Repair				
Diagnose short and long arm suspension system noises, body sway, and uneven ride height concerns; determine needed action.	P-1	P-1		
Diagnose strut suspension system noises, body sway, and uneven ride height concerns; determine needed action.	P-1	P-1		
Inspect, remove, and/or replace upper and lower control arms, bushings, shafts, and rebound bumpers.	P-3	P-3	P-1	
Inspect, remove, and/or replace strut rods and bushings.	P-3	P-3		
Inspect, remove, and/or replace upper and/or lower ball joints (with or without wear indicators).	P-2	P-2	P-1	
Inspect, remove, and/or replace steering knuckle assemblies.	P-3	P-3		
Inspect, remove and/or replace short and long arm suspension system coil springs and spring insulators.	P-3	P-3		
Inspect, remove, and/or replace torsion bars and mounts.	P-3	P-3	P-1	
Inspect, remove, and/or replace front/rear stabilizer bar (sway bar) bushings, brackets, and links.	P-3	P-3	P-1	
Inspect, remove, and/or replace strut cartridge or assembly, strut coil spring, insulators (silencers), and upper strut bearing mount.	P-3	P-3	P-2	
Inspect, remove, and/or replace track bar, strut rods/radius arms, and related mounts and bushings.	P-3	P-3	P-1	
Inspect rear suspension system leaf spring(s), spring insulators (silencers), shackles, brackets, bushings, center pins/bolts, and mounts.	P-1	P-1	P-1	

Task List	MAST	AST	MLR	Chapter
Inspect front strut bearing and mount.			P-1	
Inspect rear suspension system lateral links/arms (track bars), control (trailing) arms.			P-1	
Related Suspension and Steering Service				
Inspect, remove, and/or replace shock absorbers; inspect mounts and bushings.	P-1	P-1	P-1	
Remove, inspect, service and/or replace front and rear wheel bearings.	P-1	P-1		
Describe the function of suspension and steering control systems and components, (i.e. active suspension and stability control).	P-3	P-3	P-3	
Wheel Alignment Diagnosis, Adjustment, and Repair				
Diagnose vehicle wander, drift, pull, hard steering, bump steer, memory steer, torque steer, and steering return concerns; determine needed action.	P-1	P-1		
Perform prealignment inspection; measure vehicle ride height; determine needed action.	P-1	P-1	P-1	
Describe alignment angles (camber, caster, and toe).			P-1	
Prepare vehicle for wheel alignment on alignment machine; perform four-wheel alignment by checking and adjusting front and rear wheel caster, camber and toe as required; center steering wheel.	P-1	P-1		
Check toe-out-on-turns (turning radius); determine needed action.	P-2	P-2		
Check steering axis inclination (SAI) and included angle; determine needed action.	P-2	P-2		
Check rear wheel thrust angle; determine needed action.	P-1	P-1		
Check for front wheel setback; determine needed action.	P-2	P-2		
Check front and/or rear cradle (subframe) alignment; determine needed action.	P-3	P-3		
Reset steering angle sensor.	P-2	P-2		
Wheels and Tires Diagnosis and Repair				
Inspect tire condition; identify tire wear patterns; check for correct tire size, application (load and speed ratings), and air pressure as listed on the tire information placard/label.	P-1	P-1	P-1	
Diagnose wheel/tire vibration, shimmy, and noise; determine needed action.	P-2	P-2		
Rotate tires according to manufacturer's recommendation including vehicles equipped with tire pressure monitoring systems (TPMS).	P-1	P-1	P-1	
Measure wheel, tire, axle flange, and hub runout; determine needed action.	P-2	P-2		
Diagnose tire pull problems; determine needed action.	P-1	P-1		
Dismount, inspect, and remount tire on wheel; balance wheel and tire assembly.	P-1	P-1	P-1	
Dismount, inspect, and remount tire on wheel equipped with tire pressure monitoring system sensor.	P-1	P-1	P-1	
Inspect tire and wheel assembly for air loss; perform needed action.	P-1	P-1	P-1	
Repair tire following vehicle manufacturer approved procedure.	P-1	P-1	P-1	
Identify indirect and direct tire pressure monitoring system (TPMS); calibrate system; verify operation of instrument panel lamps.	P-1	P-1	P-1	
Demonstrate knowledge of steps required to remove and replace sensors in a tire pressure monitoring system (TPMS) including relearn procedure.	P-1	P-1	P-1	
BRAKES				
General: Brake Systems Diagnosis				
Identify and interpret brake system concerns; determine needed action.	P-1	P-1		
Research vehicle service information including fluid type, vehicle service history, service precautions, and technical service bulletins.	P-1	P-1	P-1	
Describe procedure for performing a road test to check brake system operation including an anti-lock brake system (ABS).	P-1	P-1	P-1	
Install wheel and torque lug nuts.	P-1	P-1	P-1	
Identify brake system components and configuration.			P-1	

(continued)

Task List	MAST	AST	MLR	Chapter
Hydraulic System Diagnosis and Repair				
Diagnose pressure concerns in the brake system using hydraulic principles (Pascal's Law).	P-1	P-1		
Measure brake pedal height, travel, and free play (as applicable); determine needed action.	P-1	P-1		
Describe proper brake pedal height, travel, and feel.			P-1	
Check master cylinder for internal/external leaks and proper operation; determine needed action.	P-1	P-1	P-1	
Remove, bench bleed, and reinstall master cylinder.	P-1	P-1		
Diagnose poor stopping, pulling or dragging concerns caused by malfunctions in the hydraulic system; determine needed action.	P-1	P-3		
Inspect brake lines, flexible hoses, and fittings for leaks, dents, kinks, rust, cracks, bulging, wear; and loose fittings/supports; determine needed action.	P-1	P-1	P-1	
Replace brake lines, hoses, fittings, and supports.	P-2	P-2		
Fabricate brake lines using proper material and flaring procedures (double flare and ISO types).	P-2	P-2		
Select, handle, store, and fill brake fluids to proper level; use proper fluid type per manufacturer specification.	P-1	P-1	P-1	
Inspect, test, and/or replace components of brake warning light system.	P-3	P-3		
Identify components of hydraulic brake warning light system.	P-2	P-2	P-3	
Bleed and/or flush brake system.	P-1	P-1	P-1	
Test brake fluid for contamination.	P-1	P-1	P-1	
Drum Brake Diagnosis and Repair				
Diagnose poor stopping, noise, vibration, pulling, grabbing, dragging or pedal pulsation concerns; determine needed action.	P-1	P-1		
Remove, clean, and inspect brake drum; measure brake drum diameter; determine serviceability.	P-1	P-1	P-1	
Refinish brake drum and measure final drum diameter; compare with specification.	P-1	P-1	P-1	
Remove, clean, inspect, and/or replace brake shoes, springs, pins, clips, levers, adjusters/self-adjusters, other related brake hardware, and backing support plates; lubricate and reassemble.	P-1	P-1	P-1	
Inspect wheel cylinders for leaks and proper operation; remove and replace as needed.	P-2	P-2	P-2	
Pre-adjust brake shoes and parking brake; install brake drums or drum/hub assemblies and wheel bearings; perform final checks and adjustments.	P-1	P-1	P-1	
Disc Brake Diagnosis and Repair				
Diagnose poor stopping, noise, vibration, pulling, grabbing, dragging, or pulsation concerns; determine needed action.	P-1	P-1		
Remove and clean caliper assembly; inspect for leaks, damage, and wear; determine needed action.	P-1	P-1	P-1	
Inspect caliper mounting and slides/pins for proper operation, wear, and damage; determine needed action.	P-1	P-1	P-1	
Remove, inspect, and/or replace brake pads and retaining hardware; determine needed action.	P-1	P-1	P-1	
Lubricate and reinstall caliper, brake pads, and related hardware; seat brake pads; inspect for leaks.	P-1	P-1	P-1	
Clean and inspect rotor and mounting surface; measure rotor thickness, thickness variation, and lateral runout; determine needed action.	P-1	P-1	P-1	
Remove and reinstall/replace rotor.	P-1	P-1	P-1	
Refinish rotor on vehicle; measure final rotor thickness and compare with specification.	P-1	P-1	P-1	
Refinish rotor off vehicle; measure final rotor thickness and compare with specification.	P-1	P-1	P-1	
Retract and re-adjust caliper piston on an integrated parking brake system.	P-2	P-2	P-2	
Check brake pad wear indicator; determine needed action.	P-1	P-1	P-1	
Describe importance of operating vehicle to burnish/break-in replacement brake pads according to manufacturer's recommendations.	P-1	P-1	P-1	

Task List	MAST	AST	MLR	Chapter
Power-Assist Units Diagnosis and Repair				
Check brake pedal travel with and without engine running to verify proper power booster operation.	P-2	P-2	P-2	
Identify components of the brake power assist system (vacuum and hydraulic); check vacuum supply (manifold or auxiliary pump) to vacuum- type power booster.	P-1	P-1	P-1	
Inspect vacuum-type power booster unit for leaks; inspect the check-valve for proper operation; determine needed action.	P-1	P-1		
Inspect and test hydraulically-assisted power brake system for leaks and proper operation; determine needed action.	P-3	P-3		
Measure and adjust master cylinder pushrod length.	P-3	P-3		
Related Systems (i.e. Wheel Bearings, Parking Brakes, Electrical) Diagnosis and Repair				
Diagnose wheel bearing noises, wheel shimmy, and vibration concerns; determine needed action.	P-1	P-2		
Remove, clean, inspect, repack, and install wheel bearings; replace seals; install hub and adjust bearings.	P-2	P-2	P-1	
Check parking brake system and components for wear, binding, and corrosion; clean, lubricate, adjust and/or replace as needed.	P-1	P-1	P-2	
Check parking brake operation and parking brake indicator light system operation; determine needed action.	P-1	P-1	P-1	
Check operation of brake stop light system.	P-1	P-1	P-1	
Replace wheel bearing and race.	P-3	P-3	P-2	
Remove, reinstall, and/or replace sealed wheel bearing assembly.	P-1	P-1		
Inspect and replace wheel studs.	P-1	P-1	P-1	
Electronic Brake Control Systems: Antilock Brake (ABS), Traction Control (TCS), and Electronic Stability Control (ESC) Systems Diagnosis and Repair				
Identify and inspect electronic brake control system components (ABS, TCS, ESC); determine needed action.	P-1	P-1		
Identify traction control/vehicle stability control system components.			P-3	
Describe the operation of a regenerative braking system.	P-3	P-3	P-3	
Diagnose poor stopping, wheel lock-up, abnormal pedal feel, unwanted application, and noise concerns associated with the electronic brake control system; determine needed action.	P-2			
Diagnose electronic brake control system electronic control(s) and components by retrieving diagnostic trouble codes, and/or using recommended test equipment; determine needed action.	P-2			
Depressurize high-pressure components of an electronic brake control system.	P-2			
Bleed the electronic brake control system hydraulic circuits.	P-1			
Test, diagnose, and service electronic brake control system speed sensors (digital and analog), toothed ring (tone wheel), and circuits using a graphing multimeter (GMM)/digital storage oscilloscope (DSO) (includes output signal, resistance, shorts to voltage/ground, and frequency data).	P-2			
Diagnose electronic brake control system braking concerns caused by vehicle modifications (tire size, curb height, final drive ratio, etc.).	P-1			
ELECTRICAL/ELECTRONIC SYSTEMS **General: Electrical System Diagnosis**				
Research vehicle service information including vehicle service history, service precautions, and technical service bulletins.	P-1	P-1	P-1	
Demonstrate knowledge of electrical/electronic series, parallel, and series-parallel circuits using principles of electricity (Ohm's Law).	P-1	P-1	P-1	10
Use wiring diagrams to trace electrical/electronic circuits.			P-1	

(continued)

Task List	MAST	AST	MLR	Chapter
Demonstrate proper use of a digital multimeter (DMM) when measuring source voltage, voltage drop (including grounds), current flow and resistance.	P-1	P-1	P-1	7, 10
Demonstrate knowledge of the causes and effects from shorts, grounds, opens, and resistance problems in electrical/electronic circuits.	P-1	P-1	P-1	6, 7, 8, 10
Demonstrate proper use of a test light on an electrical circuit.	P-1	P-1	P-2	7, 10
Use fused jumper wires to check operation of electrical circuits.	P-1	P-1	P-2	10
Use wiring diagrams during the diagnosis (troubleshooting) of electrical/electronic circuit problems.	P-1	P-1		10, 11
Diagnose the cause(s) of excessive key-off battery drain (parasitic draw); determine needed action.	P-1	P-1		
Measure key-off battery drain (parasitic draw).			P-1	
Inspect and test fusible links, circuit breakers, and fuses; determine needed action.	P-1	P-1	P-1	8, 9, 10
Inspect, test, repair, and/or replace components, connectors, terminals, harnesses, and wiring in electrical/electronic systems (including solder repairs); determine needed action.	P-1	P-1	P-1	9, 11
Check electrical/electronic circuit waveforms; interpret readings and determine needed repairs.	P-2			12
Repair data bus wiring harness.	P-1			11
Identify electrical/electronic system components and configurations.			P-1	
Battery Diagnosis and Service				
Perform battery state-of-charge test; determine needed action.	P-1	P-1	P-1	13
Confirm proper battery capacity for vehicle application; perform battery capacity and load test; determine needed action.	P-1	P-1	P-1	
Maintain or restore electronic memory functions.	P-1	P-1	P-1	13
Inspect and clean battery; fill battery cells; check battery cables, connectors, clamps, and hold-downs.	P-1	P-1	P-1	13
Perform slow/fast battery charge according to manufacturer's recommendations.	P-1	P-1	P-1	13
Jump-start vehicle using jumper cables and a booster battery or an auxiliary power supply.	P-1	P-1	P-1	13
Identify safety precautions for high voltage systems on electric, hybrid, hybrid-electric, and diesel vehicles.	P-2	P-2	P-2	13
Identify electrical/electronic modules, security systems, radios, and other accessories that require reinitialization or code entry after reconnecting vehicle battery.	P-1	P-1	P-1	13
Identify hybrid vehicle auxiliary (12v) battery service, repair, and test procedures.	P-2	P-2	P-2	13
Starting System Diagnosis and Repair				
Perform starter current draw tests; determine needed action.	P-1	P-1	P-1	14
Perform starter circuit voltage drop tests; determine needed action.	P-1	P-1	P-1	14
Inspect and test starter relays and solenoids; determine needed action.	P-2	P-2	P-2	14
Remove and install starter in a vehicle.	P-1	P-1	P-1	14
Inspect and test switches, connectors, and wires of starter control circuits; determine needed action.	P-2	P-2	P-2	10, 14
Differentiate between electrical and engine mechanical problems that cause a slow-crank or a no-crank condition.	P-2	P-2		14
Demonstrate knowledge of an automatic idle-stop/start-stop system.	P-2	P-2	P-3	
Charging System Diagnosis and Repair				
Perform charging system output test; determine needed action.	P-1	P-1	P-1	15
Diagnose (troubleshoot) charging system for causes of undercharge, no-charge, or overcharge conditions.	P-1	P-1		15
Inspect, adjust, and/or replace generator (alternator) drive belts; check pulleys and tensioners for wear; check pulley and belt alignment.	P-1	P-1	P-1	15
Remove, inspect, and/or replace generator (alternator).	P-1	P-1	P-2	15
Perform charging circuit voltage drop tests; determine needed action.	P-1	P-1	P-2	15

Task List	MAST	AST	MLR	Chapter
Lighting Systems Diagnosis and Repair				
Diagnose (troubleshoot) the causes of brighter-than-normal, intermittent, dim, or no light operation; determine needed action.	P-1	P-1		16
Inspect interior and exterior lamps and sockets including headlights and auxiliary lights (fog lights/driving lights); replace as needed.	P-1	P-1	P-1	16
Aim headlights.	P-2	P-2	P-2	16
Identify system voltage and safety precautions associated with high-intensity discharge headlights.	P-2	P-2	P-2	16
Instrument Cluster and Driver Information Systems Diagnosis and Repair				
Inspect and test gauges and gauge sending units for causes of abnormal readings; determine needed action.	P-2	P-2		
Diagnose (troubleshoot) the causes of incorrect operation of warning devices and other driver information systems; determine needed action.	P-2	P-2		
Reset maintenance indicators as required.	P-2	P-2		
Body Electrical Systems Diagnosis and Repair				
Diagnose operation of comfort and convenience accessories and related circuits (such as: power window, power seats, pedal height, power locks, truck locks, remote start, moon roof, sun roof, sun shade, remote keyless entry, voice activation, steering wheel controls, back-up camera, park assist, cruise control, and auto dimming headlamps); determine needed repairs.	P-2	P-3		
Diagnose operation of security/anti-theft systems and related circuits (such as: theft deterrent, door locks, remote keyless entry, remote start, and starter/fuel disable); determine needed repairs.	P-2	P-3		
Diagnose operation of entertainment and related circuits (such as: radio, DVD, remote CD changer, navigation, amplifiers, speakers, antennas, and voice-activated accessories); determine needed repairs.	P-3	P-3		
Diagnose operation of safety systems and related circuits (such as: horn, airbags, seat belt pretensioners, occupancy classification, wipers, washers, speed control/collision avoidance, heads-up display, park assist, and back-up camera); determine needed repairs.	P-1	P-3		
Diagnose body electronic systems circuits using a scan tool; check for module communication errors (data communication bus systems); determine needed action.	P-2	P-3		
Describe the process for software transfer, software updates, or reprogramming of electronic modules.	P-2	P-3		
Disable and enable supplemental restraint system (SRS); verify indicator lamp operation.			P-1	
Remove and reinstall door panel.			P-1	
Describe the operation of keyless entry/remote-start systems.			P-3	
Verify operation of instrument panel gauges and warning/indicator lights; reset maintenance indicators.			P-1	
Verify windshield wiper and washer operation; replace wiper blades.			P-1	
HEATING, VENTILATION, AND AIR CONDITIONING (HVAC)				
General: A/C System Diagnosis and Repair				
Identify and interpret heating and air conditioning problems; determine needed action.	P-1	P-1		
Research vehicle service information including refrigerant/oil type, vehicle service history, service precautions, and technical service bulletins.	P-1	P-1	P-1	
Identify heating, ventilation and air conditioning (HVAC) components and configuration.			P-1	
Performance test A/C system; identify problems.	P-1	P-1		
Identify abnormal operating noises in the A/C system; determine needed action.	P-2	P-2		
Identify refrigerant type; select and connect proper gauge set/test equipment; record temperature and pressure readings.	P-1	P-1		
Leak test A/C system; determine needed action.	P-1	P-1		
Inspect condition of refrigerant oil removed from A/C system; determine needed action.	P-2	P-2		
Determine recommended oil and oil capacity for system application.	P-1	P-1		
Using a scan tool, observe and record related HVAC data and trouble codes.	P-3	P-3		

(continued)

Task List	MAST	AST	MLR	Chapter
Refrigeration System Component Diagnosis and Repair				
Inspect, remove, and/or replace A/C compressor drive belts, pulleys, tensioners and visually inspect A/C components for signs of leaks; determine needed action.	P-1	P-1	P-1	
Inspect, test, service and/or replace A/C compressor clutch components and/or assembly; check compressor clutch air gap; adjust as needed.	P-2	P-2		
Remove, inspect, reinstall, and/or replace A/C compressor and mountings; determine recommended oil type and quantity.	P-2	P-2		
Identify hybrid vehicle A/C system electrical circuits and service/safety precautions.	P-2	P-2	P-2	
Determine need for an additional A/C system filter; perform needed action.	P-3	P-3		
Remove and inspect A/C system mufflers, hoses, lines, fittings, O-rings, seals, and service valves; perform needed action.	P-2	P-2		
Inspect for proper A/C condenser airflow; determine needed action.	P-1	P-1		
Inspect A/C condenser for airflow restrictions; determine necessary action.			P-1	
Remove, inspect, and replace receiver/drier or accumulator/drier; determine recommended oil type and quantity.	P-2	P-2		
Remove, inspect, and install expansion valve or orifice (expansion) tube.	P-1	P-1		
Inspect evaporator housing water drain; perform needed action.	P-1	P-1		
Diagnose A/C system conditions that cause the protection devices (pressure, thermal, and/or control module) to interrupt system operation; determine needed action.	P-2			
Determine procedure to remove and reinstall evaporator; determine required oil type and quantity.	P-2	P-2		
Remove, inspect, reinstall, and/or replace condenser; determine required oil type and quantity.	P-2			
Heating, Ventilation, and Engine Cooling Systems Diagnosis and Repair				
Inspect engine cooling and heater systems hoses and pipes; perform needed action.	P-1	P-1	P-1	
Inspect and test heater control valve(s); perform needed action.	P-2	P-2		
Diagnose temperature control problems in the HVAC system; determine needed action.	P-2			
Determine procedure to remove, inspect, reinstall, and/or replace heater core.	P-2	P-2		
Operating Systems and Related Controls Diagnosis and Repair				
Inspect and test HVAC system blower motors, resistors, switches, relays, wiring, and protection devices; determine needed action.	P-1	P-1		
Diagnose A/C compressor clutch control systems; determine needed action.	P-2	P-2		
Diagnose malfunctions in the vacuum, mechanical, and electrical components and controls of the heating, ventilation, and A/C (HVAC) system; determine needed action.	P-2	P-2		
Inspect and test HVAC system control panel assembly; determine needed action.	P-3	P-3		
Inspect and test HVAC system control cables, motors, and linkages; perform needed action.	P-3	P-3		
Inspect HVAC system ducts, doors, hoses, cabin filters, and outlets; perform needed action.	P-1	P-1	P-1	
Identify the source of HVAC system odors.	P-2	P-2	P-2	
Check operation of automatic or semi-automatic HVAC control systems; determine needed action.	P-2	P-2		
Refrigerant Recovery, Recycling, and Handling				
Perform correct use and maintenance of refrigerant handling equipment according to equipment manufacturer's standards.	P-1	P-1		
Identify A/C system refrigerant; test for sealants; recover, evacuate, and charge A/C system; add refrigerant oil as required.	P-1	P-1		
Recycle, label, and store refrigerant.	P-1	P-1		

Task List **ENGINE PERFORMANCE** **General: Engine Diagnosis**	MAST	AST	MLR	Chapter
Identify and interpret engine performance concerns; determine needed action.	P-1	P-1		
Research vehicle service information including vehicle service history, service precautions, and technical service bulletins.	P-1	P-1	P-1	
Diagnose abnormal engine noises or vibration concerns; determine needed action.	P-3	P-3		
Diagnose the cause of excessive oil consumption, coolant consumption, unusual exhaust color, odor, and sound; determine needed action.	P-2	P-2		
Perform engine absolute manifold pressure tests (vacuum/boost); determine needed action.	P-1	P-1	P-2	
Perform cylinder power balance test; determine needed action.	P-2	P-2	P-2	
Perform cylinder cranking and running compression tests; determine needed action.	P-1	P-1	P-2	
Perform cylinder leakage test; determine needed action.	P-1	P-1	P-2	
Diagnose engine mechanical, electrical, electronic, fuel, and ignition concerns; determine needed action.	P-2	P-2		
Verify engine operating temperature; determine needed action.	P-1	P-1	P-1	
Verify correct camshaft timing including engines equipped with variable valve timing systems (VVT).	P-1	P-1		
Remove and replace spark plugs; inspect secondary ignition components for wear and damage.			P-1	
Computerized Controls Diagnosis and Repair				
Retrieve and record diagnostic trouble codes (DTC), OBD monitor status, and freeze frame data; clear codes when applicable.	P-1	P-1	P-1	
Access and use service information to perform step-by-step (troubleshooting) diagnosis.	P-1	P-1		
Perform active tests of actuators using a scan tool; determine needed action.	P-1	P-2		
Describe the use of OBD monitors for repair verification.	P-1	P-1	P-1	
Diagnose the causes of emissions or driveablility concerns with stored or active diagnostic trouble codes (DTC); obtain, graph, and interpret scan tool data.	P-1			
Diagnose emissions or driveablility concerns without stored or active diagnostic trouble codes; determine needed action.	P-1			
Inspect and test computerized engine control system sensors, powertrain/engine control module (PCM/ECM), actuators, and circuits using a graphing multimeter (GMM)/digital storage oscilloscope (DSO); perform needed action.	P-2			
Diagnose driveablility and emissions problems resulting from malfunctions of interrelated systems (cruise control, security alarms, suspension controls, traction controls, HVAC, automatic transmissions, non-OEM installed accessories, or similar systems); determine needed action.	P-2			
Ignition System Diagnosis and Repair				
Diagnose (troubleshoot) ignition system related problems such as no-starting, hard starting, engine misfire, poor driveablility, spark knock, power loss, poor mileage, and emissions concerns; determine needed action.	P-2	P-2		
Inspect and test crankshaft and camshaft position sensor(s); determine needed action.	P-1	P-1		
Inspect, test, and/or replace ignition control module, powertrain/engine control module; reprogram/initialize as needed.	P-3	P-3		
Remove and replace spark plugs; inspect secondary ignition components for wear and damage.	P-1	P-1		
Fuel, Air Induction, and Exhaust Systems Diagnosis and Repair				
Diagnose (troubleshoot) hot or cold no-starting, hard starting, poor driveablility, incorrect idle speed, poor idle, flooding, hesitation, surging, engine misfire, power loss, stalling, poor mileage, dieseling, and emissions problems; determine needed action.	P-2			
Check fuel for contaminants; determine needed action.	P-2	P-2		

(continued)

Task List	MAST	AST	MLR	Chapter
Inspect and test fuel pump(s) and pump control system for pressure, regulation, and volume; perform needed action.	P-1	P-1		
Replace fuel filter(s) where applicable.	P-2	P-2	P-2	
Inspect, service, or replace air filters, filter housings, and intake duct work.	P-1	P-1	P-1	
Inspect throttle body, air induction system, intake manifold and gaskets for vacuum leaks and/or unmetered air.	P-2	P-2		
Inspect, test, and/or replace fuel injectors.	P-2	P-2		
Verify idle control operation.	P-1	P-1		
Inspect integrity of the exhaust manifold, exhaust pipes, muffler(s), catalytic converter(s), resonator(s), tail pipe(s), and heat shields; perform needed action.	P-1	P-1	P-1	
Inspect condition of exhaust system hangers, brackets, clamps, and heat shields; determine needed action.	P-1	P-1	P-1	
Perform exhaust system back-pressure test; determine needed action.	P-2	P-2		
Check and refill diesel exhaust fluid (DEF).	P-2	P-2	P-2	
Test the operation of turbocharger/supercharger systems; determine needed action.	P-2			
Emissions Control Systems Diagnosis and Repair				
Diagnose oil leaks, emissions, and driveablility concerns caused by the positive crankcase ventilation (PCV) system; determine needed action.	P-3	P-3		
Inspect, test, service, and/or replace positive crankcase ventilation (PCV) filter/breather, valve, tubes, orifices, and hoses; perform needed action.	P-2	P-2	P-2	
Diagnose emissions and driveablility concerns caused by the exhaust gas recirculation (EGR) system; inspect, test, service and/or replace electrical/electronic sensors, controls, wiring, tubing, exhaust passages, vacuum/pressure controls, filters, and hoses of exhaust gas recirculation (EGR) systems; determine needed action.	P-2	P-3		
Inspect and test electrical/electronically-operated components and circuits of secondary air injection systems; determine needed action.		P-3		
Diagnose emissions and driveablility concerns caused by the secondary air injection system; inspect, test, repair, and/or replace electrical/electronically-operated components and circuits of secondary air injection systems; determine needed action.	P-2			
Diagnose emissions and driveablility concerns caused by the evaporative emissions control (EVAP) system; determine needed action.	P-1			
Diagnose emission and driveablility concerns caused by catalytic converter system; determine needed action.	P-2	P-3		
Interpret diagnostic trouble codes (DTCs) and scan tool data related to the emissions control systems; determine needed action.	P-2	P-2		
REQUIRED SUPPLEMENTAL TASKS				
Shop and Personal Safety				
Identify general shop safety rules and procedures.	R	R	R	2
Utilize safe procedures for handling of tools and equipment.	R	R	R	3
Identify and use proper placement of floor jacks and jack stands.	R	R	R	
Identify and use proper procedures for safe lift operation.	R	R	R	
Utilize proper ventilation procedures for working within the lab/shop area.	R	R	R	2
Identify marked safety areas.	R	R	R	
Identify the location and the types of fire extinguishers and other fire safety equipment; demonstrate knowledge of the procedures for using fire extinguishers and other fire safety equipment.	R	R	R	2
Identify the location and use of eye wash stations.	R	R	R	
Identify the location of the posted evacuation routes.	R	R	R	2
Comply with the required use of safety glasses, ear protection, gloves, and shoes during lab/shop activities.	R	R	R	2

Task List	MAST	AST	MLR	Chapter
Identify and wear appropriate clothing for lab/shop activities.	R	R	R	2
Secure hair and jewelry for lab/shop activities.	R	R	R	
Demonstrate awareness of the safety aspects of supplemental restraint systems (SRS), electronic brake control systems, and hybrid vehicle high voltage circuits.	R	R	R	
Demonstrate awareness of the safety aspects of high voltage circuits (such as high intensity discharge (HID) lamps, ignition systems, injection systems, etc.).	R	R	R	
Locate and demonstrate knowledge of safety data sheets (SDS).	R	R	R	2
Tools and Equipment				
Identify tools and their usage in automotive applications.	R	R	R	
Identify standard and metric designation.	R	R	R	3
Demonstrate safe handling and use of appropriate tools.	R	R	R	3
Demonstrate proper cleaning, storage, and maintenance of tools and equipment.	R	R	R	3
Demonstrate proper use of precision measuring tools (i.e. micrometer, dial-indicator, dial-caliper).	R	R	R	3
Preparing Vehicle for Service				
Identify information needed and the service requested on a repair order.	R	R	R	1
Identify purpose and demonstrate proper use of fender covers, mats.	R	R	R	
Demonstrate use of the three Cs (concern, cause, and correction).	R	R	R	1
Review vehicle service history.	R	R	R	1
Complete work order to include customer information, vehicle identifying information, customer concern, related service history, cause, and correction.	R	R	R	5
Preparing Vehicle for Customer				
Ensure vehicle is prepared to return to customer per school/company policy (floor mats, steering wheel cover, etc.).	R	R	R	
Workplace Employability Skills				
Personal Standards (see Standard 7.9)				
Reports to work daily on time; able to take directions and motivated to accomplish the task at hand.	R	R	R	
Dresses appropriately and uses language and manners suitable for the workplace.	R	R	R	
Maintains appropriate personal hygiene.	R	R	R	
Meets and maintains employment eligibility criteria, such as drug/alcohol-free status, clean driving record, etc.	R	R	R	
Demonstrates honesty, integrity and reliability.	R	R	R	
Work Habits/Ethic (see Standard 7.10)				
Complies with workplace policies/laws.	R	R	R	
Contributes to the success of the team, assists others and requests help when needed.	R	R	R	
Works well with all customers and coworkers.	R	R	R	
Negotiates solutions to interpersonal and workplace conflicts.	R	R	R	
Contributes ideas and initiative.	R	R	R	
Follows directions.	R	R	R	
Communicates (written and verbal) effectively with customers and coworkers.	R	R	R	
Reads and interprets workplace documents; writes clearly and concisely.	R	R	R	
Analyzes and resolves problems that arise in completing assigned tasks.	R	R	R	
Organizes and implements a productive plan of work.	R	R	R	
Uses scientific, technical, engineering and mathematics principles and reasoning to accomplish assigned tasks.	R	R	R	
Identifies and addresses the needs of all customers, providing helpful, courteous and knowledgeable service and advice as needed.	R	R	R	

GLOSSARY

3 Cs A term used to describe the repair documentation process of 1st documenting the customer concern, 2nd documenting the cause of the problem, and 3rd documenting the correction.

absorbed glass mat (AGM) Batteries that have the electrolyte absorbed within a mat of fine glass fibers.

aftermarket A company other than the original manufacturer that produces equipment or provides services.

Allen wrench A type of hexagonal drive mechanism for fasteners.

alternating current (AC) A type of current flow that flows back and forth.

American Wire Gauge (AWG) Standardized wire gauge used in North America. The higher the AWG number, the smaller the wire is, and the lower the current carrying capacity.

ammeter A device used to measure current flow.

amp An abbreviation for amperes, the unit for current measurement.

arc joint pliers Pliers with parallel slip jaws that can increase in size. Also called Channellocks.

armature The rotating wire coils in motors and generators. It is also the moving part of a solenoid or relay, as well as the pole piece in a permanent magnet generator.

attenuator A device that weakens, or attenuates, a high-level input signal.

aviation snips A scissor-like tool for cutting sheet metal.

ball-peen (engineer's) hammer A hammer that has a head that is rounded on one end and flat on the other; designed to work with metal items.

ballast Increases the voltage substantially and controls the current to the bulb.

barrier cream A cream that looks and feels like a moisturizing cream but has a specific formula to provide extra protection from chemicals and oils.

bench vice A device that securely holds material in jaws while it is being worked on.

bipolar junction transistor (BJT) A semiconductor device constructed with three doped semiconductor regions (base, collector and emitter) separated by two P-N junctions.

blind rivet A rivet that can be installed from its insertion side.

bolt A type of threaded fastener with a thread on one end and a hexagonal head on the other.

bolt cutters Strong cutters available in different sizes, designed to cut through non-hardened bolts and other small-stock material.

bottoming tap A thread-cutting tap designed to cut threads to the bottom of a blind hole.

box-end wrench A wrench or spanner with a closed or ring end to grip bolts and nuts.

butt-connectors Crimp-type connectors that can be used to electrically join two pieces of wire.

C-clamp A clamp shaped like the letter C; it comes in various sizes and can clamp various items.

capacitance (C) The ability of a capacitor to store an electrical charge.

capacitor A device that can quickly store a small amount of electrical energy, at which point it is charged.

cause Part of the 3Cs, documenting the cause of the problem. This documentation will go on the repair order, invoice, and service history.

center punch Less sharp than a prick punch, the center punch makes a bigger indentation that centers a drill bit at the point where a hole is required to be drilled.

charge (Q) In measuring capacitance, the amount of electrical energy present.

charge carrier A mobile particle that has a positive or negative electrical charge.

circuit breaker A device that trips and opens a circuit, preventing excessive current flow in a circuit. It is resettable to allow for reuse.

circuit or schematic diagram A pictorial representation or road map of the wiring and electrical components.

coarse (UNC) Used to describe thread pitch; stands for Unified National Coarse.

cold chisel The most common type of chisel, used to cut cold metals. The cutting end is tempered and hardened so that it is harder than the metals that need to be cut.

cold cranking amps (CCA) The load in amps that a battery can deliver for 30 seconds while maintaining a voltage of 1.2 volts per cell (7.2 volts for a 12-volt battery) or higher at 0°F (−18°C).

combination pliers A type of pliers for cutting, gripping, and bending.

combination wrench A type of wrench that has a box-end wrench on one side and an open end on the other.

commutator A device made on armatures of electric generators and motors to control the direction of current flow in the armature windings.

concern Part of the 3Cs, documenting the original concern that the customer came into the shop with. This documentation will go on the repair order, invoice, and service history.

conductor A material that allows electricity to flow through it easily. It is made up of atoms with one to three valance ring electrons.

connector The plastic housing on the end of a wiring harness that holds the wire terminals in place. It can also refer to a type of wire terminal that connects wires together or to a common point such as a bolt.

continuity A conductive path between two points.

control module A generic term that identifies an electronic unit that controls one or more electrical systems in the vehicle; also called a control unit.

conventional theory The theory that electrons flow from positive to negative.

correction Part of the 3Cs, documenting the repair that solved the vehicle fault. This documentation will go on the repair order, invoice, and service history.

counter-electromotive force (CEMF) Voltage generated in the armature windings as the motor armature spins in the magnetic field.

cranking amps (CA) The load in amps that a battery can deliver for 30 seconds while maintaining a voltage of 1.2 volts per cell (7.2 volts for a 12-volt battery) or higher at 32°F (0°C).

crankshaft A vehicle engine component that transfers the reciprocating movement of pistons into rotary motion.

cross-arm A description for an arm that is set at right angles or 90 degrees to another component.

cross-cut chisel A type of chisel for metal work that cleans out or cuts key ways.

current clamp Measures the magnetic field generated by current flow through a wire or cable.

current flow The flow of electrons, typically within a circuit or component.

curved file A type of file that has a curved surface for filing holes.

dead blow hammer A type of hammer that has a cushioned head to reduce the amount of head bounce.

delay circuit A combination of electrical and electronic components that provide a time delay for switching an electrical circuit.

depletion layer An area of neutral charge in semiconductors.

depth micrometer A measuring device that accurately measures the depth of a hole.

diagonal cutting pliers Cutting pliers for small wire or cable.

dial bore gauge An accurate measuring device for inside bores, usually made with a dial indicator attached to it.

dial indicator An accurate measuring device where measurements are read from a dial and needle.

die Used to cut external threads on a metal shank or bolt.

die stock A handle for securely holding dies to cut threads.

digital multimeter (DMM) A test instrument with a digital display for measuring voltage, resistance, and current. Also called a digital volt ohm meter (DVOM).

digital volt-ohm meter (DVOM) A test instrument with a digital display for measuring voltage, resistance, and current. Also called a digital multimeter (DMM).

diode A two-lead electronic component that allows current flow in one direction only.

direct current (DC) Movement of current that flows in one direction only.

display refresh rate The rate at which a GMM or oscilloscope can display new electrical information. The higher the display refresh rate, the higher the resolution of the test device.

doping The introduction of impurities to pure semiconductor materials to provide N- and P-type semiconductors.

double flare A seal that is made at the end of metal tubing or pipe.

double insulated Tools or appliances that are designed in such a way that no single failure can result in a dangerous voltage coming into contact with the outer casing of the device.

drift punch A type of punch used to start pushing roll pins to prevent them from spreading.

drill vice A tool with jaws that can be attached to a drill press table for holding material that is to be drilled.

duty cycle The percentage of one period of time in which the circuit is powered on.

ear protection Protective gear worn when the sound levels exceed 85 decibels, when working around operating machinery for any period of time, or when the equipment you are using produces loud noise.

elasticity The amount of stretch or give a material has.

electrical capacity The amount of charge a typical lead-acid battery can store; determined primarily by the total surface area of the plates.

electrical power A measurement of the rate at which electricity is consumed or created.

electrical resistance A material's property that slows down the flow of electrical current.

electrolysis A method of using electrical current to create a chemical reaction.

electrolyte A mixture of water and acid that contains free ions that make it electrically conductive.

electrolyte An electrically conductive solution.

electromagnetic induction The production of an electrical current in a conductor when it moves through a magnetic field or a magnetic field moves past it.

electromotive force An electrical pressure or voltage.

electron theory The theory that electrons, being negatively charged, repel other electrons and are attracted to positively charged objects; thus electrons flow from negative to positive.

energy The ability to do work.

engineering and work practice controls Systems and procedures required by OSHA and put in place by employers to protect their employees from hazards.

Environmental Protection Agency (EPA) Federal government agency that deals with issues related to environmental safety.

fasteners Devices that securely hold items together, such as screws, cotter pins, rivets, and bolts.

feeler gauge A thin blade device for measuring space between two objects.

female terminals Terminals that accept the protruding male terminals. These are easily damaged by improper probing with test lights and meter leads.

field-effect transistor (FET) A transistor in which most current is carried along a channel whose effective resistance can be controlled by a transverse electric field.

fine (UNF) Used to describe thread pitch; it stands for Unified National Fine.

finished rivet A rivet after the completion of the riveting process.

first aid The immediate care given to an injured or suddenly ill person.

fixed resistor A resistor that has a fixed value.

flare nut wrench A type of box-end wrench that has a slot in the box section to allow the wrench to slip through a tube or pipe. Also called a flare tubing wrench.

flasher can A mechanical device that switches the vehicle's turn signal and hazard flasher bulbs on and off.

flasher control An electronic device that switches the vehicle's turn signals on and off.

flasher unit The name given to the system that is responsible for switching the vehicle's flashers on and off.

flat blade screwdriver A type of screwdriver that fits a straight slot in screws.

flat-nose pliers Pliers that are flat and square at the end of the nose.

forcing screw The center screw on a gear, bearing, or pulley puller. Also called a jacking screw.

free electron An electron located on the outer ring, called the valence ring, that is only loosely held by the nucleus and that is free to move from one atom to another when an electrical potential (pressure) is applied.

freeze frame data Refers to snapshots that are automatically stored in a vehicle's power train control module (PCM) when a fault occurs (only available on model year 1996 and newer).

fuse A safety device that self-destructs to prevent excessive current flowing in a circuit in the event of a fault.

gallium-arsenide A semiconductor used in high-frequency circuits.

gas welding goggles Protective gear designed for gas welding; they provide protection against foreign particles entering the eye and are tinted to reduce the glare of the welding flame.

gasket scraper A broad sharp flat blade to assist in removing gaskets and glue.

gassing When gas escapes the battery; caused by overcharging or rapid charging a battery.

gear pullers A tool with two or more legs and a cross bar with a center forcing screw to remove gears.

germanium A type of semiconductor.

graphing multimeter (GMM) Used to display electrical data in graphical fashion. The base functionality of a GMM is similar to an oscilloscope, but GMMs typically have slower sampling rates and lower display resolution than oscilloscopes.

ground A point where the circuit connects to the negative side of the electrical system.

ground The return path for electrical current in a vehicle chassis, other metal of the vehicle, or dedicated wire.

halogen lamps A type of incandescent lamp that is filled with a halogen gas such as bromine or iodine.

hard rubber mallet A special-purpose tool with a head made of hard rubber; often used for moving things into place where it is important not to damage the item being moved.

hazardous material Any material that poses an unreasonable risk of damage or injury to persons, property, or the environment if it is not properly controlled during handling, storage, manufacture, processing, packaging, use and disposal, or transportation.

headgear Protective gear that includes items like hairnets, caps, or hard hats.

heat buildup A dangerous condition that occurs when the glove can no longer absorb or reflect heat, and heat is transferred to the inside of the glove.

hertz The unit for electrical frequency measurement.

high resistance A term that describes a circuit or components with more resistance than designed.

high resistance The resistance of a component or circuit relative to a low resistance. It can also refer to a faulty circuit where a section or component has excess unwanted resistance.

high-intensity discharge (HID) Headlamps that produce light with an electric arc rather than a glowing filament.

hold function A setting on a DVOM to store the present reading.

hold-in winding The winding that is responsible for holding the solenoid in the "ON" position; typically draws less current than the pull-in winding.

hole theory The theory that as electrons flow from negative to positive, holes flow from positive to negative.

hollow punch A punch with a center hollow for cutting circles in thin materials such as gaskets.

hot junction The heating point of a thermocouple.

impact driver A tool that is struck with a hammer to provide an impact turning force to remove tight fasteners.

incandescent lighting One or more filaments that heat up to approximately 5000°F (2760°C) and glow white hot.

inside micrometer A micrometer designed to measure internal diameters.

insulated-gate bipolar transistor (IGBT) A three-terminal power semiconductor device primarily used as an electronic switch.

insulator A material that has properties that prevent the easy flow of electricity. These materials are made up of atoms with five to eight electrons in the valance ring.

integrated circuit (IC) An electronic circuit formed on a small piece of semiconducting material.

intermediate tap One of a series of taps designed to cut an internal thread. Also called a plug tap.

intermittent faults A fault or customer concern that you cannot detect all of the time and only occurs sometimes.

invertor A device that changes direct current into alternating current.

ion An atom that has fewer electrons than protons (positive) or that has more electrons than protons (negative).

keep alive memory (KAM) A certain minimum amount of parasitic current draw that is used by the vehicle's electronic systems.

Kirchhoff's current law An electrical law stating that the sum of the current flowing into a junction is the same as the current flowing out of the junction.

Kirchhoff's voltage law The sum of all voltages drops in a circuit are equal to source voltage.

light-emitting diode (LED) A diode that emits light when current flows through it.

light-emitting diodes (LEDs) Diodes that produces light when current flows across the P/N junction.

locking pliers A type of pliers where the jaws can be set and locked into position.

lockout/tag-out A safety tag system to ensure that faulty equipment or equipment in the middle of repair is not used.

logic-controlled relay A relay that is turned on and off by an electronic control module.

lug wrench A tool designed to remove wheel lugs nuts and commonly shaped like a cross.

magnetic pickup tools An extending shaft, often flexible, with a magnet fitted to the end for picking up metal objects.

magnetic pickup tools Useful for grabbing items in tight spaces, it typically is a telescoping stick that has a magnet attached to the end on a swivel joint.

male terminals Terminals that protrude into the receiving terminal.

mandrel The shaft of a pop rivet.

mandrel head The head of the pop rivet that connects to the shaft and causes the rivet body to flare.

measuring tape A thin measuring blade that rolls up and is contained in a spring-loaded dispenser.

mechanical fingers Spring-loaded fingers at the end of a flexible shaft that pick up items in tight spaces.

metal–oxide–semiconductor field-effect transistor (MOSFET) A type of field-effect transistor (FET) which has an insulated gate whose voltage determines the conductivity of the device.

microcontroller A stand-alone electronic control module that can be programmed to perform various functions.

micrometer An accurate measuring device for internal and external dimensions. Commonly abbreviated as "mic."

microprocessor An electronic control unit that can process data and control one or more devices.

min/max setting A setting on a DVOM to display the maximum and minimum readings.

multimeter A test instrument used to measure volts, ohms, and amps. A digital multimeter may also be called a digital volt-ohmmeter (DVOM).

N-type semiconductor Semiconductor material with a small amount of extra electrons.

needle-nose pliers Pliers with long tapered jaws for gripping small items and getting into tight spaces.

nippers (pincer pliers) Pliers designed to cut protruding items level with the surface.

normally closed (NC) An electrical contact that is closed in the at-rest position.

normally open (NO) An electrical contact that is open in the at-rest position.

NPN transistor A transistor in which P-type material is sandwiched between two layers of N-type material.

nut A fastener with a hexagonal head and internal threads for screwing on bolts.

Occupational Safety and Health Administration (OSHA) Government agency created to provide national leadership in occupational safety and health.

offset screwdriver A screwdriver with a 90-degree bend in the shaft for working in tight spaces.

offset vice A vice that allows long objects to be gripped vertically.

ohm The unit for measuring electrical resistance.

Ohm's law A law that defines the relationship between current, resistance, and voltage.

oil filter wrench A specialized wrench that allows extra leverage to remove an oil filter when it is tight.

open A term used to describe a circuit that does not have a complete path for current to flow.

open circuit A circuit that has a break that prevents current from flowing.

open-end wrench A wrench with open jaws to allow side entry to a nut or bolt.

original equipment manufacturer (OEM) The company that manufactured the vehicle.

oscilloscope A test instrument that graphs voltage over time and displays the results on a screen.

oscilloscope Instrument used to display the waveform of electrical signals. These have much higher resolution and sample rates than GMMs.

outside micrometer A micrometer designed to measure the external dimensions of items.

overrunning clutch drive Allows the alternator to freewheel when the engine suddenly decelerates, such as during gear changes and engine shutdown.

P-type semiconductor Semiconductor material with holes where electrons are missing.

parallax error A visual error caused by viewing measurement markers at an incorrect angle.

parasitic draw The current draw that occurs once the vehicle has been turned off and the systems have shut down.

peening A term used to describe the action of flattening a rivet through a hammering action.

personal protective equipment (PPE) Safety equipment designed to protect the technician, such as safety boots, gloves, clothing, protective eyewear, and hearing protection.

phase A term used to describe one set of windings from an alternator or alternating current electric motor.

Phillips head screwdriver A type of screwdriver that fits a head shaped like a cross in screws.

photovoltaic (PV) effect The conversion of sunlight into electricity.

piezoelectric A type of electricity in which a material such as a quartz crystal produces voltage when mechanical pressure distorts it.

pin punch A type of punch in various sizes with a straight or parallel shaft.

pipe wrench A wrench that grips pipes and can exert a lot of force to turn them. Because the handle pivots slightly, the more pressure put on the handle to turn the wrench, the more the grip tightens.

pliers A hand tool with gripping jaws.

PN junction The junction between N- and P-type semiconductor materials.

PNP transistor A transistor in which N-type material is sandwiched between two layers of P-type material. This type of semiconductor material has holes, meaning it is missing electrons.

polarity sensitive A term used to describe a component that must be connected into a circuit with the correct polarity to its terminals.

policy A guiding principle that sets the shop direction.

pop rivet gun A hand tool for installing pop rivets.

potentiometer Also called a pot, a three-terminal resistive device with one terminal connected to the input of the resistor, one terminal connected to the output of the resistor, and the third terminal connected to a movable wiper arm that moves up and down the resistor.

power The rate at which work is done; electrical power is measured in watts.

power windings The current-carrying windings in an alternator or motor.

prick punch A pinch with a sharp point for accurately marking a point on metal.

primary winding The coil of wire in the low-voltage circuit that creates the magnetic field in a step-up transformer.

probing technique The way in which test probes are connected to a circuit.

procedure A list of the steps required to get the same result each time a task or activity is performed.

pry bar A high-strength carbon steel rod with offsets for levering and prying.

pull-in winding The magnetic coil in a solenoid that is responsible for creating the initial movement in the solenoid when it is powered on.

pullers A generic term to describe hand tools that mechanically assist the removal of bearings, gears, pulleys, and other parts.

pulse width modulation (PWM) A digital on/off electrical signal used as a variable control for devices such as solenoids.

punches A generic term to describe a high-strength carbon steel shaft with a blunt point for driving. Center and prick punches are exceptions and have a sharp point for marking or making an indentation.

ratchet A generic term to describe a handle for sockets that allows the user to select direction of rotation. It can turn sockets in restricted areas without the user having to remove the socket from the fastener.

ratcheting box-end wrench A wrench with an inner piece that is able to rotate within the outer housing, allowing it to be repositioned without being removed.

ratcheting screwdriver A screwdriver with a selectable ratchet mechanism built into the handle that allows the screwdriver tip to ratchet as it is being used.

rectification The process of converting the AC voltage that the alternator naturally generates into the DC voltage that is required by the battery and nearly all of the automobile systems.

rectifier bridge An arrangement of diodes that is used to convert the AC voltage produced in the automotive alternator into DC voltage.

relay An electromechanical switching device whereby the magnetism from a coil winding acts on a lever that switches a set of contacts.

repair order The document that is given to the repair technician that details the customer concern and any needed information.

reserve capacity (RC) The time in minutes that a new, fully charged 12-volt battery at 80°F (27°C) will supply a constant load of 25 amps without its voltage dropping below 10.5 volts.

resistor A component designed to have a fixed resistance.

respirator Protective gear used to protect the wearer from inhaling harmful dusts or gases. Respirators range from single-use disposable masks to types that have replaceable cartridges. The correct types of cartridge must be used for the type of contaminant encountered.

rheostat An adjustable resistor that varies current flow through a circuit.

ribbon cable Cable with many conducting wires running parallel to one another.

roll bar Another type of pry bar, with one end used for prying and the other end for aligning larger holes, such as engine motor mounts.

rotor Electromagnet that rotates freely in the alternator; responsible for creating the magnetic field necessary to generate electricity.

safety data sheet (SDS) A sheet that provides information about handling, use, and storage of a material that may be hazardous.

safety glasses Safety glasses are protective eye glasses with built-in side shields to help protect your eyes from the front and side. Approved safety glasses should be worn whenever you are in a workshop. They are designed to help protect your eyes from direct impact or debris damage.

sampling speed How many measurements can be taken in a specified measure of time.

screw extractor A tool for removing broken screws or bolts.

secondary winding The coil of wire in which high voltage is induced in a step-up transformer.

semiconductor A material used to make microchips, transistors, and diodes.

series-parallel circuit A circuit that has both a series and a parallel circuit combined into one circuit.

service advisor The person at a repair facility that is in charge of communicating with the customer.

service history A complete listing of all the servicing and repairs that have been performed on that vehicle.

short Also called a short circuit, the flow of current along an unintended route.

short circuit A condition in which the current flows along an unintended route.

short to power A condition in which current flows from one circuit into another.

silicon A material commonly used to make semiconductors.

silicon carbide A type of material used to make semiconductors.

sine wave A mathematical function that describes a repetitive waveform such as an alternating current signal.

single flare A sealing system made on the end of metal tubing.

sledgehammer A heavy hammer, usually with two flat faces, that provides a strong blow.

sliding T-handle A handle fitted at 90 degrees to the main body that can be slid from side to side.

snap ring pliers A pair of pliers for installing and removing snap rings or circlips.

socket An enclosed metal tube commonly with 6 or 12 points to remove and install bolts and nuts.

solder A metal with a low melting temperature that is used to fuse metal components.

solder-type terminals Terminals that are soldered on instead of being crimped.

solenoid An electromagnet with a moving iron core that is used to cause mechanical motion.

solid-state relay A relay that performs the function of a mechanical relay but uses only electronic components.

speed brace A U-shaped socket wrench that allows high-speed operation. Also called a speeder handle.

speed control circuit A circuit that controls the speed of a motor.

splice A point in the wiring harness where multiple wires are connected electrically.

split ball gauge A measuring device used to accurately measure small holes.

square file A type of file with a square cross section.

square thread A thread type with square shoulders used to translate rotational to lateral movement.

stator A cylindrical, laminated iron core which carries the three- (or four-) phase windings.

steel hammer A hammer with a head made of hardened steel.

steel rule An accurate measuring ruler made of steel.

step-down transformer A transformer used to reduce the voltage, such as to allow a battery charger operated on 120 volts to charge a 12-volt battery.

step-up transformer A transformer used to increase the voltage from a lower input voltage to a higher output, such as an ignition coil.

straight edge A measuring device generally made of steel to check how flat a surface is.

strategy-based diagnostic process A systematic process used to diagnose faults in a vehicle.

stud A type of threaded fastener with a thread cut on each end rather than having a bolt head on one end.

switch An electrical device with contacts that turns current flow on and off.

switch-controlled relay A relay that is controlled by a mechanical switch.

tap A term used to generically describe an internal thread-cutting tool.

tap handle A tool designed to securely hold taps for cutting internal threads.

taper tap A tap with a tapper; it is usually the first of three taps used when cutting internal threads.

technical service bulletin (TSB) Service notifications and procedures sent out by the manufacturers to dealer groups alerting technicians about common issues with a particular vehicle or group of vehicles.

telescoping gauge A gauge that expands and locks to the internal diameter of bores; a caliper or outside micrometer is used to measure its size.

tensile strength In reference to fasteners, the amount of force it takes before a fastener breaks.

terminals Metal connectors that are attached to wire ends. They are used to create electrical connections that can be disconnected and reconnected.

thermal runaway Also referred to as venting the flame; during thermal runaway, the high heat of the failing cell will propagate to neighboring cells, causing them to become thermally unstable as well. When lithium-ion batteries enter thermal runaway, extreme overheating, and in some cases, fire, can be expected.

thermocouple A temperature-sensing component that consists of two dissimilar metals that produce voltage proportional to temperature.

thermopile Several thermocouples connected in series to boost output voltage.

thread file A type of file that cleans clogged or distorted threads on bolts and studs.

thread pitch The coarseness or fineness of a thread as measured by either the threads per inch or the distance from the peak of one thread to the next. Metric fasteners are measured in millimeters.

thread repair A generic term to describe a number of processes that can be used to repair threads.

threaded fasteners Bolts, studs, and nuts designed to secure parts that are under various tension and sheer stresses. These include bolts, studs, and nuts, and are designed to secure vehicle parts under stress.

threshold limit value (TLV) The maximum allowable concentration of a given material in the surrounding air.

timer delay relay (TDR) A relay that remains on for a set period of time after power has been removed from the relay coils.

tin snips Cutting device for sheet metal, works in a similar fashion to scissors.

torque Twisting force applied to a shaft that may or may not result in motion.

torque angle A method of tightening bolts or nuts based on angles of rotation.

torque specifications Supplied by manufacturers and describes the amount of twisting force allowable for a fastener or a specification showing the twisting force from an engine crankshaft.

torque wrench A tool used to measure the rotational or twisting force applied to fasteners.

torque-to-yield A method of tightening bolts close to their yield point or the point at which they will not return to their original length.

torque-to-yield (TTY) bolts Bolts that are tightened using the torque-to-yield method.

toxic dust Any dust that may contain fine particles that could be harmful to humans or the environment.

transformer action The transfer of electrical energy from one coil to another through induction in a transformer.

transistor A semiconductor device that allows a small current in the base lead to control a larger current through the emitter collector leads.

triangular file A type of file with three sides so it can get into internal corners.

tube flaring tool A tool that makes a sealing flare on the end of metal tubing.

tubing cutter A hand tool for cutting pipe or tubing squarely.

turn signal switch A switch that turns the left and right turn signal lights on and off.

V blocks Metal blocks with a V-shaped cutout for holding shafts while working on them. Also referred to as vee blocks.

vacuum tube fluorescent (VTF) Lighting used for instrumentation displays on vehicle instrument panel clusters. This type of lighting emits a very bright light with high contrast and can display in various colors.

variable resistor A component that has a mechanism for varying resistance.

vernier calipers An accurate measuring device for internal, external, and depth measurements that incorporates fixed and adjustable jaws.

volt The unit used to measure potential difference or electrical pressure.

voltage (V) The electrical pressure that causes current to flow in a circuit.

voltage drop The amount of potential difference between two points in a circuit.

voltage regulator Component responsible for controlling the current supplied to the rotor. As the output voltage increases to the maximum regulated voltage, the voltage regulator reduces the current flow through the rotor, reducing alternator output.

wad punch A type of punch that is hollow for cutting circular shapes in soft materials such as gaskets.

warding file A type of thin, flat file with a tapered end.

watt The unit for measuring electrical power.

waveform A graphical representation of how electrical signals vary with time.

welding helmet Protective gear designed for arc welding; it provides protection against foreign particles entering the eye, and the lens is tinted to reduce the glare of the welding arc.

wire Flexible metal, usually made of copper and wrapped in insulation; used to transmit electricity within circuits.

wiring harness The network of wires, connectors, and terminals that make up an electrical circuit.

work The process by which one type of energy is transformed into another type of energy.

wrenches A generic term to describe tools that tighten and loosen fasteners with hexagonal heads.

yield point The point at which a bolt is stretched so hard that it will not return to its original length when loosened; it is measured in pounds per square inch of bolt cross section.

Zener diode A diode that forward biases when a certain voltage is reached.

INDEX

Note: The letters "f" and "t" following locators refer to figures and tables respectively.

A

absorbed glass mat (AGM), 302
active listening skills, 4
AGM. See absorbed glass mat (AGM)
Alternating current, 144
American National Standards Institute (ANSI), 29
amps, 141
arc joint pliers, 69f
aviation snips, 70

B

barrier cream, 27
battery
  charging and discharging cycles
    corrosion inhibitor spray, 306f
    DVOM being tested voltage leaks, 305f
    felt corrosion inhibitors, 305f
    maintenance, 304–305
    operating conditions, 303
    rechargeable cell, 303–304
    temperature monitoring, 303
    terminals cleaning, 306f
    wet cell, 306f
  components, 300f
  condition, 300
  jump starting, 301
  lead acid
    AGM, 309–310
    AGM battery, cutaway of, 309f
    explosion, 307f
    gel, 309
    gel cell and AGM type, 310f
    gelled electrolyte, 309f
    plate arrangement in, 307f
    plates separator, 307f
    ratings, 308
    types, 308–309
  maintenance, 301–302
  parasitic draw, 300–301
  recycling procedures, 302
  service precautions, 301
battery testing procedure
  battery cables, 312
  battery state-of-charge, 311–312
  capacity, 318–321
  charging, 316–317
  clamps, 312
  conductance, 318
  connections, 312
  corroded battery post and clamp, 315f
  corrosion, 312–315
  defective cables, testing, 314
  disconnection, 322
  DVOM hooked up measure parasitic draw, 313f
  generic top-post replacement battery clamp, 316f
  gravity of electrolyte, 311f
  hydrometer, 311f
  inspecting, cleaning, filling, and replacing, 315–316
  jump starting procedures, 317–318, 319–320
  load testing, 321
  low amps, 323f
  measuring parasitic draw, 324
  parasitic draw, 322–323
bench vice, 79f
bipolar junction transistor (BJT), 208
BJT. See bipolar junction transistor (BJT)
bolt cutters, 70
bolts
  grading of, 118
  metric, 116
  sizing, 117
  standard, 116
  strength of
    ductility, 120
    fatigue, 120
    proof load, 119–120
    shear, 119
    tensile, 119
    torsional, 120
    toughness, 120
box-end wrench, 58, 59f
broken fasteners, 128
bubble flare, 83

C

cable extension style, 63f
castellated nuts, 121
CEMF. See counter-electromotive force (CEMF)
channel locks, 67
charging systems
  alternator principles
    AC sine wave, 365f
    air-cooled alternator end frames, 366f
    alternating current, 360–361
    alternator, 364f
    alternator fan and pulley, 368f
    brush assembly, 366–367
    component failure diagnosing, 368–369
    computer-controlled voltage regulation, 361–362
    cooling fan and pulley, 367
    and drive belt, 364f
    end frames and bearings, 366
    forward direction, current flow, 363f
    liquid-cooled alternator, 366f
    mechanic's stethoscope, 368f
    overrunning alternator decoupler (OAD), 368f
    positive rectifier assembly, 367f
    rectification, 360–361
    rectifier assembly, 367
    rectifier bridge, 362f
    reverse direction, current flow, 362f
    reverse direction, single phase in, 363f
    rotor, 364, 364f
    rotor circuit, 364–365
    running clutch drive, 367
    single-phase AC signal, 361f
    slip rings, 366–367
    stator, 365–366
    three phases, not rectified, 363f
    three phases, rectified, 363f
    three-phase AC signal, 361
    warning light, 361f
  charging system procedures
    belt tension gauge, 374f
    circuit voltage drop testing, 370, 371
    diode testing, 373
    drive belt condition, 373
    drive belt tension, 373
    pulley condition/alignment, 373
    serpentine poly-V belt, 373f
    system output testing, 370–373
    undercharge and overcharge conditions, 375
    v-belt, 373f
  hybrid vehicle charging systems
    DC-DC converter operation, 369
    orange high-voltage cables, 369f
  procedures
    alternator replacing, 375–378
  theory, 360
circuit types
  fused jumper leads, 255–256
  Ohm's Law diagnosing, 251–252
  testing light, 254–255
  wiring diagrams to, 253
circuits
  Kirchhoff's current law, 174–175
  Kirchhoff's law of voltage drop, 175–176
  parallel resistance, 173
  parallel, 173
  series, 172
  series-parallel, 174f
  with unequal resistance, 173f

cold cranking amps (CCA), 308
combination pliers, 67
combination wrench, 58, 59f
conductors, 139
conventional theory, 140
counter-electromotive force (CEMF), 334
cranking amps (CA), 308
cross-cut chisel, 74
current
    DVOM readings, 234t, 237t, 238t
    exercises, 234
    magnetic fields, 232
    resistance effects on, 236–238
    resistance exercises
        measuring, 235
        rust, corrosion, and Debris effect resistance, 235–236
    and resistance exercises, 232
    switching current through resistor, 233
curved files, 77

D

dead blow hammer, 73
deep socket, 63f
depth micrometers, 91
diagnostic trouble codes (DTCs), 8
diagonal cutting pliers, 67
diagonal side cutters, 69f
dial bore gauge, 94
dial gauges, 99
die stock, 81f
digital multimeter
    analog volt-ohm meter, 216f
    DVOM lead adapters and accessories, 219f
    DVOM selector dial, 219f
    DVOM values, 219t
    DVOM/DMM testing
        circuit integrity, 227–229
        circuit voltages, explanation of, 231t
        measuring current, 232
        readings, 232t
        unequal loads, 231
        for voltage drops, 227–229, 229t
        voltage measurements, 228t, 229t
    fundamentals, 216–220
    high-voltage rubber gloves, 218
    meter CAT ratings, 217f, 217t
    uses
        back-probing techniques, 222–223
        DVOM, setting up, 221
        min/max and hold setting, 221
        piercing wire taps, 223
    volts, ohms, and amps, measuring
        auto ranging meters, 226
        low-amps probe, 216
        manual ranging meters, 225
        voltage ranges, interpreting, 225
direct current, 144
drift punch, 75

E

ear protection, 28
elasticity, 125
electrical circuits, 142
electrical components
    automotive applications, 182f
    automotive fuse box, 184f
    automotive switches, 183f
    capacitors, 198
    circuit breakers, 184
    FETs, 185
    flash can/control, 186–187
    fuses, 184, 185f
    fusible link current reading, 185t
    fusible links, 184
    motors
        compound motor, 192f
        cutaway drawing, 194f
        cutaway drawing of synchronous motor, 193f
        DC, 190
        permanent magnet motor, 191f
        series wound motor, 192f
        simplified electric motor diagram, 191f
        stepper, 192
        synchronous AC, 193f
        three-phase, 193
    PTC devices, 186
    relays and control circuits
        logic-controlled, 189
        spike-protected relays, 187f
        switched, 188
    solenoids, 189–190
    switch in series to load, 182f
    switches, 182–183
    timer delay relays, 189
    turn signal switch, 183f
    virtual fuses, 185
    wiring diagram, 183f
electrical fundamentals
    atom parts, 138f
    basic, 138–140
    conductors, 139f, 140
    electricity, heating effect of, 141
    free electrons, 140
    insulator, 139f, 140
    negative ion, 139f
    positive ion, 139f
electrical systems
    alternating current, 144–146
    amps, 141–142
    circuits, 142–144
    direct current, 144–146
    fundamentals, 138–141
    ohms, 141–142
    open, short, and high resistance, 146–148
    power and ground, 146
    semiconductors, 143–144
    volts, 141–142
electrical testing procedures
    graphing multimeters
        applications, 290–291
        for electrical, 291
    menu, 291f
    oscilloscopes
        circuit waveforms, 294–295
        digital storage oscilloscope (DSO), 295f
        GMMs, 294
    test ports, 291f
    waveforms, 292–293
        data scale, 292f
        fast refresh rate, 292f, 293f
        slow refresh rate, 292f
        voltage scales, 292f
electricity
    effects
        chemical, 161
        electromagnetic, 161–162
        heating, 160–161
        light, 161
    sources of
        automotive battery construction, 158f
        electrochemical energy, 158
        electromagnetic induction, 160, 160f
        electrostatic, 156–157
        exhaust system, thermocouple, 158f
        knock sensor, 159f
        photovoltaic effect, 159f
        photovoltaic energy, 158–159
        piezo fuel injector, 159f
        piezoelectric energy, 159
        thermocouple, 157f
        thermocouple tested with voltmeter, 158f
        thermoelectric, 157
        thermopile, 157f
electrolysis, 158
electrolyte, 158
electromotive force, 140
electron theory, 140
electronic components
    Arduino controller, 210f
    automotive control units, 210f
    control modules, 208–209
    delay circuits, 210
    diodes
        with an ohmmeter, 205f
        image, 205f
        schematic symbol, 205f
    integrated circuits, 209
    introduction, 204
    microcontrollers, 210
    microprocessors, 209–210
    PN junction, 204f
    speed control circuits
        duty cycle, 211
        pulse width modulation, 211
    transistors
        BJTs, 208
        FETs, 207

IGBTs, 208
MOSFETs, 207–208
Zener diodes
rectifier bridge, 206
electrons, 138
energy transformation, 170
engine starting system
armature and starter drive, 331f
chamfered pinion gear teeth, 332f
direct drive/gear reduction systems, 331
gear reduction starter, 331f
pinion-to-ring gear alignment, 332f
principles, 330–331
spur gear and planetary gear
reduction, 331f
starter drive one-way clutch, 332f
starter motor engagement, 332–333
starter motor solenoid, 331f
Environmental Protection Agency
(EPA), 33
eye ring terminals, 266

F

fastener standardization
metric bolts, 116
standard bolts, 116
feeler blades, 103
feeler gauges, 103
FET. See field-effect transistor (FET)
field-effect transistor (FET), 207
fire triangle, 42f
first aid principles, 46–47
flare nut wrench, 59, 59f
flare tubing wrench, 59
flat blade screwdriver, 71
flat-bottomed tap, 79
flat-nose pliers, 67
flex socket style, 63f
flexible extensions, 63f
flooded cell battery, 306
forcing screw, 82, 83
four- and 8-point sockets, 63f
free electrons, 139
freeze frame data, 8

G

gallium-arsenide, 144f
gas welding goggles, 30
gasket scraper, 75
gassing, 307
gear pullers, 82
germanium, 144f

H

hand tools
Allen wrenches, 70
chisels, 74, 74f
clamps, 78–79
cutting tools, 70
dies, 80

files
coarse bastard, 77
dead smooth, 77
rough, 77
second-cut, 77
smooth, 77
flaring, 83–86, 86–87
gasket scrapers, 76–77, 76f
gear pullers, 83, 84–85
hammers, 73–74, 74f
magnetic pickup tools, 73
mechanical fingers, 73
pliers, 67–70, 68f–69f
pry bars, 76
punches, 74–76, 75f–76f
riveting, 86–88, 89–90
screw extractors, 81–82, 82f
screwdrivers, 70–73
types, 70f–71f
sockets
anatomy of, 62f
deep, 63f
flexible extensions, 63f
impact, 62, 62f
matching drive, 62f
six- and 12-point, 63f
standard length, 63f
standard wall, 62, 62f
tools to turn, 63f
taps
bottoming, 80
intermediate, 80
taper, 80
thread, 80
torque charts, 64–65
bolt, 65f
torque wrenches, 65–66, 65f
wrenches
box-end, 58, 59f
combination, 58, 59f
flare nut, 59, 59f
oil filter, 60, 61f
open-end, 58, 59f
open-end adjustable, 60f
pipe, 60f
pipe wrench, 60
ratcheting box-end, 59, 60f
ratcheting open-end, 60f
hard rubber mallet, 74
hazardous materials safety
cleaning, 45–46
data sheets, 44–45
engine oils, 46
fluids, 46
heating element, 158
HEPA. See high-efficiency particulate air
(HEPA)
HID. See high-intensity discharge (HID)
high resistance
open circuit, 256
protection devices, 258

relays, 258
short circuit, 256
short to power, 257
solenoids, 258
switches, 258
high-efficiency particulate air
(HEPA), 46
high-intensity discharge (HID), 385
high-resistance fault, 148f
hold-in winding, 338
hole theory, 140
hollow punches, 75, 76
hot junction, 157

I

IGBT. See insulated-gate bipolar transistor
(IGBT)
ignition coils
fixed resistors, 195
potentiometer exercise, 196
potentiometer diagnosing, 198
potentiometers, 196
resistor colors, 196f
resistors, 196
rheostats, 196
rheostat diagnosing, 197
step-down transformers, 195
and transformers, 193
variable resistor, 195
impact driver, 72
impact socket, 62f
insulated-gate bi-polar-transistor (IGBT),
145, 208
insulators, 139
intermittent faults, 7
inverters, 145

L

LEDs. See light-emitting diodes (LEDs)
light-emitting diodes (LEDs), 144
lighting system fundamentals
auxiliary relay circuit, 394f
circuit construction, 396–397
driving lights, 392–393
fog lights, 394–395
HID headlight assembly, 396f
lamps
back-up, 391f
festoon-style, 388f
halogen bulb, 384f
halogen lighting, 384
halogen vs. HID lighting, 385f
HID headlamp assembly, 385f
HID system safety precautions, 386
HID systems, 385–386
incandescent lighting, 384
lamp/light bulb information, 387–388
LED lighting, 386–387
license plate, 389f
park, 388f

lighting system fundamentals (*Continued*)
    vacuum tube fluorescent, 385
    warning, 392f
    wedge-style, 388f
  projector bulb assembly, 395f
  replaceable halogen bulb, 395f
  styles and types
    automatic lighting systems, 393
    back-up lights, 390–391
    brakes lights and CHMSL, 390
    daytime running lights, 392
    hazard warning lights, 391–392
    license lights, 388
    marker, 388
    park, 388
    tail, 388
    timer and delay circuits, 393
    turn signal and cornering lights, 391
    warning lights, 392
lighting systems procedures
  aiming headlights, 397–398
  headlight brightness, 398–400
  headlight bulb, 397
  and peripheral systems, 397
line wrench, 59
locking pliers, 69, 69f
lug wrench, 64, 64f

M

magnetic pickup tools, 73
material safety data sheets (MSDS), 44
measuring tapes, 89
MSDS. *See* material safety data sheets (MSDS)
multimeter, 142f

N

needle-nose pliers, 67, 69f
negative ion, 138
nippers, 67
N-type semiconductor, 143
nuts
  castle, 121, 121f
  locking, 120
  speciality, 121–122, 121f

O

Occupational Safety and Health Administration (OSHA), 32
offset screwdriver, 72
offset vice, 79f
ohms, 141
Ohm's law
  calculation, 169
  and circuits, 168–169
  current flow, 169f
  electrical power, 170
  load device, 168f
  power equation
    solving, 171
    Watt's law triangle, 171f
  solving, 169–170
oil filter wrench, 60, 61f
open-end adjustable wrench, 60f
open-end wrench, 58, 59f
original equipment manufacturer (OEM), 2
OSHA. *See* Occupational Safety and Health Administration (OSHA)
outside micrometer, 91

P

parallel circuit exercises
  DVOM Readings, 247t, 248f, 249t, 250t–251t
  four resistors in parallel, 250f
  three unequal resistors in series, 247f
  two unequal resistors in parallel, 249f
  12-volt DC supply, 248f
  12-volt supply, 247f
  wiring diagram of parallel circuit, 247f
parallelogram steering
personal protective equipment (PPE), 24
personal safety
  breathing devices, 28–29
  clothing, protection hand, 26
  ear, 28
  eye, 29–30
  headgear, 27
  lifting, 31
Phillips head screwdriver, 71
Phillips screw and screwdriver, 71f
photovoltaic (PV) effect, 158
piezoelectric energy, 159
pin punches, 75
pipe wrench, 60, 60f
plug tap, 79, 80
PN junction, 143
POP rivet guns, 88
pop rivet guns, 88
positive ion, 139
potential difference, 172
power, 170
PPE. *See* personal protective equipment (PPE)
precision measuring tools
  dial bore gauges, 94, 97–98
  dial indicators, 99–102, 101–102
  and equipment, 105–106, 106–107
  feeler gauges, 103–105, 104–105
  micrometers, 91–94
    depth, 91
    inside, 91
    outside, 91
    parts outside of, 92f
    using, 93–94, 95–96
  split ball gauges, 94
  steel rulers, 90–91
  straight edges, 103
  tapes, 89–90
  telescoping gauges, 94
  vernier calipers, 97–99, 99–100
P-type semiconductor, 143
pull-in winding, 337
pulse width modulation (PWM), 208
push-on spade terminals, 266
PWM. *See* pulse width modulation (PWM)

R

ratcheting box-end wrench, 59, 60f
ratcheting open-end wrench, 60f
ratcheting screwdriver, 72
rat-tail file, 78
rectifier assembly, 367
rectifier bridge, 160
repair documentation
  additional service, 17
  order, 17
  three Cs
    cause, 17
    concern, 16–17
    correction, 17
repair order, 17
reserve capacity (RC), 308
resistive heating, 160
ribbon cable, 264

S

safe edge files, 77
safety
  hazardous materials, 44–46
  personal, 24–31
  shop, 31–43
safety data sheets (SDS), 44
safety glasses, 29
safety guidelines
  handling tools, 52
  handling tools and equipment, 52–56, 54–55
  lockout/tag-out, 56
  standard and metric designations, 57
  tools and equipment, 52
  tools storage, 56–57
  tools use, 52–56, 53
screw extractors, 81
screws
  machine, 124, 124f
  self-tapping, 124, 124f
  sheet metal, 124–125
  trim, 124
self-parking, 39
self-resetting circuit breakers, 185
semiconductors
  materials, 144
  NP-N transistor, 143f
  N-type material, 143f
  operation, 143–144
  P-type material, 143f

simple circuit, 143f
static electricity, 144f
series circuit exercises
  DVOM readings, 240t, 241t, 242t–243t, 244t, 245t–246t
  resistors of equal value connected in series, 240f
  two resistors of equal value connected in series, 239
  typical circuit with battery supply, fuse, switch, and resistor, 239
  unequal resistors in series, 242f
series-parallel circuit, 169, 174
service advisor, 4
sheet metal screws, 125f
shop safety
  air quality, 39–40
  electrical, 40–41
  EPA, 32–33
  equipment
    adequate ventilation, 39
    doors and gates, 39
    handrails, 38
    machinery guards, 38
    painted lines, 39
    sound-insulated rooms, 39
    temporary barriers, 39
  evacuation routes, 42–43
  fire extinguishers, 41–42
  fire preventing, 41
  housekeeping and orderliness, 35
  OSHA, 32–33
  policy, 34
  procedures, 34
  safe attitude
    accidents and injury, 37
    dangers, 36–37
  signs
    background color, 38
    pictorial message, 38
    signal word, 37
    text, 38
  ventilation, 40
short circuit, 147f
silicon carbide, 144f
sine wave, 145
six- and 12-point sockets, 63f
sizing bolts, 117
sledgehammer, 73
sliding T-handle, 62
slotted screw and screw driver, 71f
snap ring pliers, 67
solder-type terminals, 267
solid-state relay, 187
speed brace, 64
split ball gauge, 94
square file, 77
standard length socket, 63f
standard system, 92
standard wall socket, 62f
starter draw test, 343

starter motor construction
  commutation and brushes, 335–336
  and components, 335
  electrical schematic of power flow, 336f
  electromagnetic fields, 333f
  fixed pole shoe starter, 334
  magnet types, 333–334
  moveable pole shoe, 334
  permanent magnet motors, 334–335
  permanent magnetic fields, 333f
  pole shoes, 334f
  series-parallel-wound starter motor, 334f
  single-loop motor and electromagnetic fields, 336f
  starter pole shoes, 334f
  two semi-circle segments, 337f
starter motor operation
  control circuit, 339f, 340f
  crank position, 338f
  drive linkage, 339f
  electrical windings, 339f
  hybrid vehicle starting systems
    belt-alternator starter motor (BAS), 342f
    high-voltage starters, 342–343
    Honda hybrid drive motor, 342f
    hybrid transmission, 342f
    12-volt systems, 342
  pull-in winding, 338f
  and ring gear, 341
  "S" terminal on a starter solenoid, 338f
  solenoid operating principles testing, 338–339
  solenoid starter contacts, 339f
  starter drive one-way clutch, 341f
  starter drive, view of, 340f
  windings energized and solenoid plunger starting, 337f
starting system procedures
  circuit voltage drop testing, 345–346
  control circuit, 347–349
  draw testing, 343
  electrical and mechanical problems, 353–354
  removing and reinstalling a starter, 352–353
  testing relays and solenoids, 349–351
  testing starter circuit voltage drop, 346–347
  testing starter draw, 343–345
starting systems
  engine, 330–333
  motor construction, 333–336
  motor operation, 337–340
  procedures, 343–354
  ring gear, 340–343
  starter drives, 340–343
steel hammer, 73
steel ruler, 90
step-up transformer, 193

strategy-based diagnostic process, 5, 6f
  customer concern, 6–8
  faults, 8–10
  focused testing, 10–13
  information gathering, 8–10
  need for, 5–6
  repairing
    customer approval, 14
    pay attention, 15
    prior updates, 15
    service procedure, 13
    time taken, 14
    tools, 13–14
  verifying repair, 15–16

T

tap drill chart, 81f
tap handle, 81f
TDR. See timer delay relay (TDR)
technical service bulletins (TSBs), 9, 9f
telescoping gauge, 94
tensile strength, 119
tension wrench, 65
thermocouple, 157
thermopiles, 157
thread file, 78
thread pitch, 117
thread repair, 128–131, 132–133
threaded fasteners, 114
threadlocker, 123
timer delay relay (TDR), 189
tin snips, 70
toothed lock washers, 122
torque angle, 125
torque angle gauge, 126–127
torque specifications, 64
torque wrenches, 65, 65f
torque-to-yield (TTY) bolts, 125
toxic dust, 45
triangular file, 77
trim screws, 125f
TSBs. See technical service bulletins (TSBs)
T-shaped tap handle, 81f
TTY. See torque-to-yield (TTY)
tube flaring tool, 83
tubing cutter, 85
typical Allen wrench head, 71f

U

U-joint style, 63f
UNC. See Unified National Coarse Thread (UNC)
UNF. See Unified National Fine Thread (UNF)
Unified National Coarse Thread (UNC), 117
Unified National Fine Thread (UNF), 117
universal joint, 63f

V

vacuum fluorescent display (VFD), 385
vacuum tube fluorescent (VTF), 385
valence ring, 139
valve-regulated lead-acid (VRLA), 309
variable resistors, 195
vehicle service, 2
vernier caliper, 57
VFD. *See* vacuum fluorescent display (VFD)
vice grips, 69
voltage drop, 172
voltage regulator, 360
volts, 141
VRLA. *See* valve-regulated lead-acid (VRLA)
VTF. *See* vacuum tube fluorescent (VTF)

W

wad punches, 75, 76
warding file, 77
washers
    flat, 122
    grades of, 122
    lock, 122
    star, 122
welding helmet, 30
wire maintenance
    cleaning and tinned soldering iron tip, 282f
    corroded copper wire, 278f
    dirty, black, oxidized soldering iron tip, 282f
    electrical tape applied to connector, 280f
    heat-shrink tubing, applying, 280–281
    installing solderless terminal, 278–279
    insulated and non-insulated terminals, 278f
    liquid electrical tape applied to connector, 280f
    and repairing, 275
    resultant problems, 277
    selecting proper wire, 276–276
    soldering a wire connection, 283f
    soldering iron being tinned, 283f
    soldering wires and connectors, 283–284
    soldering wires, connectors, and terminals, 281–283
    solderless terminals, 277–278
    stripping insulation, 276–277
    stripping techniques, 277
wiring
    Butt connector, 267f
    connector body, removing terminal from, 269
    connector failures, diagnosing, 267
    connector types, 266–267
    crimping pliers, 267f
    diagram fundamentals
        automotive wiring harness symbols, 271f
        electrical diagram, 271f
    disassembling electrical connectors, 268
    eye ring terminal, 267f
    fundamentals
        wire, 264
    harness labels
        color, 272
        grounds, 274
        size, 272
        splice ID number, 272f
        splices, 272
        wire codes and associated names, 274t
    harnesses
        construction, 270
        splices and grounds, 270
    insulated and non-insulated crimp on terminals, 268f
    ribbon cable, 264f
    shielding
        drain lines, 265
        Mylar tape, 265
        twisted pair, 265
    sizes, 265–266
    spade terminal, 267f
    standard, 264f
    terminating resistor, 265f
    weather-tight seals on an electrical connector, 267f
    wiring harness connector, 266f
wiring harness connectors, 266
wobble extension style, 63f

X

xenon headlamps, 385

Y

yield point, 125